Advanced Series on
Directions in High Energy Physics — Vol. 26

THE STANDARD THEORY OF PARTICLE PHYSICS

Essays to Celebrate CERN's 60th Anniversary

Editors

Luciano Maiani
INFN & Sapienza University of Rome, Italy

Luigi Rolandi
CERN, Switzerland & Scuola Normale Superiore, Pisa, Italy

World Scientific

NEW JERSEY • LONDON • SINGAPORE • BEIJING • SHANGHAI • HONG KONG • TAIPEI • CHENNAI • TOKYO

Published by

World Scientific Publishing Co. Pte. Ltd.
5 Toh Tuck Link, Singapore 596224
USA office: 27 Warren Street, Suite 401-402, Hackensack, NJ 07601
UK office: 57 Shelton Street, Covent Garden, London WC2H 9HE

Library of Congress Cataloging-in-Publication Data
Names: Maiani, L. (Luciano), editor. | Rolandi, L. (Luigi), 1953– editor. |
European Organization for Nuclear Research, honouree.
Title: The standard theory of particle physics : Essays to Celebrate CERN's 60th Anniversary /
editors, Luciano Maiani (INFN, Italy & University of Rome, Italy),
Luigi Rolandi (CERN, Switzerland & Scuola Normale Superiore di Pisa, Italy).
Other titles: Advanced series on directions in high energy physics ; v. 26.
Description: Singapore ; Hackensack, NJ : World Scientific Publishing Co.
Pte. Ltd., [2016] | 2016 | Series: Advanced series on directions in high
energy physics, ISSN 1793-1339 ; vol. 26
Identifiers: LCCN 2016011545| ISBN 9789814733502 (hardcover) |
ISBN 9814733504 (hardcover)
Subjects: LCSH: Particles (Nuclear physics) | Field theory (Physics) | Higgs bosons.
Classification: LCC QC776 .S73 2016 | DDC 539.7/2--dc23
LC record available at http://lccn.loc.gov/2016011545

British Library Cataloguing-in-Publication Data
A catalogue record for this book is available from the British Library.

Typeset by Stallion Press
Email: enquiries@stallionpress.com

Contents

Preface

From its appearance, the theory of electroweak unification has been received with a certain degree of reserve by particle physicists, as shown by low profile name that physicists used to indicate it, the *Standard Model*. Indeed the theory consisted of two parts, which, paraphrasing what Einstein said about his General Relativity theory, could be seen as the two wings of a house, *one wing ...made of fine marble, but the other wing ...built of low grade wood.*[1] The marble part is, of course, the part determined by the symmetry, identified by Glashow: currents, gauge interactions, quarks and leptons. And the low grade wood: the symmetry breaking.

The choice made by Weinberg and Salam, to break the theory spontaneously, along the lines indicated by Brout and Englert and by Higgs, is of course physically and mathematically compelling. However, the choice of a single, elementary scalar doublet whose vacuum expectation value breaks the symmetry and provides altogether the masses of the vector bosons, quarks and leptons, looked as a mere expedient motivated only by the extreme simplicity of the result, to be later reconsidered and made more compelling. Only *a model*, indeed.

The striking contrast remains when Quantum Chromodynamics, another pure marble wing, is added, making the Standard Model as we know it. With the additional dissatisfaction of seeing, side by side, two totally independent symmetries, which call for further unification.

A further indication that the scheme may have to be revised is that the mass of an elementary scalar is not *a priori* protected from being driven to large values by quantum corrections, the *hierarchy problem*. Suggested remedies are Supersymmetry, which links the scalar to chiral fermions, or compositeness, advocating that the scalar is composite by a fermion pair, in analogy with the Cooper pairs of superconductivity.

Although these possibilities remain open at the moment, it is a fact that the 125 GeV particle discovered at CERN came with physical characteristics as advertised, within present errors, and marked a point in favor of the scheme as it is.

Time has come to consider the *Standard Theory* as a closed structure in itself, like the Maxwell's equations that it supersedes so well. Like all theories we know of, it will be transformed and improved by the new physics that may manifest itself at higher energies.

The book gives a quite complete and up-to-date picture of the Standard Theory with a collection of articles written by some of the protagonists of present particle physics. The theoretical developments that led to the theory, the understanding of basic field theory we have acquired along the way and the possible future developments are described together with the most up-to-date experimental tests, including the discovery of the Higgs boson and the measurement of its properties and the most precise measurements of the top mass, to give the reader a complete description of present particle physics.

The book is written on the occasion of the sixty years of CERN, a laboratory that has given crucial contributions to the advancement of particle physics, from the discovery of the neutral currents, in 1973, the observation of the intermediate vector bosons, in 1982, and the long hunt of the Higgs boson, started with LEP2 and carried to a successful end in 2012.

We thank the authors of the articles for their generous efforts and dedicate the book to the community of scientists, engineers, administratives, young students and senior people, who have dedicated their ingenuity and important parts of their lifes to build and test the heroic, admirable construction known today as the Standard Theory.

Luciano Maiani and Luigi Rolandi

Reference

1. A. Einstein, *Out of My Later Years* (Philosophical Library, New York, 1950), p. 84.

Chapter 1

The Evolution of Quantum Field Theory: From *QED* to Grand Unification

Gerard 't Hooft
Institute for Theoretical Physics, Utrecht University
and
Spinoza Institute, Postbox 80.195
3508 TD Utrecht, the Netherlands
g.thooft@uu.nl

In the early 1970s, after a slow start, and lots of hurdles, Quantum Field Theory emerged as the superior doctrine for understanding the interactions between relativistic sub-atomic particles. After the conditions for a relativistic field theoretical model to be renormalizable were established, there were two other developments that quickly accelerated acceptance of this approach: first the Brout–Englert–Higgs mechanism, and then asymptotic freedom. Together, these gave us a complete understanding of the perturbative sector of the theory, enough to give us a detailed picture of what is now usually called the Standard Model. Crucial for this understanding were the strong indications and encouragements provided by numerous experimental findings. Subsequently, non-perturbative features of the quantum field theories were addressed, and the first proposals for completely unified quantum field theories were launched. Since the use of continuous symmetries of all sorts, together with other topics of advanced mathematics, were recognised to be of crucial importance, many new predictions were pointed out, such as the Higgs particle, supersymmetry, and baryon number violation. There are still many challenges ahead.

1. The Early Days, Before 1970

Before 1970, the particle physics community was (unequally) divided concerning the relevance of quantised fields for the understanding of subatomic particles and their interactions. On hindsight, one can see clearly why the experts were negative about this approach. Foremost was the general feeling that this theory was ugly, requiring various fixes to cover up its internal mathematical inconsistencies. The first inconsistency, as it was generally perceived, was the fact that the corrections to the particle interaction properties, generated by higher order quantum effects, invariably appeared to be infinitely strong. The energy contents of a field surrounding a particle, would clearly add an infinitely large correction to its mass, and also electric charge and other interaction parameters would receive infinite corrections by vacuum fluctuations in the vicinity of a particle.

Now, it was true that a remedy had been proposed to this particular disease, first perhaps by Hans Kramers[1] around 1933, and more precisely by Julian Schwinger,[2] Freeman Dyson,[3] Sin-Itiro Tomonaga,[4] Richard Feynman[5] and others, which was that the 'original' masses and interaction strengths of a particle are ill-defined, so that these could be adjusted to cancel out the unwanted infinities, which were now replaced to experimentally inaccessible regions near the cores of these particles. A systematic application of this procedure, called *renormalization*, turned out to be quite successful in the study of electromagnetic forces between particles.[6] The anomalous magnetic moment of the electron thus obtained agreed extremely well with experimental determinations, and other successes of this theory, called *Quantum Electrodynamics* (QED), soon followed.[7]

Yet there were reasons to mistrust these results. The mathematical rigour of the calculations was lacking, it looked as if the difficulties had been swept under the rug. Perhaps these arguments were approximately right for QED, but what were the principles lying behind the other interactions? And how can we understand the renormalization procedure from a more formal point of view? Indeed, if one attempted to understand the small-distance limit of QED, a new difficulty showed up: the interactions due to virtual particles in the surrounding vacuum accumulate there, and in spite of renormalization, the effective interaction strengths eventually tend to infinity. Today, we know that that is because, in the small distance limit, you are looking again at the bare masses and coupling strengths, and these had just been agreed upon to be infinite. Lev Landau saw that this infinity would behave like a physically unacceptable 'ghost particle', now called 'Landau ghost'.[8] Today, we know how to handle the Landau ghost, but for Landau this clearly implied that you had to abandon quantised field theories altogether.

In the West, investigators were a bit bolder. Murray Gell-Mann an Francis Low had proposed that there could be an ultraviolet fixed point.[9] That, however, did not help very much, because this fixed point would be in a domain where accurate calculations are impossible. It looked as if Nature were telling us that the real particle spectrum is more subtle, and to understand that, you would have to start from scratch. Stay away from quantum field theory.

Indeed, experimental results were not encouraging at that time. The weak force appeared to be non-renormalizable for simple mathematical reasons (Enrico Fermi's fundamental interaction constant carried the wrong dimension if you simply considered how it was defined). If you would try to replace Fermi's original theory by a theory of exchanged intermediate particles (now known as the W bosons), you would end up with fundamental spin-1 force carriers, particles that appeared to require a totally different approach as well. And, most of all, the strong force seemed to resist any rational approach altogether. The hadron spectrum suggested the existence of sub-units called "quarks" by Gell-Mann,[10] and 'aces' by George Zweig,[a,11] but how a field theory could be responsible for that strange situation was beyond us.

It is always dangerous to combine several non-convincing arguments to reach a conclusion with certainty, but this is what almost happened. Quantum field theory was not *bon ton*. Indeed, there were alternatives. You could start at a more basic level. When an experiment with subatomic particles is carried out, one begins with beams of particles directed towards one another, the so-called *in*-situation, or in-state. After the particles collide, you end up with different beams of particles going *out*, the out-state. The out-state depends on the in-state chosen. This dependence is described by a matrix called S-matrix (S for *Streuung*, or *scattering*).[12] One can derive mathematical equations that this S-matrix must obey. By demanding that no signal can ever travel faster than light, one finds the so-called *dispersion relations*, relations between frequencies and wavelengths,[13] and more general features in multi-particle scattering.

It was hoped that the S-matrix could be derived from such relations, if combined with some general symmetry properties. To this end, *current algebras* were constructed.[14] What investigators tried to avoid is to talk of operators acting at single space–time points. That would have been helpful in the current algebras, but it smelled too much like field theory.

Searching for totally different principles, Steven Frautschi and Tullio Regge[15] attempted to consider amplitudes as functions of angular momentum, which could be analytically continued to the complex plane. This yielded the famous 'Regge trajectories', curves that could be extended to include all resonances, giving useful but ill-understood relations between spin and energy.

Only a handful of researchers resisted the mainstream thought. First of all, QED was further refined, and appeared to work beautifully.[16] Quantities such as the anomalous magnetic moment $g - 2$ of the electron and the muon, could be calculated and compared with experiment up to an incredible precision. It would be great if anything like that could be constructed to describe any of the other forces.

Suppose we forget those negative preconceptions about field theory, forget even the experimental data, and instead, just ask the question: what shape could a fundamental interaction possibly have?

Maybe an axiomatic approach would help. We had Arthur Wightman's famous axioms,[17] idealising the demands any quantised field theory should obey. On hindsight, one may say that these demands were too strict; even today's Standard Model formally does not obey them, but as its breakdown occurs somewhere beyond the Planck scale, no one cares about that anymore.

---

[a]Gell-Mann thought of three fundamental quarks, but Zweig, as he would explain later, assumed that there should be four, thus anticipating the idea of charm. There are four aces in a deck of cards.

In a more practical vein, a 'toy model' had been coined by Gell-Mann and Maurice Lévy, to describe strongly interacting pions in agreement with their symmetry structure, a spontaneously broken global symmetry called *chiral symmetry.* At that time, this was phrased in terms of a "partly conserved axial vector current" (PCAC).[18] The model worked qualitatively well, certainly in connection with the famous Goldberger–Treiman relation,[19] but, being an ill-understood strong interaction theory, it could not be expected to be very accurate. Ingenious resummation techniques were attempted, but such attempts, as would also be demonstrated at several occasions later, are fruitless if one does not understand the underlying physics.

And then there was the Yang–Mills theory.[20] What Frank (C. N.) Yang and Robert Mills had done, way back in 1954, would turn out to be extremely important: Indeed, if the only two force theories that are really successful, being electrodynamics and Einstein's General Relativity, are both based on some fundamental local symmetry, are there other ways to employ symmetries in a similar way, to describe different forces? When I was M. Veltman's undergraduate student, he would already point out this paper to us. "This you must know", he said, "this is very important". When I asked why, he said, "I don't know, just read it."

But the theory they came up with seemed to make no sense. Yang–Mills theory required the existence of massless spin-1 particles, much like photons, except that, unlike photons, they carry charges themselves. Such particles were known not to exist, that is, if charged spin-1 particles exist, they must have mass. Yang and Mills were wise enough nevertheless to publish their result. That result was a new kind of quantum field theory.

Without understanding the physics, mathematics does not answer your questions. Without understanding the mathematics, your physics theories will not work successfully either, is what we had to discover (and we keep forgetting time and again).

What Veltman had seen, was that there seemed to be a deep connection between the experimental data concerning the weak force, and Yang–Mills theory. Martin (Tini) Veltman[21] commenced his own personal battle to make sense out of these strange observations. This was a quantum field theory, it had infinities that had to be renormalized, and Nature appeared to be telling him that these ideas should work. Nobody really understood the physics, but he did understand which mathematical equations had to apply. These were so complicated that he decided to construct a computer program to address lengthy equations. "Schoonschip", was the name of his program, a word that only Dutch citizens can pronounce, so that his property rights would be guaranteed. Schoonschip told Veltman that, indeed, there was something wrong with the physics of the Yang–Mills theory.

Another obscure corner was investigated by Peter Higgs,[22] Robert Brout and François Englert.[23] They enjoyed little attention when they argued that the symmetry employed by Yang and Mills had to be spontaneously broken. The reason for that was that this alley had also been closed by the "experts". There was the famous 'Goldstone Theorem':[24] *Whenever a symmetry is spontaneously broken, at least one particle must become massless.* Indeed, in the Gell-Mann–Lévy Model, the pion behaves as a massless particle. The weak interaction, however, did not seem to involve massless objects. Higgs, Brout and Englert saw no massless particles in their models either, but a major fraction of the community did not believe them. So they were mainly ignored. Veltman paid no attention at all to formal mathematics, so he believed neither Higgs, Brout and Englert, nor Jeffrey Goldstone. He only believed the experiments, and his computer.

It was in this atmosphere that, independently, three people did foresee models for the electric and weak forces that would later turn out to be the precursors of the Standard Model. Abdus Salam[25] gave some general talks advocating theories resembling what is now called the BEH mechanism to understand these forces. Shelley Glashow[26] saw how Yang–Mills photons, slightly modified to make them massive, could generate quite neatly weak forces as observed in the experiments, and Steven Weinberg[27] wrote down the most detailed theory for the entire lepton sector,[b] including the effects of the Higgs particle. They were mostly ignored, and even the authors themselves continued working on other subjects. The unsolved problem was how to renormalize these theories.

2. The New Ideas of the 1970s

Historians often talk of a 'crisis' that precedes one or more revolutions of thought for the realisation of new breakthroughs. I don't think that applies here. There was no crisis, new experimental results were coming in, the nature of our problems was clearly identified, and there were plentiful ideas. Yet, we had no advance warning that new landslides were ahead, and these came, in a very quick succession. Problems that at one time had looked hopelessly complex, were solved with unexpected elegancy, and when the dust settled, we had a beautiful and relatively simple "Standard Model" for all known subatomic particles.

It is also not true that our work on Yang–Mills theories was motivated by our wish to put the Standard Model on a proper mathematical footing, as the story is now often told. The Standard Model wasn't there yet, the only existing theories that had a more or less proper mathematical footing were QED, and models that included purely scalar fields, which did not seem to apply to anything. Landau's difficulty was still there, in both these systems. We wanted to understand how

[b]Weinberg left out all hadronic weak interactions, and this was for a very good reason: the hadrons did not seem to fit in his model. Weinberg understood that his model would predict strangeness changing neutral current interactions, while these were not observed in the experiments. The GIM mechanism, only to be discovered later, would turn out to be the explanation of this apparent contradiction.

to deal with infinities when there are fundamental vector particles (particles with spin-1). Veltman had seen correctly that the infinities are particularly mild when the interactions have a Yang–Mills structure. Two more things had to be done.

First, we needed to understand how the original Yang–Mills theory would have to be treated as a quantum field theory, and how its interactions would have to be renormalized, without jeopardising the local symmetry structure. This meant that one cannot simply say that $\infty - \infty =$ something finite, but one has to establish how these finite expressions reflect the correct symmetry structure and dispersion relations.[c] The local symmetry is a gauge symmetry, just as in electromagnetism, and so, these theories were also called gauge theories.

What are the Feynman rules? Feynman had discovered that the mathematical equations for field theories can be framed in terms of neat sets of rules. In the new theories, however, Feynman's rules could be phrased in many ways, and they did not seem to be equivalent. The original idea was that the propagation of particles (real or virtual ones) is represented by lines, called 'propagators', in Feynman's diagrams, but now we seemed to get lines that did not describe a particle at all. Worse still, early investigations by Bryce DeWitt,[28] Ludwig Faddeev, Victor Popov,[29] Richard Feynman[30] and Stanley Mandelstam[31] all seemed to produce *different* rules for the propagators!

We had to understand what the new rules for these 'ghost particles' would be. This problem was solved,[32] the hard way, meaning that we analysed diagram by diagram. We found that there are many different ways to generate Feynman rules, so that DeWitt, Faddeev, Popov and Mandelstam all were using correct rules, except Feynman himself: he had used the rules for *massive* Yang–Mills theories, which are not the same, because there is a third spin direction that does not go away when you send the mass to zero. This is why Feynman was unable to go beyond one loop: he used the wrong theory (massive Yang–Mills theory without a physically observable Higgs particle, was later found not to be reormalisable).

To see how unitarity and dispersion relations work out in a gauge theory, we selected the proper equations that should be obeyed by the renormalised diagrams. These equations were the non-Abelian generalisations of the older Ward–Takahashi identities for QED, which we needed only when the external lines are on their mass shells. The identities looked like symmetry relations, but we could not identify the symmetry, because of all sorts of curious minus signs everywhere. We did not bother to work out how our relations have to be modified when we go off mass shell, although our combinatorial proofs did use off-shell diagrams. This omission was quickly corrected when, independently, Andrei Slavnov[34] and John C. Taylor[35] wrote down the more complete expressions.

Although we thought that our proofs worked just fine, not everyone was happy with our diagrammatic formalism. It was a few years later when Carlo Becchi, Alain

[c]Indeed, the old ideas to use dispersion relations and symmetries were still quite useful!

Rouet, Raymond Stora and Igor Tyutin[36] found a more elegant procedure to handle the Feynman rules: they discovered a curious, apparently unphysical supersymmetry, now called BRST symmetry, which holds for theories where gauge conditions have been chosen, while they ensure the possibility to transform to different gauge choices. BRST made extensive use of the Slavnov–Taylor identities. So these *were* symmetry relations, and our curious minus signs were trying to tell us that this was a supersymmetry!

With these problems out of the way, we seemed to be ready to renormalize the theory. All one had to keep in mind was that the BRST symmetry should not be disturbed. From where we were then, it was also not difficult to give mass to the Yang–Mills bosons.[33] The next step that we planned to take, turned out to have been analysed earlier: the older papers by Englert, Brout, Higgs and Weinberg[22,23,27] were quickly unearthed and found to be relevant. The result of our mathematical excursion could almost have been guessed (as Salam and Weinberg had done): *If you write the Lagrangian density for the theory, you read off all dynamical variables and all interaction parameters. They must all have a strict canonical form; in that case, a systematic perturbation expansion can be set up, and if the interaction parameters are not too big, you get a very accurate theory.*

However, another difficulty showed up: anomalies. It could easily happen that when a symmetry property is imposed on one aspect of an interaction amplitude, a violation of a similar symmetry property elsewhere pops up, as had been noted by Steve Adler, John Bell and Roman Jackiw.[37] These anomalies resemble a lid that does not properly fit onto a jar. One such anomaly causes neutral pions to decay into two photons, while chiral symmetry would have forbidden such a decay. Do we have non-Abelian gauge anomalies? If so, these would be standing in the way of our renormalization procedure.

The problem would be solved if we could find a 'gauge invariant regulator'. A regulator is some procedure, invoking some hidden physical phenomenon, that makes the theory finite. We wanted such a regulator that respects local gauge symmetry. We knew the regulator that was often used for QED. It had been found by Wolfgang Pauli and Felix Villars[38] that one can introduce 'very heavy ghost fermions' that do the desired trick. We did not succeed in finding such ghost particles that work in the Yang–Mills case. One trick was a procedure that would later become popular as the 'Kaluza–Klein theory': Theodor Kaluza and Oskar Klein[39] had proposed to employ an extra 'fifth dimension'. Particles moving in the fifth dimension could act as our regulator, but only for the first quantum corrections, the ones described by Feynman diagrams with only one closed loop in them. In spite of vigorous attempts, we could not tame the diagrams with more than one, overlapping, loops. Six or seven dimensions perhaps? To no avail.

The answer turned out to be that one has to use a continuously varying number n of dimensions, choose $n = 4 + \varepsilon$, where ε is only infinitesimally small, and take the limit $\varepsilon \to 0$ sufficiently carefully.[40] That worked! This method, to be called 'dimensional regularisation and renormalisation', would later turn out to have the additional advantage that the algebra stays really simple, so that calculations can be done quickly and efficiently. This way we learned how to renormalize the pure Yang–Mills theory up to all orders in the quantum corrections. This theory would now be at least as good as QED!

It soon turned out that dimensional regularisation had been discovered independently and practically simultaneously by C. G. Bollini and J. J. Giambiagi[41] and also by J. F. Ashmore.[42] They applied it to QED, showing that gauge-invariance is maintained there.

However, the anomalies give an extra twist to the story: the axial vector current is special to a theory in 4 space–time dimensions, so that dimensional renormalization does not directly work for that kind of current. This is why anomalies can occur in its conservation law. If gauge fields couple to axial vector currents, one must check explicitly whether these currents are properly conserved. *The anomalies must all cancel out, which requires a special relation between left helicity and right helicity fermionic particles.* Quite generally, this check needs to be performed only for the 1-loop diagrams, as was already understood by Adler and Bardeen.[43] If the chiral anomaly cancels out here, it will cancel out at all higher loop levels as well.

The importance of the discovery how to renormalize Yang–Mills theories with BEH mechanism, was almost immediately realised by a majority in the particle community. Benjamin W. Lee, Kurt Symanzik, Jean-Loup Gervais and Pronob Mitter had been lecturers in the 1970 Cargèse School of Subnuclear Physics, discussing the Gell-Mann–Lévy sigma model, so that they knew the importance of spontaneous symmetry breaking. Sidney Coleman, as a guru of mathematical physics at the time, embraced the new and important role of mathematical group theory. Also, all were delighted to see that now the rules of the game had been made clear. One can write down models, generalisations of, or alternatives to, the Standard Model, and immediately read off their main predictions.

One of the predictions was due to the practically unavoidable presence of a neutral component of the carriers of the weak force. Now, the effects of this *neutral current interaction* could be precisely calculated. It was predicted that the behaviour of neutrinos would be affected, as now they could scatter elastically against electrons, which could be confirmed with ingenious experiments.[44] Also, it was now strongly suspected that the quark spectrum known at the time could not be complete. The *charmed quark*, proposed by Glashow, John Iliopoulos and Luciano Maiani[45] in 1970 was needed to accommodate for the left–right asymmetry in the weak interactions; it explained the absence of flavor changing neutral current effects, and it was also needed to cancel out the chiral anomaly there.

Why were all these findings so remarkable? We now know that the resulting model could not be *infinitely* accurate. The canonical conditions on the parameters of the theory just happen to guarantee that perturbation expansions can be carried out *up to any order of the expansion*. The only thing that can go wrong — and it does — is that the expansion will not converge. In practice, however, this is only a formal difficulty; one can calculate the consequences of the theory much more precisely, in principle, than any measurement — provided the couplings are not too strong. So, paradoxically, the successes of the renormalization program for perturbative Yang–Mills theories are due to the fact that our theory cannot be the ultimate theory of Nature.[d]

3. The Strong Interactions

In those early days, there were other, more urgent difficulties. The *strong* interactions of course require strong interaction parameters, so by their very nature, these would not allow for a perturbative treatment. We had the important suggestion, inspired by experimental observations, and the ensuing phenomenological theories, that hadrons are formed of quarks, but these were never observed. It seemed that a fundamental new principle was at work here.

The resolution to this difficulty came again quickly and unexpectedly. Some mathematical features of the renormalisation procedure had been investigated, way back in the 1950s, by Stueckelberg and Petermann.[46] The freedom of choosing which part of the interaction parameters should be used to cancel infinities at higher orders, the core of the renormalization program, should manifest itself in a certain group property of the theory, which they called the 'renormalization group'. Actually, only that part of the renormalization group that gets involved in *scale transformations* leads to novel features that would otherwise be difficult to understand. So it happened that the renormalization group became practically synonymous to the group of scale transformations.

Invariance under renormalization group transformations was cast into equations by Curtis Callan[47] and Kurt Symanzik.[48] These equations contain new functions α, β, and γ, all depending on the coupling parameters, and describing what happens under scaling. Of these, particularly the function $\beta(\vec{g})$ became important, where $\vec{g}$ stands for the set of coupling parameters.[e] β tells us where the Landau ghost is. The crucial thing to find out is its sign.

Theorists thought they knew everything about that sign. The sign is positive. At least, this is so for QED as well as the theories with scalars, and it was almost proven to be a universal fact. This would mean that *every* quantum field theory should have its Landau ghosts, and these would cripple the theory. Moreover, if

[d]Here, we talk of the *purely theoretical* argument that the Standard Model is not infinitely precise. Besides that, we would later encounter other evidence from experimental observations (dark matter) and phenomenology (the failure of exact grand unification) that the Standard Model is incomplete.

[e]There was also a, less significant, dependence on mass parameters $\vec{m}$.

there were strong interactions, the Landau ghosts would be close by and nasty. In fact, this had been Landau's reason to dismiss quantum field theories altogether.

The story of the sign of the β (beta) function is one full of misunderstandings and miscommunications. How do vector particles (particles with spin 1) contribute to beta? How can renormalization counterterms switch sign? Why did everyone believe that the sign had to be positive?

The latter mistake is actually quite understandable. The renormalization of charge-like parameters is due to clouds of virtual particles hovering near a charged particle, due to vacuum fluctuations. A spinning particle is just like a spinless particle of which there are several species. Why should their effects depend so much on spin at all? Why should the *sign* change due to spin? Rough estimates, instead of accurate calculations, would indeed suggest a universal sign. Around 1970, David Gross was firmly committed to prove the universal sign of the beta functions, and he was about to publish his proof.[49]

In 1965, Vladimir Vanyashin and Mikhail Terentyev[50] found a negative sign in the charge renormalization of charged vector mesons, but they attributed this 'absurd' result to the non-renormalizability of the theory. In 1969, Iosif Khriplovich[51] correctly calculated the charge renormalization of Yang–Mills theories in the Coulomb gauge, where there are no ghosts. He found the unusual sign, but his important result was not noticed.

The story of the calculations of the β function is given in more detail by Misha Shifman.[52] He concludes that asymptotic freedom was not noticed before 1973, when David Politzer, David Gross, and Frank Wilczek published their results,[53,54] but actually the story told says something altogether different: asymptotic freedom was discovered three times[f] before 1973, but not recognized as a new discovery. This is just one of those cases of miscommunication. The "experts" were so sure that asymptotic freedom was impossible, that signals to the contrary were not heard, let alone believed. In turn, when I did the calculation I found it difficult to believe that the result was still not known.

In the mean time, in a very different topic of research, James (BJ) Bjorken[55] had found that scaling properties of hadrons may be easy to explain if, at very high energies, constituent particles of hadrons are weakly interacting. This was called Bjorken scaling. Now this would require a β function that is negative or vanishing. It was suspected that Bjorken scaling should therefore be the strongest argument *against* quantum field theories.

The author had done his calculations on how Yang–Mills theories scale in 1971,[g] found the negative sign, and decided he understood nothing of the arguments people had against field theories for the strong interactions. He did not see any Landau

---

[f]I am sure of the third case.

[g]A brief remark in Ref. 33, at the bottom of the first page, refers to this result, which he expected to be known to the other investigators.

ghost there, and started to investigate strong interactions his own way. The obvious candidate was a pure $SU(3)$ Yang–Mills theory *without* any BEH mechanism. Now, how does one *prove* that the quarks do not emerge as free particles? To advance such a theory, one would have to come with a credible explanation of the *quark confinement mechanism.* I was told that I should not make a fool of myself by publishing a theory of the strong force if *that* problem could not be addressed. This held me up. In June 1972, I did communicate my result about the negative β function, orally at a conference in Marseille. One of the attendants was Symanzik, who strongly urged me to publish it. "If you don't, someone else will", he said, and right he was.

Politzer,[53] a student of Coleman, and independently Gross and Wilczek,[54] were the first to publish their finding that $\beta(g)$ is negative for Yang–Mills fields, in 1973. They also understood its significance for understanding Bjorken scaling, and how it could help understanding quark confinement (in a qualitative way). This was the beginning of a more precise understanding of the strong force. The basis for the new theory had already been proposed by Harald Fritzsch, Murray Gell-Mann and Heinrich Leutwyler[56] in 1973, but in their earliest ideas they still had to struggle with the confinement problem, and the problem of the high-energy behaviour — the Landau ghost still seemed to be there. In a paper with Peter Minkowski,[57] the new name for this theory was coined: *Quantum Chromodynamics* (QCD).

4. The First Years of the Standard Model. Quantum Chromodynamics

By now, we had some glimpses of a new synthesis. In total, the local gauge symmetry structure appeared to be $SU(3)_{\text{strong}} \otimes SU(2)_{\text{weak}} \otimes U(1)_{\text{EM}}$. The leptons known at the time came in two $2_{\text{left}} \oplus 1_{\text{right}}$ representations, the quarks in two $3 \otimes 2_{\text{left}} \oplus (3 \oplus 3)_{\text{right}}$ representations (the third family would come somewhat later). The BEH mechanism involved a complex scalar doublet field. Three of its four real components provide mass to the charged W bosons and the Z boson. One component, the radial one, is gauge-invariant and should have observable energy quanta: the Higgs particle. It was known that the model we had here was not yet complete; there was no mechanism yet for CP violation, and it was not known whether neutrinos had mass. If they did, we would need extra neutral ("sterile") fermionic fields for them. It had been proposed by Makoto Kobayashi and Toshihide Maskawa[58] that a third family of quarks and leptons would provide for a natural mechanism for CP violation; the various members of this third family were discovered over the decades that followed.

An urgent theoretical problem was the better understanding of QCD. Just before the gauge theory revolution another development had taken place: the dual resonance models. Gabriele Veneziano's phenomenological expression for the elastic scattering of two mesons, yielding realistic descriptions of sequences of higher spin resonances, had been generalised to encompass the n-particle amplitude by Ziro Koba and Holger-Bech Nielsen.[59] They, along with Yoichiro Nambu and Lenny Susskind, realised how these expressions could be interpreted physically: these mesons are pieces of a quantised, relativistic string.

Now this could be reconciled with the quark picture of QCD if the forces holding quarks together can be described in terms of narrow vortex lines connecting the quarks. If these vortices may also be assumed to expand somewhat in the two transverse dimensions, then a clear picture arises of the behaviour of the QCD-gluon fields: they condense into vortex structures. The question was then, how to understand this behaviour starting from the Yang–Mills Lagrangian for QCD.

Here, an insightful observation by Nielsen and Poul Olesen[60] was very helpful. If one considers an Abelian Higgs theory, then this Higgs field is a single complex scalar. Such a field allows for a topologically stable string-like configuration, the Nielsen–Olesen vortex: this vortex occurs whenever the complex Higgs field makes a full rotation in the complex plane. This feature is well-known in the BCS theory for superconductors. A material that is infinitely conducting cannot contain a magnetic field; magnetic fields are shielded. If however, a magnetic field gets stronger than some limit, the superconductor becomes unstable, temporarily loses its superconductivity, and is forced to allow the magnetic field in. One then finds that such a field takes the shape of a vortex, the Abrikosov vortex, the solid-state analog of the Nielsen–Olesen vortex.

The Nielsen–Olesen vortex, and the Abrikosov vortex, should be stable, unless they can break in pieces; in the latter case, one should be able to describe what an end point looks like. In superconducting material, this is clear: the vortex carries magnetic flux, so end points can only occur if we have magnetic monopoles. In ordinary physical systems, magnetic monopoles do not exist, and so the Nielsen–Olesen vortex is stable. Incidentally, since the total magnetic flux in a vortex must be quantised in units of $2\pi/e$, where e is the electric charge quantum, this lead to a simple way to see that the magnetic charge g_m of magnetic monopoles must also be quantised the same way. This was already known by Paul Dirac.[61]

Curiously, Julian Schwinger[62] had arrived at $4\pi/e$ as the unit of magnetic charge. He had a problem with Dirac's value, which is half of that, so he thought Dirac's value is impossible. We'll show in Section 7 that indeed something special is going on at Dirac's charge, but our theories do not forbid that.

Can Nielsen–Olesen vortices also help understanding permanent quark confinement? At first sight, no. This would require quarks to behave as magnetic monopoles, but these were difficult to describe. If, as stipulated by Bjorken, at very high energies quarks behave nearly as free particles, their magnetic monopole charges, having values such as $2\pi/e$, should be weak, so that the gauge field couplings e should be very strong — this would be a contradiction.

What would the non-Abelian analogue be of the Nielsen–Olesen vortex? Here, something interesting happens. The stability of the vortex hinges on the question whether a closed loop in the space of gauge transformations is contractible or not. If it is contractible, we say that the gauge group is simply connected and the vortex is unstable, otherwise it is stable. In mathematical terms, the quantisation of vortices is controlled by the elements of the homotopy group π_1 of the gauge group G. Now, suppose we have a Higgs mechanism, breaking a gauge group such as $SO(3)$ into a subgroup, say $SO(2)$, or equivalently, $U(1)$. For $U(1)$ or $SO(2)$, the homotopy group is $\pi_1(U(1)) = \mathbb{Z}$, the addition group of the integers. For $SO(3)$, however, π_1 is only $\mathbb{Z}(2)$, the group of additions *modulo* 2. This means that if we would add two vortices, they could annihilate each other. In terms of magnetic monopoles, if you combine two monopoles with the same magnetic charge, they could annihilate one another.

This implied that, after the BEH mechanism is switched on, the fields could carry away a magnetic monopole charge to an amount of $4\pi/e$, or: it should be possible to construct a magnetic monopole with charge $4\pi/e$ out of the gauge fields in this model!

Once we realised that the gauge fields in a BEH system can generate magnetic monopoles, it was not hard to show how the construction went.[63] The result was of high physical interest: *If you have a unified gauge theory whose covering group is compact* (so that π_1 is finite), *it allows for the construction of magnetic monopoles!* Their physical properties, such as their masses, can be calculated. In general, the monopole mass is of order M_W/e^2, where M_W is the mass of the heaviest elementary vector particle in the theory.

The monopole solution of the gauge fields in a BEH system was discovered by the author and, independently, by Aleksandr Polyakov.[64] He had been searching for what he called a 'hedgehog' configuration. When discussing this at the Landau Institute in Moscow, it was Lev Okún who noticed that the thing carries a magnetic monopole charge (according to a footnote in his paper).

Now, if we call the QCD gauge coupling constant g_3, then, if there were a BEH mechanism, there would be gauge magnetic monopoles with charges $6\pi/g_3$, with a large mass, of order $1/g_3^2$. But, there is no reason to assume a BEH mechanism in QCD at all. Without this mechanism, there is no reason to assume any lower limit to the energy required for gauge-magnetic charges in QCD. Better still, it is reasonable to suspect *a BEH mechanism for the magnetic gauge charges* in QCD.

The effect of this would be wonderful. The QCD vortex must be the electromagnetic dual of a Nielsen–Oleson vortex! In that case, it was not the magnetic monopoles but the quark charges that would be confined by the vortex forces. Since triple electric monopole charges freely roam around in the QCD vacuum — these are now simply the QCD gluons — the quark confining forces will be active *modulo* 3, which is why both baryonic and mesonic states can be formed out of quarks.

Properties of the baryonic and mesonic bound states can be calculated numerically in lattice models. The QCD lattice was first discussed by Kenneth Wilson.[65] Formally, one can perform the $1/g_3$ expansion on the lattice, to observe that not only the confinement phase is realised in that formalism, but also chiral symmetry is broken exactly in the way observed, and in the way described by the qualitatively successful Gell-Mann–Lévy sigma model.

The pieces all fell into place this way, and studying all related dynamical properties of QCD became a big industry. The QCD vacuum is still a hot topic of research. It was found to depend on a *PC* violating angle θ due to instantons (see below), besides the number of light fermion species. In model calculations, various types of transitions are found to take place as these numbers are varied. The QCD vacuum also gets highly non-trivial properties when put in a box in 4-space, with various possible topological boundary conditions.[66]

A supplementary development was the idea of *jet physics*[67]: in high-energy scattering, jets emerge whose leading particle represents a single quark or gluon, fragmenting into mesons and baryons. The properties of these jets are calculable, so that they give us a means to compare theory with experiment.

5. The Large N Limit. Planar Diagrams

Even though our understanding of QCD was greatly improving, numerical calculations continued to be laborious and voluminous. We are still looking for more efficient approximation methods, such as the highly successful loop expansion in QED. What is needed is a small parameter in the theory, in terms of which we can perform asymptotic expansions. There is one parameter that perhaps could be used for this purpose. Suppose we replace the gauge group $SU(3)$ by a group of the form $SU(N)$, would there be a $1/N$ expansion? The question was asked by Claude Itzykson and Bernard Zuber,[68] but they thought that nothing special happens in the infinite N limit, since diagrams with many loops will still dominate.

Yet something special does happen: of all diagrams, *only the planar ones* survive.[69] Using a simple topological argument, this could be proved. A planar Feynman diagram is a diagram that can be drawn on a piece of paper without lines crossing one another. These diagrams emerge in what is now called the 't Hooft limit: $N \to \infty$, $\tilde{g}^2 \equiv g^2 N$ is kept fixed. This result was very suggestive. Planar diagrams are very similar to the world sheets of strings. Could it be that confinement can be proven in this limit?

The $N \to \infty$ limit indeed simplifies the computation of Feynman diagrams considerably, but summing exactly all diagrams that contribute, even in this limit, is not possible with presently known techniques. Today, we do know a lot more about this limit. It is frequently employed in AdS/CFT transformations.

6. Grand Unification

It was soon realised that what we call the 'Standard Model' today, may well be the tip of an iceberg. Having seen that the weak force and the electromagnetic force become (partly) unified, one could ask whether unification can occur in a more drastic fashion when we look further. On the one hand, one may speculate on wilder gauge field structures at moderate energies. Direct evidence for extended gauge structures have been lacking so far, but one is free to speculate. All sorts of models were suggested. A good attempt was the speculation that many or all of the elementary particles in today's Standard Model could turn out to be composites. The simplest construction that could lead to such a picture is a repetition of QCD at roughly a thousand times higher energies: "Technicolor". Today's quarks and leptons are tomorrow's mesons and baryons of Technicolor. Estia Eichten and Kenneth Lane[70] pioneered this theory, but, as of this moment, strong supporting experimental evidence for technicolor is still lacking; many of its predictions were falsified by experiments. Multitudes of repairs were attempted, but the theory is still not in a good shape.

On the other hand, it is also very tempting to try to unify the three gauge groups seen today, $SU(3)$, $SU(2)$ and $U(1)$, into one. Abdus Salam and Jogesh Pati, John Ellis and Dimitri Nanopoulos, together with many others, investigated various possibilities. One of the early approaches[71] involved the gauge group $SU(5)$. It regards a single generation of fermions as a 5 plus a 10 representation. The sterile neutrinos form invariant singlets.

Such a theory would predict the existence of physical magnetic monopoles, although these would be very heavy. It would also imply that protons are unstable, decaying into pions, electrons, positrons, muons, neutrinos or others. The decay time could be estimated. Ingenious experiments were designed and carried out, but no such decay has been detected thus far.

$SU(5)$ is a subgroup of $SO(10)$, and this is a nicer group, as it very elegantly puts each generation of fermionic particles (including the sterile neutrinos) in a single fermionic 16 representation.[72] This representation handles the fermions just as the space–time group $SO(3,1)$ does, as if one should combine these groups into a single $SO(13,1)$.

7. Magnetic Monopoles, Solitons and Instantons

The discovery of magnetic monopoles in certain unified theories demonstrated that gauge theories have a rich topological structure. It was an interesting exercise to search for more examples of this. *Solitons* are particle-like configurations that are stable on a one-dimensional line. They typically describe a boundary between two equivalent vacuum configurations. These are stable if one has a double-well potential in a real scalar field variable, both potential wells being equally deep (because of a symmetry). Subjecting these to a rigorous quantisation procedure is an interesting exercise in mathematical physics.

The Abrikosov–Nielsen–Oleson vortex is the two-dimensional analog of that, in the sense that it can be regarded as a soliton living in two space dimensions, taking the shape of a line if we add the third space dimension. It owes its existence to the fact that the minima of the Higgs field potential form a closed loop in field space. One can lasso this loop around the line.[h]

Next comes the magnetic monopole. It is a stable, particle-like configuration in three space dimensions. Apart from its indirect presence in QCD, no signs have been observed for its existence as a particle with a $U(1)$ magnetic charge. Further research has shown that these particles would be special indeed. Bernard Julia and Tony Zee[74] observed that, besides their magnetic charge, monopoles can also carry electric charge. This charge itself needs not obey the usual charge quantisation rule. In a two-dimensional plane where we plot the allowed values of the 2-vector $(q_i, g_{m\,i})$, where q_i is the electric charge and $g_{m\,i}$ the magnetic charge of a stable monopole-like configuration i, the outer products $(q_1\, g_{m2} - q_2\, g_{m1})$ must be integer multiples of 2π. This allows for a universal arbitrary angle: $g_{m\,i} = \frac{2\pi}{e} n_i$; $q_i = k_i e + \frac{\theta}{2\pi} n_i e$, where n_i and k_i are integers. The angle θ could be related to another angle in the unbroken theory, the instanton angle θ (see below).

Another curious feature in magnetic monopoles is due to the large value of their magnetic charges. Since elementary charged fermions carry magnetic moments with a gyromagnetic ratio of about 2, one can calculate that the dipole force between a charged fermion and a monopole (if the monopole is electrically neutral) can generate an attractive or repulsive $1/r^2$ potential that is so strong that it may cancel the usual angular momentum term in Schrödinger's equation. This means that these particles reach severely modified bound state configurations, which may need extra boundary terms for the two particles coming very close. At the close positions, effects due to the underlying grand unified theory may become sizeable, even if the fermions have very low energy. One finds that *baryon number decay* may become a strong force there. In short: magnetic monopoles could behave as very strong catalysts for proton decay. This discovery was made and worked out by Valery Rubakov and C. G. Callan.[75]

There is even more to be learned from magnetic monopoles. Careful analysis shows that, if we take an electric charge q orbiting a magnetic monopole with charge g_m, the total angular momentum takes values $q\, g_m/4\pi$ + integers. This means that a bosonic electric charge e and a bosonic magnetic monopole with magnetic charge $2\pi/e$ will produce fermionic bound states[76]! The calculation of these states is straightforward. The fact that also the statistical properties of the bound state are those of fermions was first understood by Alfred Goldhaber.[77]

Having seen soliton-like structures in one, two, and three dimensions, we can ask whether there is anything interesting going on in four dimensions. Are there topological structures that are stable in four dimensions? Such objects would be

[h]Sidney Coleman[73] gave famous lectures about this at the Erice Summer School of Subnuclear Physics. He demonstrated the notion of topology by winding the cord of his microphone around his neck. Students got worried that he might strangle himself.

space–time points, or, *event*-like. For that reason, they are called *instantons*. The simplest classically stable instanton can be seen to occur in a scalar field theory if the self-interaction is a curve with two (or more) minima, while the vacuum state is chosen to be in the 'wrong' minimum, the minimum that is locally stable but does not represent the lowest energy possible. This vacuum state would then not be absolutely stable: *quantum tunnelling* could cause the vacuum to decay spontaneously into the true vacuum. The tunnelling could initiate at one spot in space and time: the instanton event. Careful analysis of the mathematical physical question how to calculate the decay probability of the false vacuum, shows that the probability is dominated by the exponent of the total action of an $SO(4)$ invariant field configuration in Euclidean space–time, the scalar field instanton.

Non-Abelian gauge theories also have instantons. Alexander Belavin, A. Polyakov, Albert Schwarz, and Yu. S. Tyupkin[78] described a four-dimensional topological solution of the classical Yang–Mills field equations in Euclidean space, calling it a four-dimensional 'pseudo-particle'. They also made the interesting observation that the parity-odd expression $\int \mathrm{d}^4x\, F_{\mu\nu}\tilde{F}_{\mu\nu}$ takes values $32\,n\,\pi^2/g^2$ if you have n of such instantons. Now it was known that, due to the chiral anomaly, the axial current J^A_μ for massless fermions is not exactly conserved. Instead, it obeys

$$\partial_\mu J^A_\mu = \frac{g^2\,L}{16\pi^2} F_{\mu\nu}\tilde{F}_{\mu\nu}, \tag{1}$$

where L is the number of flavors. This means that exactly $2L$ units of chiral charge are annihilated or produced by this BPST instanton. Indeed, for every flavor one elementary fermion flips from left to right, or they all flip from right to left. The fact that the chirality of chiral fermions is not conserved in a BPST instanton can be understood by carefully inspecting the boundary conditions for fermions there, and noticing that these allow for a four-dimensional, zero action Jackiw Rebbi bound state.[79]

The BPST instanton describes tunnelling between different vacuum states in a gauge theory. These vacuum states form an infinite sequence of states, each connected to one another by topologically non-trivial gauge transformations. Classically, these quantum states are disconnected; quantum mechanically, they can tunnel to one another. In the quantum tunnelling amplitude, a phase rotation $e^{i\theta}$ may take place, where the angle θ emerges as a new constant of nature. The intensity of this instanton-induced process can be calculated from the exponent of the action of this instanton. The fact that there are zero energy, or action, solutions for fermions in monopoles and instantons, gives them very special physical properties. In mathematics, the numbers of such solutions are controlled by index theorems.[80] The bound state solutions furthermore play a role in establishing the nature and degeneracies of the instanton solutions themselves.

In effective field models for the mesons in QCD, there had been several problems. One was the decay $\pi^0 \to 2\gamma$. According to the theory of spontaneous breakdown of axial symmetry when light quarks are present, this decay should be strongly suppressed. The fact that it is not suppressed had been attributed to the Adler–Bell–Jackiw anomaly, Eq. (1). This was understood fairly well. However, the same symmetry should dictate that the η particle is the Goldstone boson of chiral symmetry, hence it should have a low mass value, comparable to the pions. The fact that the η mass is much bigger, presented us with the η *problem*. How do we compute this mass? We now found that the BPST instanton is the culprit. It generates an effective action where the number of chiral charges is not conserved, such as the η mass term. This effective action is strong, and it may also play a role in different systems. The admixture of biquark mesons with tetra-quark mesons, found in more recent experiments, is an example of a force that appears to be induced by the same instanton.[81]

These are instantons associated with the strong $SU(3)$ interaction, whose effects are highly visible in QCD, but we also have instantons in the electroweak $SU(2) \otimes U(1)$ sector of the Standard Model. Being weak, these instantons carry a very *large* action, and as such, the quantum tunnelling effects they describe are *extremely* weak. Now it so happens that, in the electroweak theory, baryon number is unevenly spread over the left-handed sector and the right-handed sector of the fermion spectrum. Consequently, the instanton induces baryon number violating interactions. Each generation of fermions hands in one unit of baryon number, so that a single electroweak instanton produces a change of three units in the total baryon number: $\Delta B = \pm 3$.

In fact, one can now understand why, in the perturbative regime, the gauge anomalies have to cancel out. If they hadn't cancelled out, left helicity and right helicity particles would have been produced in such a way that electric charge is not conserved; this would be inadmissible in any gauge theory.

This instanton tunnelling effect will be exorbitantly weak, yet it is thought perhaps to play a role in the genesis of baryon- and lepton-asymmetry in the early universe.[82] This is because, at very high temperatures, the transition can occur classically, without tunnelling. The transition goes *via* a classical, unstable field configuration called *sphaleron*, pioneered by Frans Klinkhamer and Nick Manton.[83]

It is a fact that instantons induce the kind of CP violation needed to understand the matter–antimatter asymmetry in the universe. Quantitative calculations however, show that there is still a problem: the matter–antimatter imbalance in our universe seems to be greater than one would expect from these calculations.

8. Supersymmetry and Gravity

While all these developments took place, and many new insights developed, there were numerous other, related researches that greatly affected our understanding of quantised fields. Without the fantastic experimental efforts over almost a century, we would not have been able to answer any of our questions, or even ask them.

Advances in numerical techniques, such as lattice theories using supercomputers, enabled us to investigate the QCD vacuum structure.

But a more peculiar feature turned out to yield all sorts of information that shines a new light on many of our findings: supersymmetry. It started when various coincidences were discovered in models that related fermionic and bosonic fields and their properties. It seemed more and more obvious that fermions and bosons, somehow, are related. In the early 1970s, Julius Wess and Bruno Zumino wrote down their first models that exhibited supersymmetry, first for lower-dimensional models, then for chiral theories with scalars and chiral Dirac fermions, later for gauge theories with fermionic and scalar fields, which displayed higher forms of supersymmetry, called $\mathcal{N} = 2$ and $\mathcal{N} = 4$ supersymmetry, and when Sergio Ferrara and Peter van Nieuwenhuizen hit upon some peculiar coincidences in perturbative quantum gravity, also the gravitational field was found to have supersymmetric connections with other fields. Here, we could continue all the way to $\mathcal{N} = 8$ supergravity.

Supersymmetric versions of QCD gave new insights into the various confinement modes. Furthermore, these theories appeared to display rich duality structures, relating one model to a different one. Supersymmetry also turned out to serve as an essential ingredient of the quantised relativistic string theories, originally designed to help us understand quark confinement, but later seen to function even better when used as a super unifying theory, connecting the gravitational force to all other forces in Nature. Without supersymmetry, these systems cannot be quantised in a completely satisfactory manner, so that this subject is now known as *superstring theory*. This theory is being advocated as an extremely promising approach towards quantising gravity, and it has indeed deeply transformed the topic of reconciling General Relativity with Quantum Mechanics. To what extent superstring theories have become more than promising, is discussed in other chapters of this book.

There is one thing theoreticians do not seem to be able to do. Being so ambitious as to attempt to produce a *Theory of Everything*, did not help us answer in a satisfactory manner, a very basic question: *Why is our universe so complex?* What we mean here is the obvious observation that our world is controlled by large and small numbers. There are numerous examples of this. Not only is Sommerfeld's fine-structure constant α fairly small, $\alpha \approx 1/137.036$, and m_p/m_e is fairly large, ≈ 1836.15, while the mass ratio for the electron and its neutrino is somewhere in the range 10^5 to 10^8, but we have much more extreme numbers:

$$\frac{M_{\rm Planck}}{m_{\rm proton}} = 1.301 \times 10^{19}; \quad \Lambda\, L^2_{\rm Planck} \approx 6 \times 10^{-122}. \tag{2}$$

The universe is so big and complex because its laws are based on very large and very small dimensionless numbers. Where do these numbers come from? This is the *hierarchy problem*. The only answer given today is the anthropic principle: *these numbers have the values they have because these would be the conditions for having life in the universe.* We wouldn't be there to ask the question, if these numbers had different values.

This argument would have been acceptable if the theory gave us a list of discrete numbers from which we could have chosen the constants. No theory today can give us such a list.

9. Calculations

In theoretical physics, models are often referred to as theories (Yang–Mills theory, String theory, ...) while theories are often called models. The Standard Model is a case in point. It now appears to describe the world of the fundamental particles so well, that it has become much more than a model. The time has come to refer to it as the *Standard Theory*, as in the title of this book.

We have learned how to do the perturbation expansion for this theory to all orders (in principle), we identified various non-perturbative features such as soliton-like solutions, confinement mechanisms, tunnelling through instantons, and effective descriptions of confined particles such as quarks and gluons in terms of *jets*. Now the question is how to do calculations more efficiently and accurately.

The perturbation expansion is formally divergent, but when the expansion parameter is not too big, the first two or three terms already give quite accurate results, and sometimes, such as in the case of $g-2$ for the electron or muon, impressive precision can be reached by including even higher terms. In many cases, however, this expansion is not good enough. Instanton effects are fundamentally non-perturbative, being dominated by exponential terms such as $e^{-8\pi^2/g^2}$, but subsequently, these can, in principle, be supplemented by ordinary perturbation expansions.

A persistent feature is that individual Feynman diagrams yield logarithms of mass/momentum ratios that can become rather large. One is then interested in identifying these 'leading logarithms' and to resum them. The leading logarithms can be understood by renormalization group arguments, and resumming them, in particular for strongly interacting particles such as quarks and gluons, has become an important industry.

An equally challenging question is how the perturbation expansion may behave at very high orders, so as to understand how their contributions can be rearranged and combined. Here again, instantons may play a role. Combining instantons and anti-instanton in pairs, such that these have vanishing topological winding numbers, their contribution is found to be $\mathcal{O}(e^{-16\pi^2/g^2})$, and this indicates that the perturbation series may diverge as

$$C\sum_{n=0}^{\infty} n!\, g^{2n}/(16\pi^2)^n, \tag{3}$$

a kind of behaviour that can be understood by performing Borel resummation methods. Indeed, as was argued by Lev Lipatov,[84] it is well-known from ordinary quantum mechanics, that tunnelling phenomena cause perturbation expansions to diverge this way. In terms of Feynman diagrams, one can observe that this divergence is related to the fact that the total number of Feynman diagrams diverges as $A^n n!$ at nth order, for some number A.

There is, however, another source of divergent behaviour in perturbation expansions. This has nothing to do with tunnelling, but is due to renormalization. If we consider a self-energy diagram that requires renormalization, it develops a logarithmic dependence of the momentum k^2. When a propagator with such a self-energy correction included, in turn occurs inside a loop diagram, one finds that integration over the momentum k generates $n!$-dependence because of that logarithm. Now this observation can be generalised by including renormalization group arguments to understand the k-dependence of sub-diagrams, and so get a more systematic understanding of $n!$ divergences of higher loop diagrams. The numbers of such diagrams increase geometrically with higher orders, that is, with powers A^n but without $n!$. The $n!$ dependence now comes from renormalization. Because of the similarity between these effects and the instanton effects, we attribute the new $n!$ dependence to *renormalons.* Should the contributions of renormalons and instantons be computed separately and just be added into the final expressions? No, by carefully studying the *instanton–anti-instanton gas*, and the way the instantons here behave under scaling, it was found that the instanton and renormalon effects blur in a more interesting way.[85]

Equipped with a better understanding of the perturbation expansion, could we now ask the question whether we can really understand quantum field theory beyond this expansion? Are there instances where the perturbation expansion can be made absolutely convergent? Should we not consider the axiomatic construction of all quantum field theory amplitudes, relying exclusively on convergent summations, as a mandatory step in developing true understanding of nature?

Attempts in that direction have been made. If, due to renormalization group behaviour, the coupling strengths of a theory run to infinity at high energies, then there is little chance that the theory can be rigorously defined, because the Landau ghost simply destroys it. Such theories can at best be *effective* theories, valid as approximative models only at energies well below the Landau ghost. Beyond the Landau ghost, the couplings of the original theory would run in the wrong direction, showing that unitarity, locality, and/or causality have broken down. This means that the theory in that domain would have to be replaced by something altogether different. In asymptotically free theories, there is much more reason for hope that they can be rigorously constructed, but even here, the mathematical proof has not been given.[i]

[i]In fact, the question of how to give a mathematically sound formulation of such theories, in particular proving that they have a mass gap, is one of the "Millennium Problems" formulated by the Clay Mathematics Institute, in Providence, RI.

What could be done is that we proved the existence of a planar limit, or the limit $N \to \infty$, with g^2N fixed, for an asymptotically-free theory, when both the far ultraviolet and the far infrared regions are controlled by the ordinary coupling constant expansion. From a physical point of view, such theories are utterly uninteresting, but, here at least, we could prove their mathematical existence.[86]

10. Conclusions and Outlook

Resurfacing as a phoenix from its ashes, in the 1970s, Quantum Field Theory has become a major discipline in elementary particle physics. Paradoxically, it owes its strengths to the fact that we could indicate where its limits are. Usually, we restrict ourselves to the cases where the theory is renormalizable. The real reason for doing this is that if an interaction with coupling strength g is renormalizable, we can perform a perturbation expansion in powers of g, so that very high accuracies can be reached, and very detailed comparisons with experiment can be made in a highly non-trivial manner.

Eventually, such theories will be plagued either by Landau ghosts or by very difficult infrared divergences such as in quantum chromodynamics. This implies that uncertainties of the order $e^{-C/g}$ or e^{-C/g^2} may become inevitable. If we wish to deal with these difficulties, we either must search for alternative calculation methods, such as lattice simulations in QCD, or search for altogether different theories that replace Landau ghosts by more acceptable descriptions.

In contrast, non-renormalizable theories may also still be useful as effective models if their coupling parameters g are so small that higher-order loop corrections can be omitted altogether, such as is the case in the older Fermi theory for the weak interactions, or in quantum gravity in practice today.

These observations have to be kept in mind, in order not to make the mistake to regard the demand of renormalizability as an absolute one in particle physics. It is a pragmatic demand. At first sight, quantum chromodynamics (QCD) seems to be an exception: the theory is renormalizable, and by using lattice simulations one can address its infrared behaviour. Here, however, we have to keep in mind that mathematical proofs for the internal consistency of this theory are still lacking. Most of us believe without doubt that the theory will work fine under all circumstances, with unlimited precision in principle, and we have good reasons for this belief, but we cannot be sure.

When theoreticians had exactly found out what the rules are for quantum field theories, it was soon realised that our theories form an infinite class of possible algebraic structures. Why stop at $U(1)$, $SU(2)$, $SU(3)$ and a rather small unifying group such as $SU(5)$ or $SO(10)$? In principle, we can imagine quantum field theories with unbounded complexity; after all, why don't we consider theories that describe endless sequences of elementary particles, with continuously increasing masses, and unending complexities in their algebra? We were suspecting such complications to come our way, long before LHC was switched on.

Here, however, it appears as if Nature confronts us with another surprise. The complications have still not arrived. Without any sign either of new technicolor forces, or supersymmetry, it looks as if the conventional Standard Model will be all that is needed. It seems as if elementary particles follow the textbooks on quantised relativistic field theories much better than we should have expected in all reasonableness. True, the indications for dark matter in the universe, found by astronomers, are very strong, so we know that there *is* more, but what is it, and why did our particle accelerators not reveal what dark matter is?

With the Standard Model working so well, the question what will bring it down, how and when, is becoming a more vexing one. We can only imagine two scenarios for the future paths of research that may bring us the answers: one, we persevere in building more energetic particle accelerators, that could unlash new physics by checking the standard theories with more precision, or two, we persevere in constructing more satisfactory theories, ones that address the naturalness problems, the hierarchy problem, the questions concerning dark energy and dark matter up front. Besides these two major alleys, we have our investigations of cosmological models, and also less conventional experiments such as the direct searches for dark matter, which all can help to direct us further.

At all these fronts, research must continue, and the role of renormalizable quantum field theories will be decisive. Will there be a new revolution? We all hope for new and interesting developments.

References

1. H. A. Kramers, *Die Grundlagen der Quantentheorie* (Akademische Verlagsgesellschaft, Leipzig, 1938) [transl. D. ter Haar, *Quantum Mechanics* (North-Holland Pub. Co. Amsterdam, 1957)].
2. J. Schwinger, *Phys. Rev.* **73**, 416 (1948); *Phys. Rev.* **74**, 1439 (1948).
3. F. J. Dyson, *Phys. Rev.* **75**, 486 (1949).
4. S-I, Tomonaga, *Prog. Theoret. Phys.* **1**, 27 (1946).
5. R. P. Feynman, *Rev. Mod. Phys.* **20**, 367 (1948); *Phys. Rev.* **74**, 939, 1430 (1948).
6. J. Mehra and K. A. Milton, Schwinger, Tomonaga, Feynman, and Dyson: The Triumph of Renormalization, in *Climbing the Mountain: The Scientific Biography of Julian Schwinger* (Oxford Univ. Press, 2003).
7. S. Laporta and E. Remiddi, The analytical value of the electron (g-2) at order a^3 in QED, *Phys. Lett. B* **379**, 283 (1996).
8. L. Landau, in *Niels Bohr and the Development of Physics*, ed. W. Pauli (Pergamon Press, London, 1955).
9. M. Gell-Mann and F. Low, Quantum electrodynamics at small distances, *Phys. Rev.* **95**, 1300 (1954).
10. M. Gell-Mann, *The Eightfold Way*, Caltech report CTSL-20 (1961), unpub.; *Phys. Lett.* **8**, 214 (1964); Y. Ne'eman, *Nucl. Phys.* **26**, 222 (1961).

11. G. Zweig, (1964), "An SU(3) model for strong interaction symmetry and its breaking", in *Developments in the Quark Theory of Hadrons*, Vol. **1**, 22–101, D. B. Lichtenberg, (Ed.), S. P. Rosen, (Ed.) and CERN Geneva — TH. 401 (REC.JAN. 64).
12. W. Heisenberg, Die 'beobachtbaren grössen' in der theorie der elementarteilchen-physik, *Z. Phys.* **120**, 513–38 (1943).
13. N. G. van Kampen, S matrix and causality condition. II. Nonrelativistic particles, *Phys. Rev.* **91**, 1267 (1953).
14. C. F. Chew, *S-Matrix Theory of Strong Interactions* (Benjamin Publ., New York, 1961).
15. T. Regge, *Nuovo Cim.* **14**, 951 (1959); G. F. Chew and S. C. Frautschi, *Phys. Rev. Lett.* **7**, 394 (1961).
16. R. P. Feynman, Mathematical formulation of the quantum theory of electromagnetic interaction, *Phys. Rev.* **80**(3), 440 (1950).
17. R. F. Streater and A. S. Wightman, *PCT, Spin and Statistics, and All That* (Benjamin, New York, 1964), latest ed. Princeton, 2000; A. S. Wightman, Quantum Field Theory in Terms of its Vacuum Expectation Values, *Phys. Rev.* **101**, 860 (1956).
18. M. Gell-Mann and M. Lévy, *Nuovo Cim.* **16**, 705 (1960).
19. M. L. Goldberger and S. B. Treiman, Decay of the pi meson, *Phys. Rev.* **110**, 1178 (1958); Form factors in β decay and μ capture, *ibid.* **111**, 354 (1958).
20. C. N. Yang and R. L. Mills, Conservation of isotopic spin and isotopic gauge invariance, *Phys. Rev.* **96**, 191 (1954); R. Shaw, Cambridge Ph.D. Thesis, unpublished.
21. M. Veltman, Divergence conditions and sum rules, *Phys. Rev. Lett.* **17**, 553 (1966); Perturbation theory of massive Yang-Mills fields, *Nucl. Phys. B* **7**, 637 (1968); J. Reiff and M. Veltman, Massive Yang-Mills fields, *Nucl. Phys. B* **13**, 545 (1969); M. Veltman, Generalized Ward identities and Yang-Mills fields, *Nucl. Phys. B* **21**, 288 (1970).
22. P. W. Higgs, Broken symmetries, massless particles and gauge fields, *Phys. Lett.* **12**, 132 (1964); Broken symmetries and the masses of gauge bosons, *Phys. Rev. Lett.* **13**, 508 (1964).
23. F. Englert and R. Brout, Broken symmetry and the mass of gauge vector mesons, *Phys. Rev. Lett.* **13**, 321 (1964).
24. J. Goldstone, Field theories with superconductor solutions, *Nuovo Cim.* **19**, 154 (1961); Y. Nambu and G. Jona-Lasinio, Dynamical model of elementary particles based on an analogy with superconductivity. I, *Phys. Rev.* **122**, 345 (1961); II, *ibid.* **124**, 246 (1961).
25. A. Salam and J. C. Ward, On a gauge theory of elementary interactions, *Nuovo Cim.* **19**, 165 (1961).
26. S. L. Glashow, Partial-symmetries of weak interactions, *Nucl. Phys.* **22**, 579 (1961).
27. S. Weinberg, A model of leptons, *Phys. Rev. Lett.* **19**, 1264 (1967).
28. B. S. DeWitt, Theory of radiative corrections for non-Abelian gauge fields, *Phys. Rev. Lett.* **12**, 742 (1964); Quantum theory of gravity. I. The canonical theory, *Phys. Rev.* **160**, 1113 (1967); II. The manifestly covariant theory, *ibid.* **162**, 1195 (1967); III. Applications of the covariant theory, *ibid.* **162**, 1239 (1967).
29. L. D. Faddeev and V. N. Popov, Feynman diagrams for the Yang-Mills field, *Phys. Lett. B* **25**, 29 (1967); L. D. Faddeev, The Feynman integral for singular Lagrangians, *Theor. and Math. Phys.* **1**, 1 (1969).
30. R. P. Feynman, The quantum theory of gravitation, *Acta Phys. Polonica* **24**, 697 (1963).
31. S. Mandelstam, Feynman rules for electromagnetic and Yang-Mills fields from the gauge-independent field-theoretic formalism, *Phys. Rev.* **175**, 1580 (1968).

32. G. 't Hooft, Renormalization of massless Yang-Mills fields, *Nucl. Phys. B* **33**, 173 (1971).
33. G. 't Hooft, Renormalizable Lagrangians for massive Yang-Mills fields, *Nucl. Phys. B* **35**, 167 (1971).
34. A. Slavnov, Ward identities in gauge theories, *Theor. Math. Phys.* **10**, 153 (1972) (in Russian), *Theor. Math. Phys.* **10**, 99 (1972) (Engl. Transl.).
35. J. C. Taylor, Ward identities and charge renormalization of the Yang-Mills field, *Nucl. Phys. B* **33**, 436 (1971).
36. C. Becchi, A. Rouet and R. Stora, Renormalization of the abelian Higgs-Kibble model, *Commun. Math. Phys.* **42**, 127 (1975); Renormalization of gauge theories *Ann. Phys. (N.Y.)* **98**, 287 (1976); I. V. Tyutin, Gauge invariance in field theory and statistical physics in operator formalism (In Russian), Lebedev Prepr. FIAN 39 (1975), unpubl.
37. S. L. Adler, Axial vector vertex in spinor electrodynamics, *Phys. Rev.* **177**, 2426 (1969); J. S. Bell and R. Jackiw, A PCAC puzzle: $\pi^0 \to \gamma\gamma$ in the sigma model, *Nuovo Cim. A* **60**, 47 (1969).
38. W. Pauli and F. Villars, Renormalizable Lagrangians for massive Yang-Mills fields, *Rev. Mod. Phys.* **21**, 434 (1949).
39. Th. Kaluza, Zum unitätsproblem in der physik, *Sitzungsber. Preuss. Akad. Wiss. Berlin. Math. Phys.* 966 (1921); O. Klein, Quantentheorie und fünfdimensionale Relativitätstheorie, *Z. Physik* **37**, 895–906 (1926).
40. G. 't Hooft and M. Veltman, Regularization and renormalization of gauge fields, *Nucl. Phys. B* **44**, 189 (1972).
41. C. G. Bollini and J. J. Giambiagi, Dimensional renormalization: The number of dimensions as a regularizing parameter, *Nuovo Cim. B* **12**, 20 (1972).
42. J. F. Ashmore, A method of gauge-invariant regularization, *Nuovo Cim. Lett.* **4**, 289 (1972).
43. S. L. Adler and W. A. Bardeen, Absence of higher order corrections in the anomalous axial vector divergence equation, *Phys. Rev.* **182**, 1517 (1969).
44. D. Haidt, *Discovery of Weak Neutral Currents*, Talk presented at the International XXVI Conference on Neutrino and Astrophysics at Boston USA, June 2, 2014.
45. S. L. Glashow, J. Iliopoulos and L. Maiani, Weak interactions with lepton-hadron symmetry, *Phys. Rev. D* **2**, 1285 (1970).
46. E. C. G. Stueckelberg and A. Petermann, La renormalisation des constants dans la theorie de quanta, *Helv. Phys. Acta.* **26**, 499 (1953).
47. C. G. Callan, Broken scale invariance in scalar field theory, *Phys. Rev. D* **2** 1541 (1970).
48. K. Symanzik, Small distance behavior in field theory and power counting, *Commun. Math. Phys.* **18**, 227 (1970).
49. D. J. Gross, in *The Rise of the Standard Model* (Cambridge Univ. Press, 1997), ISBN 0-521-57816-7 (pbk) p. 199.
50. V. S. Vanyashin and M. V. Terentyev, *Zh. Eksp. Teor. Phys.* **48**, 565 (1965) [*Sov. Phys. JETP* **21**, 375 (1965)].
51. I. B. Khriplovich, Green's functions in theories with non-abelian gauge group, *Yad. Fiz.* **10**, 409 (1969) [*Sov. J. Nucl. Phys.* **10**, 235 (1969)].
52. M. Shifman, Historical curiosity: How asymptotic freedom of the Yang-Mills theory could have been discovered three times before Gross, Wilczek, and Politzer, but was not, in *At the Frontier of Particle Physics, Handbook of QCD* (World Scientific, Singapore, 2001), Vol. 1, p. 126.

53. H. D. Politzer, Reliable perturbative results for strong interactions?, *Phys. Rev. Lett.* **30**, 1346 (1973); Asymptotic freedom: An approach to strong interactions, *Phys. Rep.* **14**, 129 (1974).
54. D. J. Gross and F. Wilczek, Ultraviolet behavior of non-abelian gauge theories, *Phys. Rev. Lett.* **30**, 1343 (1973).
55. J. D. Bjorken, Asymptotic Sum rules at infinite momentum, *Phys. Rev.* **179**, 1547 (1969).
56. H. Fritzsch, M. Gell-Mann and H. Leutwyler, Advantages of the color octet gluon picture, *Phys. Lett.* **47B**, 365 (1973).
57. H. Fritzsch, M. Gell-Mann and P. Minkowski, Vector-like weak currents and new elementary fermions, *Phys. Lett. B* **59**, 256 (1975).
58. M. Kobayashi and T. Maskawa, CP violation in the renormalizable theory of weak interaction, *Prog. Theor. Phys.* **49**, 652 (1973).
59. Z. Koba and H. B. Nielsen, Reaction amplitude for n-mesons, a generalization of the Veneziano-Bardakçi-Ruegg-Virasoro model, *Nucl. Phys. B* **10**, 633 (1969); Generalized Veneziano model from the point of view of manifestly crossing-invariant parametrization, *Z. Phys. A Hadrons and Nuclei* **229**, 243 (1969).
60. H. B. Nielsen and P. Olesen, Vortex-line models for dual strings, *Nucl. Phys. B* **61**, 45 (1973).
61. P. A. M. Dirac, Quantised singularities in the electromagnetic field, *Proc. Roy. Soc. A* **133**, 60 (1934); The theory of magnetic poles, *Phys. Rev.* **74**, 817 (1948).
62. J. Schwinger, Magnetic charge and quantum field theory, *Phys. Rev.* **144**, 1087 (1966).
63. G. 't Hooft, Magnetic monopoles in unified gauge theories, *Nucl. Phys. B* **79**, 276 (1974).
64. A. M. Polyakov, Particle spectrum in quantum field theory, *Zh. Eksp. Teor. Fiz. Pis'ma. Red.* **20**, 430 (1974) [*JETP Lett.* **20**, 194 (1974)].
65. K. G. Wilson, Confinement of quarks, *Phys. Rev. D* **10**, 2445 (1974).
66. P. van Baal, Some results for SU(N) gauge fields on the hypertorus, *Commun. Math. Phys.* **85**, 529 (1985).
67. G. Sterman and S. Weinberg, Jets from quantum chromodynamics, *Phys. Rev. Lett.* **39**, 1436 (1977).
68. C. Itzykson, personal communication.
69. G. 't Hooft, A planar diagram theory for strong interactions, *Nucl. Phys. B* **72**, 461 (1974).
70. E. Eichten and K. Lane, Dynamical breaking of weak interaction symmetries, *Phys. Lett. B* **90**, 125 (1980).
71. H. Georgi and S. Glashow, Unity of all elementary-particle forces, *Phys. Rev. Lett.* **32**, 438 (1974).
72. H. Georgi, The state of the art — gauge theories, in *Particles and Fields 1974*, ed. Carl E. Carlson, AIP Conference Proceedings Vol. 23 (1975), pp. 575–582.
73. S. Coleman, The uses of instantons, in *The Whys of Subnuclear Physics*, Erice 1977, ed. A. Zichichi (Plenum New York and London, 1979), p. 805.
74. B. Julia and A. Zee, Poles with both magnetic and electric charges in non-Abelian gauge theory, *Phys. Rev. D* **11**, 2227 (1975).
75. V. A. Rubakov, Superheavy magnetic monopoles and proton decay, *JETP Lett.* **33**, 644 (1981) [*Pisma Zh. Eksp. Teor. Fiz.* **33**, 658 (1981)]; Adler-Bell-Jackiw anomaly and fermion number breaking in the presence of a magnetic monopole, *Nucl. Phys. B* **203**, 311 (1982); C. G. Callan, Disappearing dyons, *Phys. Rev. D* **25**, 2141 (1982); Dyon-fermion dynamics, *Phys. Rev. D* **26**, 2058 (1982).

76. R. Jackiw and C. Rebbi, Solitons with fermion number ½, *Phys. Rev. D* **13** 3398 (1976); Vacuum periodicity in a Yang-Mills quantum theory, *Phys. Rev. Lett.* **37**, 172 (1976).
77. A. Goldhaber, Connection of spin and statistics for charge-monopole composites, *Phys. Rev. Lett.*, **36**, 1122 (1976).
78. A. A. Belavin, A. M. Polyakov, A. S. Schwarz and Yu. S. Tyupkin, Pseudoparticle solutions of the Yang-Mills equations, *Phys. Lett. B* **59**, 85 (1975).
79. G. 't Hooft, Symmetry breaking through Bell-Jackiw anomalies, *Phys. Rev. Lett.* **37**, 8 (1976); Computation of the quantum effects due to a four-dimensional pseudoparticle, *Phys. Rev. D* **14**, 3432 (1976) [Erratum *ibid.* **18**, 2199 (1978)].
80. M. F. Atiyah, N. J. Hitchin and I. M. Singer, Deformations of instantons, *Proc. Natl. Acad. Sci. U.S.A.*, **74**, 2662 (1977); M. F. Atiyah, N. J. Hitchin, V. G. Drinfeld and Yu. I. Manin, Construction of instantons, *Phys. Lett. A* **65**, 185 (1978).
81. G 't Hooft, G. Isidori, L. Maiani, A. D. Polosa and V. Riquer, *A theory of scalar mesons*, *Phys. Lett. B* **662**, 424 (2008).
82. M. E. Shaposhnikov and G. R. Farrar, Baryon asymmetry of the universe in the minimal standard model, *Phys. Rev. Lett.* **70**, 2833 (1993).
83. F. R. Klinkhamer and N. S. Manton, A saddle-point solution in the Weinberg-Salam theory, *Phys. Rev. D* **30**, 2212 (1984).
84. L. N. Lipatov, Divergence of the perturbation-theory series and the quasi-classical theory, *Zh. Eksp. Teor. Fi.* **72**, 411 (1977).
85. E. V. Shuryak, The role of instantons in quantum chromodynamics (i): Physical vacuum, *Nucl. Phys. B* **203**, 93 (1982).
86. G. 't Hooft, Rigorous construction of planar diagram field theories in four-dimensional Euclidean space, *Comm. Math. Phys.* **88**, 1 (1983), Reprinted in: *The large N Expansion in Quantum Field Theory and Statistical Physics: From Spin Systems to 2-dimensional Gravity*, eds. E. Brézin and S. R. Wadia (World Scientific, Singapore, 1993), pp. 157–181.

Chapter 2

The Making of the Standard Theory

John Iliopoulos
Laboratoire de Physique Théorique, École Normale Supérieure,
24 rue Lhomond, 75005 Paris, France

1. Introduction

The construction of the Standard Model, which became gradually the Standard Theory of elementary particle physics, is, probably, the most remarkable achievement of modern theoretical physics. In this Chapter we shall deal mostly with the weak interactions. It may sound strange that a revolution in particle physics was initiated by the study of the weakest among them (the effects of the gravitational interactions are not measurable in high energy physics), but we shall see that the weak interactions triggered many such revolutions and we shall have the occasion to meditate on the fundamental significance of "tiny" effects. We shall outline the various steps, from the early days of the Fermi theory to the recent experimental discoveries, which confirmed all the fundamental predictions of the Theory. We shall follow a phenomenological approach, in which the introduction of every new concept is motivated by the search of a consistent theory which agrees with experiment. As we shall explain, this is only part of the story, the other part being the requirement of mathematical consistency. Both went in parallel and the real history requires the understanding of both. In fact, as we intend to show, the initial motivation was not really phenomenological. It is one of these rare cases in which a revolution in physics came from theorists trying to go beyond a simple phenomenological model, not from an experimental result which forced them to do so. This search led to the introduction of novel symmetry concepts which brought geometry into physics. It is this exciting story which will be presented here.

2. Prehistory

2.1. *The electron spectrum in β-decay*

Chadwick vs Hahn and Meitner — Ellis and Wooster settle the issue.

In the early years of the 20th century the only known weak interaction was nuclear β-decay, which was believed to be a two-body decay of the form $N_1 \to N_2 + e^-$. The first revolution came from the study of the electron spectrum, which, for a two-body decay, is expected to be mono-energetic. The measurements were performed by two groups: (i) Otto Hahn and Lise Meitner[1] in Berlin were measuring the electron energy by looking at the penetration depth in various materials. The precision was not very good and their assumptions were rather crude, but the results could be interpreted as compatible with mono-energetic rays. (ii) James Chadwick, working in the group of Hans Geiger, also in Berlin, used a magnetic spectrometer and an electron counter. His results,[2] first published in 1914, showed instead a continuous spectrum, incompatible with a two-body decay. This was the first *energy crisis* in β-decay.

Hahn and Meitner attempted to explain the continuous spectrum by the re-scattering of the electron in nuclear matter, before its ejection from the nucleus. (Remember, they were still using the old nuclear model of a nucleus being a bound state of protons and electrons.) The issue was settled by a calorimetry experiment performed by Charles Drummond Ellis and William Alfred Wooster in 1927. They measured the total energy released during a certain number of decays N and they found[3] that $E_{\text{tot}} = NE_{\text{mean}}$, the *mean* energy of the electron spectrum, while, if Hahn and Meitner were right, they should have found $E_{\text{tot}} = NE_{\text{max}}$. The electrons were indeed emitted with a continuous spectrum. This was the second, and the most serious, *energy crisis.* Meitner, who declared having felt "a great shock" reading the paper, repeated the experiment and confirmed the result,[4] but she proposed no explanation. Resolving the energy crisis was left to the theorists.

2.2. *Enter the neutrino*

Pauli vs Bohr — Bohr looses the battle.

In the meantime the crisis was getting worse with new evidence showing that not only energy, but also angular momentum was not conserved and the Pauli exclusion principle was violated. Faced with such challenges many prominent physicists were ready to abandon the validity of all conservation laws in the new physics. The most important among them was no lesser man than Niels Bohr who, already in 1924,[5] had a scheme in which energy was conserved only *in the mean.* Heisenberg, and even Einstein and Dirac among others, toyed with this idea for a while.

This confusion brings us to December 1930. A Conference was organised in Tübingen to debate all the relevant issues. Pauli was invited but he decided not to attend. He sent a letter instead,[6] written in an inimitable style, in which he makes a bold suggestion: nuclear β-decay *is not a two-, but a three-body decay!* The electron is accompanied by a light, neutral, weakly interacting particle which carries away part of the energy. If, in addition, it is assumed to have spin 1/2, all problems are solved. Bohr did not give up immediately and was considering energy violating theories as late as 1936, but, by and large, the new particle was generally accepted. Pauli had called it "neutron", but when Chadwick discovered our neutron in 1932, Fermi coined the name "neutrino". In Fermi's theory we shall see next, the neutrino is a particle like any other. The first direct observation of the neutrino had to wait until 1956[7] with Frederick Reines and Clyde Lorrain Cowan. Their first announcement of the discovery was a telegram to Pauli.[8]

2.3. *Fermi's Tentativo*

Quantum Field Theory becomes the language of particle physics.

Already in 1926, before the introduction of the Schrödinger wave equation, Fermi had published two papers with the rules of counting particles which established the Fermi quantum statistics and gave fermions their name.[9] In 1933 he came back with one of the most influential papers in particle physics in which he proposes a field theory model for the β-decay of neutrons. Even today, when this theory has been superseded by the Standard Model of weak interactions, Fermi's theory is still used as a good low energy approximation.

This paper contains many revolutionary ideas. Fermi was one of the first physicists who believed in the physical existence of the neutrino. Contrary to Heisenberg, in the Bohr–Pauli controversy Fermi sided clearly with the latter. But he went further and broke completely with the prevailing philosophy, according to which particles which come out from a nucleus ought to be present inside it.[a] In his paper he formulated the full quantum field theory for fermion fields and introduced for them the formalism of creation and annihilation operators. It was the first time that quantised fermion fields appeared in particle physics. The paper appeared at the beginning of 1934[11] in Italian[b] under the title *Tentativo di una teoria della emissione di raggi* β.

Fermi's starting point was an analogy with the electromagnetic interactions in which the current j_μ produced by the charged particles acts as the source of the electromagnetic potential A^μ. This simple idea influenced all further developments we shall review in this book. He introduced quantised fermion fields for the electron, $\psi_{(e)}(x)$, and the neutrino, $\psi_{(\nu)}(x)$. He did not do the same for the nucleons because there was still a confusion regarding the magnetic moment of the proton and it

[a]Similar ideas had been expressed before by D. Iwanenko, in 1932 and Francis Perrin in 1933. The latter wrote: "... The neutrino ... does not preexist in atomic nuclei, it is created when emitted, like the photon",[10] but Fermi was the first to show how such a thing could actually happen.

[b]An english version had been submitted earlier to *Nature*, but it was rejected "because it contained speculations too remote from reality to be of interest to the reader".

was not clear whether the Dirac equation was applicable to them. He bypassed this difficulty by considering a static density for the nucleons and used the isospin operators $\tau_\pm$ Heisenberg had introduced earlier for the nuclear forces. The result was the following expression for the β-decay interaction Hamiltonian:

$$H_I = \frac{G_F}{\sqrt{2}}[\tau_- \psi^\dagger_{(e)}(x)\psi_{(\nu)}(x) + \tau_+ \psi^\dagger_{(\nu)}(x)\psi_{(e)}(x)] \tag{1}$$

where $\dagger$ means "hermitian adjoint" and $\frac{G_F}{\sqrt{2}}$ is a coupling constant, F stands for Fermi.[c] As we see, the effort to understand the nuclear forces went in parallel to that of the weak interactions and they influenced each other considerably. By 1936 the use of Dirac fields quantised *à la* Fermi for the nucleons became common in describing all nuclear interactions and the β-decay Hamiltonian of Eq. (1) took the form:

$$H_I = \frac{G_F}{\sqrt{2}} \sum_{i=1}^{5} \bar{\Psi}_{(p)}(x) O_i \Psi_{(n)}(x) \bar{\psi}_{(e)}(x) O_i \psi_{(\nu)}(x) \tag{2}$$

where the sum runs over the five Dirac invariants. It is the form under which the Fermi theory is known.[d]

2.4. *The high energy behaviour*

Infinities are never good.

An important property of the Fermi Hamiltonian was discovered in 1936 by Markus Fierz,[13] who computed the cross section for neutrino scattering and found that, at high energies, it increases with the neutrino energy:

$$d\sigma(\bar{\nu} + p \to n + e^+) = \frac{G_F^2}{2\pi^2} p_\nu^2 d\Omega \tag{3}$$

where p_ν is the neutrino momentum in the centre-of-mass system and $d\Omega$ is the element of the solid angle of the positron momentum. Similar conclusions were reached also by Heisenberg for the inelastic cross sections. One could guess these results by a simple dimensional analysis, taking into account the fact that the coupling constant G_F has the dimensions of inverse mass square. It became immediately obvious that such a behaviour is unacceptable because, at sufficiently high energies, the higher order terms will exceed this lowest order result, which means that an expansion in powers of G_F is meaningless. This problem haunted weak interactions for many years and led to the formulation of the new theory whose exposition is the subject of this book.

---

[c]This is the modern notation. Fermi used simply the symbol g.

[d]It is not clear whether Fermi ever wrote this form. It is possible that he had introduced it in one of his lectures. It appeared for the first time in a review article by H. A. Bethe and R. F. Bacher in 1936.[12]

3. Thirty Years of Unconcern, Thirty Years of Doubt

3.1. *Fermi's theory as the most successful phenomenology*

Elegance is the name of the game.

Following an impressive series of experimental and theoretical investigations, the form (2) was gradually reduced to a superposition of only the vector and pseudo-vector parts in the V-A combination which violates maximally the invariance under space inversions. Even a superficial history of the subject should include some of the early experiments, which count among the most significant and beautiful ones in physics. They include the discovery of parity violation in the β-decay of ^{60}Co by Chien-Shiung Wu in 1956[14] and the measurement of the neutrino helicity by Maurice Goldhaber in 1957.[15] On the other hand, the main theoretical ideas were: (i) The β-decay $\leftrightarrow$ μ-decay universality. (ii) The Conserved Vector Current (CVC) hypothesis. (iii) The Partial Conservation of the Axial Current (PCAC) hypothesis. (iv) The Cabibbo extension to $SU(3)$ currents and the generalised universality condition. All these gave rise to the current$\times$current theory, with the current being a sum of a leptonic and a hadronic part, of the form:

$$H_I = \frac{G_F}{\sqrt{2}} J^\mu(x) J^\dagger_\mu(x) \quad ; \quad J^\mu(x) = l^\mu(x) + h^\mu(x). \tag{4}$$

The leptonic part was written, as Fermi had done it in the early thirties, in terms of the fields of known leptons:[e]

$$l^\mu(x) = \bar{\nu}_{(e)}(x)\gamma^\mu(1+\gamma_5)e(x) + \bar{\nu}_{(\mu)}(x)\gamma^\mu(1+\gamma_5)\mu(x) \tag{5}$$

while the explicit form of the hadronic part depended on the assumptions regarding the strong interactions, in some sense coming back to the original Fermi formula (1). This simple and elegant form, not only described all weak interaction phenomena known at the time, but also led to the discovery of several fundamental symmetry properties in particle physics. It was a very satisfactory model, especially if one compared it with the situation in strong interactions, for which we had neither a successful phenomenology, nor an elegant form.

3.2. *Fermi's theory as the most inspiring model*

Chiral symmetry — Current algebra.

The leptonic part of the weak interaction current being determined, the effort was concentrated on the hadronic part $h^\mu(x)$. In a typical semi-leptonic weak interaction one needs the matrix elements $\langle a|h^\mu|b\rangle$, where $|a\rangle$ and $|b\rangle$ are hadronic states. So, when we say "to determine h^μ", we really mean "to identify it with a known operator of the strong interactions". The only operators whose properties are supposed

[e] Today we should add a third term for the τ-lepton.

to be known independently of the details of a particular dynamical model, are the currents of whichever symmetries one assumes for strong interactions. This way the effort to understand the structure of the weak interaction Hamiltonian helped discovering fundamental symmetries of the strong interactions. The important steps were the following:

- CVC. The vector part of the strangeness conserving weak current is the charged component of the isospin current.[16]
- The Cabibbo universality condition. Nicola Cabibbo generalised the universality and the CVC conditions to include the strangeness changing currents.[17] He wrote the weak hadronic current as:

$$h_\mu(x) = \cos\theta h_\mu^{(\Delta S=0)}(x) + \sin\theta\; h_\mu^{(\Delta S=1)}(x) \tag{6}$$

  with θ "the Cabibbo angle". This way the weak current became a member of an $SU(3)$ octet. Cabibbo checked that this hypothesis fits the experimental results on the decays of strange particles.
- PCAC, or Partially Conserved Axial Current. The CVC hypothesis answered the question of determining the vector part of the weak hadronic current by identifying it with the $SU(3)$ symmetry current. Could an analogous determination be extended to the axial part?[18] At first sight the answer is no, because no approximate axial symmetry seems to be present in the spectrum of hadrons. Yoichiro Nambu[19] gave the correct explanation: strong interaction dynamics is approximately invariant under chiral transformations, but the symmetry is spontaneously broken. The pions, or the octet of 0^- bosons if we extend the idea to $SU(3)$, are the corresponding Nambu–Goldstone bosons.[20]
- The Algebra of Currents. These considerations led Murray Gell-Mann[21] to postulate an algebraic scheme which translated the approximate symmetries of strong interactions. The symmetry group is assumed to be:

$$U(3) \times U(3) \sim U(1) \times U(1) \times SU(3) \times SU(3). \tag{7}$$

  To this symmetry correspond 18 conserved, or approximately conserved, currents out of which we can construct 18 charges, the generators of the group transformations. It is convenient to write them as follows: (i) Q_V for the vector $U(1)$. (ii) Q_A for the axial $U(1)$. (iii) Q_R^a for the right-hand $SU(3)$ and (iv) Q_L^a for the left-hand part, $a = 1, 2, \ldots, 8$. In the limit of exact symmetry they satisfy the commutation relations:

$$[Q_R^a\,, Q_R^b] = if^{abc}Q_R^c\;;\;\; [Q_L^a\;,\; Q_L^b] = if^{abc}Q_L^c\;;\;\; a, b, c = 1, 2, \ldots, 8 \tag{8}$$

  with all other commutators vanishing. f^{abc} are the structure constants of $SU(3)$ and a sum over repeated indices is understood. It is instructive to see the fate of the various factors in (7): (i) The vector $U(1)$ remains as an exact symmetry and the corresponding conservation law is that of baryon number. (ii) The axial

$U(1)$ puzzled people for a long time and it took some years before it was finally understood that, at the quantum level, the symmetry is broken by a phenomenon we shall see later, called "the axial anomaly". (iii) The $SU(3) \times SU(3)$ part is spontaneously broken as:

$$SU(3)_R \times SU(3)_L \rightarrow SU(3)_V \tag{9}$$

with $SU(3)_V$ being the diagonal subgroup of the chiral $SU(3)_R \times SU(3)_L$ which contains only vector currents. It is called "the flavour group". As we said before, the corresponding Nambu–Goldstone bosons form the 0^- octet of flavour $SU(3)$. (iv) There was a hierarchy of strong interactions: the "very strong interactions" were invariant under the full $SU(3)_V$. The "medium strong interactions" were assumed to break explicitly $SU(3)_V$ and leave invariant only the isospin subgroup.

3.3. *Fermi's theory as a an effective field theory*

Where is the cut-off, or the vital importance of precision measurements.

Fermi's theory cannot be viewed as a fundamental theory because, in technical terms, it is *non-renormalisable.* In practical terms this means that, if we write any physical amplitude as a power series in the Fermi coupling constant G_F, every term in the expansion requires the introduction of a cut-off parameter Λ. In a renormalisable theory, such as quantum electrodynamics, there exists a well-defined prescription to take the limit $\Lambda \rightarrow \infty$ and obtain unambiguous results, but to a non-renormalisable theory the prescription does not apply. The cut-off must remain finite and its value determines the energy scale above which the theory cannot be trusted. This is the definition of *an effective theory.*

Can we estimate an order of magnitude for the cut-off? A very simple method is the following: Ordinary dimensional analysis tells us that a physical quantity $\mathcal{A}$, for example a weak decay amplitude, can be written in a series expansion as:

$$\mathcal{A} = A_1 G_F \left(1 + \sum_{n=2}^{\infty} A_n (G_F \Lambda^2)^{n-1} \right) \tag{10}$$

where, in every order of the expansion, we have kept only the highest power in Λ. We see that the expression $g_{\text{eff}} = G_F \Lambda^2$ acts as an effective, dimensionless coupling constant. The expansion will become meaningless when $g_{\text{eff}} \sim 1$, which, for the numerical value of G_F, gives $\Lambda \sim 300\,\text{GeV}$, a value which, for the accelerators of the 1960s, was essentially infinite.

It was B. L. Ioffe and E. P. Shabalin,[23] from the Soviet Union, who first remarked that, in fact, one can do much better. Let us go back to the expansion (10) and consider also the sub-dominant terms in powers of Λ. We can rephrase their argument and write any physical quantity as a double expansion in g_{eff} and G_F:

$$\mathcal{A} = \sum_{n=0}^{\infty} A_n^{(0)} g_{\text{eff}}^n + G_F M^2 \sum_{n=0}^{\infty} A_n^{(1)} g_{\text{eff}}^n + (G_F M^2)^2 \sum_{n=0}^{\infty} A_n^{(2)} g_{\text{eff}}^n + \cdots \tag{11}$$

where the quantities $A_n^{(i)}$ may contain powers of the logarithm of Λ. M is some mass parameter, which, for a typical quantity in particle physics, is of the order of 1 GeV. The first series contains the terms with the maximum power of Λ for a given power of G_F, they are called *the leading divergences.* Similarly, the second series contains all the *next-to-leading divergences*, the third the *next-to-next-to-leading divergences*, etc. Following Ioffe and Shabalin, let us choose for $\mathcal{A}$ a quantity in strong interactions, for example the energy levels in a nucleus. The leading divergences represent the weak interaction corrections to this quantity. But weak interactions violate parity and/or strangeness, therefore the high precision with which such effects are known to be absent in nuclear physics gives a much more stringent bound for Λ, of the order of 2–3 GeV. Similarly the next-to-leading divergences contribute to "forbidden" weak interaction processes, such as ΔS=2 transitions (the $K_L^0 - K_S^0$ mass difference), or $K_L^0 \to \mu^+\mu^-$ decays. Again, the precision measurements of such quantities give the same 2–3 GeV limit for Λ.

4. Gauge Theories

4.1. *Gauge invariance in classical physics*

From electrodynamics to general relativity.

Classical electrodynamics is traditionally formulated in terms of the electric and magnetic fields which form a redundant set of variables. A first step towards a more reduced system was the introduction of the vector potential during the first half of the nineteenth century, either implicitly or explicitly, by several authors independently. It appears in some manuscript notes by Carl Friedrich Gauss as early as 1835 and it was fully written by Gustav Kirchhoff in 1857, following some earlier work by Franz Neumann.[22] It was soon noticed that it still carried redundant variables and several "gauge conditions" were used. The condition, which in modern notation is written as $\partial_\mu A^\mu = 0$, was proposed by the Danish mathematical physicist Ludvig Valentin Lorenz in 1867. However, the profound geometric interpretation of gauge invariance was not noticed until much later.

At the beginning of the twentieth century the development of the General Theory of Relativity offered a new paradigm for a gauge theory. The fact that it can be written as the theory invariant under local translations was certainly known to Hilbert, hence the name of *Einstein–Hilbert action.* The two fundamental forces known at that time, namely electromagnetism and gravitation, were thus found to obey a gauge principle. It was, therefore, tempting to look for a unified theory. Today we know the attempt by Theodor Kaluza, completed by Oscar Benjamin

Klein, which is often used in supergravity and superstring theories. These authors consider a theory of General Relativity formulated in a five-dimensional space–time (1+4). They remark that if the fifth dimension is compact the components of the metric tensor along this dimension may look to a four-dimensional observer as those of an electromagnetic vector potential. What is less known is that the idea was introduced earlier by the Finnish Gunnar Nordström who had constructed a scalar theory of gravitation. In 1914 he wrote a five-dimensional theory of electromagnetism[24] and showed that, if one assumes that the fields are independent of the fifth coordinate, the assumption made later by Kaluza, the electromagnetic vector potential splits into a four-dimensional vector and a four-dimensional scalar, the latter being identified to his scalar field of gravitation, in some sense the mirror theory of Kaluza and Klein.

4.2. *Gauge invariance in quantum mechanics*

The phase of the wave function.

The transformations of the vector potential in classical electrodynamics are the first example of an *internal symmetry* transformation, namely one which does not change the space–time point x. However, the concept, as we know it today, belongs really to quantum mechanics. It is the phase of the wave function, or that of the quantum fields, which is not an observable quantity and produces the internal symmetry transformations. The local version of these symmetries are the gauge theories we study here. The first person who realised that the invariance under local transformations of the phase of the wave function in the Schrödinger theory implies the introduction of an electromagnetic field was Vladimir Aleksandrovich Fock in 1926, just after Schrödinger wrote his equation. Fock noticed[25] that Schrödinger's equation, together with the normalisation condition of the wave function,

$$i\frac{\partial\Psi(\mathbf{x},t)}{\partial t} = -\frac{1}{2m}\Delta\Psi(\mathbf{x},t)\ ;\quad \int|\Psi(\mathbf{x},t)|^2 d\mathbf{x} = 1 \tag{12}$$

are invariant under the transformation $\Psi(\mathbf{x},t) \to e^{i\theta}\Psi(\mathbf{x},t)$, with θ a constant phase. Fock asked the question of what happens if the transformation becomes *local*, i.e. if we replace the constant θ by an arbitrary function of space and time. The answer is that we can restore invariance if we introduce a vector and a scalar potential $\mathbf{A}(\mathbf{x},t)$ and $A_0(\mathbf{x},t)$ and replace in the Eq. (12) the derivative operators by the covariant derivatives:

$$\left(\frac{\partial}{\partial t}\right)_{\text{cov}} = \frac{\partial}{\partial t} + ieA_0(\mathbf{x},t)\ ;\quad (\nabla)_{\text{cov}} = \nabla - ie\mathbf{A}(\mathbf{x},t) \tag{13}$$

where e is a constant introduced only for convenience. The new equation:

$$i\frac{\partial\Psi(\mathbf{x},t)}{\partial t} = \left[-\frac{1}{2m}(\nabla - ie\mathbf{A}(\mathbf{x},t))^2 + eA_0(\mathbf{x},t)\right]\Psi(\mathbf{x},t) \tag{14}$$

is invariant under local phase transformations, provided the potentials transform as:

$$A_0(\mathbf{x},t) \to A_0(\mathbf{x},t) + \frac{1}{e}\frac{\partial\theta(\mathbf{x},t)}{\partial t} \quad ; \quad \mathbf{A}(\mathbf{x},t) \to \mathbf{A}(\mathbf{x},t) - \frac{1}{e}\nabla\theta(\mathbf{x},t). \qquad (15)$$

This equation describes the motion of a charged particle in an external electromagnetic field. The electromagnetic interactions are generated by a gauge principle. In 1929 Hermann Klaus Hugo Weyl[f] extended this work to the Dirac equation.[26] In this work he introduced many concepts which have become classic, such as the Weyl two-component spinors and the vierbein and spin-connection formalism. Although the theory is no more scale invariant, he still used the term *gauge invariance*, a term which has survived ever since.

4.3. *From general relativity to particle physics*

The direct road is not always obvious.

Naturally, one would expect non-Abelian gauge theories to be constructed following the same principle immediately after Heisenberg introduced the concept of isospin in 1932. But here history took a totally unexpected route.

The first person who tried to construct the gauge theory for $SU(2)$ is Oskar Klein[27] who, in an obscure conference in 1938, he presented a paper with the title: *On the theory of charged fields.* The most amazing part of this work is that he follows an incredibly circuitous road: He considers general relativity in a five-dimensional space and compactifies *à la* Kaluza–Klein. Then he takes the limit in which gravitation is decoupled. In spite of some confused notation, he finds the correct expression for the field strength tensor of $SU(2)$. He wanted to apply this theory to nuclear forces by identifying the gauge bosons with the new particles which had just been discovered, (in fact the muons), misinterpreted as the Yukawa mesons in the old Yukawa theory in which the mesons were assumed to be vector particles. He considered massive vector bosons and it is not clear whether he worried about the resulting breaking of gauge invariance.

The second work in the same spirit is due to Wolfgang Pauli[28] who, in 1953, in a letter to Abraham Pais, developed precisely this approach: the construction of the $SU(2)$ gauge theory as the flat space limit of a compactified higher-dimensional theory of general relativity. He had realised that a mass term for the gauge bosons breaks the invariance and he had an animated argument during a seminar by Yang in the Institute for Advanced Studies in Princeton in 1954.[29] What is surprising is that Klein and Pauli, fifteen years apart one from the other, decided to construct the $SU(2)$ gauge theory for strong interactions and both choose to follow this totally counter-intuitive method. It seems that the fascination

[f]He is more known for his 1918 unsuccessful attempt to enlarge diffeomorphisms to local scale transformations.

which general relativity had exerted on this generation of physicists was such that, for many years, local transformations could not be conceived independently of general coordinate transformations. Yang and Mills[30] were the first to understand that the gauge theory of an internal symmetry takes place in a fixed background space which can be chosen to be flat, in which case general relativity plays no role.

4.4. *Yang–Mills and weak interactions*

With, or without, electromagnetism?

In particle physics we put the birth of non-Abelian gauge theories in 1954, with the fundamental paper of Chen Ning Yang and Robert Laurence Mills. It is the paper which introduced the $SU(2)$ gauge theory and, although it took some years before interesting physical theories could be built, it is since that date that non-Abelian gauge theories became part of high energy physics. It is not surprising that they were immediately named *Yang–Mills theories*. Although the initial motivation was a theory of the strong interactions, the first semi-realistic models aimed at describing the weak and electromagnetic interactions. In fact, following the line of thought initiated by Fermi, the theory of electromagnetism has always been the guide to describe the weak interactions.

Already in 1957, Julian Schwinger had conjectured[31] that the theory (4) should be modified with the introduction of an *intermediate vector boson* (IVB) $W_\mu^\pm$:

$$H_I = gJ^\mu(x)W_\mu^-(x) + \text{hc} \tag{16}$$

with g a new dimensionless coupling constant. This way weak interactions looked pretty much like the electromagnetic ones, a vector boson coupled to a current, but with some very important differences: (i) The photon is massless and the electromagnetic interactions are long ranged. The weak interactions are known to be short ranged, so the W's must be massive. (ii) The photon is neutral, the W's are charged. (iii) The electromagnetic current is conserved, the weak current is not. It was soon clear that these differences implied that the theory (16) was in fact as hopelessly non-renormalisable as (4).

The early attempts to use Yang–Mills theories to describe the weak interactions followed immediately the IVB hypothesis. Schwinger assumed the existence of a triplet of intermediate bosons, which he called $Z^{\pm,0}$, the two charged ones mediating the weak interactions and the neutral one being the photon. A year later, in 1958, S. A. Bludman[32] built the first $SU(2)$ Yang–Mills theory for weak interactions in which all three gauge bosons were coupled to $V - A$ currents. No connection with electromagnetism was assumed.

The most important contribution from this period dates from 1961 and it is due to Sheldon Lee Glashow.[33] He uses an $SU(2) \times U(1)$ gauge group, thus having two neutral gauge bosons. He is the first to propose a unified description for weak and electromagnetic interactions and introduces the idea of a mixing between the two neutral bosons. The photon field is a linear combination of the fields associated with $U(1)$ and the third generator of $SU(2)$ with an angle which he called θ (today it is called θ_W).

In the same year we have another important paper by Gell-Mann and Glashow.[34] This paper extends the Yang–Mills construction, which was originally done for $SU(2)$, to arbitrary Lie algebras. The well-known result of associating a single coupling constant with every simple factor in the algebra appeared for the first time in this paper. They even introduced the idea of a grand unified theory, as we shall explain later.

4.5. *A model for leptons*

The synthesis.

Gauge invariance requires the conservation of the corresponding currents and a zero masse for the Yang–Mills vector bosons. None of these properties seemed to be satisfied for the weak interactions. People were aware of the difficulty,[g] but had no means to bypass it. The mechanism of spontaneous symmetry breaking was invented a few years later,[35] in 1964. It is presented in a different Chapter in this Book, so here we shall continue the story after it.

The synthesis of Glashow's 1961 model with the mechanism of spontaneous symmetry breaking was made in 1967 by Steven Weinberg, followed a year later by Abdus Salam.[36,37] It is the work which gave rise to the Standard Model. The group is $U(1) \times SU(2)$ and has four gauge bosons, two charged ones and two neutral. At that time people did not yet know how to avoid the appearance of strangeness changing neutral currents, or $\Delta S = 2$ transitions, so the model applied only to leptons. The mechanism which allowed the extension to hadrons was found in 1970 and will be presented in the next section.

Many novel ideas have been introduced in this paper, mostly connected with the use of the spontaneous symmetry breaking which became the central point of the theory. They include:

- Its use for the weak interactions. We remind that the initial motivation was the breaking of flavour $SU(3)$.
- The fact that the same mechanism is the origin of the weak–electromagnetic mixing which had been postulated by Glashow.
- It is also the mechanism which gives masses to the fermions.

We shall present the general form of the model in a subsequent section.

[g]Glashow talks about *partially gauge invariant theories.*

5. Fighting the Infinities

5.1. *The phenomenology front*

A two-front battle.

A cut-off as low as 2–3 GeV was clearly unacceptable, it meant that, at least for some processes, Fermi's theory should be corrected already at low energies. The fact that the Fermi theory was non-renormalisable was known since the early years, but I believe it is fair to say that it was the work of Ioffe and Shabalin which showed that the problem was not only mathematical but also physical. A long and painful struggle against the infinities started. Although it was fought by few people,[h] it has been an epic battle given in two fronts: The first, the phenomenology front, aimed at finding the necessary modifications to the theory in order to eliminate the disastrous leading and next-to-leading divergences. The second, the field theory front,[38] tried to find the conditions under which a quantum field theory involving massive, charged, vector bosons is renormalisable. It took the success in both fronts to solve the problem. In this Chapter we shall describe the efforts in the first front.

5.2. *Early attempts*

Can we determine the Cabibbo angle? Are we ready to sacrifice elegance?

In the early attempts the effort was not focused on a particular physical problem, but aimed instead at eliminating the divergences, at least from physically measurable quantities. Some were very ingenious, but lack of space does not allow us to present them in any detail. A very incomplete list contains:

- The *physical* Hilbert space contains states with negative metric.[39] The introduction of negative metric states is considered unacceptable because it implies violation of the unitarity condition. However, Tsung Dao Lee and Gian Carlo Wick observed that, if the corresponding "particles", in this case the weak vector bosons, are very short lived, the resulting unitarity violations could be confined into very short times and be undetectable.
- The V-A form of the Fermi theory is an illusion and, in reality, the intermediate bosons mediating weak interactions are scalars.[40] By a Fierz transformation, the effective Lagrangian could look like a vector theory for some processes. This way the theory is renormalisable, but at the price of loosing all insight into the fundamental role of the weak currents.
- The theory (16) is an approximation and the real theory contains a large number of intermediaries with couplings arranged to cancel the most dangerous divergences.[41] The idea was simple: divergences arise in perturbation theory

[h]Most people doubted about the physical significance of the problem because of widespread mistrust towards field theory in general and higher order diagrams in particular. Since we had no theory, why bother about its higher order effects?

because a massive vector boson has a propagator which behaves like a constant at large momenta. This behaviour cannot be improved without violating unitarity. However, for a matrix valued field, we can obtain cancelations for some matrix elements. With a clever arrangement of the couplings, we can hide all bad divergences from the physically relevant quantities. A simple idea whose implementation turned out to be very complicated.[i]

- The weak interaction divergences and the value of the Cabibbo angle.[42,43] The idea was to compute the coefficient of the divergent term, for example in a loop expansion, for both the weak and the electromagnetic contributions. Setting this coefficient equal to zero gives an equation for the Cabibbo angle. The work by itself has today only a historical interest, but, as by-products, two interesting results emerged, summarised in the following two relations:

$$\tan\theta = \sqrt{\frac{m_d}{m_s}} \; ; \; \frac{|m_d - m_u|}{m_d + m_u} \sim \mathcal{O}(1) \tag{17}$$

where the masses are those of the three quarks. The first is in good agreement with experiment and relates the Cabibbo angle with the medium strong interactions which break $SU(3)$. The second, obtained by Cabibbo and Maiani, is more subtle: The prevailing philosophy was that isospin is an exact symmetry for strong interactions broken only by electromagnetic effects. In this case one would expect the mass difference in a doublet to be much smaller than the masses themselves. The second relation of (17) shows instead that isospin is badly broken in the quark masses and the approximate isospin symmetry in hadron physics is accidental, due to the very small values, in the hadronic mass scale, of m_u and m_d.

5.3. *The leading divergences*

The breaking of $SU(3) \times SU(3)$.

The leading divergences in the series (11) raised the spectrum of strangeness and parity violation in strong interactions. The first step was to find the conditions under which this disaster could be avoided.[44] The argument is based on the following observation: at the limit of exact $SU(3) \times SU(3)$ one can perform independent right- and left-handed rotations in flavour space and diagonalise whichever matrix would multiply the leading divergent term. As a result, any net effect should depend on the part of the interaction which breaks $SU(3) \times SU(3)$. In particular, one can prove that, under the assumption that the chiral $SU(3) \times SU(3)$ symmetry breaking term transforms as a member of the $(3, \bar{3}) \oplus (\bar{3}, 3)$ representation, the matrix multiplying the leading divergent term is diagonal in flavour space, i.e. it does not connect states with different quantum numbers, strangeness and/or parity. Therefore, all its effects could be absorbed in a redefinition of the parameters of the strong interactions and

[i]S. L. Glashow's remark: "Few would concede so much sacrifice of elegance to expediency".

no strangeness or parity violation would be induced. This was first found for the one-loop diagrams and then extended to all orders. This particular form of the symmetry breaking term has a simple interpretation in the formalism of the quark model: it corresponds to an explicit quark mass term and it was the favourite one to most theorists, so it was considered a welcome result.

5.4. *The next-to-leading divergences*

Lepton–hadron symmetry — Charm.

The solution of the leading divergence problem was found in the framework of the commonly accepted theory at that time. On the contrary, the next-to-leading divergences required a drastic modification, although, in retrospect, it is a quite natural one.[45]

Let us first state the problem. A firmly established experimental fact is that flavour changing weak processes obey certain selection rules: One of them, known as the $\Delta S = 1$ rule, states that the flavour number, in this case strangeness S, changes by at most one unit. A second rule is that the allowed ΔFlavour $= 1$ processes involve only charged currents. It follows that $\Delta S = 2$ transitions, as well as Flavour Changing Neutral Current processes (FCNC), must be severely suppressed. The best experimental evidence for the first is the measured $K_L - K_S$ mass difference which equals 3.48×10^{-12} MeV and, for the second, the branching ratio $B_{\mu^+\mu^-} = \Gamma(K_L \to \mu^+\mu^-)/\Gamma(K_L \to all)$ which equals 6.87×10^{-9}. It was this kind of tiny effects which led to the small value of the cut-off we mentioned earlier. In fact, this problem can be addressed at two levels. They are both easier to visualise in the framework of the quark model. At the limit of exact flavour symmetry, quark quantum numbers, such as strangeness, are not well defined. Any basis in quark space is as good as any other. By breaking this symmetry the medium strong interactions choose a particular basis, which becomes the privileged one. Weak interactions, however, define a different direction, which forms an angle θ with respect to the first one. Having only three quarks to play with, one can form only one charged current of the form postulated by Cabibbo:

$$J_\mu(x) = \bar{u}(x)\gamma_\mu(1+\gamma_5)[\cos\theta \; d(x) + \sin\theta \; s(x)]. \tag{18}$$

The expression (18) can be interpreted as saying that the u quark is coupled to a certain linear combination of the d and s quarks, $d_C = \cos\theta \; d + \sin\theta \; s$. The orthogonal combination, namely $s_C = -\sin\theta \; d + \cos\theta \; s$ remains uncoupled. Notice the difference with the leptonic current. We have four leptons, two neutrals, the $\nu_{(e)}$ and the $\nu_{(\mu)}$ and two negatively charged ones, the electron and the muon. They are all coupled and the weak current (5) has two pieces.

The first level of the problem is to consider a theory satisfying a current algebra. The neutral component of the current will be related to the commutator of J_μ and $J_\mu^\dagger$ and will contain terms like $\bar{d}_C d_C$, thus having flavour changing pieces. Notice again that this does not happen with the leptonic current. The commutator of the current (5) with its hermitian adjoint has no terms violating the two lepton flavour numbers. Phrased this way, the solution is almost obvious: we must use the s_C combination, but, in order to do so, we must have a second up-type quark. If we call it c, for *charm* the charged weak current (18) will have a second piece:

$$J_\mu(x) = \bar{u}(x)\gamma_\mu(1+\gamma_5)d_C(x) + \bar{c}(x)\gamma_\mu(1+\gamma_5)s_C(x) \tag{19}$$

or, in a matrix notation,

$$J_\mu(x) = \bar{U}(x)\gamma_\mu(1+\gamma_5)CD(x) \tag{20}$$

with

$$U = \begin{pmatrix} u \\ c \end{pmatrix}; \quad D = \begin{pmatrix} d \\ s \end{pmatrix}; \quad C = \begin{pmatrix} \cos\theta & \sin\theta \\ -\sin\theta & \cos\theta \end{pmatrix}. \tag{21}$$

The important point is that, now, a current J_3, given by the commutator of J and $J^\dagger$, is diagonal in flavour space.

This solves the first level of the problem, but it is not enough to explain the observed rates. For example, the $K_L \to \mu^+\mu^-$ decay can be generated by the box diagram of Fig. 1 which, although of higher order in the weak interactions, it is quadratically divergent and contributes a term proportional to $G_F g_{\text{eff}}$.

Here comes the second ingredient of the mechanism. With a fourth quark, there is a second diagram, with c replacing u, Fig. 2.

In the limit of exact flavour symmetry the two diagrams cancel. The breaking of flavour symmetry induces a mass difference between the quarks, so the sum of the two diagrams is of order $g^4(m_c^2 - m_u^2)/m_W^2 \sim G_F(G_F m_c^2)$. Therefore, Ioffe and Shabalin's estimations can be translated into a limit for the new quark mass and

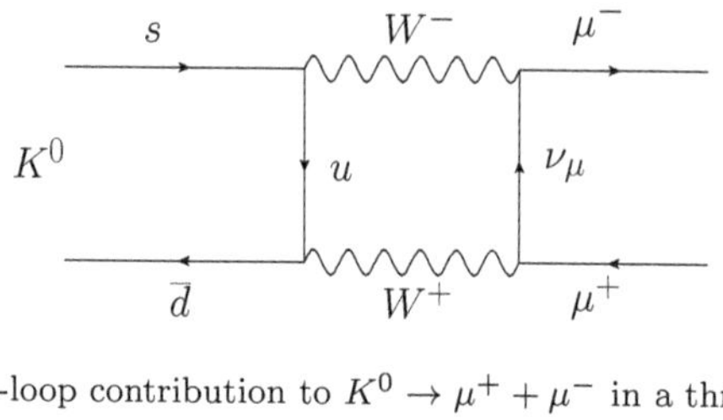

Fig. 1. The one-loop contribution to $K^0 \to \mu^+ + \mu^-$ in a three quark theory.

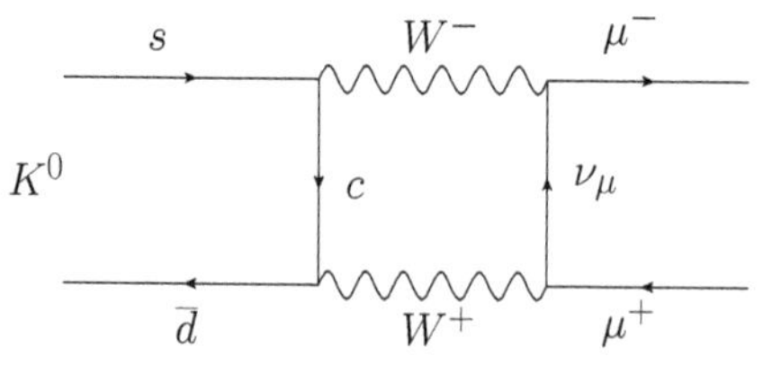

Fig. 2. The charm quark contribution.

yield an upper bound of a few GeV for the masses of the new hadrons. This fact is very important. A prediction for the existence of new particles is interesting only if they cannot be arbitrarily heavy.

In the early days of the Fermi theory there was a kind of symmetry between hadrons and leptons: proton–neutron *vs* neutrino–electron. The first discovery of heavy flavours appeared to break this symmetry: we had two new leptons, the muon and its associated neutrino, but only one new hadron, the strange quark. The introduction of the charmed quark restored this symmetry. By doing so, the mechanism obtained two important results: (i) It solved the technical problem of the low value of the Ioffe and Shabalin cut-off by replacing it with the masses of new hadrons. (ii) It opened the way to a formulation of the theory in terms of current operators which satisfy algebraic properties. It is this second result which allowed the use of Yang–Mills theories for the entire weak interactions, leptonic as well as hadronic.

6. The Standard Model

With the work done in the field theory front, and especially that of Gerard 't Hooft and Martinus Veltman, which is presented in a special Chapter in this Book, it became clear that weak *and* electromagnetic interactions are described by a gauge theory. The ball now was again in the phenomenology camp to decide which one.

6.1. *Which model?*

Do-it-yourself guide for gauge models.

We want to apply all the powerful machinery of gauge theories to the real world. The essential steps are the following:

- Choose a gauge group G.
- Choose the fields of the "elementary" particles whose interactions you want to describe and assign them to representations of G. Include scalar fields to allow for the Brout–Enblert–Higgs (BEH) mechanism.
- Write the most general renormalisable Lagrangian invariant under G. At this stage gauge invariance is still exact and all gauge vector bosons are massless.
- Choose the parameters of the BEH potential so that spontaneous symmetry breaking occurs. In practice, this often means to choose a negative value for a parameter μ^2.
- Translate the scalars and rewrite the Lagrangian in terms of the translated fields. Choose a suitable gauge and quantise the theory.

A remark: Gauge theories provide only the general framework, not a detailed model. The latter will depend on the particular choices made in steps 1) and 2).

6.1.1. *No neutral currents*

Today all experimental evidence points unmistakably to a single model, but this was not the case in the early days. In particular there was no evidence for the existence of weak neutral currents, so some models tried to avoid them. We mention two among them:

- The $SO(3)$ model.[46] The photon is the only neutral gauge boson, so the spontaneous symmetry breaking is $SO(3) \to U(1)$. The leptons belong to a triplet of $O(3)$, so we need heavy, positively charged partners of the electron and the muon. The hadronic sector is much more complicated. Even after the discovery of the weak neutral currents, this model has survived as a toy model because it has one interesting feature: the photon is a gauge boson of a simple Lie algebra. In this sense it is an elementary version of what we shall call *grand-unified theories* in the last section.
- A model without neutrino induced neutral currents.[47] This model had a short lifetime and it was proposed as a possible theoretical answer to a confusion regarding the existence of neutral currents in the early neutrino experiments.

6.1.2. *The $U(1) \times SU(2)$ model*

This is *the Standard Model.* It has four gauge bosons $W^{\pm}$, Z^0 and the photon. Following the notation which was inspired by the hadronic physics, we call T_i, $i = 1, 2, 3$ the three generators of $SU(2)$ and Y that of $U(1)$. Then, the electric charge operator Q will be a linear combination of T_3 and Y. By convention, we write: $Q = T_3 + \frac{1}{2}Y$. The coefficient in front of Y is arbitrary and only fixes the normalisation of the $U(1)$ generator relatively to those of $SU(2)$.

This ends our discussion of the first step. Regarding the choice of the matter fields, the model assumes a modular structure with the basic unit being a *family* of spin-1/2 chiral fermions. We start with the leptons for which the assignment in $SU(2)$ representations is as follows (the index i denotes the three families, the electron, the muon and the tau):

$$\Psi_L^i(x) = \frac{1}{2}(1 + \gamma_5) \begin{pmatrix} \nu_i(x) \\ \ell_i^-(x) \end{pmatrix}; \quad i = 1, 2, 3. \tag{22}$$

The right-handed components are assigned to singlets of $SU(2)$:

$$\nu_{iR}(x) = \frac{1}{2}(1 - \gamma_5)\nu_i(x) \ \ (?) \quad ; \quad \ell_{iR}^-(x) = \frac{1}{2}(1 - \gamma_5)\ell_i^-(x). \tag{23}$$

The question mark next to the right-handed neutrinos means that the presence of these fields is not confirmed by the data. We shall drop them in this Chapter, but there is a special one devoted to the neutrino masses. Notice that, with this assignment and in the absence of ν_R, the neutrinos will be massless and individual lepton numbers will be separately conserved.

For the hadrons, the model is written in terms of elementary quark fields. In order to explore the lepton–hadron universality property, it uses also doublets and singlets, but with some novel features, as compared to leptons:

- All quarks appear to have non-vanishing Dirac masses, so we must introduce both right-handed singlets for each family.
- Quark numbers are not individually conserved, so the formalism must allow for mixings among them, but not in the neutral current sector.
- We know that each quark appears in three species, called *colours*, so we have three times as many fields for each family.

Since left and right fields belong to different representations of $SU(2)$, all fermions are massless

$$Q_L^i(x) = \frac{1}{2}(1+\gamma_5)\begin{pmatrix} U^i(x) \\ D^i(x) \end{pmatrix}; \quad U_R^i(x); \quad D_R^i(x) \tag{24}$$

with the index i running over the three families as $U^i = u, c, t$ and $D^i = d, s, b$ for $i = 1, 2, 3$, respectively. An additional index a, running also through 1, 2 and 3 and denoting the colour, is understood.

We still have to introduce scalar fields for the mechanism of spontaneous symmetry breaking and the simplest choice is to have a doublet Φ containing a ϕ^+ and a ϕ^0 with the conjugate fields ϕ^- and ϕ^{0*} forming $\Phi^\dagger$.

These choices fully determine the model. What follows is straightforward algebra. We write the most general, renormalisable, Lagrangian, invariant under gauge transformations of $SU(2) \times U(1)$. The requirement of renormalisability implies that all terms in the Lagrangian are monomials in the fields and their derivatives and their canonical dimension is smaller or equal to four. The presence of the scalar fields generate Yukawa interactions with the fermions. The physical consequences of the model are obtained after spontaneous symmetry breaking and translation of the scalar field. We give a list of the most important ones:

- *Gauge boson mass terms.* The breaking is $U(1) \times SU(2) \to U(1)_{\rm em}$ with the generator of the electromagnetic group Q obtained as a superposition of T_3 and Y. As a result three gauge bosons become massive and the fourth one, the photon, remains massless. Let us call $\vec{W}$ and B the gauge fields associated to $SU(2)$ and $U(1)$ respectively and g and g' the corresponding coupling constants. After the breaking we obtain two charged bosons $W^\pm$ and two neutral ones, Z^0 and A, orthogonal combinations of W^3 and B:

$$Z_\mu = \sin\theta_W B_\mu - \cos\theta_W W_\mu^3; \quad A_\mu = \cos\theta_W B_\mu + \sin\theta_W W_\mu^3 \tag{25}$$

with $\tan\theta_W = g'/g$. They correspond to the mass eigenvalues

$$m_W = \frac{vg}{2}; \quad m_Z = \frac{v(g^2 + g'^2)^{1/2}}{2} = \frac{m_W}{\cos\theta_W}; \quad m_A = 0 \tag{26}$$

where v is the vacuum expectation value of the scalar field.

• *Fermion masses.* They come from the Yukawa couplings between the fermions and the scalar field. They are given by:

$$\mathcal{L} = \sum_{i=1}^{3}\left[-G_i(\bar{\Psi}_L^i R_i \Phi + \text{h.c.}) + G_u^i(\bar{Q}_L^i U_R^i \tilde{\Phi} + \text{h.c.})\right] + \sum_{i,j=1}^{3}\left[(\bar{Q}_L^i G_d^{ij} D_R^j \Phi + \text{h.c.})\right] \tag{27}$$

where the summation runs over the three families. A further summation for the three quark colours is understood. $\tilde{\Phi}$ is the doublet made out of ϕ^{0*} and ϕ^-. It has the same transformation properties under $SU(2)$ as Φ, but the opposite Y charge.

Two remarks on this expression: (i) We have assumed the absence of right-handed neutrinos and this explains the fact that we have only one term for leptons. (ii) In the two terms for the quarks we have chosen to diagonalise the one referring to the up-quarks. This is a convention. As a result, the coupling of the second term involving the down-quarks is non-diagonal in flavour space.

After translation of the scalar field we obtain masses for all fermions with the exception of the neutrinos. The model does not predict their values, but allows to fit them. They are all proportional to the corresponding Yukawa coupling constant G. In addition, after diagonalisation of the down-quark masses, we obtain the Cabibbo–Kobayashi–Maskawa matrix[48] for the weak couplings:

$$\text{CKM} = \begin{pmatrix} c_1 & s_1c_3 & s_1s_3 \\ -s_1c_3 & c_1c_2c_3 - s_2s_3e^{i\delta} & c_1c_2s_3 + s_2c_3e^{i\delta} \\ -s_1s_2 & c_1s_2c_3 + c_2s_3e^{i\delta} & c_1s_2s_3 - c_2c_3e^{i\delta} \end{pmatrix} \tag{28}$$

with the notation $c_k = \cos\theta_k$ and $s_k = \sin\theta_k$, $k = 1, 2, 3$. The novel feature is the possibility of introducing the phase δ. This means that a six-quark model has a natural source of CP, or T, violation, while a four-quark model does not.

- $W^{\pm}$*-fermion couplings.* The charged W's couple to the usual $V - A$ weak current. The hadronic part involves the CKM matrix. The model was designed for that.
- *Photon-fermion couplings.* Again, we obtain the usual electromagnetic current.
- Z^0*-fermion couplings.* This is a new feature of the model. The weak neutral current is a particular mixture of right- and left-components given by the same parameter θ_W which enters in the gauge boson masses. For example, the Z^0-lepton

coupling is given by:

$$-\frac{e}{\sin(2\theta_W)}\left[\bar{\nu}_L\gamma^\mu\nu_L - \cos(2\theta_W)\bar{e}_L\gamma^\mu e_L + 2\sin^2\theta_W\bar{e}_R\gamma^\mu e_R\right]Z_\mu. \quad (29)$$

- *Vector boson self-couplings.* They are characteristic of the Yang–Mills structure of the theory. In particular, the coupling of the photon with the charged W's involve a single coupling constant e and gives very specific predictions concerning the electromagnetic parameters of the W's. The gyromagnetic ratio equals two and the quadrupole moment equals $-em_W^{-2}$.
- *Scalar field-fermion couplings.* They come from the Yukawa term (27) and give the most important signature of the model: the strength of the coupling is proportional to the fermion mass.
- *Scalar self-coupling.* Another prediction of the model is that the value of the mass of the physical scalar particle determines the strength of its self-coupling. With the value of 126 GeV we find $\lambda \sim 1/6$, appreciable, but according to our experience, still in the perturbative regime.

The agreement of the theory with experiment has been spectacular.

6.2. *A problem of anomalies*

An obscure higher-order effect determines the structure of the world.

Gauge invariant quantum field theories present a special feature as compared to other field theories, to wit gauge invariance. In order to define the theory one needs to *fix the gauge*, i.e. to impose some condition to eliminate redundant degrees of freedom. A change of gauge produces a completely new theory. All these theories which look, and in many respects are, very different, must give the same answer for physical quantities. This is achieved because they are linked together through a set of relations called *Ward identities.* They are the results, at the level of the Green functions, of the conservation equations for the symmetry currents of the theory. But this raises a new problem:

Let us consider the example of quantum electrodynamics and, for simplicity, neglect the electron mass. At the classical level the Lagrangian is invariant under separate phase transformations of the right- and left-components of the electron field. This $U(1) \times U(1)$ symmetry yields two currents, a vector and an axial:

$$j_\mu(x) = \bar{\psi}(x)\gamma_\mu\psi(x); \quad j^5_\mu(x) = \bar{\psi}(x)\gamma_\mu\gamma^5\psi(x) \quad (30)$$

and, using the classical equations of motion, we see that they are both conserved. The problem arrives at the quantum level where we can prove that the two currents cannot be simultaneously conserved. It is the famous Adler–Bell–Jackiw[49] triangle anomaly[j] which tells us that, if we enforce the conservation of the vector current $J_\mu(x)$, the equation for the axial one becomes:

$$\frac{\partial}{\partial x_\mu} j^5_\mu(x) = \frac{e^2}{8\pi^2} \epsilon_{\nu\rho\sigma\tau} F^{\nu\rho}(x) F^{\sigma\tau}(x) \tag{31}$$

where e is the charge of the electron, $\epsilon_{\nu\rho\sigma\tau}$ is the completely anti-symmetric four index tensor which equals 1 if the indices form an even permutation of (0,1,2,3) and $F^{\nu\rho}$ is the electromagnetic field strength given by $F^{\nu\rho}(x) = \frac{\partial A^\rho(x)}{\partial x_\nu} - \frac{\partial A^\nu(x)}{\partial x_\rho}$ with $A^\nu(x)$ the electromagnetic vector potential. The r.h.s. of Eq. (31) is called *the axial anomaly*, which is a fancy way to say that the axial current of massless quantum electrodynamics is not conserved, contrary to what the classical equations of motion indicate.

This result has important physical consequences in particle physics but here we shall present only its implications for the electroweak theory. For quantum electrodynamics the non-conservation of the axial current can be considered as a curiosity because this current does not play any direct physical role. However, in the electroweak theory both vector and axial currents are important and in deriving the Ward identities we need the conservation of both. The axial anomaly breaks this conservation and the entire program collapses. As a result, the purely leptonic model, the one which was first constructed, is mathematically inconsistent.

The solution was first found in 1972.[50] The important observation is that the anomaly is independent of the fermion mass. Every fermion of the theory, light or heavy, contributes the same amount and we must add all contributions in order to get the right answer. For the electroweak theory this means that we need both the leptons and the quarks. A simple calculation shows that the total anomaly produced by the fermions of each family will be proportional to $\mathcal{A}$ given by:

$$\mathcal{A} = \sum_i Q_i \tag{32}$$

where the sum extends over all fermions in a given family and Q_i is the electric charge of the ith fermion. Since $\mathcal{A} = 0$ is a necessary condition for the mathematical consistency of the theory, we conclude that each family must contain the right amount of leptons and quarks to make the anomaly vanish. This condition is satisfied by the three colour model with charges 2/3 and $-1/3$, but also by other models such as the old Han–Nambu model which assumes three quark doublets with integer charges given by (1,0), (1,0) and (0,$-$1). In fact, the anomaly cancellation condition (32) has a wider application. The Standard Model could have been invented after the Yang–Mills theory was written, much before the discovery of the quarks. At that

[j] The term is slightly misleading, as it may give the impression that something contrary to common sense has happened. The real reason is that going from the classical equations to the quantum theory involves a series of steps which often include a limiting procedure, for example the limit of some parameter, the cut-off, going to infinity. This limit, although well defined, may not respect some of the symmetries of the classical equations.

time the "elementary" particles were thought to be the electron and its neutrino, the proton and the neutron, so we would have used one lepton and one hadron doublet. The condition (32) is satisfied. When quarks were discovered we changed from nucleons to quarks. The condition is again satisfied. If tomorrow we find that our known leptons and/or quarks are composite, the new building blocks will be required to satisfy this condition again. Since the contribution of a chiral fermion to the anomaly is independent of its mass, it must be the same no matter which mass scale we are using to compute it.

The moral of the story is that families must be complete.[k] Thus, the discovery of a new lepton, the tau, implied the existence of two new quarks, the b and the t, prediction which was again verified experimentally.

The above discussion was confined to the $SU(2) \times U(1)$ gauge theory but the principle of anomaly cancellation should be imposed in any gauge theory in order to ensure mathematical consistency. This includes models of strong interactions and grand-unified theories. H. Georgi and S.L. Glashow[51] found the generalisation of the anomaly equation (32) for a gauge theory based on any Lie algebra. It takes a surprisingly simple form:

$$\mathcal{A}_{abc} = \mathrm{Tr}\left(\gamma^5 \{\Gamma_a, \Gamma_b\} \Gamma_c\right) \tag{33}$$

where Γ_a denotes the Hermitian matrix which determines the coupling of the gauge field W_a^μ to the fermions through the interaction $\bar{\Psi}\gamma_\mu \Gamma_a \Psi W_a^\mu$. As we see, Γ_a may include a γ^5. Georgi and Glashow showed that the anomaly is always a positive multiplet of $\mathcal{A}_{abc}$, so this quantity should vanish identically for all values of the Lie algebra indices a, b and c.

Since gauge theories are believed to describe all fundamental interactions, the anomaly cancellation condition plays an important role not only in the framework of the Standard Model, but also in all modern attempts to go beyond, from grand unified theories to superstrings. It is remarkable that this seemingly obscure higher order effect dictates, to a certain extent, the structure of the world.

6.3. *The Standard Model becomes the Standard Theory*

The Standard Model wins all the battles.

The detailed comparison of the Standard Model with experiment will be shown in several Chapters of this Book. Obviously, in computing the theoretical predictions, one should include also the strong interactions, so the model is really the gauge theory of the group $U(1) \times SU(2) \times SU(3)$. Here we shall present only a list of the most spectacular successes in the electroweak sector.

[k]The title of the GIM paper was *Weak interactions with lepton–hadron symmetry*. With this work we showed that the title was indeed correct.

- The discovery of weak neutral currents by Gargamelle in 1972. Both, their strength and their properties were predicted by the Model.
- The discovery of charmed particles at SLAC in 1974–1976. Their characteristic property is to decay predominantly in strange particles.
- A necessary condition for the consistency of the Model is that $\sum_i Q_i = 0$ inside each family. When the τ lepton was discovered the b and t quarks were predicted with the right electric charges.
- The discovery of the W and Z bosons at CERN in 1983 was a remarkable achievement of experimental physics and accelerator technology. The characteristic relation of the Standard Model with an isodoublet scalar $m_Z = m_W / \cos\theta_W$ is checked with very high accuracy (including radiative corrections).
- The t-quark was *seen* at LEP through its effects in radiative corrections before its actual discovery at Fermilab.
- An impressive series of experiments have tested the Model at a level such that the weak interaction radiative corrections are important.
- The final touch: The recent discovery of the Brout–Englert–Higgs scalar.

All these successes give us full confidence that we have THE STANDARD THEORY of strong, electromagnetic and weak interactions of elementary particles.

7. Beyond the Standard Model

7.1. *Why and how*

We know why — we do not know how.

In spite of its enormous success, there are several reasons to suspect that the gauge theory of the Standard Model cannot be considered as a truly fundamental theory. Let us only mention some of its shortcomings.

- The family problem: why do we observe three, apparently similar, families of elementary fermions?
- The problem of masses. It is hard to believe that all these widely spread mass values are arbitrary parameters in a fundamental theory. This problem existed already with mass values such as m_e and m_t. It is accentuated with the values of the neutrino masses. We expect in a fundamental theory to be able to compute mass ratios.
- $U(1) \times SU(2) \times SU(3)$ is not a unified theory at all. Each group factor comes with its own coupling strength. Even worse is the presence of $U(1)$ because it allows for any number of coupling constants. We have already explained that in a non-Abelian group the coupling constant is fixed, but for $U(1)$ this is not so. In other words, the present theory does not explain why electric charge appears to be quantised and we do not see particles with charge πe. For the standard

model the observed very precise equality (up to one part in 10^{20}) of the electric charges of the proton and the positron, seems to be accidental.

- With the discovery of the scalar boson the Standard Model is complete. We can compute any quantity at any given order in the perturbation expansion. Following K. Wilson, we can fix a scale Λ and imagine that we integrate over all degrees of freedom with energies above Λ. We thus obtain an effective field theory describing the light, meaning lighter than Λ, degrees of freedom. Even without computations, we can guess the form of this effective theory by dimensional analysis. Integrating over the heavy degrees of freedom does not break any symmetry, so the effective theory will be a sum over all operators built out of the light fields consistent with the symmetries of the Standard Model. We can distinguish three classes of operators: (i) Operators whose dimension d_i is larger than four. Their contribution decreases as a power of Λ, so they become irrelevant for large Λ. (ii) Operators with $d_i = 4$. They are precisely the operators appearing in the Standard Model Lagrangian. (iii) Operators with $d_i < 4$. Their coefficients grow like positive powers of Λ, so they become dominant at large scales. In the Standard Model there are only two such operators: the unit operator **1** with dimension equal to zero and the operator Φ^2, where Φ is the scalar field, with dimension equal to two. The first contributes only to the induced cosmological constant which, in the absence of gravitational interactions, is not observable. We conclude that *the only dominant operator of the Standard Model is the mass term of the scalar boson.* It receives corrections which grow quadratically with the energy scale. This problem is often referred to as *the hierarchy problem* and it is a genuine instability of all generic quantum field theories involving scalar fields. This argument allows us to introduce the concept of *naturalness.* The underlying idea is that all physical theories are effective theories valid up to a certain scale, because we can never assume that we know physics at all scales. A quantum field theory will be called natural if the values of its parameters depend only logarithmically on this large energy scale. According to this definition, the Standard Model is not natural. It must be replaced by a different theory above a certain scale Λ.
- Last, but not least, the Standard Model leaves out the gravitational forces. Although, at present energies, the latter are very weak, we expect a fundamental theory to describe *all* fundamental interactions.

7.2. *The most beautiful speculations*

A personal choice.

If there are plenty of answers to the question *Why,* we are still in the dark concerning the question *How.* An old theoretical prejudice states that a better theory is a more symmetric one, so it is not surprising that most theoretical speculations aim at increasing the symmetry of the Standard Model. We give a short selection of theories, each one of which tries to address some of the problems mentioned above. None solves them all, which probably means that none is *the* correct theory.

7.2.1. *Grand unified theories*

The seed for grand unification can be found in the 1961 paper of Gell-Mann and Glashow.[34] In a footnote they write: The remarkable universality of the electric charge would be better understood were the photon not merely a singlet, but a member of a family of vector mesons comprising a simple partially gauge invariant theory. The first "realistic" grand-unified theories were proposed in the early seventies, just after the Standard Model was complete.[52]

The basic hypothesis of grand unification states that $U(1) \times SU(2) \times SU(3)$ is the remnant of a larger, simple or semi-simple group G, which is spontaneously broken at very high energies. The scheme looks like:

$$G \xrightarrow{M} U(1) \times SU(2) \times SU(3) \xrightarrow{m_W} U(1)_{e.m.} \times SU(3) \tag{34}$$

where the breaking of G may be a multistage one and M is one (or several) characteristic mass scale(s). As Gell-Mann and Glashow had observed, in order to explain electric charge quantisation, the charge operator should be a generator of G. Therefore it must be represented by a traceless matrix, which implies that the sum of the charges of the particles belonging to an irreducible representation must be equal to zero. This property is not true for either the known leptons, or the known quarks considered separately. We conclude that, unless we assume the existence of unknown exotic particles, an irreducible representation of G must include both leptons *and* quarks. This means that there exist gauge bosons of G which can change a lepton into a quark, or vice versa. Therefore, a generic prediction of GUT's can be stated as an alternative: new exotic particles, and/or violation of the separate conservation of baryons and leptons. An intense experimental effort has been devoted to the detection of a possible proton decay. The amplitude for such a decay is given by the exchange of the corresponding gauge boson and therefore, it is of order M^{-2}, where M is the gauge boson's mass. The resulting proton life-time will be of order:

$$\tau_p \sim \frac{M^4}{{m_p}^5}. \tag{35}$$

Using the experimental limit (for particular decay modes), of a few times 10^{33} years, we can put a lower limit on M:

$$M \geq 10^{16}\,\text{GeV}. \tag{36}$$

Grand unification is not a low-energy phenomenon!

Another general feature of grand unification concerns the three coupling constants of the Standard Model. At present energies they have very different numerical values. We can use the renormalisation group equations and follow their evolution as a function of the energy scale.[53] For the grand unification idea to be correct, they must reach roughly the same value at a scale of M. This property can be used to test each particular model and the result is not completely satisfactory. We shall come back to this in the next section.

A last general remark is that grand unified theories predict the existence of magnetic monopoles,[54] although their masses are of order M.

Several groups have been used for grand unification and an incomplete list includes:

- $SU(5)$. It is the simplest possible choice. The group of the Standard Model $U(1) \times SU(2) \times SU(3)$ is of rank four and we can prove that the only simple, or semi-simple, group of rank four which could be used for grand unification is $SU(5)$. The leptons of each family fill a $(10 + \bar{5})$ reducible representation. If a right-handed neutrino exists, it should belong to a singlet. There are 24 gauge bosons, 12 belonging to the Standard Model group and 12 new ones which, since they may mediate proton decay, they must be superheavy with masses of order M. The two step spontaneous symmetry breaking requires at least two distinct scalar field representations, the simplest system consisting of dimensions 24 and 5. We can immediately see the hierarchy problem we mentioned above: Let H and h denote the scalar fields of the 24 and 5 dimensional representations, respectively. They both get non-zero vacuum expectation values, the first $V \sim 10^{16}\,\text{GeV}$ and the second $v \sim 10^2\,\text{GeV}$. For this, the scalar potential must have terms M^2H^2 and μ^2h^2, with $M \sim V$ and, presumably, $\mu \sim v$. But if these terms are present, the term λH^2h^2 must also be present. Translating the field H by V will generate an induced mass term for h equal to λV^2. So, unless the coupling constant λ is of order 10^{-28}, the "light" field h will be pushed to the high mass scale, in other words the system is not capable to sustain naturally this hierarchy of widely separated mass scales. This is a problem with practically all grand unified theories.
- $SO(10)$. The simplest $SU(5)$ model does not fit the data very well and people have looked for higher groups. An attractive choice is the rank five group $SO(10)$. It has a sixteen dimensional spinorial irreducible representation capable of accommodating all chiral spinors of a family, including a right-handed neutrino. $SO(10)$ contains $SU(5)$ as a subgroup and the 16-plet decomposes under $SU(5)$ into $16 = 10 \oplus \bar{5} \oplus 1$. The proton decay prediction is similar to that of $SU(5)$. The main experimental prediction of $SO(10)$, which differs substantially from that of $SU(5)$, concerns the neutrino mass, but this problem will be addressed in another Chapter.
- The exceptional groups E_6 and E_8 have also been used. They offer many theoretical advantages but they have quite large representations (for example, the adjoint representation of E_8 has 248 dimensions), which means that a large number of up to now unknown particles are predicted.

7.2.2. *Supersymmetry*

Gauge theories contain three independent worlds. The world of radiation with the gauge bosons, the world of matter with the fermions and the world of BEH scalars. In the framework of gauge theories these worlds are essentially unrelated to each other. Given a group G the world of radiation is completely determined, but we have no way to know a priori which and how many fermion representations should be introduced; the world of matter is, to a great extent, arbitrary.

This arbitrariness is even more disturbing if one considers the world of BEH scalars. Not only their number and their representations are undetermined, but their mere presence introduces a large number of arbitrary parameters into the theory. Notice that this is independent of our computational ability, since these are parameters which appear in our fundamental Lagrangian. What makes things worse, is that these arbitrary parameters appear with a wild range of values. From the theoretical point of view, an attractive possibility would be to connect the three worlds with some sort of symmetry principle. Then the knowledge of the vector bosons will determine the fermions and the scalars and the absence of quadratically divergent counterterms in the fermion masses will forbid their appearance in the scalar masses. We shall call such transformations supersymmetry transformations and we see that a given irreducible representation will contain both fermions and bosons.[55] It is not *a priori* obvious that such supersymmetries can be implemented consistently, but in fact they can. The generators of the algebra contain operators Q which are fermionic with spin 1/2. The algebra closes using both commutators and anticommutators and, in its simplest version, takes the form:

$$[Q_\alpha, \bar{Q}_\beta]_+ = -2\gamma^\mu_{\alpha\beta} P_\mu \tag{37}$$

where P_μ are the generators of space–time translations.

There is a special Chapter in this Book devoted to supersymmetry,[56] so we will not go into any details here. We shall see there that supersymmetric field theories have remarkable renormalisation properties[57] which make them unique. In particular, they offer the only field theory solution of the hierarchy problem. Another attractive feature refers to grand unification. The presence of the supersymmetric particles modifies the renormalisation group equations and the effective coupling constants meet at high scales.

An interesting extension consists of considering gauge supersymmetry transformations, i.e. transformations whose infinitesimal parameters — which are anticommuting spinors — are also functions of the space–time point x. There are several reasons to go from global to local supersymmetry[58]:

- We have learned in the last years that all fundamental symmetries in nature are local (or gauge) symmetries.
- The supersymmetry algebra contains the translations. So local supersymmetry transformations imply local translations and we know that invariance under local

translations leads to general relativity which, at least at the classical level, gives a perfect description of the gravitational interactions.

- In the last section we saw that in a supersymmetric grand unified theory the unification scale approaches the Planck mass (10^{19} GeV) at which gravitational interactions can no more be neglected.
- The miraculous cancelation of divergences we find in supersymmetry theories raises the hope that the supersymmetric extension of general relativity will give a consistent quantum field theory. In fact local supersymmetry, or "supergravity", is the only field theoretic extension of the Standard Model which addresses the issue of quantum gravity.

Since the supersymmetry generators have spin 1/2, when applied to a state with spin projection s_z, they transform it into one with $s_z \pm 1/2$. A well-established theoretical prejudice is that in Nature there are no elementary particles with spin higher than 2. It follows that the maximum number of independent supersymmetry generators we can consider is $N = 8$.[59] The irreducible representation of one-particle states contains:

$$\begin{aligned}&\text{1 spin-2 graviton}\\&\text{8 spin-3/2 Majorana gravitini}\\&\text{28 spin-1 vector bosons}\\&\text{56 spin-1/2 Majorana fermions}\\&\text{70 spin-0 scalars}\end{aligned} \tag{38}$$

$N = 8$ supergravity promised to give us a truly unified theory of all interactions, including gravitation and a description of the world in terms of a single fundamental multiplet. The main question is whether it defines a consistent field theory. At the moment we have no clear answer to this question, although it sounds rather unlikely. In some sense $N = 8$ supergravity can be viewed as the end of a road, the road of local quantum field theory. The usual response of physicists whenever faced with a new problem was to seek the solution in an increase of the symmetry. This quest for larger and larger symmetry led us to the standard model, to grand unified theories and then to supersymmetry, to supergravity and, finally, to the largest possible supergravity, that with $N = 8$. In the traditional framework we are working, that of local quantum field theory, there exists no known larger symmetry scheme. The next step had to be a very radical one. The very concept of point particle, which had successfully passed all previous tests, was abandoned. During the last decades the theoretical investigations have moved towards the theory of interacting extended objects.

This Chapter has touched so many subjects that a complete list of references is impossible. The selection is arbitrary and I have chosen either the original articles, or some which are not so well-known. I apologise for the numerous omissions.

References

1. O. Hahn and L. Meitner, *Phys. Zeitschr.* **9**, 321 (1908); **9**, 697 (1908); **10**, 741 (1909).
2. J. Chadwick, *Verh. d. Deutsch. Phys. Ges.* **16**, 383 (1914).
3. C. D. Ellis and W. A. Wooster, *Proc. Roy. Soc. A* **117**, 109 (1927).
4. L. Meitner and W. Orthmann, *Zeitschr. f. Phys.* **60**, 143 (1930).
5. N. Bohr, H. A. Kramers and J. C. Slater, *Phil. Mag.* **47**, 745 (1925).
6. W. Pauli, in *Collected Scientific Papers*, eds. R. Kronig and V. Weisskopf (Interscience, New York, 1964), Vol 2, p. 1313.
7. C. L. Cowan *et al.*, *Science* **124**, 103 (1956).
8. F. Reines, in *Proc. Int. Colloq. History of Particle Phys, J. de Phys.* **43**, suppl to No. 12, p. C8–237 (1982).
9. E. Fermi, *Rend. Acc. Lincei* **3**, 145 (1926); *Zeitschr. f. Phys.* **36**, 902 (1926).
10. D. Iwanenko, *Comptes Rendus* **195**, 439 (1932); F. Perrin, *Comptes Rendus* **197**, 1625 (1933).
11. E. Fermi, *Ric. Scient.* **4**, 491 (1934).
12. H. A. Bethe and R. F. Bacher, *Rev. Mod. Phys.* **8**, 82 (1936).
13. M. Fierz, *Helv. Phys. Acta* **9**, 245 (1936).
14. C. S. Wu *et al.*, *Phys. Rev.* **105**, 1413 (1957).
15. M. Goldhaber, L. Grodzins and A. W. Sunyar, *Phys. Rev.* **109**, 1015 (1958).
16. S. S. Gershtein and Ya. B. Zeldovich, *JETP* **29**, 698, (1955) [*Soviet Physics JETP* **2**, 576 (1958)]; R. P. Feynman and M. Gell-Mann, *Phys. Rev.* **109**, 193 (1958).
17. N. Cabibbo, *Phys. Rev. Lett.* **10**, 531 (1963).
18. M. L. Goldberger and S. B. Treiman, *Phys. Rev.* **110**, 1178 (1958); **111**, 354 (1958).
19. Yoichiro Nambu *Phys. Rev. Lett.* **4**, 380 (1960).
20. J. Goldstone *Nuov. Cim.* **19**, 154 (1961).
21. M. Gell-Mann, *Phys.* **1**, 63 (1964).
22. O. Darrigol, *Electrodynamics from Ampère to Einstein* (Oxford University Press, 2000).
23. B. L. Ioffe and E. P. Shabalin, *Yad. Fiz.* **6**, 828 (1967) [*Sov. Jour. Nucl. Phys.* **6**, 603 (1968)]; see also R. N. Mohapatra, J. Subba Rao and E. Marshak, *Phys. Rev. Lett.* **20**, 1081 (1968).
24. G. Nordström, *Phys. Z.* **15**, 504 (1914).
25. V. Fock, *Z. Phys.* **39**, 226 (1926) [*Physics-Uspekhi* **53**, 839 (2010)].
26. H. Weyl, *Z. Phys.* **56**, 330 (1929).
27. O. Klein, in *Les Nouvelles Théories de la Physique*, Paris 1939, p. 81. Report in a Conference organised by the "Institut International de Coopération Intellectuelle", Warsaw 1938.
28. W. Pauli, Unpublished. It is summarised in a letter to A. Pais dated July 22–25 1953. A. Pais, *Inward Bound* (Oxford Univ. Press, 1986), p. 584.
29. C. N. Yang, *Selected Papers 1945–1980 with Commentary* (Freeman, San Francisco), p. 525.
30. C. N. Yang and R. L. Mills, *Phys. Rev.* **96**, 191 (1954); It seems that similar results were also obtained by R. Shaw in his thesis at the University of Cambridge.
31. J. Schwinger, *Ann. of Phys.* **2**, 407 (1957).
32. S. A. Bludman, *Nuovo Cim.* **9**, 433 (1958).
33. S. L. Glashow, *Nucl. Phys.* **22**, 579 (1961).
34. S. L. Glashow and M. Gell-Mann, *Ann. of Phys.* **15**, 437 (1961); see also: R. Utiyama, *Phys. Rev.* **101**, 1597 (1956).

35. F. Englert and R. Brout *Phys. Rev. Lett.* **13**, 321 (1964); P. Higgs *Phys. Lett.* **12**, 132 (1964); P. Higgs *Phys. Rev. Lett.* **13**, 508 (1964); G. S. Guralnik, C. R. Hagen and T. W. B. Kibble *Phys. Rev. Lett.* **13**, 585 (1964); T. W. B. Kibble *Phys. Rev.* **155**, 1554 (1967).
36. S. Weinberg *Phys. Rev. Lett.* **19**, 1264 (1967).
37. A. Salam, Lecture given at *Eighth Nobel Symposium*, ed. N. Svartholm, p. 367 (1968).
38. G. 't Hooft, Chapter 1 in this Book.
39. T. D. Lee and G. C. Wick, *Nucl. Phys. B* **9**, 209 (1969).
40. W. Kummer and G. Segré, *Nucl. Phys.* **64**, 585 (1965).
41. M. Gell-Mann, M. L. Goldberger, N. M. Kroll and F. Low, *Phys. Rev.* **179**, 1518 (1969).
42. R. Gatto, G. Sartori and M. Tonin, *Phys. Lett. B* **28**, 128 (1968); *Nuov. Cim. Lett.* **1**, 1 (1969).
43. N. Cabibbo and L. Maiani, *Phys. Lett. B* **28**, 131 (1968); *Phys. Rev. D* **1**, 707 (1970).
44. Cl. Bouchiat, J. Iliopoulos and J. Prentki, *Nuov. Cim. A* **56**, 1150 (1968); J. Iliopoulos, *Nuov. Cim. A* **62**, 209 (1969).
45. S. L. Glashow, J. Iliopoulos and L. Maiani, *Phys. Rev. D* **2**, 1285 (1970).
46. H. Georgi and S. L. Glashow, *Phys. Rev. Lett.* **28**, 1494 (1972).
47. J. Prentki and B. Zumino, *Nucl. Phys. B* **47**, 99 (1972); B. W. Lee, *Phys. Rev. D.* **6**, 188 (1972).
48. M. Kobayashi and T. Maskawa, *Progr. in Theor. Phys.* **49**, 652 (1973).
49. S. L. Adler, *Phys. Rev.* **177**, 2426 (1969); J. S. Bell and R. Jackiw, *Nuovo Cim. A* **60**, 47 (1969).
50. Cl. Bouchiat, J. Iliopoulos and Ph. Meyer, *Phys. Lett. B* **38**, 519 (1972); D. J. Gross and R. Jackiw, *Phys. Rev. D* **6**, 477 (1972).
51. H. Georgi and S. L. Glashow, *Phys. Rev. D* **6**, 429 (1972).
52. H. Georgi and S. L. Glashow, *Phys. Rev. Lett.* **32**, 438 (1974); J. C. Pati and A. Salam, *Phys. Rev. D* **8**, 3633 (1973).
53. H. Georgi, H. R. Quinn and S. Weinberg, *Phys. Rev. Lett.* **33**, 451 (1974).
54. G. 't Hooft, *Nucl. Phys. B* **79**, 276 (1974); A. M. Polyakov, *JETP Lett.* **20**, 194 (1974); *Pisma Zh. Eksp. Teor. Fiz.* **20**, 430 (1974); C. P. Dokos and T. N. Tomaras, *Phys. Rev. D* **21**, 2940 (1980).
55. J. Wess and B. Zumino, *Nucl. Phys. B* **70**, 39 (1974); Yu. A. Gol'fand and E. P. Likhtman, *JETP Lett.* **13**, 323 (1971); D. V. Volkov and V. P. Akulov, *Phys. Lett. B* **46**, 109 (1973).
56. P. Fayet, Chapter 20 in this Book.
57. J. Iliopoulos and B. Zumino, *Nucl. Phys. B* **76**, 310 (1974); S. Ferrara, J. Iliopoulos and B. Zumino, *Nucl. Phys. B* **77**, 413 (1974).
58. D. Z. Freedman, P. van Nieuwenhuizen and S. Ferrara, *Phys. Rev. B* **13**, 3214 (1976); S. Deser and B. Zumino, *Phys. Lett. B* **62**, 335 (1976).
59. E. Cremmer, B. Julia, Joel Scherk, *Phys. Lett. B* **76**, 409 (1978); E. Cremmer, B. Julia, *Phys. Lett. B* **80**, 48 (1978); *Nucl. Phys. B* **159**, 141 (1979).

Chapter 3

Quantum Chromodynamics and Deep Inelastic Scattering

R. Keith Ellis

Institute for Particle Physics Phenomenology
Physics Department, Durham University
Durham DH1 3LE, United Kingdom
keith.ellis@durham.ac.uk

This article first describes the parton model which was the precursor of the QCD description of hard scattering processes. After the discovery of QCD and asymptotic freedom, the first successful applications were to Deep Inelastic lepton–hadron scattering. The subsequent application of QCD to processes with two initial state hadrons required the understanding and proof of factorization. To take the fledgling theory and turn it into the robust calculational engine it has become today, required a number of technical and conceptual developments which will be described. Prospects for higher loop calculations are also reviewed.

1. Hard Scattering before QCD

In the preface to his 1972 book, entitled *Photon–hadron interactions*,[1] Richard Feynman indicates how his book was stimulated by his participation in the *International Symposium on Electron and Photon Interactions at High Energies* held at Cornell[2] the previous year. By the time of that conference the approximate scaling behaviour in Deep Inelastic scattering (DIS) was established and the search for a theoretical framework to interpret it was underway. Scaling in the context of DIS is the observation that the structure functions, F_i, that describe the differential cross section for lepton–hadron scattering, are functions of a single dimensionless variable, x. For a hadron of momentum P interacting with an exchanged virtual photon of momentum q,

$$x = \frac{-q^2}{2P \cdot q}. \tag{1}$$

The theoretical explanations in vogue were the light cone expansion (that somehow produced free field behaviour on the light cone, but was strongly interacting off the light cone) and various formulations of the parton model (that all required an *ad hoc* bound on the transverse momentum of the partons, to explain the observed scaling behaviour). At the Cornell conference, gluons were mentioned in the talk of Bjorken (but they were labelled as speculative and no details of their couplings were given). The fractional charges of quarks were still controversial and were probed by the large x behaviour of the Deep Inelastic scaling functions.

In his parton model, Feynman envisaged that a proton of momentum P was made up of partons of momenta $\xi_i P$ all sharing the longitudinal momentum of the proton in various proportions, ξ_i. By analysing the data on deep inelastic scattering in the parton model, and in particular the scaling behaviour, the relative strengths of the longitudinal and transverse virtual photons, and the fraction of the parton momentum carried by charged constituents, he deduced that the charge carrying partons should have spin 1/2 and at least some fraction of the momentum of the proton should be carried by electrically neutral partons. As a further development of the theory he considered the possibility that partons should be identified with quarks.

Just one year later the property of asymptotic freedom was discovered[3,4] and a candidate Lagrangian for the theory of strong interactions, called Quantum Chromodynamics (QCD) was written down. The Lagrangian contained fractionally charged quarks, which were triplets under the $SU(3)$ colour degree of freedom, and eight coloured gluons in the adjoint representation of SU(3). The expression for the classical Lagrangian density of QCD is

$$\mathcal{L}_{\text{classical}} = -\frac{1}{4} F^A_{\alpha\beta} F^{\alpha\beta}_A + \sum_{\text{flavours}} \bar{q}_a (i \not{D} - m)_{ab} q_b. \qquad (2)$$

The Lagrangian describes the interaction of spin-$\frac{1}{2}$ quarks with masses specified by the diagonal mass matrix m, and massless spin-1 gluons. $\not{D}$ in Eq. (2) is a symbolic notation for $\gamma_\mu D^\mu$, where $D^\mu = \partial^\mu + igt \cdot \mathcal{A}^\mu$ is the covariant derivative, and the matrices t are the eight generators of the $SU(3)$ colour group in the fundamental representation. The spinor indices of γ_μ and q_a have been suppressed. $F^A_{\alpha\beta}$ is the field strength tensor derived from the gluon field $\mathcal{A}^A_\alpha$,

$$F^A_{\alpha\beta} = \partial_\alpha \mathcal{A}^A_\beta - \partial_\beta \mathcal{A}^A_\alpha - g f^{ABC} \mathcal{A}^B_\alpha \mathcal{A}^C_\beta. \qquad (3)$$

The indices A, B, C run over the eight colour degrees of freedom of the gluon field, whereas the indices a, b run over the three colour degrees of freedom of the quark field. f^{ABC} are the structure constants of the $SU(3)$ colour group.

Since QCD purports to be a complete theory of the strong interactions it should describe both high energy and low energy strong interactions. In the present Chapter we shall be concerned with the high energy manifestations of QCD, which are analysable using the techniques of perturbation theory, because of the property of asymptotic freedom. It can be argued that the discovery of QCD was the best of all possible worlds. On the one hand, the high energy behaviour of the theory can be analysed using the techniques of perturbation theory, that were so successfully used in Quantum Electrodynamics. On the other hand, the theory presented many different features, in particular a confining phase where free quarks and gluons are not observed, that presented significant intellectual challenges in the application to real world situations. The theoretical physics community has risen to this challenge and provided astonishing advances in calculational technique to elaborate the consequences of QCD. The most significant of these advances for the high energy regime will be described in this review.

2. The Discovery of Asymptotic Freedom

Politzer,[3] Gross and Wilczek[4,5] discovered that non-Abelian gauge theories are asymptotically free, viz. that the coupling of the theory decreases at high energy, according to the β function equation.[a] The running of the coupling constant α_S is determined by the renormalization group equation,

$$Q^2 \frac{\partial \alpha_S}{\partial Q^2} = \beta(\alpha_S), \tag{4}$$

where Q denotes the energy scale. In QCD, the β function has the perturbative expansion[6–9]

$$\beta(\alpha_S) = -b\alpha_S^2(1 + b'\alpha_S + b''\alpha_S^2 + b'''\alpha_S^3 + O(\alpha_S^4)), \tag{5}$$

where the one-loop coefficient of the β-function is,

$$b = \frac{(11C_A - 2n_{\rm lf})}{12\pi} = \frac{(33 - 2n_{\rm lf})}{12\pi}, \tag{6}$$

and $n_{\rm lf}$ is the number of active light flavours. The coefficients b', b'', b''' are also known. The form of the beta function utilizing all the known coefficients is shown in Fig. 1. Since the beta function is negative the coupling constant decreases as the energy increases. We may solve Eq. (4) for the dependence of the coupling on the scale Q^2. In the lowest approximation the result is that the running coupling vanishes as an inverse power of a logarithm of the scale,

$$\alpha_S(Q^2) = \frac{1}{b \ln(Q^2/\Lambda^2)}. \tag{7}$$

Λ is a constant of integration that determines the scale at which the coupling constant becomes large. Because the coupling constant vanishes at large scale we may expand perturbative quantities as a series in α_S. For a physical quantity f, the relationship between the perturbative series evaluated at two different scales, Q_0 and Q_1 is,

$$\begin{aligned} f &= c_1\alpha_S(Q_0^2) + c_2\alpha_S^2(Q_0^2) + O(\alpha_S^3) \\ &\equiv c_1\alpha_S(Q_1^2) + (c_2 - bc_1 \ln(Q_0^2/Q_1^2))\alpha_S^2(Q_1^2) + O(\alpha_S^3). \end{aligned} \tag{8}$$

[a]It is often wrongly stated that the coupling decreases at small distances, but since the theory only introduces a logarithmic modification of the Coulomb potential, $1/r$, this cannot be the case.

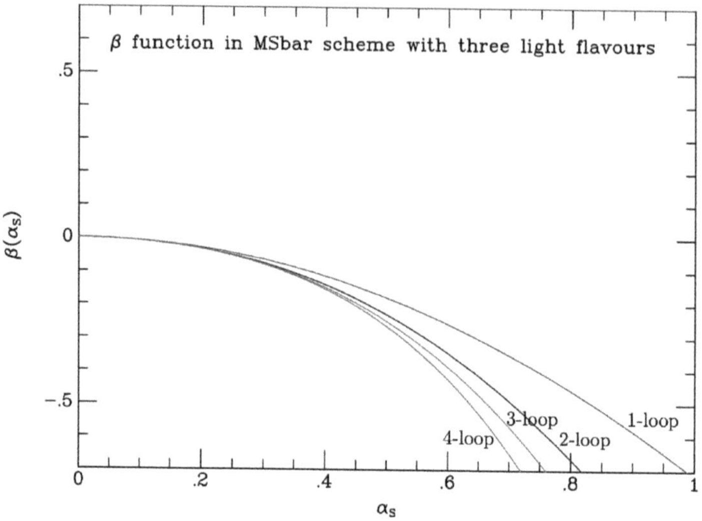

Fig. 1. Perturbative expansion of the QCD beta function as a function of α_S in the one-loop approximation,[3,4] two-loop approximation,[6] three-loop approximation,[7,8] and four-loop approximation.[9] The one- and two-loop approximations are scheme independent. The three- and four-loop approximations are in the Minimal Subtraction (MS) scheme.

This illustrates the special character of QCD perturbation series, where the difference between two choices of scales, is itself a non-leading effect. Therefore a meaningful discussion about the choice of scale requires at least calculation of next-to-leading (NLO) effects.

3. Deep Inelastic Scattering

The first application of QCD was to the Deeply inelastic scattering of a charged lepton off a proton target by virtual photon exchange, $lp \to lX$. With four quark flavours, the result for the *QCD improved parton model* is

$$F_2^{\rm em}(x,Q^2) = x\left[\frac{4}{9}(u(x,Q^2)+\bar{u}(x,Q^2)+c(x,Q^2)+\bar{c}(x,Q^2)) + \frac{1}{9}(d(x,Q^2)+\bar{d}(x,Q^2)+s(x,Q^2)+\bar{s}(x,Q^2))\right]. \quad (9)$$

This is in contradistinction with the *naive parton model* in which the parton distribution functions are independent of the scale.[b] For compactness of notation we shall use the variable t to represent Q^2.

[b]Feynman would bristle at the use of the word *naive* to qualify his parton model. Several times I heard him remonstrate, saying ‘Was it so naive?’

The scale dependent parton distribution functions satisfy sum rules that count the valence quarks in the proton, and the total momentum of the partons

$$\int_0^1 dx(u(x,t)-\bar{u}(x,t)) = 2,$$

$$\int_0^1 dx(d(x,t)-\bar{d}(x,t)) = 1,$$

$$\int_0^1 dx\, x\left[g(x,t) + \sum_{i=d,u,s,c} (f_i(x,t)+\bar{f}_i(x,t))\right] = 1. \tag{10}$$

The parton distributions themselves must be determined by experiment, but their t-dependence is determined in QCD. The t-dependence is simplest for (flavour) non-singlet combinations of quark distributions, $V = q_i - q_j$. A first formulation of the evolution equation was in terms of the moments of the parton distributions:

$$f(j,t) = \int_0^1 dx\, x^{j-1} f(x,t), \quad f = q_i, g. \tag{11}$$

In terms of moments, the evolution equation is

$$t\frac{\partial}{\partial t}V(j,t) = \frac{\alpha_S(t)}{2\pi}\gamma_{qq}^{(0)}(j)V(j,t), \tag{12}$$

where $\gamma_{qq}^{(0)}(j)$ are anomalous dimensions given in Eq. (13). In general the evolution equation is a matrix equation in the space of quarks and gluons. The complete set of leading-order anomalous dimensions in the matrix equation is,

$$\gamma_{qq}^{(0)}(j) = C_F\left[-\frac{1}{2}+\frac{1}{j(j+1)} - 2\sum_{k=2}^{j}\frac{1}{k}\right],$$

$$\gamma_{qg}^{(0)}(j) = T_R\left[\frac{(2+j+j^2)}{j(j+1)(j+2)}\right],$$

$$\gamma_{gq}^{(0)}(j) = C_F\left[\frac{(2+j+j^2)}{j(j^2-1)}\right],$$

$$\gamma_{gg}^{(0)}(j) = 2C_A\left[-\frac{1}{12}+\frac{1}{j(j-1)}+\frac{1}{(j+1)(j+2)} - \sum_{k=2}^{j}\frac{1}{k}\right] - \frac{2}{3}n_f T_R, \tag{13}$$

where $C_F = 4/3, C_A = 3$ and $T_R = 1/2$. A graphical illustration of the j-dependence of these functions in the LO and NLO approximations is shown in Fig. 2. Note that

$$\gamma_{qq}^{(0)}(1) = 0,$$

$$\gamma_{qq}^{(0)}(2) + \gamma_{gq}^{(0)}(2) = 0,$$

$$2n_{lf}\gamma_{qg}^{(0)}(2) + \gamma_{gg}^{(0)}(2) = 0. \tag{14}$$

Inserting the lowest-order form Eq. (7) for the running coupling, we obtain the solution of Eq. (12) for the moments of the non-singlet distribution,

$$V(j,t) = V(j,t_0)\left(\frac{\alpha_S(t_0)}{\alpha_S(t)}\right)^{d_{qq}(j)}, \quad d_{qq}(j) = \frac{\gamma_{qq}^{(0)}(j)}{2\pi b}, \tag{15}$$

where b is given in Eq. (6). Finally, the distribution in x space can be obtained using the inverse Mellin transform integral,

$$V(x,t) = \frac{1}{2\pi i}\int_C dj\; x^{-j}\; V(j,t), \tag{16}$$

where the integration contour in the complex j plane is parallel to the imaginary axis and to the right of all singularities of the integrand. Except in very special cases, the inverse transformation can only be performed by numerical integration.

It is straightforward to show that $d_{qq}(1) = 0$ and that $d_{qq}(j) < 0$ for $j > 1$, cf. Fig. 2. This in turn implies that as t increases the non-singlet distribution function *decreases* at large x and *increases* at small x in accord with experimental observation.

The description of Deep Inelastic scattering beyond the leading order requires higher-order calculation of the DIS anomalous dimensions[10–15] as well as the calculation of higher order corrections to the coefficient functions.[16,17]

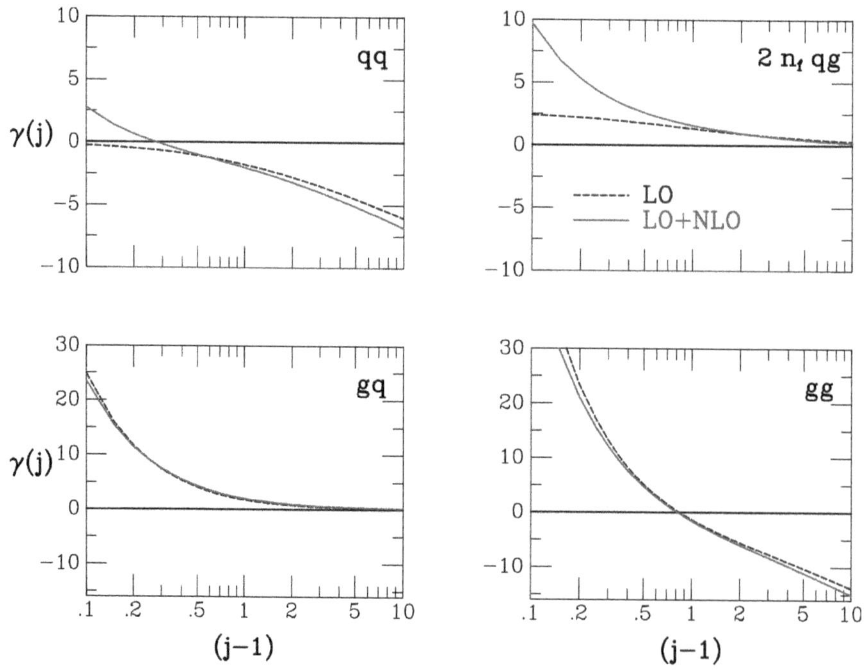

Fig. 2. Anomalous dimensions in moment space.

4. Factorization and the QCD Improved Parton Model

The treatment of Deep Inelastic scattering outlined above was done in the context of operator product expansion, building on the work of Wilson.[18] At the time it was not widely known which class of diagrams was resummed by including the effects of the anomalous dimensions. This became clear after the papers of Altarelli and Parisi[19] and Dokshitzer.[20] Earlier papers by Gribov and Lipatov[21,22] were also important. These papers made clear that the diagrams summed by the leading order anomalous dimension were ladder diagrams, in any physical gauge in which the vanishing of the helicity amplitudes in the forward direction was manifest. These papers also effectively separated the ladder sum from any particular physical process, opening the door to generalisations to processes other than deep inelastic scattering.

For example, for the flavour non-singlet combination, the DGLAP equation at lowest order reads,

$$t\frac{d}{dt}V(x,t) = \frac{\alpha_S(t)}{2\pi}\int_x^1 \frac{dz}{z}P_{qq}\left(z,\alpha_S(t)\right)V\left(\frac{x}{z},t\right). \tag{17}$$

The leading-order splitting functions are

$$\begin{aligned}
P_{qq}^{(0)}(z) &= C_F\left[\frac{1+z^2}{(1-z)_+}+\frac{3}{2}\delta\left(1-z\right)\right],\\
P_{qg}^{(0)}(z) &= T_R\left[z^2+(1-z)^2\right], \quad T_R=\frac{1}{2},\\
P_{gq}^{(0)}(z) &= C_F\left[\frac{1+(1-z)^2}{z}\right],\\
P_{gg}^{(0)}(z) &= 2C_A\left[\frac{z}{(1-z)_+}+\frac{1-z}{z}+z\left(1-z\right)\right]\\
&\quad +\delta(1-z)\frac{(11C_A-4n_fT_R)}{6},
\end{aligned} \tag{18}$$

where the 'plus prescription' on the singular parts of the functions is defined in Eq. (19):

$$\int_0^1 dz\; f(z)[g(z)]_+ = \int_0^1 dz(f(z)-f(1))g(z). \tag{19}$$

We shall refer to the splitting functions without the plus prescription and the end-point contributions $\delta(1-z)$, as unregularised splitting functions, $\hat{P}$.

It took some time for the whole community to accept that the DGLAP equation was a systematically improvable approximation, rather than just a leading log statement. This became clear with the work of Refs. 23, 24 and the work of Curci, Furmanski and Petronzio.[12,13] The latter authors used the method of Ellis *et al.*[23] and calculated the previously known NLO anomalous dimensions in a compact form, without reference to a particular hard process.

The first NLO application of QCD to hadron–hadron processes[25] was the calculation of higher-order corrections to muon pair production in hadronic collisions, the so-called Drell–Yan process. At fixed target energies it was found that the corrections were large, so large that the applicability of perturbation theory was called into question. The ratio of the NLO to LO predictions is often referred to as the K-factor, a term introduced by Altarelli.[26] In the parton model with a colour degree of freedom, there is a factor of $1/N_c$ expressing the fact that only a red and anti-red quark, etc. can annihilate to give a colour singlet final state. Since the experimental results in 1979 were larger than the expected in the coloured quark model without NLO corrections, the introduction of the colour degree of freedom was not universally accepted by experimentalists. Nowadays the validity of QCD in this process is beautifully confirmed by the data on W and Z production[27] which is compared with NNLO QCD.[28]

5. Parton Shower Monte Carlo

The sequential emission of partons in a high energy collision will give rise to a shower of emitted partons, just an energetic hadron hitting the atmosphere will give rise to an extended air shower. The formulation of the DGLAP equation given previously is not the most useful when we want to study the detailed features of a parton shower. We therefore consider an equivalent formulation. In the simplest case, in which we have only one species of parton, the DGLAP equation may be rewritten by introducing the Sudakov form factor,

$$\Delta(t) \equiv \exp\left[-\int_{t_0}^{t} \frac{dt'}{t'} \int dz \, \frac{\alpha_S}{2\pi} \hat{P}(z)\right]. \tag{20}$$

We shall see shortly that this function has a simple physical interpretation. Now we can write

$$t\frac{\partial}{\partial t} f(x,t) = \int \frac{dz}{z} \frac{\alpha_S}{2\pi} \hat{P}(z) f(x/z,t) + \frac{f(x,t)}{\Delta(t)} t \frac{\partial}{\partial t} \Delta(t) \tag{21}$$

and hence

$$t\frac{\partial}{\partial t}\left(\frac{f}{\Delta}\right) = \frac{1}{\Delta} \int \frac{dz}{z} \frac{\alpha_S}{2\pi} \hat{P}(z) f(x/z,t). \tag{22}$$

Notice that this is similar in form to the Altarelli–Parisi equation (17), except that f is replaced by f/Δ and the regularised splitting function P is replaced by the unregularised one $\hat{P}$. This equation can be integrated to give an integral equation for $f(x,t)$ in terms of the initial parton distribution $f(x,t_0)$:

$$f(x,t) = \Delta(t) f(x,t_0) + \int_{t_0}^{t} \frac{dt'}{t'} \frac{\Delta(t)}{\Delta(t')} \int \frac{dz}{z} \frac{\alpha_S}{2\pi} \hat{P}(z) f(x/z,t'). \tag{23}$$

The first term on the right-hand side is the contribution from paths that do not branch between scales t_0 and t. Thus the Sudakov form factor $\Delta(t)$ is simply the probability of evolving from t_0 to t without branching. The second term is the contribution from all paths which have their last branching at scale t'. The factor of $\Delta(t)/\Delta(t')$ represents the probability of evolving from t' to t without branching. This is consistent with the above interpretation of $\Delta(t)$ because we see from Eq. (20) that $\Delta(t_0) = 1$.

Equation (23) can be used to develop a parton shower in a Monte Carlo program by a sequential series of branchings. Each individual branching is characterised by a lower scale t_2 at which the branching occurs, and a value of z which characterises the fractional longitudinal momenta carried away in the branching. Both of these quantities can be generated with the appropriate probabilities specified by Eq. (23) by solving the following equations for t_2 and x_2/x_1,

$$\frac{\Delta(t_2)}{\Delta(t_1)} = \mathcal{R},$$

$$\int_\epsilon^{x_2/x_1} dz \frac{\alpha_S}{2\pi} P(z) = \mathcal{R}' \int_\epsilon^{1-\epsilon} dz \frac{\alpha_S}{2\pi} P(z), \tag{24}$$

where $\mathcal{R}$ and $\mathcal{R}'$ are random numbers. The effects of coherence can be approximately introduced in the parton shower by imposing angular ordering, so that successive branchings occur at ever smaller angles.[29]

6. Jet Cross Sections

At collider energies, the event displays of a small fraction of events, show conglomerations of tracks or energy deposits, which one is tempted to interpret as the fragmentation products of individual partons. At the high energy the existence of jets in these events is evident to the naked eye. To put this simple idea on a more rigorous footing one must introduce a jet algorithm. For both experimenters and theorists a jet algorithm defines a procedure for including a set of particles into a jet and for defining the momentum of the resulting jet. A good algorithm can be applied to both real experimental data and simulated Monte Carlo data. In the latter case it should be applicable at both the hadronic and partonic level. The necessity for this consensus between experimenters and theorists has the result that, according to Tannenbaum,[30] "large sections of the publications read more like legal contracts between experimentalists and theorists than like scientific papers". The paper of Sterman and Weinberg[31] established that any infrared finite quantity would be calculable in perturbation theory. Therefore the primary legal requirement for a theorist is that the jet algorithm should be infrared finite, i.e. it should be insensitive to the emission of soft or collinear parton radiation. Within this overarching requirement, many algorithms are legal and no one algorithm or jet size will be suitable in all circumstances.

Jet algorithms come in two main types: cone algorithms and sequential recombination jet algorithms. Historically, many proposed cone algorithms have had issues of infrared safety. A good modern example of an infrared finite cone algorithm is the Seedless, infrared safe cone (Siscone) algorithm.[32] Sequential recombination jet algorithms proceed by defining a distance measure between particles or clusters of energy. If, according to the distance measure, two particles are closer than some jet resolution size, they are combined by some well-defined prescription into a pseudo-particle. The whole procedure is re-iterated until no further recombinations can occur. Two famous examples are the longitudinally invariant k_T clustering algorithm[33] which has a clustering pattern, approximating the QCD branching structure, and the anti-k_T jet algorithm,[34] a sequential recombination jet algorithm giving rise to approximately conical jets. For a full review, see the article by Salam.[35]

7. Technical Advances

The method of dimensional regularisation[36] was introduced in order to have a regularisation for non-Abelian theories that would respect the Ward identities of the theory. Ultraviolet loop divergences are regulated by reducing the number of space–time dimensions, d, to $d < 4$. Thus the integration measure becomes,

$$\frac{d^4k}{(2\pi)^4} \longrightarrow (\mu)^{2\epsilon} \frac{d^{4-2\epsilon}k}{(2\pi)^{4-2\epsilon}}, \tag{25}$$

where $\epsilon = 2 - \frac{d}{2}$. For a detailed discussion of renormalisation and dimensional regularisation we refer the reader to the book by Collins.[37] It was quickly realised that dimensional regularization could also be used to regulate both infrared singularities coming from soft emission and mass singularities present in the region of collinear emission.[38,39] Essentially all higher loop QCD calculations are now done using dimensional regularisation.

Because of the gauge structure of QCD, the final results of the calculation of an amplitude are often simpler than the intermediate results, (e.g. the result from an individual Feynman diagram). In this circumstance symbolic manipulation programs, such as Veltman's *Schoonschip*[40] and its successor Vermaseren's *Form*[41,42] are invaluable. Commercial products such as *Maple* and *Mathematica* are also widely used. This software (and corresponding improvements in computer hardware) have greatly extended the complexity of the calculations that can be successfully undertaken.

The spinor-helicity method has also led to great advances in understanding the structure of QCD amplitudes. This method starts from the observation that the polarisation vector of an external gluon field can be represented as a bispinor. Thus a polarisation vector of a gluon with momentum k (and gauge vector b) can be written as,

$$\varepsilon^{\mu}_{\pm}(k,b) = \pm \frac{\langle k \pm |\gamma^{\mu}| b\pm \rangle}{\sqrt{2}\langle b \mp | k\pm \rangle}. \tag{26}$$

Physical results for gauge invariant (sub-)amplitudes will be independent of the choice for b. The function $u_\pm(k_i) \equiv |k_i\pm\rangle$ is a massless Weyl spinor of momentum k_i and positive or negative chirality. In terms of these solutions of the free Dirac equation, the spinor products are defined by,

$$\langle i\, j\rangle = \langle i^-|j^+\rangle = \bar{u}_-(k_i)u_+(k_j), \tag{27}$$

$$[i\, j] = \langle i^+|j^-\rangle = \bar{u}_+(k_i)u_-(k_j). \tag{28}$$

The spinor products are given, up to a phase, by the square root of a dot product, $\langle i\, j\rangle\, [j\, i] = 2k_i \cdot k_j$. Spinor products are the natural language to represent gauge theory amplitudes, because the amplitudes contain square root singularities. The spinor-helicity formalism,[43,44] has been used to great effect for tree graphs, and for one-loop calculations.[45,46] The result for one four-gluon helicity amplitude expressed in terms of spinor products is

$$m(1_g^-, 2_g^-, 3_g^+, 4_g^+) = -i\frac{\langle 12\rangle^4}{\langle 12\rangle\langle 23\rangle\langle 34\rangle\langle 41\rangle}. \tag{29}$$

This formula, which generalises to an arbitrary number of gluons,[47] is an example of a maximum helicity violating amplitude (MHV). The amplitudes with less helicity violation $m(1_g^+, 2_g^+, 3_g^+, 4_g^+)$ and $m(1_g^-, 2_g^+, 3_g^+, 4_g^+)$ are equal to zero at tree graph level.

7.1. *One-loop calculations*

Scalar integrals are loop integrals with no powers of the loop momentum in the numerator; they are classified by the number of external lines, $N = 1, 2, 3, 4$, corresponding to scalar tadpoles, bubbles, triangles and boxes. Any one-loop amplitude can be expressed as a sum of scalar tadpoles, bubbles, triangles and box integrals. The existence of this decomposition is one of the most important results for one-loop calculations; the result relies on simple Lorentz invariance, which allows a decomposition of tensor integrals to invariant form factors, and on the four-dimensional nature of space–time, which allows scalar higher-point integrals to be reduced to sums of boxes. Using the property of Lorentz invariance, Passarino and Veltman[48] demonstrated how to reduce the tensor integrals (with $N \leq 4$ and with k powers of the loop momentum in the numerator) to scalar integrals. In a renormalisable theory, tensor integrals occur with a limited number of powers of the loop momentum in the numerator, $k \leq N$. Higher point scalar integrals can also be reduced to lower point functions. Thus a scalar pentagon in d dimensions can be written as a sum of the five box diagrams obtainable by removing one propagator, if we neglect terms of order ϵ.[49–51] The general one-loop N-point integral in $d = 4-2\epsilon$ dimensions for $N \geq 6$ can be recursively obtained as a linear combination of pentagon integrals,[49,50] provided that the external momenta are restricted to four dimensions.

Thus any integral I_N can be written as a linear combination of one-loop scalar integrals, that include four-, three-, two- and one-point functions and a remnant of the dimensional regularisation procedure that is called *the rational part* $\mathcal{R}$,

$$I_N = c_{4;j}I_{4;j} + c_{3;j}I_{3;j} + c_{2;j}I_{2;j} + c_{1;j}I_{1;j} + \mathcal{R} + \mathcal{O}(D-4). \tag{30}$$

The index j labels the combination of momenta that enter into each particular scalar integral $I_{n;j}$. The significance of Eq. (30) is that once the scalar one-loop integrals are tabulated for $n \leq 4$, *any* one-loop calculation is reduced to the determination of both the coefficients $c_{n;j}$ and the rational part $\mathcal{R}$.

Complete results for finite scalar one-loop integrals were given by 't Hooft and Veltman.[52] A corresponding set of complete results for integrals containing soft or collinear divergences, regulated using dimensional regularisation, were provided by the program QCDLoop.[53] As far as one-loop diagrams are concerned, the scalar integrals are completely known. Therefore the calculation of one-loop amplitudes is reduced to the determination of the coefficients of these known one-loop integrals.

The reduction coefficients and the rational part, can be obtained by efficient numerical methods based on an algebraic understanding of the structure of the integrand, as first noted by Ossola, Papadopoulos and Pittau.[54] This procedure exploits the general parameterisation of the structure of the one-loop *integrand* involving a mix of terms that give rise to the wanted coefficients, $c_{n;j}$, and evanescent terms that vanish upon integration. By repeated evaluation of the integrand at differing values of the loop momentum (typically values at which some subset of the denominators vanish), we can determine the coefficients of both the physical and the evanescent terms. A detailed description of this method, including its extension to d dimensions to determine the rational terms, can be found in the *Physics Report* of Ellis *et al.*[55]

8. The Age of the Automation

8.1. *Tree graphs*

Although completely specified by the Feynman rules, the calculation of tree graphs can lead to considerable algebraic complexity. Here we document examples of attempts to deal with this complexity. Helas[56] is a set of Fortran subroutines for helicity amplitudes which enable the calculation of an arbitrary tree diagram using building blocks supplied by a number of subroutine calls. Two years later it was realized by Stelzer and Long,[57] working in Madison, that a script, dubbed Madgraph, could be written to assemble the needed subroutine calls for an arbitrary process. This has subsequently been further developed into a sophisticated tool that can generate both tree diagrams and one-loop diagrams, for both the standard model and BSM models, and that can sum over parton distributions and generate events for further processing by parton showers.[58]

Subsequently there have been many further developments in the automatic generation of Feynman graphs. As an example of these programs we quote *qgraf* by Nogueira,[59] which can handle both tree diagrams and loop diagrams. Automatic graph generation is an essential feature for higher-order calculations. In addition, much effort has been put into automatic Feynman rule generation (for models different than QCD). A standard interface for the Feynman rule output that can serve as the input for other amplitude generators, called Universal Feynman-rule Output (UFO)[60] has been proposed.

Also of great importance is the generation of Feynman amplitudes recursively. The earliest of these methods, due to Berends and Giele,[61] generates colour-ordered off-shell currents which are further used as the input for higher-point currents by attaching all possible branching processes to the off-shell line. The computer evaluation of amplitudes using off-shell recursion has been compared with analytic formula for QCD amplitudes and the conclusion is that a hybrid approach may be the most efficient.[62] Thus for a large number of legs, and for helicity choices far from the MHV, Berends–Giele recursion is computationally superior to analytic formulas.

Berends–Giele recursion relations perform recursion with off-shell currents. By contrast BCFW recursion[63] performs recursion with on-shell colour-ordered amplitudes. The proof of BCFW recursion relies heavily on the analyticity properties of the colour-ordered amplitudes. Thus for example, using BCFW recursion we can show that a four-point MHV gluon amplitude can be calculated as a product of two three-point amplitudes with shifted external momenta, connected by a scalar propagator. For an elementary introduction to BCFW recursion, both at tree graph level and at one-loop level I refer the reader to the lectures of Lance Dixon.[64]

8.2. *NLO calculations*

With the factorisation theorem in place, and advances in the theory of jet algorithms, the conceptual framework needed to perform NLO one-loop calculations of physical quantities was in place. If the Born diagram for a given process is a tree graph, the NLO calculation will require the calculation of the Born diagram with one additional parton, and interference of the one-loop virtual contribution with the Born diagram. We shall refer to these as the real and virtual contributions respectively. In QCD both of these contributions are individually singular because of the emission of a soft gluon or because of the collinear emission of effectively massless partons. The form of the singular terms is completely understood.[65,66] For suitable inclusive quantities there is a cancellation between real and virtual contributions.

The subtraction method, first used for 3-jet structure in e^+e^- annihilation,[67] introduces a subtraction term that cancels the behaviour of the real diagrams in the singular limit,

$$d\sigma_{ab,\mathrm{NLO}} = \int_{n+1} \left[d\sigma^R_{ab,\mathrm{NLO}} - d\sigma^S_{ab,\mathrm{NLO}}\right] + \int_n \left[d\sigma^V_{ab,\mathrm{NLO}} + d\sigma^S_{ab,\mathrm{NLO}}\right]. \tag{31}$$

The subtraction terms are chosen to be sufficiently simple that they can be analytically integrated (in d dimensions) over the kinematic variables of the additional parton. The resultant integrated counterterm contains poles in ϵ that, when added to the virtual term, yield a finite result. The subtraction procedure was turned into a precise algorithm by Catani and Seymour[68] who also introduced a procedure for mapping the n-parton phase space onto the $(n+1)$-parton phase space, both for initial and final radiation. Also worthy of note is the subtraction procedure of Frixione *et al.*[69] which uses energy and angle variables to define the subtraction term.

An alternative to the subtraction procedure is the slicing method which separates the $(n+1)$-parton phase space into two regions. In the first region, $(n+1)$-partons are resolved, whereas in the second region n-partons are resolved. If the invariant mass of two partons, s_{ij}, is smaller than some resolution parameter s_{min} only n partons are resolved. All of the divergences from the $(n+1)$-parton final state occur in regions where only n-partons are resolved. The resolution parameter s_{min} must be judiciously chosen, small enough that the sub-leading effects can be ignored in the n-parton phase space, but not so small that the divergences in the $(n+1)$-parton phase space are numerically unstable. A review of the slicing method, with references to the original literature is given in Ref. 70.

9. Outlook for NNLO

Since at high energy the coupling constant $\alpha_S \approx 0.1$, it might be expected that NLO calculations would give about 10% precision. This is not the case because of the special nature of asymptotically-free perturbation theory where the coupling decreases only logarithmically at large energy. Leading order calculations give little idea of the correct choice of scale, cf. Eq. (8). Thus the NLO calculation is the first serious approximation. With this viewpoint it is clear that NNLO has an important role to play. NNLO gives the first information about the behaviour of the asymptotic series.

The pioneers of these calculations were van Neerven and collaborators[28,71] who performed NNLO calculations of the $2 \to 1$ processes such as muon pair production in hadron–hadron collisions and Higgs boson production by gluon–gluon fusion. There are two major challenges in these calculations, the calculation of the two-loop matrix elements and the treatment and isolation of the singular regions in the real radiation contributions.

The NNLO mass factorised partonic cross section has three contributions: the double real, real-virtual and double virtual corrections,

$$d\sigma_{ab,\mathrm{NNLO}} = \int_{n+2} d\sigma^{RR}_{ab,\mathrm{NNLO}} + \int_{n+1} \left[d\sigma^{RV}_{ab,\mathrm{NNLO}} + d\sigma^{MF,1}_{ab,\mathrm{NNLO}}\right] + \int_{n} \left[d\sigma^{VV}_{ab,\mathrm{NNLO}} + d\sigma^{MF,2}_{ab,\mathrm{NNLO}}\right]. \tag{32}$$

Each of these terms is separately divergent, either containing singularities in regions of single or double unresolved phase space or explicit IR poles in ϵ. The sum of all three contributions can be rearranged, so that all singularities (poles in ϵ) cancel,

$$d\sigma_{ab,\mathrm{NNLO}} = \int_{n+2} \left[d\sigma^{RR}_{ab,\mathrm{NNLO}} - d\sigma^{S}_{ab,\mathrm{NNLO}}\right] + \int_{n+1} \left[d\sigma^{RV}_{ab,\mathrm{NNLO}} - d\sigma^{T}_{ab,\mathrm{NNLO}}\right] + \int_{n} \left[d\sigma^{VV}_{ab,\mathrm{NNLO}} - d\sigma^{U}_{ab,\mathrm{NNLO}}\right]. \tag{33}$$

We mention three methods for choosing the subtractions. First, the antenna subtraction method[72] was developed for e^+e^- annihilation, but has subsequently been adapted and successfully used for hadron–hadron collisions. Second, the sector improved subtraction technique of Czakon[73] has been applied to several physical processes. Lastly, the NNLO corrections for processes involving the production of colourless particles, such as W, Z and Higgs bosons are also known.[74,75]

Two loop (and higher) calculations generate families of integrals with differing denominators and numerators. Although it would in principle be possible to calculate all of these integrals individually, it is more efficient to note that there are relations between the integrals that allow reduction to a set of master integrals. The most significant of these relationships are the integration by parts (IBP) identities.[76,77] Here I provide a simple example, which illustrates the power of the method. Consider an integral that is a total derivative and that vanishes in dimensional regularisation,

$$0 = \int \frac{d^dp d^dq}{(2\pi)^{2d}} \frac{\partial}{\partial p^\mu} \frac{1}{p^2(p-k)^2q^2(q-k)^2} \frac{(p-q)^\mu}{(p-q)^2}. \tag{34}$$

Performing the differentiation it follows that in $d = 4 - 2\epsilon$ dimensions,

$$\epsilon \int \frac{d^dp\, d^dq}{(2\pi)^{2d}} \frac{1}{p^2(p-k)^2q^2(q-k)^2(p-q)^2} = \int \frac{d^dp\, d^dq}{(2\pi)^{2d}} \frac{1}{p^2(p-k)^4q^2(p-q)^2} - \int \frac{d^dp\, d^dq}{(2\pi)^{2d}} \frac{1}{p^2(p-k)^4q^2(q-k)^2}. \tag{35}$$

Even with IBP identities in hand, it is quite onerous to find the correct reduction to a set of master integrals. The Laporta Algorithm[78] provides a systematic lexicographic ordering of the integrals, that allows a decision on the most complicated integral that should be eliminated using the IBP and other identities. Publicly available programs implement this algorithm,[79,80] however it remains a compute-intensive problem. The calculation of phenomenologically relevant two-loop amplitudes is going through a period of rapid development at the time of writing. The current state of the art can assessed by looking at recent work.[81,82]

10. Epilogue

There can be no doubt that the enormous human effort invested into high-energy QCD, has had a great impact on the interpretation of the experiments at e^+e^- and hadron–hadron colliders throughout the world. Although this support role is important, the continuing challenge of understanding the workings of QCD, an interacting quantum field theory, is no less important.

References

1. R. Feynman, *Photon-hadron interactions*, Addison-Wesley, (1988).
2. N. Mistry, *International Symposium on Electron and Photon Interactions at High Energies*, Ithaca, New York. (1971).
3. H. D. Politzer, *Phys. Rev. Lett.* **30**, 1346–1349 (1973).
4. D. Gross and F. Wilczek, *Phys. Rev. D* **8**, 3633–3652 (1973).
5. D. Gross and F. Wilczek, *Phys. Rev. D* **9**, 980–993 (1974).
6. W. E. Caswell, *Phys. Rev. Lett.* **33**, 244 (1974).
7. S. Larin and J. Vermaseren, *Phys. Lett. B* **303**, 334–336 (1993).
8. O. Tarasov, A. Vladimirov, and A. Y. Zharkov, *Phys. Lett. B* **93**, 429–432 (1980).
9. T. van Ritbergen, J. Vermaseren, and S. Larin, *Phys. Lett. B* **400**, 379–384, (1997).
10. E. Floratos, D. Ross, and C. T. Sachrajda, *Nucl. Phys. B* **152**, 493 (1979).
11. E. Floratos, D. Ross, and C. T. Sachrajda, *Nucl. Phys. B* **129**, 66–88 (1977).
12. G. Curci, W. Furmanski, and R. Petronzio, *Nucl. Phys. B* **175**, 27 (1980).
13. W. Furmanski and R. Petronzio, *Phys. Lett. B* **97**, 437 (1980).
14. S. Moch, J. Vermaseren, and A. Vogt, *Nucl. Phys. B* **688**, 101–134, (2004).
15. A. Vogt, S. Moch, and J. Vermaseren, *Nucl. Phys. B* **691**, 129–181, (2004).
16. W. A. Bardeen, A. Buras, D. Duke, and T. Muta, *Phys. Rev. D* **18**, 3998 (1978).
17. J. Vermaseren, A. Vogt, and S. Moch, *Nucl. Phys. B* **724**, 3–182 (2005).
18. K. G. Wilson, *Phys. Rev. D* **3**, 1818 (1971).
19. G. Altarelli and G. Parisi, *Nucl. Phys. B* **126**, 298 (1977).
20. Y. L. Dokshitzer, *Sov. Phys. JETP.* **46**, 641–653, (1977).
21. V. Gribov and L. Lipatov, *Sov. J. Nucl. Phys.* **15**, 675–684 (1972).
22. V. Gribov and L. Lipatov, *Sov. J. Nucl. Phys.* **15**, 438–450 (1972).
23. R. K. Ellis, H. Georgi, M. Machacek, H. D. Politzer, and G. G. Ross, *Nucl. Phys. B* **152**, 285 (1979).
24. D. Amati, R. Petronzio, and G. Veneziano, *Nucl. Phys. B* **140**, 54 (1978).
25. G. Altarelli, R. K. Ellis, and G. Martinelli, *Nucl. Phys. B* **157**, 461 (1979).

26. G. Altarelli, *International Conference on High-Energy Physics*, Geneva, **C790627**, 727–741 (1979).
27. S. Chatrchyan *et al.*, *Phys. Rev. Lett.* **112**, 191802 (2014).
28. R. Hamberg, W. van Neerven, and T. Matsuura, *Nucl. Phys.* **B359**, 343–405 (1991).
29. G. Marchesini and B. Webber, *Nucl. Phys. B* **310**, 461, (1988).
30. M. Tannenbaum, *Hard-scattering and Jets from RHIC to LHC: A Critical review*, *PoS.* **LHC07**, 004 (2007).
31. G. F. Sterman and S. Weinberg, *Phys. Rev. Lett.* **39**, 1436 (1977).
32. G. P. Salam and G. Soyez, *JHEP.* **0705**, 086 (2007).
33. S. Catani, Y. L. Dokshitzer, M. Seymour, and B. Webber, *Nucl. Phys. B* **406**, 187–224 (1993).
34. M. Cacciari, G. P. Salam, and G. Soyez, *JHEP.* **0804**, 063 (2008).
35. G. P. Salam, *Eur. Phys. J. C* **67**, 637–686 (2010).
36. G. 't Hooft and M. Veltman, *Nucl. Phys. B* **44**, 189–213 (1972).
37. J. C. Collins, *Renormalization: an introduction to renormalization, the renormalization group, and the operator product expansion*, Cambridge (1984).
38. W. Marciano and A. Sirlin, *Nucl. Phys. B* **88**, 86 (1975).
39. W. Marciano, *Phys. Rev. D* **12**, 3861 (1975).
40. H. Strubbe, *Manual for Schoonschip: A CDC 6000/7000 program for symbolic evaluation of algebraic expressions*, *Comput. Phys. Commun.* **8**, 1–30 (1974).
41. J. Vermaseren, The Symbolic manipulation program FORM, (1992).
42. J. Kuipers, T. Ueda, J. Vermaseren, and J. Vollinga, *FORM version 4.0*, *Comput. Phys. Commun.* **184**, 1453–1467 (2013).
43. Z. Xu, D.-H. Zhang, and L. Chang, *Nucl. Phys. B* **291**, 392 (1987).
44. L. J. Dixon, *Calculating scattering amplitudes efficiently.* (hep-ph/9601359, 1996).
45. Z. Bern, L. J. Dixon, and D. A. Kosower, *Ann. Rev. Nucl. Part. Sci.* **46**, 109–148, (1996).
46. Z. Bern, L. J. Dixon, and D. A. Kosower, *Annals Phys.* **322**, 1587–1634 (2007).
47. S. J. Parke and T. Taylor, *Phys. Rev. Lett.* **56**, 2459 (1986).
48. G. Passarino and M. Veltman, *Nucl. Phys. B* **160**, 151, (1979).
49. D. Melrose, *Nuovo Cim.* **40**, 181–213 (1965).
50. W. van Neerven and J. Vermaseren, *Phys. Lett. B* **137**, 241 (1984).
51. Z. Bern, L. J. Dixon, and D. A. Kosower, *Phys. Lett. B* **302**, 299–308 (1993).
52. G. 't Hooft and M. Veltman, *Nucl. Phys. B* **153**, 365–401 (1979).
53. R. K. Ellis and G. Zanderighi, *JHEP.* **0802**, 002 (2008).
54. G. Ossola, C. G. Papadopoulos, and R. Pittau, *Nucl. Phys. B* **763**, 147–169 (2007).
55. R. K. Ellis, Z. Kunszt, K. Melnikov, and G. Zanderighi, *Phys. Rept.* **518**, 141–250 (2012).
56. H. Murayama, I. Watanabe, and K. Hagiwara, *HELAS: HELicity Amplitude Subroutines for Feynman diagram evaluations*, (1992).
57. T. Stelzer and W. Long, *Comput. Phys. Commun.* **81**, 357–371, (1994).
58. J. Alwall, R. Frederix, S. Frixione, V. Hirschi, F. Maltoni, *et al.*, *JHEP.* **1407**, 079 (2014).
59. P. Nogueira, *J. Comput. Phys.* **105**, 279–289 (1993).
60. C. Degrande, C. Duhr, B. Fuks, D. Grellscheid, O. Mattelaer, *et al.*, *Comput. Phys. Commun.* **183**, 1201–1214 (2012).
61. F. A. Berends and W. Giele, *Nucl. Phys. B* **306**, 759 (1988).
62. S. Badger, B. Biedermann, L. Hackl, J. Plefka, T. Schuster, *et al.*, *Phys. Rev. D* **87**(3), 034011 (2013).
63. R. Britto, F. Cachazo, B. Feng, and E. Witten, *Phys. Rev. Lett.* **94**, 181602 (2005).

64. L. J. Dixon, *A brief introduction to modern amplitude methods*, pp. 31–67, (2014).
65. S. Catani, *Phys. Lett. B* **427**, 161–171 (1998).
66. S. Catani, S. Dittmaier, and Z. Trocsanyi, *Phys. Lett. B* **500**, 149–160 (2001).
67. R. K. Ellis, D. Ross, and A. Terrano, *Nucl. Phys. B* **178**, 421 (1981).
68. S. Catani and M. Seymour, *Nucl. Phys. B* **485**, 291–419, (1997).
69. S. Frixione, Z. Kunszt, and A. Signer, *Nucl. Phys. B* **467**, 399–442, (1996).
70. B. Harris and J. Owens, *Phys. Rev. D* **65**, 094032 (2002).
71. V. Ravindran, J. Smith, and W. L. van Neerven, *Nucl. Phys. B* **665**, 325–366 (2003).
72. A. Gehrmann-De Ridder, T. Gehrmann, and E. N. Glover, *JHEP.* **0509**, 056 (2005).
73. M. Czakon, *Phys. Lett. B* **693**, 259–268 (2010).
74. C. Anastasiou and K. Melnikov, *Nucl. Phys. B* **646**, 220–256 (2002).
75. S. Catani and M. Grazzini, *Phys. Rev. Lett.* **98**, 222002 (2007).
76. K. Chetyrkin and F. Tkachov, *Nucl. Phys. B* **192**, 159–204, (1981).
77. F. Tkachov, *Phys. Lett. B* **100**, 65–68 (1981).
78. S. Laporta, *Int. J. Mod. Phys. A* **15**, 5087–5159 (2000).
79. A. von Manteuffel and C. Studerus, *Reduze 2 — Distributed Feynman Integral Reduction*, (2012).
80. A. Smirnov, *JHEP.* **0810**, 107 (2008).
81. T. Gehrmann, A. von Manteuffel, L. Tancredi, and E. Weihs, *JHEP.* **1406**, 032 (2014).
82. F. Caola, J. M. Henn, K. Melnikov, A. V. Smirnov, and V. A. Smirnov, *JHEP.* **1411**, 041 (2014).

Chapter 4

Electroweak Corrections

Riccardo Barbieri

Scuola Normale Superiore and INFN, Pisa, Italy

The test of the electroweak corrections has played a major role in providing evidence for the gauge and the Higgs sectors of the Standard Model. At the same time the consideration of the electroweak corrections has given significant indirect information on the masses of the top and the Higgs boson before their discoveries and important orientation/constraints on the searches for new physics, still highly valuable in the present situation. The progression of these contributions is reviewed.

1. Introduction

In spite of its apparently limited scope the study of the ElectroWeak Corrections (EWC) has played a crucial role in three different directions:

- Provide evidence for the consistency at the quantum level of the gauge and the Higgs sectors of the Standard Model (SM);
- Give significant information, although indirect, on the existence and the mass of the top quark and the Higgs boson before their direct discovery;
- Constrain and orient the search for possible new physics beyond the SM.

Although not making justice to important earlier works on the subject, in order to recall the progressive role of the EWC it makes sense, as we do in the following, to set the $t = 0$ time in 1971, when 't Hooft proved the renormalisability of the electroweak sector of the SM.[1,2] Since then, this means more than forty years of progress until the discovery of the Higgs boson, which offers new opportunities for precision physics.

2. The Pioneering Works

A few papers of Veltman in 1977 played a key role both in establishing the program of computing the EWC to physical observables in the SM and in pointing out their possible sensitivity to physics in the 100 GeV range or above, even by measurements of lower energy processes.[3,4]

The relevant physical observables are determined at tree level by three parameters: the two gauge couplings in the electroweak sector, g and g', and the vacuum expectation value of the Higgs boson doublet v. To fix these parameters, three basic observables may be chosen, which is equivalent to a renormalization procedure. Since the beginning of the electroweak correction program, two such observables have emerged: the electromagnetic fine structure constant, α, as measured by the Josephson effect or the electron $(g-2)$, and the Fermi constant, G_F, as determined from the muon lifetime. In the seventies the third observable was more often coming from the ν_μ–e or the ν_μ–hadron cross sections, in particular the ratio R_ν between the neutral and the charged current processes, with a determination, at that time, of the weak mixing angle $\sin^2\theta_W$ at about 10–20% level. The emergence from ν and $\bar{\nu}$ processes of a ρ parameter close to unity, as predicted at tree level by the SM, was recognized by Veltman as evidence for the isodoublet nature of the Higgs boson and later, more generally, as a consequence of an approximate *custodial symmetry*.[5] At the same time the finding of ρ close to unity motivated to establish the EWC program.

The absence of any other precisely measured observable, as the same relatively large error on $\sin^2\theta_W$ itself from neutrino scattering, could not allow any significant quantitative comparison with the EWC. Yet, as already mentioned, the sensitivity to heavy non-decoupled particles was identified and relevant one-loop effects calculated both due to a split fermion doublet or to the Higgs boson. A first calculated one-loop effect was the correction to the ρ-parameter, the ratio $\rho = G_{NC}/G_F$ between the neutral-current Fermi constant G_{NC}, defined in analogy with G_F from the elastic electron-neutrino scattering amplitude at $q^2 = 0$, and the charged-current Fermi constant itself. The deviation from unity of ρ was calculated from a charged lepton of mass $m_l \gg M_{W,Z}$, split from a massless neutrino in the same isodoublet[3]

$$\rho - 1 \equiv \Delta\rho = \frac{G_F m_l^2}{8\pi^2\sqrt{2}}. \tag{1}$$

This is the same correction as for a heavy top, of mass $m_t = m_l$, and a quasi massless bottom up to the extra colour factor of 3 that has to be introduced. The neutrino data mentioned above allowed to set an upper bound on m_l of 800 GeV, or — knowing now that the top is also heavy — of about 450 GeV on the top mass.

Somewhat later came the calculation in the SM of the two independent one-loop corrections growing like the logarithm of the Higgs mass, m_H.[6–11] As recalled later on, these corrections affect in a precisely defined way every observable. For example one has

$$\Delta\rho = -\frac{3\alpha}{8\pi\cos^2\theta_W}\log\frac{m_H}{M_Z}, \tag{2}$$

$$\frac{\sqrt{2}G_F M_W^2}{\pi\alpha}\left(1-\frac{M_W^2}{M_Z^2}\right) \equiv 1+\Delta r = 1+\frac{11\alpha}{24\pi\sin^2\theta_W}\log\frac{m_H}{M_Z}. \tag{3}$$

Veltman called "screening theorem" the observation of the mild logarithmic dependence on the Higgs mass, taken much larger than M_Z, of all the one-loop corrections. The root of this theorem is in the absence in the unitary gauge of any coupling of the W and the Z to the Higgs boson proportional to the quartic Higgs coupling λ.

3. Constraining m_t and m_H

Although clear evidence for the EWC[a] had to wait for the e^+e^- data at the Z-pole, new more precise measurements came in the eighties from various sources: the direct measurements of the W and Z masses at hadron colliders, ν_μ–e and ν_μ–hadron scattering, Atomic Parity Violation and e^+e^- collisions at various energies below the Z-boson mass. This quickly led the focus on the constraints that could be inferred from these data on the top quark mass and, at least indirectly, on the Higgs boson mass, which was rapidly recognized as relevant to the constraint on the top mass itself. At the summer conferences in 1989, with the LEP operation just about to start, it was possible to quote the following allowed ranges for the top mass:[12,13]

$$40\,\text{GeV} < m_t < 210\,\text{GeV}\,(90\%\text{C.L.})\,\text{for}\,m_H < 1\,\text{TeV} \tag{4}$$

$$m_t = 140^{+43}_{-52}\,\text{GeV for}\,m_H = 100\,\text{GeV}. \tag{5}$$

The fit leading to these constraints had already incorporated as well the fresh first measurement of the Z mass by SLC, $m_Z = 91.17 \pm 0.18\,\text{GeV}$, close to *per mille* precision.[14] At the same summer conferences UA1, UA2[15] and CDF[16] were quoting lower bounds on the top mass in the 60–70 GeV range.

[a]Here and in the following by EWC we mean corrections not of pure electromagnetic nature (nor *a fortiori* of strong nature). For the issues discussed here, this separation has an operative meaning at a sufficient level of approximation. Needless to say the pure electromagnetic (strong) corrections have a very important role in extracting from the experimental data the physical observables (sometimes called *pseudo-observables* for this very reason). As such these corrections have to be taken into account to a sufficient level of precision.

With the same motivations in mind several calculations were also made, among which: (i) the one-loop heavy-top correction to the $Z \to b\bar{b}$ coupling[17]

$$\delta V_\mu(Z \to b\bar{b}) = -2x \frac{g}{2\cos\theta_W} Z_\mu \bar{b}_L \gamma_\mu b_L, \quad x = \frac{G_F m_t^2}{8\pi^2\sqrt{2}}; \tag{6}$$

(ii) the two-loop heavy-top corrections to the ρ-parameter[18] and (iii) the corrections, again to the ρ-parameter, growing linearly with m_H^2 and occurring at two loops.[19] A comprehensive and efficient calculation of all these effects was made possible by the observation that they come from the so-called *gaugeless limit* of the SM, through a loop expansion controlled only by the top Yukawa coupling, λ_t, and by the quartic Higgs boson coupling.[b,20] As an example the corrections to the previously defined quantities, the only ones that grow like powers of m_t^2, take a simple form up to two-loop order

$$\Delta\rho = 3x(1 + x(22 - 2\pi^2)) \tag{7}$$

$$\delta V_\mu(Z \to b\bar{b}) = -2x(1 + x(9 - \pi^3/3)) \frac{g}{2\cos\theta_W} Z_\mu \bar{b}_L \gamma_\mu b_L. \tag{8}$$

These equations hold for $m_H^2/m_t^2 \ll 1$, but simple closed forms can be given for any value of m_H and m_t. To get these corrections the W and the Z bosons only enter in loops through their Goldstone boson components with couplings controlled by λ_t and λ. Similarly observables with W and Z bosons as external legs are immediately related, via simple Ward identities, to amplitudes for the corresponding Goldstone bosons.

These and other two-loop calculations,[21,22] quickly incorporated in programs like ZFITTER, TOPAZ0 or similar, made possible a comparison at the *per mille* level of the SM with a number of precision observables measured at the Z pole in the early nineties. Consequently the constraint on the top mass became increasingly precise. In the winter conferences of 1994 an indirect determination of the top mass was quoted as[23,24]

$$m_t = 177 \pm 13^{+18}_{-19}\,\text{GeV} \tag{9}$$

with the second error due to the Higgs mass scanning the 60–1000 GeV range. Alternatively one had

$$m_t = 158 \pm 11\,\text{GeV}\,(m_H = 65\,\text{GeV}), \quad m_t = 194 \pm 10\,\text{GeV}\,(m_H = 1\,\text{TeV}). \tag{10}$$

These numbers can be compared with the value of

$$m_t = 174 \pm 10^{+13}_{-12}\,\text{GeV} \tag{11}$$

in the CDF paper, published a few months later, that provided the first evidence for direct $t\bar{t}$ production.[25]

---

[b] A numerically relevant exception comes from two loop effects of order $\alpha_S G_F m_t^2$, but they also have nothing to do with the gauging of the electroweak interactions and, as such, are relatively simple to calculate.

A pretty analogous story for the Higgs boson mass had to wait twenty years more to be accomplished. In July 2012 the Higgs boson discovery was announced with a mass $m_H = 126.0 \pm 0.4 \pm 0.4$ GeV (ATLAS[26]) and $m_H = 125.3 \pm 0.4 \pm 0.5$ GeV (CMS[27]), against an indirect determination from the electroweak fit of $m_H = 97^{+23}_{-17}$ GeV.

4. Indirect Constraints and Orientation on New Physics

The indirect sensitivity of the ElectroWeak Precision Tests (EWPT) to physics at higher energies has not only played a role in constraining m_t and m_H, but has also been seen from the beginning as a possible search for signals of new physics. As a matter of fact, until the discovery of at least the top, the separation between the two issues was to some extent not easy to achieve. There are three different ways that can be used to try to identify the sensitivity of the EWPT to new physics, with relations to each other: (i) the examination of effects in specific models; (ii) the identification of effective parameters or of effective observables relevant in specific contexts; (iii) the use of Effective Field Theory (EFT). For clarity, to describe them it is better not to follow a strict chronological order.

4.1. *Oblique parameters*

Inspired by the early work on heavy Higgs and top in the SM, as already described, and in preparation for the e^+e^- colliders working at the Z pole, attention was drawn to the contributions to precision observables from the electroweak vacuum polarisations amplitudes[28,29]

$$\Pi_{ij}^{\mu\nu}(q^2) = -i\big[A_{ij}(0) + q^2 F_{ij}(q^2)\big]\eta^{\mu\nu} + (q^\mu q^\nu\text{-terms}) \tag{12}$$

with $i, j = W, Z, \gamma$ or possibly $i, j = 0, 3$ for the B or the W_3 bosons respectively. Of particular relevance in this context are two quantities $(s = \sin\theta_W, c = \cos\theta_W)$[30]

$$\hat{T} = \frac{1}{m_W^2}(A_{33}(0) - A_{WW}(0)); \quad \hat{S} = \frac{c}{s}F_{30}(0). \tag{13}$$

Whenever an expansion in q^2 is meaningful, $\hat{T}$ is the leading parameter describing "custodial symmetry" breaking, since $\hat{T} = \Delta\rho$, whereas $\hat{S}$ is the leading term that respects it. We shall see in a moment how $\hat{S}$ and $\hat{T}$ relate to physical observables.

Although $\hat{S}$ and $\hat{T}$ are normally defined as due to new physics only, in the SM they are the only terms receiving the one loop logarithms in a large m_H expansion that we have already encountered,

$$\hat{T} = -\frac{3\alpha}{8\pi c^2}\log\frac{m_H}{M_Z}; \quad \hat{S} = \frac{\alpha}{24\pi s^2}\log\frac{m_H}{M_Z}. \tag{14}$$

After electroweak symmetry breaking there are four non-vanishing vacuum polarisation amplitudes, $\Pi_{WW}, \Pi_{33}, \Pi_{00}$ and Π_{03}, which, expanded up to first order in q^2, are determined by eight coefficients. Three of them are absorbed in the definition of the parameters g, g', v and two are related by gauge invariance of unbroken electromagnetism: $\Pi_{\gamma\gamma}(0) = \Pi_{\gamma Z}(0) = 0$. As a consequence, other than $\hat{S}$ and $\hat{T}$, it is useful to introduce a third quantity[32]

$$\hat{U} = F_{WW}(0) - F_{33}(0), \tag{15}$$

which, however, while also breaking "custodial symmetry", is subleading with respect to $\hat{T}$ in a q^2 expansion and, as such, does not contain any "large" $\log m_H$ term. At second order in q^2 four more parameters appear.[31] Unlike $\hat{S}, \hat{T}, \hat{U}$, two of these four extra parameters respect the electroweak gauge group, but, as $\hat{S}, \hat{T}, \hat{U}$, they are also calculable in the SM, since they are related to operators of dimension 6.

Let us postpone for a while the determination of $\hat{S}, \hat{T}, \hat{U}$ from experiments.

4.2. *Effective parameters at the Z pole*

As already mentioned, observables at the Z pole have played and are still playing a major role in the comparison between theory and experiment. This rapidly led to realize that every new physics respecting quark-lepton and flavour universality could affect the Z pole observables and the W-mass only via three effective parameters $\epsilon_i, i = 1, 2, 3$, defined in terms of Δr in Eq. (3) and of the effective couplings $g^f_{V,A}$ to the Z boson of a fermion f of non-zero charge Q_f and third component of the weak isospin T^f_{3L} as[32,33]

$$g^f_A = T^f_{3L}\left(1 + \frac{\epsilon_1}{2}\right), \tag{16}$$

$$\frac{g^f_V}{g^f_A} = 1 - 4|Q_f|s^2\left(1 + \frac{\epsilon_3 - c^2\epsilon_1}{c^2 - s^2}\right), \tag{17}$$

$$\Delta r = \frac{1}{s^2}(-c^2\epsilon_1 + (c^2 - s^2)\epsilon_2 + 2s^2\epsilon_3), \tag{18}$$

where

$$s^2c^2 = \frac{\pi\alpha(M_Z)}{\sqrt{2}G_F M_Z^2}. \tag{19}$$

A few comments on these parameters are useful to make.

Equations (16) and (17) refer to any light charged fermion except the b-quark, where already in the SM there is the flavour non-universal correction (6) due to top loops. To incorporate it together with any other possible new physics effect of the same kind, thus breaking flavour universality, one has to include another parameter

ϵ_b in the effective couplings of the b quark:[34]

$$g_A^b = -\frac{1}{2}\left(1 + \frac{\epsilon_1}{2}\right)(1 + \epsilon_b), \tag{20}$$

$$\frac{g_V^b}{g_A^b} = \frac{1}{1+\epsilon_b}\left(1 - \frac{4}{3}s^2\left(1 + \frac{\epsilon_3 - c^2\epsilon_1}{c^2 - s^2}\right) + \epsilon_b\right). \tag{21}$$

In the parameters ϵ_i there is no reference to a q^2 expansion nor any restriction to vacuum polarisation corrections only. An explicit expression for the vacuum polarisation contributions to $\epsilon_{1,2,3}$ is[35]

$$\begin{aligned}
\epsilon_1 &= e_1 - e_5 + \text{non-oblique},\\
\epsilon_2 &= e_2 - c^2 e_5 - s^2 e_4 + \text{non-oblique},\\
\epsilon_3 &= e_3 - c^2 e_5 + c^2 e_4 + \text{non-oblique},
\end{aligned} \tag{22}$$

where

$$\begin{aligned}
e_1 &= \frac{1}{M_W^2}(A_{33}(0) - A_{WW}(0)), \quad e_2 = F_{WW}(M_W^2) - F_{33}(M_Z^2),\\
e_3 &= \frac{c}{s}F_{30}(M_Z^2),
\end{aligned} \tag{23}$$

$$e_4 = F_{\gamma\gamma}(0) - F_{\gamma\gamma}(M_Z^2), \quad e_5 = M_Z^2 F'_{ZZ}(M_Z^2). \tag{24}$$

With new physics in oblique corrections only and a characteristic scale M_{NP} such that an expansion in $M_{W,Z}^2/M_{NP}^2$ is meaningful, a comparison of these equations with Eqs. (13)–(15) shows that $\epsilon_{1,2,3}$ may be approximated by $\hat{T}, \hat{U}, \hat{S}$ respectively.[c]

Note finally that with the definitions (16)–(21) the ϵ_i include every electroweak correction, both from the SM and from possible new physics, so that it is useful to define the separation $\epsilon_i = \epsilon_i^{SM}(m_t, m_H) + \delta\epsilon_i$. Now that both m_t and m_H are known with significant accuracy, so are the ϵ_i^{SM}[38]:

$$\begin{aligned}
\epsilon_1^{SM} &= 5.21 \cdot 10^{-3}, \quad \epsilon_2^{SM} = -7.37 \cdot 10^{-3}, \quad \epsilon_3^{SM} = 5.28 \cdot 10^{-3}\\
\epsilon_b^{SM} &= -6.9 \cdot 10^{-3}.
\end{aligned} \tag{25}$$

This makes possible to plot the two-dimensional probability distributions from current data for $\delta\epsilon_{1,3}$ in Fig. 1, when all four $\delta\epsilon_{i,b}$ are allowed to float or fixing some of them to zero.[39] It depends on a specific model of new physics which one among Fig. 1 has to be used to constrain it.

As one can see from these figures (Fig. 1), the size of the error on the ϵ_i is currently at the 10^{-3} level, which means, comparing with (25), that the electroweak radiative corrections in the SM have been seen at about the $(15-20)\%$ level of their

---

[c]In the SM as an example, although not of new physics, while $\hat{S}$ and $\hat{T}$ contain the leading $\log M_Z/m_H$ terms, the full dependence of the EWC on M_Z/m_H is only included in the ϵ_i.[36,37]

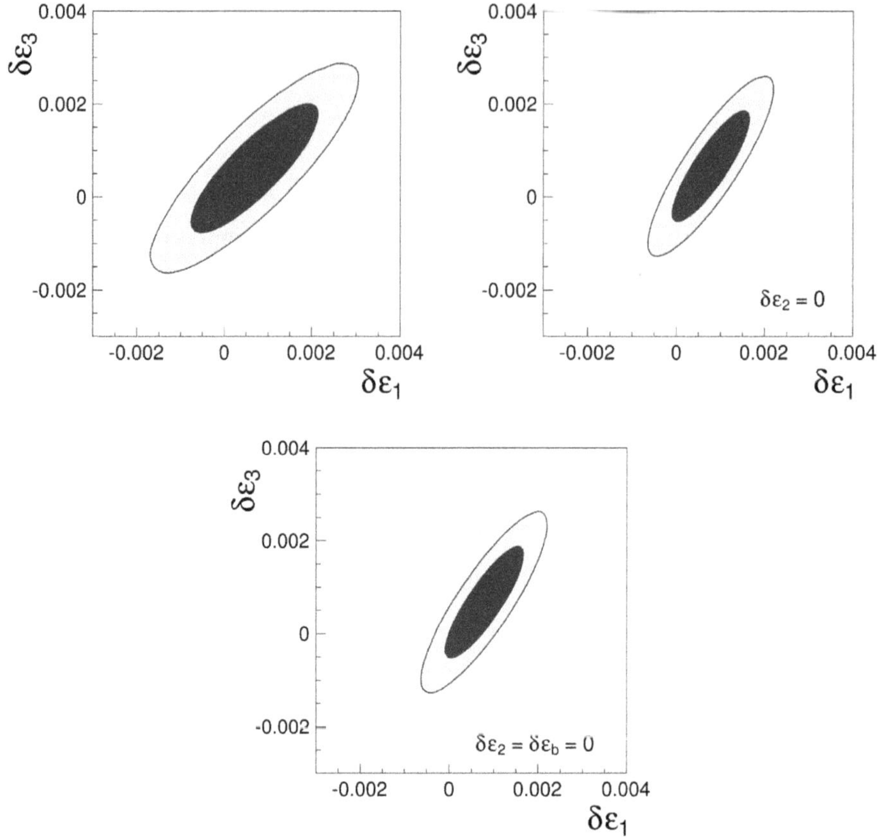

or $\delta\epsilon_2 = \delta\epsilon_b = 0$ (lower). From Ref. 39.

Fig. 1. Two-dimensional probability distributions for $\delta\epsilon_1$ and $\delta\epsilon_3$ at 68% (the dark region) and 95% (the light region) with all four $\delta\epsilon_{i,b}$ floating (upper left) or fixing $\delta\epsilon_2 = 0$ (upper right).

typical value. Interestingly this is about the same level of precision at which the loops in Flavour Changing Neutral Current processes are currently tested. So far no deviation from the SM has emerged in either case.

4.3. *Effective operators*

It is customary to view the SM as an effective low energy approximation of a more fundamental theory involving the same low energy degrees of freedom and already valid at shorter distances. As a consequence a more complete Lagrangian describing physics at any scale below a cut-off Λ is

$$\mathcal{L}_{\text{eff}}(E < \Lambda) = \mathcal{L}_{SM} + \Sigma_{i,p>0} \frac{c_{i,p}}{\Lambda^p} \mathcal{O}_i^{(4+p)}, \tag{26}$$

in terms of any gauge invariant operator $\mathcal{O}_i^{(4+p)}$ of dimension $(4+p)$, made of the SM fields. While the presence of these extra operators has in itself nothing to

Table 1. 95% lower bounds on Λ/TeV for the individual operators. The first two columns are from the year 2000 data,[43,44] whereas the third and fourth columns are from the currently available data.[42]

	$c_i = -1$	$c_1 = +1$	$c_i = -1$	$c_i = +1$
$(H^+\tau^a H)W^a_{\mu\nu}B_{\mu\nu}$	9.7	10	11.1	18.4
$\lvert H^+D_\mu H\rvert^2$	4.6	5.6	6.3	15.4
$i(H^+D_\mu\tau^a H)(\bar{L}\gamma_\mu\tau^a L)$	8.4	8.8	9.8	14.8
$i(H^+D_\mu\tau^a H)(\bar{Q}\gamma_\mu\tau^a Q)$	6.6	6.8	9.6	8.7
$i(H^+D_\mu H)(\bar{L}\gamma_\mu L)$	7.3	9.2	14.8	9.2

do with the EWC, the study of their effects has proven effective in describing the implications of the EWPT. In turn such effects typically arise from the interference of these operators with radiatively corrected SM amplitudes. This approach has indeed been pursued already in the nineties during the first decade of LEP operation and is receiving further recent attention through the measurements of the Higgs boson properties.[d]

The relative drawback of this approach is in the large number of independent operators that are involved and can affect many different observables. Nevertheless it is interesting to recall the most significant bounds on the scale Λ that were first obtained after a decade of LEP operation[43,44] and compare them with the corresponding current bounds.[42] They are shown in Table 1 at 95% C.L. The bounds are obtained by considering one operator at a time with the relative coefficient $c_{i,p} = \pm 1$, since what counts is the interference with the radiatively corrected SM amplitudes. Furthermore, in the columns from the data in the year 2000, the Higgs boson mass was taken at 115 GeV, consistent with the direct bound and corresponding to the indirect indication of the EWPT at that time. The first operator in Table 1 contributes as well to correct the Higgs boson width into a pair of photons, which is loop suppressed in the SM. From current data on this width one obtains the bounds[42]

$$\Lambda > 12.5\,\mathrm{TeV}\ (c = -1), \quad \Lambda > 7.1\,\mathrm{TeV}\ (c = 1). \tag{27}$$

These are the strongest bounds obtainable to date from the Higgs boson coupling measurements.

The bounds on Λ in the year 2000, compared with the "naturalness" estimate of the radiative corrections to the Higgs boson mass

$$\delta m_H^2 \approx (100\,\mathrm{GeV})^2 \left(\frac{\Lambda}{400\,\mathrm{GeV}}\right)^2, \tag{28}$$

[d]See, among others, Refs. 40–42.

and the already mentioned indication for a light Higgs boson in the same year 2000, motivated the title "LEP paradox" of Ref. 44. Needless to say, the identification of the two Λ parameters entering in Eqs. (26) and (28) has to be taken with a grain of salt, since in suitable extensions of the SM the squares of these parameters are related to each other by a loop factor.

4.4. *Examples in specific models*

The Effective Operator approach just described is useful in selecting particular observables not already constrained in an indirect way by other measurements. Nevertheless the large number of independent operators involved and the neglect, inherent to this approach, of any correlation among them may limit the possibility to understand the relevance of the EWC in a particular model of new physics, where, on the other hand, the EWC have been and can be explicitly computed.[e] Let us briefly recall some illustrative examples.

In composite Higgs models or in fact in any model where the standard Higgs boson mixes with a scalar that has anomalous or even vanishing couplings to the intermediate vector bosons, the coupling g_{HVV} of the newly found Higgs boson to the W and the Z will itself deviate from the SM coupling g_{HVV}^{SM}. As a consequence, new contributions appear in $\delta\epsilon_{1,3}$ or in $\hat{T}, \hat{S}$ of universal character[46,47]

$$\delta\epsilon_1 = -\frac{3\alpha}{8\pi c^2}(1 - k_V^2)\log\frac{\Lambda}{m_H}, \quad \delta\epsilon_3 = \frac{\alpha}{24\pi s^2}(1 - k_V^2)\log\frac{\Lambda}{m_H}, \tag{29}$$

in terms of

$$k_V = \frac{g_{HVV}}{g_{HVV}^{SM}}, \tag{30}$$

normally expected to be less than unity, and of a model dependent cut-off Λ, that roughly measures where VV scattering gets unitarised. The appearance in Eqs. (29) of the same coefficients as in Eq. (14) is not accidental. Barring other new physics contributions to $\delta\epsilon_{1,3}$ of suitable size and sign, which may well be there, a significant constraint is implied on k_V for $\Lambda \gtrsim 1$ TeV by comparing these equations with Fig. 1 (in the more appropriate version, depending on the model in question).

The third squark generation, in particular the partners of the left handed top and bottom, are looked for with particular attention since they play a special role

[e] In fact caution has to be used in drawing general conclusions about the relative importance of precision in Higgs coupling measurements versus the EWPT, since in specific models of new physics significant correlations may arise between various effective operators that escape a one-by-one operator analysis. See Ref. 45 for some relevant examples.

in controlling the largest radiative correction to the Higgs mass. Hence the interest in possible radiative correction effects due to virtual exchanges of these particles. An early calculation gave[48,49]

$$\delta\epsilon_1 = \frac{G_F m_t^4}{\sqrt{2}\, 8\pi^2 m_{\tilde{Q}_3}^2} \approx 10^{-3}\left(\frac{300\,\text{GeV}}{m_{\tilde{Q}_3}}\right)^2, \tag{31}$$

where $m_{\tilde{Q}_3}^2$ is the supersymmetry-breaking squared mass for the $(\tilde{t}_L, \tilde{b}_L)$ doublet. Even though Eq. (31) receives corrections from $\tilde{t}_L - \tilde{t}_R$ mixing, the sensitivity of $\delta\epsilon_1$ does not exceeds significantly the one in the same Eq. (31) with $m_{\tilde{Q}_3}$ replaced by the lightest stop mass, while it can become weaker. ϵ_3 (or the $\hat{S}$ parameter) receives corrections from the third squark generation which are about one order of magnitude lower than for ϵ_1. Radiative corrections from stop exchanges to $Z \to b\bar{b}$ and to Higgs couplings may as well play some role in future precision experiments (see Ref. 50 for a recent analysis).

The issue of the corrections to the $g-2$ factor of the muon, with its 3σ discrepancy between theory and experiment, is discussed in Ref. 51.

5. High Precision in the Standard Model

All the content of Sections 3 and 4 rests on the high precision that has been achieved so far in computing the EWC in the SM. Some key quantities like the W mass (or Δr in Eq. (3)) and the effective leptonic mixing angle

$$\sin^2\theta_{\text{eff}}^l = \frac{1}{4}\left(1 - \frac{g_V^l}{g_A^l}\right) \tag{32}$$

have been computed in the SM to full two-loop order, which consists of many different contributions[52–58] due to the several coupling constants and particles involved, and including as well some relevant three loops, going e.g. like $\alpha_S G_F^2 m_t^4$ or $G_F^3 m_t^6$.[59] As a result the current theoretical uncertainty from the missing higher order terms, as inferred for example from comparing the calculations in different renormalisation schemes, is estimated for M_W at about 5 MeV and on $\sin^2\theta_{\text{eff}}^l$ at about $5\cdot 10^{-5}$. To these uncertainties one has to add the parametric errors of 9 MeV for M_W and of $12\cdot 10^{-5}$ for $\sin^2\theta_{\text{eff}}^l$, obtained with $\Delta m_t = 1\,\text{GeV}$, $\Delta\alpha(M_Z) = 3.3\cdot 10^{-4}$ and $\Delta\alpha_S(M_Z) = 7\cdot 10^{-4}$.[60] This has to be compared with the current experimental errors of 15 MeV for M_W and $16\cdot 10^{-5}$ for $\sin^2\theta_{\text{eff}}^l$. All these numbers are collected in Table 2. While the theoretical uncertainties look adequate to compare with current precision measurements, significant improvements will be necessary to make full use of the precision foreseen at future facilities like, e.g., FCC-ee,[61] also shown in Table 2.

Table 2. Theoretical, parametric and experimental errors, present (third line) and future (fourth line), on M_W and $\sin^2\theta^l_{\rm eff}$

	δM_W/MeV	$\delta\sin^2\theta^l_{\rm eff}/10^{-5}$
Higher orders	5	5
Parametric	9	12
Exp. current	15	16
Exp. FCC-ee	0.5	0.3

Acknowledgments

It is a pleasure to thank G. Altarelli, M. Beccaria, B. Bellazzini, F. Caravaglios, P. Ciafaloni, G. Curci, M. Frigeni, S. Jadach, L. Maiani, A. Pomarol, R. Rattazzi, S. Rychkov, A. Strumia, A. Varagnolo, A. Vicere', for their collaborations on the subject treated in this work. I am indebted with S. Mishima and L. Silvestrini for providing me with Fig. 1 and to G. Degrassi for help on the uncertainties in Table 2. This work is supported in part by the European Programme "Unification in the LHC Era", contract PITN-GA-2009-237920 (UNILHC) and by MIUR under the contract 2010YJ2NYW-010.

References

1. G. 't Hooft, *Nucl. Phys. B* **33**, 173 (1971).
2. G. 't Hooft, *Nucl. Phys. B* **35**, 167 (1971).
3. M. J. G. Veltman, *Nucl. Phys. B* **123**, 89 (1977).
4. M. J. G. Veltman, *Phys. Lett. B* **70**, 253 (1977).
5. P. Sikivie, L. Susskind, M. B. Voloshin and V. I. Zakharov, *Nucl. Phys. B* **173**, 189 (1980).
6. G. Passarino and M. J. G. Veltman, *Nucl. Phys. B* **160**, 151 (1979).
7. F. Antonelli, M. Consoli and G. Corbo, *Phys. Lett. B* **91**, 90 (1980).
8. M. J. G. Veltman, *Phys. Lett. B* **91**, 95 (1980).
9. A. Sirlin, *Phys. Rev. D* **22**, 971 (1980).
10. A. C. Longhitano, *Phys. Rev. D* **22**, 1166 (1980).
11. W. J. Marciano and A. Sirlin, *Phys. Rev. D* **22**, 2695 (1980) [Erratum *ibid.* **31**, 213 (1985)].
12. J. R. Ellis and G. L. Fogli, *Phys. Lett. B* **232**, 139 (1989).
13. P. Langacker, *Phys. Rev. Lett.* **63**, 1920 (1989).
14. G. S. Abrams, C. Adolphsen, R. Aleksan, J. P. Alexander, M. A. Allen, W. B. Atwood, D. Averill and J. Ballam *et al.*, *Phys. Rev. Lett.* **63**, 724 (1989).
15. J. M. Gaillard, *Nucl. Phys. Proc. Suppl.* **16**, 30 (1990).
16. L. J. Nodulman, *Nucl. Phys. Proc. Suppl.* **16**, 40 (1990).
17. J. Bernabeu, A. Pich and A. Santamaria, *Phys. Lett. B* **200**, 569 (1988).
18. J. J. van der Bij and F. Hoogeveen, *Nucl. Phys. B* **283**, 477 (1987).
19. J. van der Bij and M. J. G. Veltman, *Nucl. Phys. B* **231**, 205 (1984).

20. R. Barbieri, M. Beccaria, P. Ciafaloni, G. Curci and A. Vicere, *Phys. Lett. B* **288**, 95 (1992) [Erratum *ibid.* **312**, 511 (1993)] [hep-ph/9205238]; *Nucl. Phys. B* **409**, 105 (1993).
21. J. Fleischer, O. V. Tarasov, F. Jegerlehner and P. Raczka, *Phys. Lett. B* **293**, 437 (1992).
22. J. Fleischer, O. V. Tarasov and F. Jegerlehner, *Phys. Lett. B* **319**, 249 (1993).
23. M. Koratzinos, in *La Thuile 1994, Results and perspectives in particle physics*, 319–332.
24. R. Barbieri, in *La Thuile 1994, Results and perspectives in particle physics*, 375–386.
25. F. Abe *et al.* [CDF Collaboration], *Phys. Rev. Lett.* **73**, 225 (1994). [hep-ex/9405005].
26. G. Aad *et al.* [ATLAS Collaboration], *Phys. Lett. B* **716**, 1 (2012). [arXiv:1207.7214 [hep-ex]].
27. S. Chatrchyan *et al.* [CMS Collaboration], *Phys. Lett. B* **716**, 30 (2012). [arXiv:1207.7235 [hep-ex]].
28. M. Consoli, S. Lo Presti and L. Maiani, *Nucl. Phys. B* **223**, 474 (1983).
29. D. C. Kennedy and B. W. Lynn, *Nucl. Phys. B* **322**, 1 (1989).
30. M. E. Peskin and T. Takeuchi, *Phys. Rev. Lett.* **65**, 964 (1990); *Phys. Rev. D* **46**, 381 (1992).
31. R. Barbieri, A. Pomarol, R. Rattazzi and A. Strumia, *Nucl. Phys. B* **703**, 127 (2004). [hep-ph/0405040].
32. G. Altarelli and R. Barbieri, *Phys. Lett. B* **253**, 161 (1991).
33. G. Altarelli, R. Barbieri and S. Jadach, *Nucl. Phys. B* **369**, 3 (1992) [Erratum *ibid.* **376**, 444 (1992)].
34. G. Altarelli, R. Barbieri and F. Caravaglios, *Nucl. Phys. B* **405**, 3 (1993).
35. R. Barbieri, M. Frigeni and F. Caravaglios, *Phys. Lett. B* **279**, 169 (1992).
36. V. A. Novikov, L. B. Okun and M. I. Vysotsky, *Nucl. Phys. B* **397**, 35 (1993).
37. A. Orgogozo and S. Rychkov, *JHEP* **1306**, 014 (2013). [arXiv:1211.5543 [hep-ph]].
38. M. Ciuchini, E. Franco, S. Mishima and L. Silvestrini, *JHEP* **1308**, 106 (2013). arXiv:1306.4644 [hep-ph].
39. M. Ciuchini, E. Franco, S. Mishima, M. Pierini, L. Reina and L. Silvestrini, arXiv:1410.6940 [hep-ph].
40. A. Falkowski, F. Riva and A. Urbano, *JHEP* **1311**, 111 (2013), arXiv:1303.1812 [hep-ph].
41. A. Pomarol and F. Riva, *JHEP* **1401**, 151 (2014), arXiv:1308.2803 [hep-ph].
42. J. de Blas, M. Ciuchini, E. Franco, D. Ghosh, S. Mishima, M. Pierini, L. Reina and L. Silvestrini, arXiv:1410.4204 [hep-ph].
43. R. Barbieri and A. Strumia, *Phys. Lett. B* **462**, 144 (1999). [hep-ph/9905281].
44. R. Barbieri and A. Strumia, hep-ph/0007265.
45. R. Barbieri and A. Tesi, *Phys. Rev. D* **89**, 5, 055019 (2014). [arXiv:1311.7493 [hep-ph]].
46. R. Barbieri, T. Gregoire and L. J. Hall, hep-ph/0509242.
47. R. Barbieri, B. Bellazzini, V. S. Rychkov and A. Varagnolo, *Phys. Rev. D* **76**, 115008 (2007). [arXiv:0706.0432 [hep-ph]].
48. R. Barbieri and L. Maiani, *Nucl. Phys. B* **224**, 32 (1983).
49. L. Alvarez-Gaume, J. Polchinski and M. B. Wise, *Nucl. Phys. B* **221**, 495 (1983).
50. J. Fan, M. Reece and L. T. Wang, arXiv:1412.3107 [hep-ph].
51. A. Hoecker and W. J. Marciano, in K. A. Olive *et al.* (Particle Data Group), *Chin. Phys. C* **38**, 090001 (2014).
52. A. Djouadi and C. Verzegnassi, *Phys. Lett. B* **195**, 265 (1987).

53. G. Degrassi, P. Gambino and A. Vicini, *Phys. Lett. B* **383**, 219 (1996). [hep-ph/9603374].
54. G. Degrassi, P. Gambino and A. Sirlin, *Phys. Lett. B* **394**, 188 (1997). [hep-ph/9611363].
55. A. Freitas, W. Hollik, W. Walter and G. Weiglein, *Phys. Lett. B* **495**, 338 (2000) [Erratum *ibid.* **570**, 260 (2003)] [hep-ph/0007091]; *Nucl. Phys. B* **632**, 189 (2002) [Erratum *ibid.* **666**, 305 (2003)] [hep-ph/0202131].
56. M. Awramik and M. Czakon, *Phys. Rev. Lett.* **89**, 241801 (2002). [hep-ph/0208113]; *Phys. Lett. B* **568**, 48 (2003). [hep-ph/0305248].
57. M. Awramik, M. Czakon, A. Onishchenko and O. Veretin, *Phys. Rev. D* **68**, 053004 (2003). [hep-ph/0209084].
58. M. Awramik, M. Czakon, A. Freitas and G. Weiglein, *Phys. Rev. Lett.* **93**, 201805 (2004). [hep-ph/0407317].
59. M. Faisst, J. H. Kuhn, T. Seidensticker and O. Veretin, *Nucl. Phys. B* **665**, 649 (2003). [hep-ph/0302275].
60. G. Degrassi, P. Gambino and P. P. Giardino, arXiv:1411.7040 [hep-ph].
61. M. Bicer *et al.* [TLEP Design Study Working Group Collaboration], *JHEP* **1401**, 164 (2014), arXiv:1308.6176 [hep-ex].

Chapter 5

Lattice Quantum Chromodynamics

C. T. Sachrajda

School of Physics and Astronomy,
University of Southampton, Southampton SO17 1BJ, UK
cts@soton.ac.uk

I review the the application of the lattice formulation of QCD and large-scale numerical simulations to the evaluation of non-perturbative hadronic effects in Standard Model Phenomenology. I present an introduction to the elements of the calculations and discuss the limitations both in the range of quantities which can be studied and in the precision of the results. I focus particularly on the extraction of the QCD parameters, i.e. the quark masses and the strong coupling constant, and on important quantities in flavour physics. Lattice QCD is playing a central role in quantifying the hadronic effects necessary for the development of precision flavour physics and its use in exploring the limits of the Standard Model and in searches for inconsistencies which would signal the presence of new physics.

1. Introduction

Quantum Chromodynamics (QCD) is now well established as the theory of the strong nuclear force. This has been possible largely as a result of the property of *asymptotic freedom* which states that the force between quarks and gluons becomes weak at distances much less than 1 fm. At such short distances, the standard analytical tool of perturbation theory can be applied and the results compared with experimental results as discussed, e.g. in the chapter by R. K. Ellis in this Book.[1] However at the typical hadronic scale of about 1 fm, the strong coupling constant α_s is too large for perturbation theory to be applied and *non-perturbative* methods are required to make quantitative determinations of hadronic effects. Of these *Lattice QCD*, i.e. the use of lattice formulations of QCD in large scale numerical simulations has emerged in recent years as a precise *ab initio* technique which can be applied to a wide range of processes and physical quantities.

As the name suggests, the evaluation of hadronic effects in lattice QCD is performed by approximating space–time by a discrete lattice of points in each space–time direction in Euclidean space. A schematic sketch is shown in Fig. 1(a).

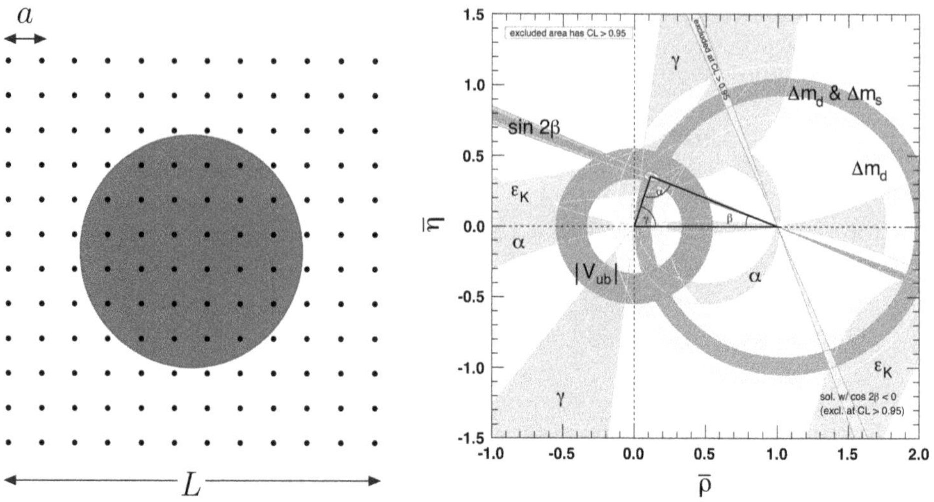

(a) (b)

Fig. 1. (a) Schematic illustration of a Euclidean lattice where a represents the lattice spacing and L its spacial extent. The green circle represents the hadron which is being studied. (b) Determination of the vertex $(\bar{\rho}, \bar{\eta})$ of the unitarity triangle from a variety of weak decay processes.[42]

The quark fields $\psi(x)$ are placed at the lattice sites, whereas gluon fields $A_\mu(x)$ are introduced in terms of *links* $U_\mu(x) = e^{iA_\mu(x+a\hat{\mu}/2)}$, where a is the spacing between neighbouring points and $\hat{\mu}$ is the unit vector in the μ direction. (Throughout this paper the lattice spacing will be denoted by a.) The links transform covariantly under SU(3) gauge transformations $U_\mu(x) \to g(x)U_\mu(x)g^\dagger(x+\hat{\mu})$ $(\psi(x) \to g(x)\psi(x))$, making it possible to construct lattice QCD actions which are exactly gauge invariant. Here we do not review the different discretisations of QCD which are used in actual simulations, but refer the reader to some of the many excellent textbooks on the subject.[2–5]

The applications of lattice QCD are numerous and it is not possible to review them all here. For example at the most recent annual symposium on lattice field theory[6] there were parallel sessions on hadron spectroscopy and interactions; hadron structure; standard model parameters and renormalisation; physics beyond the standard model; weak decays and matrix elements; QCD at nonzero temperature and density; chiral symmetry; vacuum structure and confinement, as well as on new theoretical developments, on algorithms and machines and on applications beyond QCD. In this review I will focus on some of the applications to particle physics phenomenology in general and to flavour physics in particular (see the chapters by G. Isidori[7] and F. Teubert[8] in this Book). In flavour physics we explore the limits of the standard model and search for signatures of new physics by overdetermining the four parameters of the Cabibbo–Kobayashi–Maskawa (CKM) matrix using numerous different physical processes and checking for inconsistencies. The central role of lattice simulations here is in quantifying the hadronic effects, without which the

CKM elements in general cannot be determined, and several examples are given below. This is illustrated in Fig. 1(b) in which the status of the vertex $(\bar{\rho}, \bar{\eta})$ of the unitarity triangle is shown.[42] Lattice results are used in extracting information about the CKM matrix elements when using the measured values of the indirect CP-violation parameter ϵ_K, the mass differences of neutral B mesons (Δm_d and Δm_s) and the determination of $|V_{ub}|$ from exclusive decays.

Numerical results from lattice simulations are constantly improving and I will not attempt to provide an independent detailed compilation of all the results and uncertainties. The *Flavour Physics Lattice Averaging Group* FLAG, performs detailed critical analyses of the computations and results and so far has published two editions of its reviews with a third one scheduled for early in 2016.[11,12] Throughout this paper I illustrate the discussion by quoting the averages from Ref. 12.

The plan of the remainder of this review is as follows. The following section contains a brief introduction to lattice phenomenology with the aim of providing the non-specialist reader with some intuition as to which quantities are calculable and what the limitations on precision are. In Section 3 I discuss the determination of the parameters of QCD, i.e. the quark masses and the strong coupling constant. This is followed by a discussion of a selection of important quantities in flavour physics, including the leptonic decay constants of pseudoscalar mesons, the B-parameters of neutral meson mixing, semileptonic decays as well as nonleptonic kaon decays whose amplitudes were computed for the first time very recently. Prospects for extending the range of lattice computations in flavour physics are briefly discussed in Section 5 and the review concludes with a brief summary.

2. Introduction to Lattice Phenomenology

Lattice phenomenology starts with the evaluation of correlation functions of the form:

$$\langle 0|\, O(x_1, x_2, \ldots, x_n)\, |0\rangle = \frac{1}{Z} \int [dA_\mu]\, [d\psi]\, [d\bar{\psi}]\, e^{iS}\, O(x_1, x_2, \ldots, x_n), \tag{1}$$

where $O(x_1, x_2, \ldots, x_n)$ is a multilocal operator composed of quark and gluon fields and Z is the partition function:

$$Z = \int [dA_\mu]\, [d\psi]\, [d\bar{\psi}]\, e^{iS}\ . \tag{2}$$

In lattice simulations, the infinite-dimensional functional integrals in Eq. (1) are performed by discretising space–time, and using Monte Carlo integration in Euclidean space. The physics which can be studied by computing correlation functions depends on the choice of the multilocal operator O. To illustrate this, consider two-point correlation functions of the form:

$$C_2(t) = \int d^3x\ e^{i\vec{p}\cdot\vec{x}}\ \langle 0|\, \phi_H(\vec{x}, t)\, \phi_H^\dagger(\vec{0}, 0)\, |0\rangle, \tag{3}$$

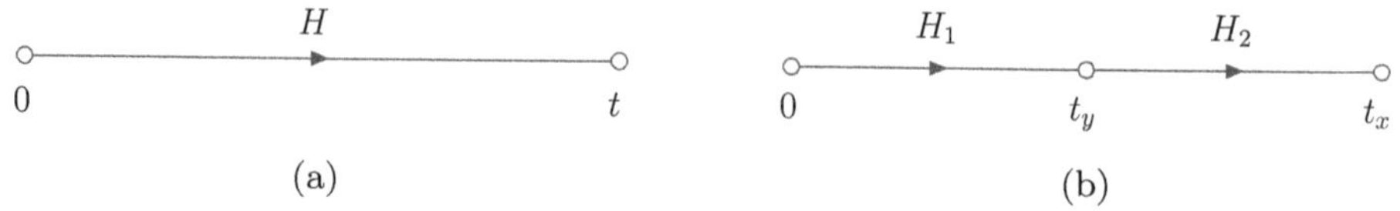

Fig. 2. Schematic illustrations of two- and three-point correlation functions.

where ϕ_H is any interpolating operator for the hadron H whose properties we wish to determine and the time t is taken to be positive. We assume that H is the lightest hadron which can be created by $\phi_H^\dagger$ and that $t > 0$. Inserting a complete set of states between the operators in Eq. (3) and exploiting translational invariance one obtains for sufficiently large t in Euclidean space

$$C_2(t) = \frac{1}{2E_H}\, e^{-E_H t}\, |\langle 0|\phi_H(0)|H(p)\rangle|^2 + \cdots , \tag{4}$$

where $E_H = \sqrt{m_H^2 + \vec{p}^{\,2}}$. A schematic sketch of $C_2(t)$ is shown in Fig. 2(a). The ellipsis in Eq. (4) represents contributions from heavier states with the same quantum numbers as H and which fall more rapidly with t. By fitting $C_2(t)$ as a function of t (at sufficiently large t so that the excited states in Eq. (4) can be neglected) we obtain both the mass m_H and the matrix element $|\langle 0|\phi_H(0)|H(p)\rangle|$. For example, if ϕ_H is the axial current $\bar{q}_1\gamma^\mu\gamma^5 q_2$, where $q_{1,2}$ are the valence quarks of the pseudoscalar meson P, then we obtain the leptonic decay constant f_P as discussed in Section 4.1.

To obtain matrix elements of the form $\langle H_2\,|\,O\,|\,H_1\rangle$, where $H_{1,2}$ are single-hadron states and O is a composite operator of quark and gluon fields, we evaluate three-point correlation functions of the form

$$C_3(t_x, t_y) = \int d^3x\, d^3y\; e^{i\vec{p}\cdot\vec{x}}\; e^{i\vec{q}\cdot\vec{y}}\; \langle 0|\phi_{H_2}(\vec{x}, t_x) O(\vec{y}, t_y)\phi_{H_1}^\dagger(\vec{0}, 0)|0\rangle \tag{5}$$

$$= \frac{e^{-E_{H_1} t_y} e^{-E_{H_2}(t_x - t_y)}}{(2E_{H_1})(2E_{H_2})} \langle 0|\phi_{H_2}(0)|H_2(\vec{p})\rangle$$

$$\times \langle H_2(\vec{p})|O(0)|H_1(\vec{p}+\vec{q})\rangle\langle H_1(\vec{p}+\vec{q})|\phi_{H_1}^\dagger(0)|0\rangle, \tag{6}$$

where ϕ_{H_1,H_2} are interpolating operators for H_1, H_2 and we assume that $t_x > t_y > 0$. All the factors in (6) can be obtained from two-point functions as described above with the exception of the matrix element $\langle H_2(\vec{p})|O(0)|H_1(\vec{p}+\vec{q})\rangle$ which is therefore determined from the computation of the three-point functions. In this way we can obtain for example, weak and electromagnetic form factors, the amplitudes for neutral meson mixing (such as the B_K parameter of K–$\bar{K}$ mixing) and the moments of hadronic structure functions.

By computing correlation functions of the form C_2 and C_3, the determination of the spectrum and of matrix elements of local operators O between single hadron states is now standard. We briefly discuss the status of the (non-standard as yet) evaluation of matrix elements with multi-hadron states and of non-local operators below.

2.1. *Uncertainties in lattice simulations*

Before presenting results, I briefly discuss some of the main contributions to the uncertainties. The evaluation of the correlation functions C_2 and C_3 is performed using Monte Carlo sampling and has an associated statistical error which is estimated by observing how the results vary as additional field configurations are added or removed. More problematic is a reliable determination of the systematic uncertainties. Some of these are specific to the particular quantities being computed; here I mention those which are common to most computations.

2.1.1. *Unphysical light-quark masses*

For much of the period since the 1980's, when lattice computations were in their infancy, the simulations were performed in the *quenched* approximation in which vacuum polarisation effects (sea-quark loops) are neglected. In practice this is implemented by setting the fermionic determinant $\mathrm{Det}(D[A]) = 1$, where $D[A]$ is the Dirac operator in the presence of the background gluon field configuration $\{A(x)\}$. This determinant arises from the integration over the quark fields in Eqs. (1) and (2); it is non-local and its presence makes the generation of the gluon configurations significantly more expensive. Whilst results for known quantities were typically correct at the 10–20% level, the problem with the quenched approximation is that the associated errors cannot be reliably determined.

From around 2000, unquenched simulations became possible albeit with unphysically heavy u- and d-quark masses m_u and m_d; generally the corresponding pion masses were in the range of 0.5–1 GeV. To obtain physical quantities it was therefore necessary to extrapolate in the light-quark masses, frequently using chiral perturbation theory (ChPT) to guide the extrapolation. Indeed as the quark masses were reduced the lattice simulations provided tests of the range of applicability of ChPT and a determination of the corresponding low-energy constants.[13,14] Today we are in the early years of simulations performed with physical values of m_u and m_d (or at least their average $(m_u + m_d)/2$). The next challenge is to include and control isospin breaking effects (including electromagnetic effects). An interesting recent example is the determination of the neutron–proton mass difference by the BMW collaboration.[9] For electromagnetic corrections to amplitudes, one has to control the cancelation of infrared divergences and some initial thoughts were presented earlier this year.[10]

2.1.2. *Lattice spacings and volumes*

Since the cost of the simulations is largely proportional to the number of points in each direction, the choice of the lattice spacing is a compromise between the two conflicting requirements of a fine lattice to minimise lattice artefacts (discretisation errors) and simultaneously one with a large physical volume. At present, depending on the discretisation of QCD being used and on the values of the light quark masses, typical values of the lattice spacing are in the range 0.05 − 0.1 fm and typical volumes are of the order of a few fm, perhaps 5 fm or so. A natural approach to quantifying the errors due to the finite lattice spacing a and volume V is to perform several simulations at different values of a and V and to perform the extrapolations to the continuum and infinite-volume limits and indeed this is done in many cases. In addition one can choose an *improved* discretisation of QCD in which the lattice artefacts decrease as higher powers of a. For example, the artefacts in results obtained using the original Wilson formulation of lattice fermions fall linearly with a whereas with many other formulations they decrease quadratically. The finite-volume corrections are dominated by the propagation of the lightest particles, the pseudo-Goldstone bosons of chiral symmetry breaking (pions and kaons) and so ChPT can be used to estimate these effects. In a finite volume the momentum spectrum is discrete and so infinite-volume momentum integrals are replaced by finite-volume sums and the Poisson summation formula is a powerful tool for calculating the difference between the two.

2.2. *Renormalisation*

A lattice formulation of QCD can be considered as a bare quantum field theory with a playing the role of the ultraviolet cut-off. Quantities computed directly in lattice simulations therefore generally require renormalisation. These might be the QCD parameters (see Section 3) or composite operators which appear when describing physical quantities using the Operator Product Expansion (OPE). With $a^{-1} \leq 4\,\text{GeV}$ it is not possible to simulate the entire standard model and so we have to rely on the OPE and effective theories, writing physical quantities ϕ in the schematic form:

$$\phi = \sum_i C_i(\mu^2)\langle f|O_i^{\rm R}(\mu^2)|i\rangle, \tag{7}$$

where the $O_i^{\rm R}(\mu^2)$ are renormalised composite local operators and μ^2 is the renormalisation scale. $|i\rangle$ and $|f\rangle$ denote the initial and final states respectively. The long-distance non-perturbative physics is contained in the operator matrix elements. The Wilson coefficients $C_i(\mu^2)$ contain the short-distance physics and are calculated in perturbation theory, generally in the $\overline{\rm MS}$ scheme which is convenient for perturbative calculations.

In lattice simulations we compute directly the matrix elements of the bare operators $O_i^B(a)$ in the discretisation of QCD which is being used. For sufficiently large ultraviolet cut-off a^{-1} and renormalisation scale μ it is possible to relate the bare and renormalised operators using perturbation theory and this was done in the early days of lattice QCD. Calculating higher order perturbative calculations in lattice QCD is challenging however, and frequently leads to large corrections. It is now common practice instead to impose the renormalisation conditions non-perturbatively.[15,16] In this way the renormalisation matrix $Z_{ij}(\mu a)$ relating the bare and renormalised operators,

$$O_i^R(\mu^2) = \sum_j Z_{ij}(\mu a)\, O_j^B(a), \tag{8}$$

is determined numerically and often with excellent precision. Some element of perturbation theory is needed however, since the perturbative calculations of the Wilson coefficients C_i are generally performed in schemes based on dimensional regularisation (such as $\overline{\text{MS}}$) which cannot be simulated. A continuum perturbative calculation is therefore needed to relate the operators renormalised in a scheme which can be imposed in a lattice simulation and that used in the calculation of the Wilson coefficients.

2.3. *Heavy quarks*

Weak decays of charm and bottom hadrons provide a particularly rich source of information with which to perform precision studies in flavour physics. They are sufficiently light to be produced copiously and yet heavy enough to have a huge number of possible decay channels, many of which are very rare within the standard model and which can therefore be used to search for evidence of new physics. It appears that the lattice spacings currently being used are sufficiently fine for the charm quark to be simulated using a discrete version of the corresponding terms in the QCD action, but for the b-quark $m_b a > 1$ and such simulations are inappropriate. Most approaches to B-physics rely on effective theories which then have to be matched to QCD. The most common ones use (i) the Heavy Quark Effective Theory which is an expansion in Λ_{QCD}/m_B (see Ref. 17 for a recent review), (ii) nonrelativistic QCD which is an expansion in the heavy quark's velocity[18,19] and (iii) the relativistic heavy quarks approach of the Fermilab group[20] and its extensions.[21] Some groups also extrapolate results from the charm to the bottom region, using scaling laws where applicable and possibly using results in the static limit in which the heavy quark is treated as being infinitely heavy. There are far fewer calculations in heavy-quark physics than of light-quark quantities (although this is currently changing) and so there has been less opportunity to check for consistency of the different approaches.

3. Determination of α_s and the Quark Masses

The strong coupling α_s and the quark masses are input parameters into QCD. They are not measurable directly, but have to be inferred from their effect on measurable quantities such as hadronic masses or, as we shall see below, on other quantities which can be computed in lattice simulations.

To illustrate the procedure consider simulations with $N_f = 2+1$ flavours, i.e. with a sea of u, d, s quarks with $m_u = m_d$. This is a typical situation, although increasingly charm quarks are also introduced into the sea. In each simulation numerical values of the bare strong coupling constant $g(a)$ and the bare quark masses are entered into the computation. If computing resources were unlimited, we would vary the bare quark masses until two dimensionless quantities agreed with their physical values; e.g. a popular choice is m_π/m_K and m_π/m_Ω. Given that the input bare quark masses are chose *a priori* and that resources are in fact limited, such a procedure is adapted to include extrapolations or interpolations of results obtained from several simulations and/or with the use of ChPT. To determine the lattice spacing a, we take the lattice result for a dimensionful quantity obtained in "lattice units", e.g. am_Ω, and write

$$a^{-1} = \frac{1.672\,\text{GeV}}{(am_\Omega)}, \tag{9}$$

where the physical value of $m_\Omega = 1.672\,\text{GeV}$. Having determined the bare quark masses $m_f(a)$, where f denotes the flavour, we need to renormalise them into a standard scheme, such as the $\overline{\text{MS}}$ scheme. For the light-quark masses the FLAG collaboration[12] quotes from $N_f = 2+1$ simulations, $m_{ud}^{\overline{\text{MS}}}(2\,\text{GeV}) = 3.42(6)(7)\,\text{MeV}$ and $m_s^{\overline{\text{MS}}}(2\,\text{GeV}) = 93.8(1.5)(1.9)\,\text{MeV}$.

In order to obtain m_u and m_d separately, rather than just their average, isospin breaking effects must be included and this is beginning to be done. In the meantime the FLAG results shown in Table 1 combine additional ChPT/Current Algebra phenomenological input with lattice results obtained from isospin-symmetric computations.

The traditional determination of the strong coupling constant α_s without using lattice QCD inputs relies on comparing experimental results for some short distance quantities (such as hadronic τ decays, $e^+e^- \to$ hadrons, deep inelastic lepton-hadron scattering and electroweak precision measurements) with the corresponding perturbative expansion. This is reviewed by G. Dissertori in this Book.[22] The PDG[42] quote $\alpha_S^{(5)}(M_Z) = 0.1183 \pm 0.0012$ for the $\overline{\text{MS}}$ coupling in the 5-flavour theory renormalised at M_Z obtained without using lattice inputs. The use of lattice simulations has the advantage that the short-distance quantities do not have to be physically measurable. The procedure is therefore to determine such a quantity ϕ^{SD} nonperturbatively in a lattice computation and to compare the result with the perturbation series:

$$\phi^{\text{SD}} = \sum_i c_i(\mu)\, \alpha_S^i(\mu) + \cdots, \tag{10}$$

where SD reminds us that the quantity must be a short-distance one and the ellipsis represents power corrections which are sometimes modelled and included in the fits. Choices for ϕ^{SD} include the heavy-quark potential, the correlation functions C_2 evaluated at short-distances or large momenta, small loops composed of products of

Table 1. Results taken from the summary table from the FLAG compilation,[12] grouped in terms of N_f, the number of dynamical quark flavours in lattice simulations. The quark masses are given in the $\overline{\rm MS}$ scheme at running scale $\mu = 2\,{\rm GeV}$. The columns marked indicate the number of results that enter the averages for each quantity having satisfied the quality criteria. Full details of the analyses for each quantity can be found in the corresponding sections of Ref. 12. The f_P are the leptonic decay constants of the pseudoscalar meson P (normalised so that $f_\pi \simeq 131\,{\rm MeV}$), the $\hat{}$ denotes the renormalisation group invariant definition of the B-parameters and $\xi = f_{B_s}\sqrt{B_{B_s}}/f_{B_d}\sqrt{B_{B_d}}$.

Quantity		$N_f = 2+1+1$		$N_f = 2+1$		$N_f = 2$
m_s(MeV)			3	93.8(1.5)(1.9)	2	101(3)
m_{ud}(MeV)			3	3.42(6)(7)	1	3.6(2)
m_s/m_{ud}			3	27.46(15)(41)	1	28.1(1.2)
m_d(MeV)				4.68(14)(7)		4.80(23)
m_u(MeV)				2.16(9)(7)		2.40(23)
m_u/m_d				0.46(2)(2)		0.50(4)
$f_+^{K\pi}(0)$			2	0.9661(32)	1	0.9560(57)(62)
f_{K^+}/f_{π^+}	2	1.194(5)	4	1.192(5)	1	1.205(6)(17)
f_K(MeV)			3	156.3(0.9)	1	158.1(2.5)
f_π(MeV)			3	130.2(1.4)		
$\hat{B}_K$			4	0.766(10)	1	0.729(25)(17)
$B_K^{\overline{\rm MS}}(2\,{\rm GeV})$			4	0.560(7)	1	0.533(18)(12)
f_D(MeV)			2	209.2(3.3)	1	208(7)
f_{D_s}(MeV)			2	248.6(2.7)	1	250(7)
f_{D_s}/f_D			2	1.187(12)	1	1.20(2)
$f_+^{D\pi}(0)$			1	0.666(29)		
$f_+^{DK}(0)$			1	0.747(19)		
f_B(MeV)	1	186(4)	3	190.5(4.2)	1	189(8)
f_{B_s}(MeV)	1	224(5)	3	227.7(4.5)	1	228(8)
f_{B_s}/f_B	1	1.205(7)	2	1.202(22)	1	1.206(24)
$f_{B_d}\sqrt{\hat{B}_{B_d}}$(MeV)			1	216(15)		
$f_{B_s}\sqrt{\hat{B}_{B_s}}$(MeV)			1	266(18)		
$\hat{B}_{B_d}$			1	1.27(10)		
$\hat{B}_{B_s}$			1	1.33(6)		
ξ			1	1.268(63)		
$\hat{B}_{B_s}/\hat{B}_{B_d}$			1	1.06(11)		
$\alpha_{\overline{\rm MS}}^{(5)}(M_Z)$			4	0.1184(12)		

gauge links, as well as quark–gluon vertices at large external momenta. Although the perturbative coefficients are frequently known to impressive orders of perturbation theory, the unavoidable truncation of the series in Eq. (10) is one of the main sources of systematic error. A related limitation is the size of the typical lattice spacing $a^{-1} \simeq 2$–$4\,{\rm GeV}$ at which the coupling constant is still fairly large. This can be overcome in principle, and increasingly frequently in practice, by the use of *step scaling*. While lattices on which hadrons are studied necessarily have spacial extents of a least a few fermi, this is not the case for the short-distance quantities used in determining α_S, or in performing renormalisation in general. Step scaling is the successive matching of one lattice with a finer one until we end up with a lattice with a sufficiently small lattice spacing allowing for a reliable perturbation series. The FLAG collaboration have critically reviewed the current lattice determinations of α_S and (conservatively) quote

$$\alpha_S^{(5)}(M_Z) = 0.1184 \pm 0.0012, \tag{11}$$

as their combined result. It should be noted that the result in (11) has a larger uncertainty than those quoted in several of the publications analysed in arriving at this result; the reasons for this are explained in Ref. 12.

4. Selected Quantities in Flavour Physics

It is not possible in this brief review to discuss every physical quantity and to analyse every lattice computation. For many quantities in flavour physics this task has been undertaken by the FLAG collaboration.[12] I will comment on a number of the quantities studied by FLAG, but start this section by reproducing in Table 1 part of the summary table of Ref. 12. The reader who is interested in specific quantities will find a critical analysis of each computation and references to the original literature in Ref. 12.

4.1. *Leptonic decays of mesons*

Among the simplest quantities for which the nonperturbative QCD effects can be computed are the amplitudes for the leptonic decays $P^+ \to \ell^+\nu_\ell$ where P is a pseudoscalar meson and ℓ a lepton. Lorentz and parity symmetries imply that all the hadronic effects are contained in a single *decay constant*, f_P, defined by:

$$\langle 0|\, \bar{q}_2\gamma_\mu\gamma_5 q_1\, |P^+\rangle = i f_P p_\mu\,, \tag{12}$$

where q_1 and q_2 are the charge $2/3$ and $-1/3$ valence quark fields of P^+ respectively. In terms of f_P the decay rate is written as

$$\Gamma(P^- \to \ell^-\bar{\nu}_\ell) = \frac{G_F^2\, |V_{q_1q_2}|^2 f_P^2}{8\pi}\, m_P\, m_\ell^2 \left(1 - \frac{m_\ell^2}{m_P^2}\right)^2 . \tag{13}$$

The decay constants are obtained from a calculation of two-point correlation functions with suitable interpolating operators for the mesons. A recent compilation of the results is presented in Table 1,[12] from which I wish to make two points. The first is to underline the remarkable progress in lattice calculations, which can be seen in the very small errors, approaching 1% precision or even better. Secondly, if the improved precision is to be reflected in an improved determination of the CKM matrix elements $V_{q_1q_2}$ isospin breaking effects, including electromagnetic corrections must be included.[10,28]

4.2. *Neutral-meson mixing and semileptonic decays of pseudoscalar mesons*

We have seen that from three-point correlation functions we can obtain matrix elements of the form $\langle f|O(0)|i\rangle$ where $O(0)$ is a local composite operator. As illustrations of the physics which can be studied this way, we will consider the mixing of neutral pseudoscalar mesons $P^0 \leftrightarrow \bar{P}^0$ and the semileptonic decays $P_1 \to P_2\ell\nu_\ell$, where $P_{1,2}$ are pseudoscalar mesons and ℓ is a lepton.

Figure 3(a) shows the quark-flow diagram for neutral kaon mixing. In the standard model, the non-perturbative hadronic effects in the dominant contribution to the indirect CP-violation parameter ϵ_K are contained in the matrix element of a single $\Delta S = 2$ four-quark operator

$$\langle \bar{K}^0|\left(\bar{s}\gamma^\mu(1-\gamma^5)d\right)\left(\bar{s}\gamma_\mu(1-\gamma^5)d\right)|K^0\rangle \equiv \frac{8}{3}f_K^2 B_K(\mu), \tag{14}$$

where f_K is the leptonic decay constant of the kaon and it is conventional to parametrise the matrix element in terms of B_K. μ represents the scale at which the operator has been renormalised. A summary of the lattice results for B_K and the corresponding quantities B_{B_d} and B_{B_s} for B-meson mixing is given in Table 1. The impressive precision of these results is now such that subdominant contributions need also to be evaluated and in Sec. 5.2 I briefly discuss the prospects for the evaluation of the long-distance contributions to ϵ_K which are expected to be $O(5\%)$.

Another important class of quantities which can be obtained from the evaluation of three-point functions are electromagnetic and weak form factors of both mesons and baryons. Within flavour physics, lattice evaluations of the weak transition form factors combined with experimental measurements of the decays rates are used to determine the corresponding CKM matrix elements. Here we illustrate this by considering semileptonic $B \to \pi\ell\nu_\ell$ decays from which the CKM matrix element V_{ub} can be determined. As is frequently the case, the main limitation on the precision in the determination of V_{ub} is due to that with which we can compute the hadronic effects. These are contained in two invariant form factors $f_{0,+}$

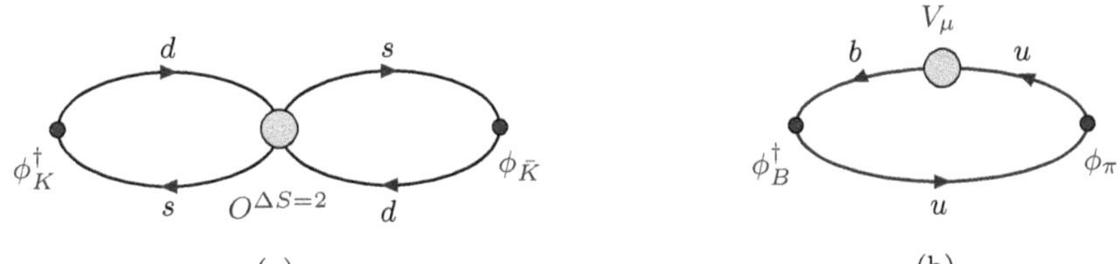

Fig. 3. Schematic illustrations of the correlation functions from which the B_K parameter of K^0–$\bar{K}^0$ mixing and the semileptonic $B \to \pi$ form-factors are obtained.

defined by

$$\langle \pi(p_\pi)|\bar{b}\gamma^\mu u|B(p_B)\rangle = f_0(q^2)\frac{m_B^2 - m_\pi^2}{q^2} + f_+(q^2)\left[(p_\pi + p_B)_\mu - \frac{m_B^2 - m_\pi^2}{q^2}q_\mu\right], \tag{15}$$

where $q = p_B - p_\pi$. The form factors are obtained from the computation of the three-point function in (5) with $\phi^\dagger_{H_1} = \phi^\dagger_B$, an interpolating operator with the quantum numbers to create a B-meson, $O(\vec{y}, t_y) = V_\mu(\vec{y}, t_y)$ the $b \to u$ weak vector current and $\phi_{H_2} = \phi_\pi$ an interpolating operator which can annihilate the pion. The evaluation of such form factors can readily be generalised to other pseudoscalar mesons in the initial and final states and extended to form factors of other operators and particles (e.g. to vector particles as in $B \to \rho$ semileptonic decays). In the case of $B \to \pi\ell\nu_\ell$ decays, for much of the phase space the pions have momenta which are large enough to resolve the discrete nature of the lattice. The calculations are therefore restricted to small pion momenta, which corresponds to large values of q^2. V_{ub} is obtained by combining the lattice results with a subset of the experimental data:

$$\Delta\zeta(q_1^2, q_2^2) \equiv \frac{1}{|V_{ub}|^2}\int_{q_1^2}^{q_2^2} dq^2\, \frac{d\Gamma}{dq^2}.$$

(The lattice results at large q^2 can also be combined with theoretically motivated parametrisations for the form factors at lower q^2, including perhaps constraints from analyticity and other general properties of field theory, to extend the range of the predictions, but this is not discussed here.) The two longstanding results for the form-factors are from the FNAL/MILC[23] and HPQCD collaborations[24] are these are combined in the FLAG compilation to give[12]

$$\Delta\zeta(16\,\mathrm{GeV}^2, q^2_{\mathrm{max}}) = 2.16(50)\,\mathrm{ps}^{-1}\,. \tag{16}$$

Combining this result with the experimental data from the BaBar (Belle) experiments, the FLAG collaboration find $V_{ub} = 3.37(21) \times 10^{-3}$ $(3.47(22) \times 10^{-3})$.[12] There is an interesting tension between results such as these obtained from exclusive decays and those obtained from inclusive $b \to u$ semileptonic decays,[42] $|V_{ub}| = (4.41 \pm 0.15^{+0.15}_{-0.19}) \times 10^{-3}$. The determination from inclusive decays has very different systematics and cannot be studied in lattice simulations. The evaluation of $f_+(q^2)$ and $f_0(q^2)$ and the subsequent determination of V_{ub} is a major priority for several collaborations (see e.g. Refs. 25 and 29 for two recent studies).

Until this year, the exclusive determination of V_{ub} has been largely performed by studying B-meson decays. A very interesting recent development has been the determination by the LHCb collaboration of $V_{ub} = (3.27 \pm 0.23) \times 10^{-3}$ from the baryonic decay $\Lambda_b^0 \to p\mu^-\bar{\nu}_\mu$[26] using form factors computed in lattice simulations.[27]

4.3. *Hadronic decays*

Up to now we have considered matrix elements of the form $\langle f|O(0)|i\rangle$ where $|i\rangle$ and $|f\rangle$ are single-particle states or the vacuum. Much of standard model phenomenology, whether involving decays or scattering, concerns multi-hadron states and it turns out that these are considerably harder to deal with in Euclidean finite-volume computations. Before studying the decays however, consider the propagation of a two-pion state with energy below the inelastic threshold. In a series of pioneering papers Lüscher showed how the corresponding energy levels depend on the finite volume and that they are given in terms of the physical scattering $\pi\pi$ phase-shifts.[30–32] For example, assuming that the s-wave $\pi\pi$ scattering is dominant, the Lüscher quantisation condition for the (discrete) two-pion energies in a finite-volume ($E_{\pi\pi}$) takes the form $\phi(p^*) + \delta_s(p^*) = n\pi$ where n is an integer, the relative momentum $p^* = \frac{1}{2}\sqrt{E_{\pi\pi}^2 - 4m_\pi^2}$, ϕ is a known kinematic function and δ_s is the s-wave phase-shift in the appropriate isospin channel. The quantisation condition can be generalised to include higher partial waves. This remarkable formula allows for a determination of the elastic scattering phase-shifts from the measured two-pion energy levels in a finite Euclidean volume. The formalism applies also to other two-body systems, including pion–nucleon states below the inelastic threshold.

The generalisation to states with higher multiplicities is much more difficult and so far has been restricted to theoretical studies of three-particle states with the recent completion of a formalism to relate the finite-volume three-body spectrum to the three-to-three scattering amplitude for a scalar quantum field theory.[33]

4.3.1. *Two-body decay amplitudes*

The finite-volume approach has been extended to the study of two-body decays. A particularly important example is that of nonleptonic $K \to \pi\pi$ decays in which both indirect and direct CP-violation were first discovered. Bose symmetry implies that the total isospin I of the final two-pion state is either 0 or 2. The evaluation of the $K \to \pi\pi_{I=2}$ amplitude A_2 has recently been achieved with physical kaon and pion masses.[34–36] Whereas for single-hadron states the finite-volume corrections decrease exponentially with the volume, for two-hadron states they only fall as powers of the volume. These effects are also given in terms of the phase-shifts (or more precisely of the derivatives of the phase-shifts with respect to the centre-of-mass momentum) and can be corrected by a multiplicative factor, the *Lellouch–Lüscher factor*.[38–40]

The evaluation of the $K \to (\pi\pi)_{I=0}$ amplitude A_0 is much more involved and complicated and it has only been very recently that the first calculation, performed with physical masses and kinematics, has been completed.[41] The evaluation of the $K \to \pi\pi$ amplitudes, in particular A_0, has for several decades been a seemingly unattainable target for lattice computations. It is therefore particularly satisfying that this target has now been reached, albeit still with significant uncertainties[41]:

$$\mathrm{Re}\left(\frac{\epsilon'}{\epsilon}\right) = (1.38 \pm 5.15 \pm 4.43) \times 10^{-4}, \tag{17}$$

where the first error is statistical and the second is systematic. The Particle Data Group compilation[42] of the experimental results is $\mathrm{Re}(\epsilon'/\epsilon) = 1.66(0.23) \times 10^{-3}$. The matrix elements which contribute to the amplitudes are obtained with better relative accuracy but, as expected, there is a significant partial cancellation between the QCD penguin contribution to $\mathrm{Im}(A_0)/\mathrm{Re}(A_0)$ and the electroweak penguin contribution to $\mathrm{Im}(A_2)/\mathrm{Re}(A_2)$ which amplifies the relative error in $\mathrm{Re}(\epsilon'/\epsilon)$.

4.3.2. *On the difficulty of studying exclusive nonleptonic B decays*

So far in this review we have discussed some of the many physical quantities which can be studied using lattice simulations. It is instructive to examine the main difficulties in the evaluation of $K \to \pi\pi$ amplitudes, because they underline why the calculations cannot be extended to a very important set of processes, two-body B-decays. A huge amount of data has been provided by the B-factories and is being provided by the LHCb experiment on decay rates and CP-asymmetries of processes such as $B \to \pi\pi$ and $B \to \pi K$ and yet without new ideas we cannot compute the corresponding hadronic effects and hence to use this data in studies of the unitarity triangle.

We start the discussion by considering the evaluation of the amplitude A_2 of $K \to (\pi\pi)_{I=2}$ decays illustrated in Fig. 4. We envisage creating a kaon by placing an interpolating operator $\phi_K^\dagger$ at time $t = t_K$ and taking a Fourier transform to project to $\vec{p}_K = 0$. The operators of the strangeness-changing $\Delta S = 1$ effective weak Hamiltonian are placed at t_H, with $t_H - t_K$ sufficiently large to suppress the propagation of states heavier than the kaon between t_K and t_H. By integrating the position of the operators over space, we ensure that the final state also has zero three momentum. Finally at time $t = t_\pi$ we place an operator which can annihilate the two pions. A natural choice for the two-pion interpolating operator is a product of two single-pion annihilation operators, one with momentum $\vec{q}$ and the other with momentum $-\vec{q}$ (see Fig. 4). Among the difficulties in the calculation is that by using

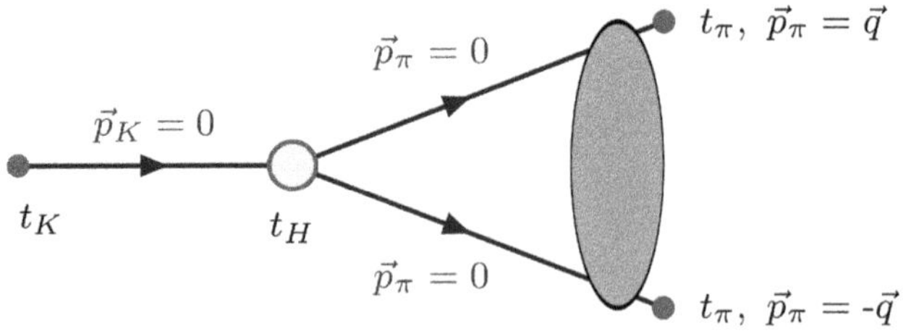

Fig. 4. Schematic illustrations of the correlation functions from which the $K \to (\pi\pi)_{I=2}$ decay amplitude A_2 (see text).

time to isolate the lightest state, energy is not conserved and so $E_{\pi\pi} \neq m_K$ unless we engineer this very carefully. The correlation function sketched in Fig. 4 can be written schematically as a function of $t_\pi - t_H$ in the form:

$$C = C_0 e^{-E^0_{\pi\pi}(t_\pi - t_H)} + C_1 e^{-E^1_{\pi\pi}(t_\pi - t_H)} + \cdots , \tag{18}$$

where $E^0_{\pi\pi}$ and $E^1_{\pi\pi}$ are the energies of the two-pion ground state and first excited state respectively corresponding to the finite volume and boundary conditions used in the simulation. The ellipsis represents the contributions of excited states with energies greater than $E^1_{\pi\pi}$. The lack of energy conservation means that even if the input momentum $\vec{q} \neq 0$, C is dominated by the term with the lowest energy, $E^0_{\pi\pi} = 2m_\pi$ (up to finite volume corrections). The grey oval in Fig. 4 represents the energy non-conserving $\pi\pi$ scattering from $E^0_{\pi\pi}$ to an excited state. In the discussion so far we have implicitly assumed the use of periodic boundary conditions. In such simulations the determination of the physical $K \to \pi\pi$ amplitude requires the study of an excited state with the volume chosen such that $E^1_{\pi\pi} = m_K$ and the determination of C_1; this is very difficult indeed. In the recent computations of A_2, antiperiodic boundary conditions were used instead and the volume was chosen such that the ground state has energy equal to m_K.[34–36]

The s-wave two-pion state $|(\pi\pi)_{I=0}\rangle$ has vacuum quantum numbers and so in addition to the terms on the right-hand side of Eq. (18) there is a constant term corresponding to the vacuum intermediate state which does not fall with $t_\pi - t_H$ and which therefore dominates the correlation function. This complication is unavoidable in the evaluation of A_0 and although the constant is calculable (it is the product of two vacuum expectation values of the form in Eq. (1)) the subtraction of the dominant term leads to a loss of precision. In addition of course, we would like the lowest-energy two-pion state to be the one with $E_{\pi\pi} = m_K$ and this requires the imposition of isospin-conserving G-parity boundary conditions.[43] The vacuum subtraction and the use of G-parity boundary conditions to ensure that the lowest energy of an $I = 0$ two-pion state is m_K were the major technical breakthroughs allowing for the evaluation of ϵ'/ϵ.[41]

Apart from the intrinsic importance of evaluating ϵ'/ϵ from first principle, the main reason for this discussion is to explain a significant limitation of lattice computations; our present inability to calculate exclusive nonleptonic decays in general. Imagine trying to evaluate $B \to \pi\pi$ decay amplitudes and consider Fig. 4 but with the kaon replaced by a B-meson. The pions in the final state each have momenta of about 2.6 GeV. In addition to the requirement of a very fine lattice to accommodate such large momenta, a more serious difficulty is that this corresponds to a highly excited two-pion state with total momentum zero, i.e. one which which is hidden well inside the ellipsis in Eq. (18) and hence is very highly suppressed. One would also need to deal with the power corrections in the volume with many possible intermediate states. Without some novel ideas it is very unlikely that calculations of nonleptonic B and D exclusive decays will be possible in the foreseeable future.

5. New Directions

In the preceding section I described some calculations in flavour physics starting with quantities which can be obtained by computing *standard* two and three-point correlation functions and proceeding to the evaluation of $K \to \pi\pi$ decay amplitudes. In this section I briefly mention some new developments which are likely to extend the physics reach of lattice phenomenology.

5.1. *Hadronic effects in the muon's electric dipole moment*

There is an intriguing 3–4σ discrepancy between the experimental measurement of the anomalous magnetic moment of the muon

$$a_\mu^{\text{exp}} = (11659208.9 \pm 6.3) \times 10^{-10} \tag{19}$$

and the standard model prediction

$$a_\mu^{\text{SM}} = (11659180.1 \pm 4.9) \times 10^{-10}, \tag{20}$$

(see for example Ref. 44 for a review and references to the original literature). The experimental precision will be significantly improved by new muon $g-2$ experiments at Fermilab and J-PARC. On the theoretical side, in addition to the dominant QED perturbative contributions (and the small weak contributions) there are hadronic effects through the *hadronic vacuum polarisation* (HVP) contribution, estimated using e^+e^- experimental data to be $(682.5 \pm 4.2) \times 10^{-10}$ and the *hadronic light-by-light contribution* (HLbL), estimated using phenomenological techniques to be $(10.5 \pm 2.6) \times 10^{-10}$. It is these hadronic effects, both HVP[45–49] and even HLbL effects[50] which can be estimated using lattice QCD, although it is a challenge to compete with the above precision.

5.2. *Long-distance contributions to hadronic processes*

In Section 4 we have discussed the evaluation of matrix elements of composite local operators. More recently calculations of long-distance effects have begun to

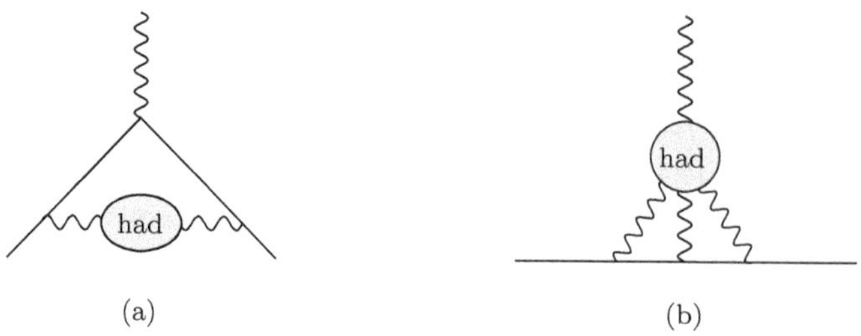

Fig. 5. (a) Hadronic vacuum polarisation and (b) hadronic light-by-light contributions to the anomalous magnetic moment of the muon. The curly lines represent photons and the straight lines the muon.

be evaluated, which involve the determination of non-local matrix elements of the form

$$\int d^4x \int d^4y \langle h_2 | T\{O_1(x) O_2(y)\} | h_1 \rangle, \tag{21}$$

where $O_{1,2}$ are composite operators and $| h_{1,2}\rangle$ are single-hadron states (one can also exploit translational invariance and set either x or y to be the origin for example). An important example is the long-distance contribution to the ϵ_K parameter.[51] This has been estimated to be at the level of a few percent,[52] so since the precision of the hadronic effects in the leading contribution is approaching this level of precision (see the result for B_K in Table 1) it becomes necessary to evaluate the long distance contributions and we can look forward to results in the next few years. Other important quantities for which the long-distance contributions are beginning to be evaluated include the K_L–K_S mass difference[53,54] and rare kaon decays $K \to \pi\nu\bar{\nu}$ and $K \to \pi\ell^+\ell^-$, where ℓ represents a charged lepton.[55,56]

5.3. *R*(*D*) *and R*(*D**)

An intriguing tension between experimental observations and standard model predictions is provided by the quantities

$$R(D) = \frac{\mathrm{Br}(\bar{B} \to D\tau^-\bar{\nu}_\tau)}{\mathrm{Br}(\bar{B} \to D\ell^-\bar{\nu}_\ell)} \quad \text{and} \quad R(D^*) = \frac{\mathrm{Br}(\bar{B} \to D^*\tau^-\bar{\nu}_\tau)}{\mathrm{Br}(\bar{B} \to D^*\ell^-\bar{\nu}_\ell)} \tag{22}$$

where $\ell = e$ or μ. Combining the results from Babar[57] and Belle[58] one obtains $R(D) = 0.388 \pm 0.047$ to be compared to the standard model prediction of $R(D) = 0.300 \pm 0.010$. Similarly, combining the Babar, Belle and LHCb[59] results gives $R(D^*) = 0.321 \pm 0.021$ compared to the predicted value of 0.252 ± 0.005. The role of lattice calculations is to evaluate $B \to D^{(*)}$ form factors with good precision. So far this has been done for $B \to D$ form factors with the recent result $R(D) = 0.299 \pm 0.011$[60] and we can anticipate improved precision and also a calculation of $R(D^*)$ in the near future.

6. Summary and Future Prospects

At the 1989 annual symposium on Lattice Field Theory held in Capri, Italy, Ken Wilson the father of the subject, made the prediction that it would take about 30 years before Lattice QCD computations were sufficiently precise and reliable to be useful to Standard Model phenomenology. Although this prediction appeared to be too pessimistic at the time, for much of the intervening period, the systematic uncertainties, and in particular the large values of the u and d sea quark masses in the simulations, meant that extrapolations and model input was necessary to obtain physical results from the computations. This has now changed! Theoretical and algorithmic developments, combined with increased computing resources have led to enormous progress in recent years in the applications of lattice

QCD to Standard Model phenomenology. In this review I have summarised these developments focussing on the determination of the parameters of the standard model and on the applications to precision flavour physics. In the near future, isospin-breaking effects (including electromagnetic corrections) will be included more frequently taking the precision of lattice phenomenology beyond 1%. In addition to continuing to improve the precision of the results for *standard* quantities, the community in also increasing the range of physical quantities and effects which can be studied in lattice simulations. A selection of such extensions was presented in Section 5. We can therefore look forward to an exciting period of time in which new experimental results from the LHC, Belle-II, J-PARC and other facilities are combined with improved lattice computations to explore the limits of the standard model with ever increasing rigour and to help unravel the underlying theoretical framework of the new physics once it is discovered.

References

1. R. K. Ellis, Chapter 3 in this Book.
2. I. Montvay and G. Münster, *Quantum Fields on a Lattice*, Cambridge Monographs on Mathematical Physics (Cambridge University Press, 1994).
3. J. Smit, Cambridge *Lect. Notes Phys.* **15**, 1 (2002).
4. T. DeGrand and C. E. Detar, *Lattice Methods for Quantum Chromodynamics* (World Scientific, 2006).
5. C. Gattringer and C. B. Lang, *Lect. Notes Phys.* **788**, 1 (2010).
6. http://www.aics.riken.jp/sympo/lattice2015/
7. G. Isidori, Chapter 17 in this Book.
8. F. Teubert, Chapter 18 in this Book.
9. S. Borsanyi *et al.*, *Science* **347**, 1452 (2015) [arXiv:1406.4088 [hep-lat]].
10. N. Carrasco, V. Lubicz, G. Martinelli, C. T. Sachrajda, N. Tantalo, C. Tarantino and M. Testa, *Phys. Rev. D* **91**, 7, 074506 (2015) [arXiv:1502.00257 [hep-lat]].
11. G. Colangelo, S. Durr, A. Jüttner, L. Lellouch, H. Leutwyler, V. Lubicz, S. Necco and C. T. Sachrajda *et al.*, *Eur. Phys. J. C* **71**, 1695 (2011) [arXiv:1011.4408 [hep-lat]].
12. S. Aoki, Y. Aoki, C. Bernard, T. Blum, G. Colangelo, M. Della Morte, S. Dürr and A. X. El Khadra *et al.*, *Eur. Phys. J. C* **74**, 2890 (2014) [arXiv:1310.8555 [hep-lat]].
13. S. Dürr *et al.* [Budapest-Marseille-Wuppertal Collaboration], *Phys. Rev. D* **90**, 11, 074506 (2014) [arXiv:1310.3626 [hep-lat]].
14. S. Dürr, PoS LATTICE **2014**, 006 (2015) [arXiv:1412.6434 [hep-lat]].
15. M. Luscher, R. Narayanan, P. Weisz and U. Wolff, *Nucl. Phys. B* **384**, 168 (1992) [hep-lat/9207009].
16. G. Martinelli, C. Pittori, C. T. Sachrajda, M. Testa and A. Vladikas, *Nucl. Phys. B* **445**, 81 (1995) [hep-lat/9411010].
17. R. Sommer, *Nucl. Part. Phys. Proc.* **261–262**, 338 [arXiv:1501.03060 [hep-lat]].
18. B. A. Thacker and G. P. Lepage, *Phys. Rev. D* **43**, 196 (1991).
19. G. P. Lepage, L. Magnea, C. Nakhleh, U. Magnea and K. Hornbostel, *Phys. Rev. D* **46**, 4052 (1992) [hep-lat/9205007].
20. A. X. El-Khadra, A. S. Kronfeld and P. B. Mackenzie, *Phys. Rev. D* **55**, 3933 (1997) [hep-lat/9604004].
21. N. H. Christ, M. Li and H. W. Lin, *Phys. Rev. D* **76**, 074505 (2007) [hep-lat/0608006].

22. G.Dissertori, Chapter 6 in this Book.
23. J. A. Bailey, C. Bernard, C. E. DeTar, M. Di Pierro, A. X. El-Khadra, R. T. Evans, E. D. Freeland and E. Gamiz *et al.*, *Phys. Rev. D* **79**, 054507 (2009) [arXiv:0811.3640 [hep-lat]].
24. E. Dalgic, A. Gray, M. Wingate, C. T. H. Davies, G. P. Lepage and J. Shigemitsu, *Phys. Rev. D* **73**, 074502 (2006) [*Phys. Rev. D* **75**, 119906 (2007)] [hep-lat/0601021].
25. J. A. Bailey *et al.* [Fermilab Lattice and MILC Collaborations], *Phys. Rev. D* **92**, 014024 (2015), arXiv:1503.07839 [hep-lat].
26. R. Aaij *et al.* [LHCb Collaboration], *Nature Phys.* **11**, 743 (2015), arXiv:1504.01568 [hep-ex].
27. W. Detmold, C. Lehner and S. Meinel, *Phys. Rev. D* **92**, 034503 (2015), arXiv:1503.01421 [hep-lat].
28. G. M. de Divitiis *et al.*, *JHEP* **1204**, 124 (2012) [arXiv:1110.6294 [hep-lat]].
29. J. M. Flynn, T. Izubuchi, T. Kawanai, C. Lehner, A. Soni, R. S. Van de Water and O. Witzel, *Phys. Rev. D* **91**, 7, 074510 (2015) [arXiv:1501.05373 [hep-lat]].
30. M. Luscher, *Commun. Math. Phys.* **105**, 153 (1986).
31. M. Luscher, *Nucl. Phys. B* **354**, 531 (1991).
32. M. Luscher, *Nucl. Phys. B* **364**, 237 (1991).
33. M. T. Hansen and S. R. Sharpe, *Phys. Rev. D* **92**, 114509 (2015), arXiv:1504.04248 [hep-lat].
34. T. Blum, P. A. Boyle, N. H. Christ, N. Garron, E. Goode, T. Izubuchi, C. Jung and C. Kelly *et al.*, *Phys. Rev. Lett.* **108**, 141601 (2012) [arXiv:1111.1699 [hep-lat]].
35. T. Blum, P. A. Boyle, N. H. Christ, N. Garron, E. Goode, T. Izubuchi, C. Jung and C. Kelly *et al.*, *Phys. Rev. D* **86**, 074513 (2012) [arXiv:1206.5142 [hep-lat]].
36. T. Blum, P. A. Boyle, N. H. Christ, J. Frison, N. Garron, T. Janowski, C. Jung and C. Kelly *et al.*, *Phys. Rev. D* **91**, 7, 074506 (2015), [arXiv:1502.00263 [hep-lat]].
37. P. A. Boyle *et al.* [RBC and UKQCD Collaborations], *Phys. Rev. Lett.* **110**, 15, 074506 (2013), [arXiv:1212.1474 [hep-lat]].
38. L. Lellouch and M. Luscher, *Commun. Math. Phys.* **219**, 31 (2001) [arXiv:hep-lat/0003023].
39. C. J. D. Lin, G. Martinelli, C. T. Sachrajda and M. Testa, *Nucl. Phys. B* **619**, 467 (2001) [arXiv:hep-lat/0104006].
40. C. h. Kim, C. T. Sachrajda and S. R. Sharpe, *Nucl. Phys. B* **727**, 218 (2005) [hep-lat/0507006].
41. Z. Bai *et al.* [RBC and UKQCD Collaborations], *Phys. Rev. Lett.* **115**, 212001 (2015), arXiv:1505.07863 [hep-lat].
42. K. A. Olive *et al.* [Particle Data Group Collaboration], *Chin. Phys. C* **38**, 090001 (2014).
43. C. Kim and N. H. Christ, PoS LAT **2009**, 255 (2009) [arXiv:0912.2936 [hep-lat]].
44. J. P. Miller, E. d. Rafael, B. L. Roberts and D. Stckinger, *Ann. Rev. Nucl. Part. Sci.* **62**, 237 (2012).
45. C. Aubin and T. Blum, *Phys. Rev. D* **75**, 114502 (2007) [hep-lat/0608011].
46. X. Feng, K. Jansen, M. Petschlies and D. B. Renner, *Phys. Rev. Lett.* **107**, 081802 (2011) [arXiv:1103.4818 [hep-lat]].
47. P. Boyle, L. Del Debbio, E. Kerrane and J. Zanotti, *Phys. Rev. D* **85**, 074504 (2012) [arXiv:1107.1497 [hep-lat]].
48. M. Della Morte, B. Jager, A. Juttner and H. Wittig, *JHEP* **1203**, 055 (2012) [arXiv:1112.2894 [hep-lat]].
49. F. Burger *et al.* [ETM Collaboration], *JHEP* **1402**, 099 (2014) [arXiv:1308.4327 [hep-lat]].

50. T. Blum, S. Chowdhury, M. Hayakawa and T. Izubuchi, *Phys. Rev. Lett.* **114**, 1 (2015), 012001 [arXiv:1407.2923 [hep-lat]].
51. N. Christ *et al.* [RBC and UKQCD Collaborations], PoS LATTICE **2013**, 397 (2014) [arXiv:1402.2577 [hep-lat]].
52. A. J. Buras, D. Guadagnoli and G. Isidori, *Phys. Lett. B* **688**, 309 (2010) [arXiv:1002.3612 [hep-ph]].
53. N. H. Christ *et al.* [RBC and UKQCD Collaborations], *Phys. Rev. D* **88**, 014508 (2013) [arXiv:1212.5931 [hep-lat]].
54. Z. Bai, N. H. Christ, T. Izubuchi, C. T. Sachrajda, A. Soni and J. Yu, *Phys. Rev. Lett.* **113**, 112003 (2014) [arXiv:1406.0916 [hep-lat]].
55. X. Feng, N. H. Christ, A. Portelli and C. Sachrajda, PoS LATTICE **2014**, 367 (2015).
56. N. H. Christ *et al.* [RBC and UKQCD Collaborations], *Phys. Rev. D* **92**, 094512 (2015), arXiv:1507.03094 [hep-lat].
57. J. P. Lees *et al.* [BaBar Collaboration], *Phys. Rev. D* **88**, 7 (2013), 072012 [arXiv:1303.0571 [hep-ex]].
58. A. Bozek *et al.* [Belle Collaboration], *Phys. Rev. D* **82**, 072005 (2010) [arXiv:1005.2302 [hep-ex]].
59. G. Ciezarek, presented on behalf of the LHCb collaboration at the conference *Flavour Physics and CP-violation 2015*, 25–29 May 2015 Nagoya University, Japan.
60. J. A. Bailey *et al.* [MILC Collaboration], *Phys. Rev. D* **92**, 034506 (2015), arXiv:1503.07237 [hep-lat].

Chapter 6

The Determination of the Strong Coupling Constant

Günther Dissertori

Institute for Particle Physics, ETH Zurich, Switzerland

The strong coupling constant is one of the fundamental parameters of the Standard Theory of particle physics. In this review I will briefly summarise the theoretical framework, within which the strong coupling constant is defined and how it is connected to measurable observables. Then I will give an historical overview of its experimental determinations and discuss the current status and world average value. Among the many different techniques used to determine this coupling constant in the context of quantum chromodynamics, I will focus in particular on a number of measurements carried out at the Large Electron–Positron Collider (LEP) and the Large Hadron Collider (LHC) at CERN.

1. Introduction

The strong coupling constant, α_s, is the only free parameter of the Lagrangian of quantum chromodynamics (QCD), the theory of strong interactions, if we consider the quark masses as fixed. As such, this coupling constant, or equivalently $g_s = \sqrt{4\pi\alpha_s}$, is one of the three fundamental coupling constants of the Standard Theory (ST) of particle physics. It is related to the $\mathrm{SU}(3)_C$ colour part of the overall $\mathrm{SU}(3)_C \times \mathrm{SU}(2)_L \times \mathrm{U}(1)_Y$ gauge symmetry of the ST. The other two constants g and g' indicate the coupling strengths relevant for weak isospin and weak hypercharge, and can be rewritten in terms of the Weinberg mixing angle $\tan\theta_\mathrm{W} = g'/g$ and the fine-structure constant $\alpha = e^2/(4\pi)$, where the electric charge is given by $e = g\sin\theta_\mathrm{W}$. Note that natural units are used throughout.

While typically denoted as *constants*, actually all these coupling strengths vary as a function of the energy scale or momentum transfer Q of the particular process looked at, as will be discussed later. In contrast to $\alpha(Q^2)$, which increases with increasing Q, the strong coupling $\alpha_s(Q^2)$ decreases for increasing scale, leading to the famous property of QCD known as asymptotic freedom. It is interesting to compare the values of these two coupling strengths at some fixed scale, such as the mass of the Z boson, $Q \approx M_\mathrm{Z} \approx 91\,\mathrm{GeV}$. We find that $\alpha(M_\mathrm{Z}^2) \approx 1/128$, whereas $\alpha_s(M_\mathrm{Z}^2) \approx 0.12$; that is, the strong coupling is still about 15 times larger than the fine-structure constant at energy scales much larger than those relevant for quark confinement into hadrons ($Q \sim 1\,\text{GeV}$). Thus, strong interactions are indeed strong compared to electroweak interactions, even at large energy scales such as those probed by CERN's past and present colliders, in particular the Large Electron–Positron Collider (LEP) or the Large Hadron Collider (LHC).

The different energy dependence of the coupling strengths triggers the immediate question if and at which exact energy scale these coupling constants become of equal strength, implying the onset of a possible *grand unification*. Obviously, the answer to this question also depends on the precision at which α and α_s have been determined by experiment, and it is instructive to realize that today $\alpha(Q^2 = 0) \approx 1/137$ is known at an accuracy of 32 parts per billion,[1] whereas the relative uncertainty of the current world average (WA) value[2] of $\alpha_s(M_Z^2) = 0.1185 \pm 0.0006$ amounts to half a percent; quite an astonishing difference.

Besides the wish to improve the accuracy of the aforementioned very high energy extrapolation, it is of general importance to know α_s at the best possible precision, since it enters the calculation of each and every process that involves strong interactions and thus ultimately limits the precision at which such processes can be predicted theoretically. As a most recent and prominent example, it is worth mentioning that the uncertainty on α_s gives a non-negligible contribution to the overall theoretical uncertainty on Higgs boson production at the LHC.[3] Correspondingly, this limits the studies looking for effects beyond the ST that could manifest themselves through deviations of the measured Higgs production cross sections from their theoretical predictions. In the following I will indicate the experimental and theoretical difficulties that limit the precision at which we know this fundamental parameter, but also highlight the dramatic improvements, which have been achieved during the last three decades.

2. Theoretical Framework

The basic elements of QCD, including a discussion of the scale dependence of $\alpha_s(Q^2)$ and the related structure of theoretical predictions obtained in perturbation theory, are summarized elsewhere in this series of reviews.[4] Further extensive descriptions of the theoretical framework can be found in, e.g., Refs. 2,5 and 6. Here I will only highlight a few important aspects of perturbative QCD (pQCD), that are relevant for the remainder of this review.

When calculating amplitudes corresponding to Feynman graphs that involve loop diagrams, ultraviolet divergences are encountered. The procedure of renormalization absorbs these divergences into a redefinition of the bare parameters and fields that appear in the lagrangian. In particular, this leads to the renormalised or so-called *running* coupling constant $\alpha_s(\mu^2)$, a function of the (unphysical) renormalisation scale μ. If μ is chosen close to the scale of the momentum transfer Q in a given process, then $\alpha_s(\mu^2 \approx Q^2)$ is indicative of the effective strength of the strong interaction in that process.[2] This explains why in the literature we often find a discussion of the running coupling constant as function of the physical scale Q, while the renormalised coupling actually is a function of the unphysical scale μ. This will also become clearer from the following discussion of the structure of perturbative predictions.

While the value of $\alpha_s(\mu^2)$ at a fixed scale μ cannot be predicted and has to be determined from experiment instead, its μ dependence is given by the renormalisation group equation,

$$\mu^2 \frac{d\alpha_s}{d\mu^2} = \beta(\alpha_s) = -\left(b_0\alpha_s^2 + b_1\alpha_s^3 + \mathcal{O}(\alpha_s^4)\right). \tag{1}$$

The first two coefficients in the perturbative expansion of the so-called β-function of QCD are $b_0 = (33 - 2n_f)/(12\pi)$ and $b_1 = (153 - 19n_f)/(24\pi^2)$, where n_f is the number of "light" quark flavours ($m_q \ll \mu$). Most importantly, for $n_f < 17$ we have $b_0 > 0$, which leads to a decreasing coupling strength for increasing scale (asymptotic freedom), as originally predicted by Politzer,[7] Gross and Wilczek.[8] Considering only the first term of the expansion on the right hand side of Eq. 1, a solution can be written as $\alpha_s(\mu^2) = \left(b_0 \ln(\mu^2/\Lambda^2)\right)^{-1}$, with $\Lambda \approx 200$ MeV defined as the scale where $\alpha_s(\mu^2)$ formally diverges. Whereas at small scales of order GeV or lower the coupling constant increases dramatically, making any perturbative approach to the solution of low-energy strong interactions and the property of confinement meaningless, it is the property of asymptotic freedom that leads to an expansion parameter α_s well below unity at large scales and thus allows perturbative methods to be applied for the calculation of scattering processes.

To second order, including the resummation of leading logarithms $\ln(\mu^2/Q^2)$, a solution of Eq. (1) allows us to relate α_s at one scale μ^2 to that at another scale Q^2,

$$\alpha_s(\mu^2) = \frac{\alpha_s(Q^2)}{w}\left(1 - \frac{b_1}{b_0}\frac{\alpha_s(Q^2)}{w}\ln w\right), \quad w = 1 + b_0\alpha_s(Q^2)\ln\frac{\mu^2}{Q^2}\,. \tag{2}$$

In the following we will use the resulting expansion

$$\alpha_s(\mu^2) \approx \alpha_s(Q^2)\left(1 - \alpha_s(Q^2) b_0 \ln\frac{\mu^2}{Q^2} + \mathcal{O}(\alpha_s^2)\right). \tag{3}$$

This shows that a change of scale only manifests itself as a non-leading effect in α_s; in other words, a meaningful determination of the *running* coupling constant necessarily has to involve a next-to-leading order (NLO) prediction. In order to highlight this even further, let us look at the perturbative structure, up to NLO, of some generic cross section that is proportional to α_s at leading order (e.g., a three-jet cross section in e^+e^- annihilations at $\sqrt{s} = Q$),

$$\sigma\left(\alpha_s(\mu^2), \frac{\mu^2}{Q^2}\right) = \alpha_s(\mu^2)A + \alpha_s^2(\mu^2)\left(B + b_0 A \ln\frac{\mu^2}{Q^2}\right) + \mathcal{O}(\alpha_s^3). \tag{4}$$

The coefficients A and B have to be calculated for the specific process at hand. Now let us first assume that only A is known for a particular process "1", i.e., only the leading order (LO) expansion is available, $\sigma_1^{\rm LO} = \alpha_s(\mu^2)A_1$. However, at the same LO we could equally well write this prediction as $\sigma_1^{\rm LO} = \alpha_s(\mu^2)A_1 = \alpha_s(Q^2)A_1$, because using the above expansion of the coupling constant, Eq. (3), we see that the scale dependence would only appear as an NLO correction, namely $\sigma_1^{\rm LO'} = \alpha_s(Q^2)A_1 - \alpha_s^2(Q^2)b_0A_1 \ln \mu^2/Q^2$. Thus, strictly sticking to the LO expression, it is clear that an experimental measurement of σ_1 and its comparison to $\sigma_1^{\rm LO}$ only allows the determination of some "effective" LO coupling constant $\alpha_s^{\rm eff,1}$, where it is unclear to which scale this really corresponds to. Furthermore, repeating the same procedure at LO for some other process "2", at some different physical energy or momentum scale, would result in a measurement $\alpha_s^{\rm eff,2}$, and most likely these two measurements of the effective coupling constant would give differing results.

Looking again at the expression $\sigma_1^{\rm LO'} = \alpha_s(Q^2)A_1 - \alpha_s^2(Q^2)b_0A_1 \ln \mu^2/Q^2$ we also see that $\sigma_1^{\rm LO'}$ strongly depends on the unphysical scale μ, since the logarithm with the explicit μ dependence already appears at NLO. Correspondingly, a determination of $\alpha_s(Q^2)$ using this prediction would result in a large uncertainty when varying the unphysical parameter μ in the fits to the measured cross section. This procedure of μ-variations, typically over a range of $0.5 < \mu/Q < 2$, has become a standard approach to estimating the possible impact of unknown higher-order contributions. In fact, the μ-dependence always appears at one order higher than the fully known perturbative expansion. More concretely, let us now assume that also the NLO coefficient B has been calculated. Then, by plugging the expansion Eq. (3) into Eq. (4) we find

$$\sigma\left(\alpha_s(\mu^2), \frac{\mu^2}{Q^2}\right) = \alpha_s(Q^2)A + \alpha_s^2(Q^2)B + \mathcal{O}\left(\alpha_s^3, \ln^2\frac{\mu^2}{Q^2}\right). \tag{5}$$

We see that there is no μ-dependence up to NLO; at this order it is thus equivalent to set $\mu = Q$ and to write $\sigma(\alpha_s(\mu^2), \frac{\mu^2}{Q^2}) = \sigma(\alpha_s(Q^2))$; i.e., we can replace the dependence of the running coupling constant on the unphysical scale μ with a dependence on the physical scale Q of the process at hand. Furthermore, we see that the explicit μ-dependence of the cross section prediction only appears at next-to-NLO (NNLO), i.e. suppressed by two powers of α_s relative to the LO term. This leads to a smaller uncertainty of the extracted $\alpha_s(Q^2)$ value when varying μ in the fit procedure. Finally, the NLO expression in Eq. (5) leads to the first non-trivial dependence of the cross section on $\alpha_s(Q^2)$ at the particular scale Q. Therefore, determinations of $\alpha_s(Q^2)$ from two different processes, at possibly different values of Q, using the NLO predictions for the cross sections and the running of α_s in order to relate the measured values to each other, should result, within uncertainties, in consistent measurements. Similarly, the value of $\alpha_s(Q^2)$ determined at NLO for some specific process can be used for predicting, at the same order, the cross section for another process at a different physical scale.

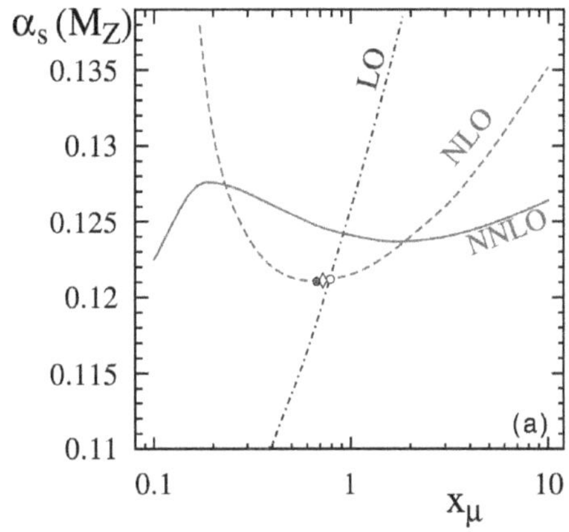

Fig. 1. $\alpha_s(M_Z^2)$ determined from the scaled hadronic width of the Z boson, in LO, NLO and NNLO QCD, as a function of the renormalisation scale $x_\mu = \mu/M_Z$; taken from Ref. 9.

The extension of this discussion to NNLO and beyond is straightforward, and easily shows that theoretical uncertainties estimated from μ-variations should decrease even further. This is nicely illustrated in Fig. 1, where the dependence of the extracted value of $\alpha_s(M_Z^2)$ is shown, when using the LO, NLO and NNLO pQCD expressions for fitting the measured hadronic width of the Z boson, normalised to its leptonic width.[9] Ultimately, for an observable known at all orders in pQCD the μ-dependence would vanish completely, as it should be for a physical observable that cannot depend on unphysical parameters. In fact, the μ-dependence of the NLO term in expression (4) could have simply been derived by imposing this requirement for a physical observable and using the renormalisation group equation.

At this stage it should have become clear, but still is worth highlighting, that the running coupling constant $\alpha_s(Q^2)$ is not a physical quantity, but simply a parameter defined in the context of a particular theoretical framework, namely pQCD up to some order in α_s. It can be determined experimentally in this context and used for making predictions for observables calculated within the same framework.

I would like to conclude these theoretical considerations by highlighting a further consequence of the particular scale behaviour of α_s: An uncertainty δ on a measurement of $\alpha_s(Q^2)$, at a scale Q, translates to an uncertainty $\delta' = (\alpha_s^2(M_Z^2)/\alpha_s^2(Q^2)) \cdot \delta$ on $\alpha_s(M_Z^2)$; that is, $\delta' < \delta$ for $Q < M_Z$. This enhances the impact of precise low-Q measurements, such as from τ decays (cf. below), in combinations performed at the M_Z scale.

3. Observables

The strong coupling constant has been measured in a large variety of physics processes, using many different observables. As depicted in Fig. 2, sensitivity to the

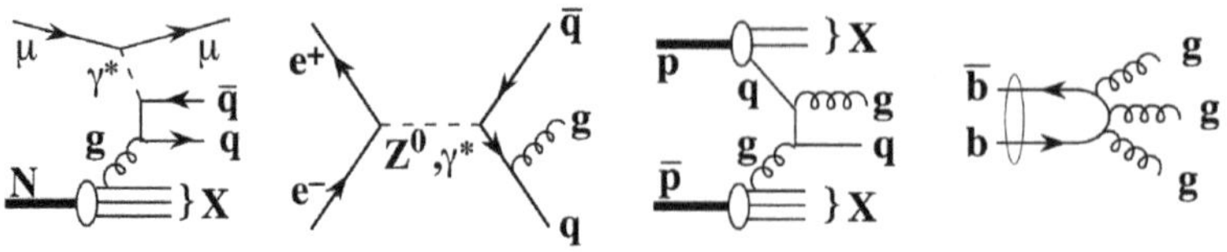

Fig. 2. Examples of Feynman diagrams describing hadronic final states in processes which are used to measure α_s; taken from Ref. 9.

coupling of quarks to gluons is obtained by studying, e.g., deep-inelastic lepton-nucleon scattering, e^+e^- annihilations, hadron collisions or resonance decays. Since we are not able to directly measure partons (quarks or gluons), but only hadrons and their decay products, a central issue is establishing a correspondence between observables obtained at the partonic and the hadronic level. The only theoretically sound correspondence is achieved by means of *infrared and collinear safe* (ICS) quantities (see e.g. Ref. 2), which allow one to obtain finite predictions at any order of perturbative QCD. ICS observables are insensitive to the addition of a soft parton, or to the splitting of one parton into two collinear ones. This guarantees that singularities, which appear in the infrared or collinear limits of diagrams involving real and/or virtual radiation, cancel order by order in perturbation theory.

Generally speaking, ICS observables can be divided into different classes, depending on how detailed the hadronic final state is being scrutinized. As for example discussed in Ref. 2, the simplest case of ICS observables are total cross sections. More generally, when measuring fully inclusive observables, the final state is not analyzed at all regarding its (topological, kinematical) structure or its composition. Basically, the relevant information consists in the rate of a process ending up in a partonic or hadronic final state.

In e^+e^- annihilation, widely used examples are the ratios of partial widths or branching ratios for the electroweak decay of particles into hadrons or leptons, such as Z or τ decays. The strong suppression of non-perturbative effects, $\mathcal{O}(\Lambda^4/Q^4)$, is one of the attractive features of such observables. However, at the same time the sensitivity to radiative QCD corrections is small, since here the perturbative expansion is of the type $1 + \sum_n c_n \alpha_s^n$, corresponding to, e.g., a 4% correction, $1 + \alpha_s(M_Z^2)/\pi \approx 1 + 0.04$, for the scaled hadronic Z width. In the case of τ decays not only the hadronic branching ratio is of interest, but also moments of the spectral functions of hadronic tau decays, which sample different parts of the decay spectrum and thus provide additional information.

Other examples of fully inclusive observables, that are used for α_s determinations, are the total top-pair production cross section in proton-proton collisions, the ratio of the hadronic to leptonic branching ratios of Υ decays, which is proportional to α_s^3 at LO (cf. the right-most diagram in Fig. 2), or structure functions (and related sum rules) in deep-inelastic scattering. Structure functions are sensitive to α_s through the Dokshitzer–Gribov–Lipatov–Altarelli–Parisi[10–13] evolution equations, e.g. $dF_2(x,Q^2)/d\ln Q^2 \propto \alpha_s(Q^2)\, P_{qg} \otimes g(x,Q^2)$, which depict in a simplified manner the scaling violation of the F_2 structure function; here x is the Bjorken scaling variable, P_{qg} is a so-called splitting function and $g(x, Q^2)$ is the parton distribution function (PDF) of the gluon. Such equations are used in global PDF fits in order to relate measurements at different Q scales to each other and to fit the PDFs at a chosen initial scale. We see that in such approaches the fit results for α_s and $g(x, Q^2)$ are strongly correlated. Similar considerations apply to the measurements of scaling violations of fragmentation functions.

Compared to inclusive observables, the particular structure, topology or composition of the hadronic final state can give enhanced sensitivity to α_s, therefore cross sections differential in one or more variables characterising this structure are of interest. The obvious example is the measurement of jet cross sections and jet rates, where the probability of producing an additional jet is directly proportional to α_s (for a general discussion of jets and jet algorithms I refer the reader to, e.g., Refs. 2, 4 and references therein). Besides jet quantities, another class of observables, so-called *event shapes*, have been widely used, in particular for measurements in e^+e^- annihilations, but also in lepton–proton and hadron collisions. The classic example of an event shape is the *Thrust*[14,15] (T or $\hat{\tau} = 1 - T$) in e^+e^- annihilations, defined as

$$T = \max_{\vec{n}_T} \frac{\sum_i |\vec{p}_i \cdot \vec{n}_T|}{\sum_i |\vec{p}_i|} , \tag{6}$$

where $\vec{p}_i$ are the momenta of the particles or the jets in the final-state and the maximum is obtained for the Thrust axis $\vec{n}_T$. In the Born limit of the production of a perfect back-to-back quark–antiquark pair the limit $\hat{\tau} \to 0$ is obtained, whereas a perfectly symmetric many-particle configuration leads to $\hat{\tau} \to 1/2$. Figure 3 (left) shows an example of measurements by the ALEPH experiment at different centre-of-mass energies.

Besides Thrust, many other similar observables such as C-parameter, Heavy Jet mass, Jet Broadening or the differential three-jet rate were proposed and used for α_s-determinations. They all provide a measure of the topology of an event, and typically are defined such that they take on small values for pencil-like (back-to-back) configurations, and large values for more spherical topologies that arise from single or multiple hard gluon radiation. This provides sensitivity to α_s at LO in perturbation theory, with normalized cross sections expressed as an expansion of the type Eq. (4). As discussed further below, predictions are known up to NNLO and complemented by the all-orders resummation of logarithms of the event-shape variable (i.e., terms of the form $\alpha_s^n \ln^m \hat{\tau}$). An important aspect of event-shape variables is their enhanced sensitivity to non-perturbative effects compared to more inclusive quantities, with power corrections of $\sim \lambda/Q$. For α_s determinations, analytical functions of these power corrections were used to complement the purely perturbative expansion, but the more widespread approach to correct for non-perturbative effects has been to use Monte Carlo simulations and their hadronization models in order to calculate the event shape at the partonic and hadronic level. As can be

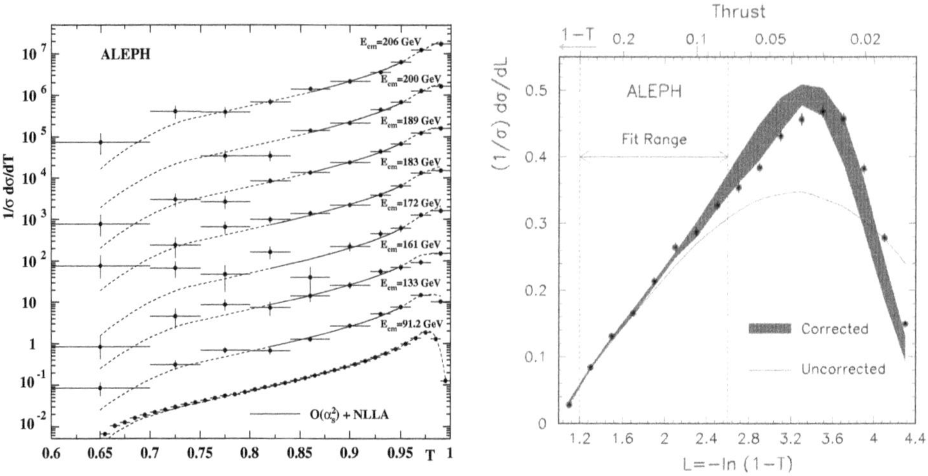

Fig. 3. Left: Thrust distribution measured by the ALEPH experiment at LEP for centre-of-mass energies between 91.2 and 206 GeV together with QCD predictions at NLO plus next-to-leading-logarithmic approximation (NLLA). The plotted distributions are scaled by arbitrary factors for presentation; taken from Ref. 16. Right: Comparison of ALEPH data for the Thrust distribution to the fitted QCD prediction (NLO+NLLA) obtained at parton level (solid line) and corrected for hadronization effects (shaded band). The width of the band covers the predictions using different hadronization models; taken from Ref. 17.

seen in Fig. 3 (right), these non-perturbative corrections can be sizeable, especially when approaching the two-jet region of the distribution, therefore the fit range has to be chosen carefully. Ultimately the model dependence of such corrections leads to systematic uncertainties on the extracted α_s values.

A completely different approach to the determination of $\alpha_s(M_Z^2)$ consists in calculating QCD predictions on the lattice for observables such as hadron mass splittings. From the comparison of data to the predictions, obtained as a function of the lattice spacing a and extrapolated to $a \to 0$, first a lattice coupling is extracted which is then converted to its perturbative counter-part $\alpha_s(M_Z^2)$. During the last decades there has been enormous progress in this field; indeed, the most precise $\alpha_s(M_Z^2)$ determinations to date are obtained from lattice QCD, though it is fair to say that the community still intensively discusses the validity of the very small systematic uncertainties, claimed by some of the involved groups. A more detailed discussion of this approach can be found in the review by Sachrajda.[18]

4. Brief Historical Overview

In the following I make an attempt to sketch some of the relevant developments that occurred during the last few decades, without any claim of being comprehensive and of covering all types of α_s studies in a balanced manner. In fact, a particular focus is put on results obtained by experiments at CERN.

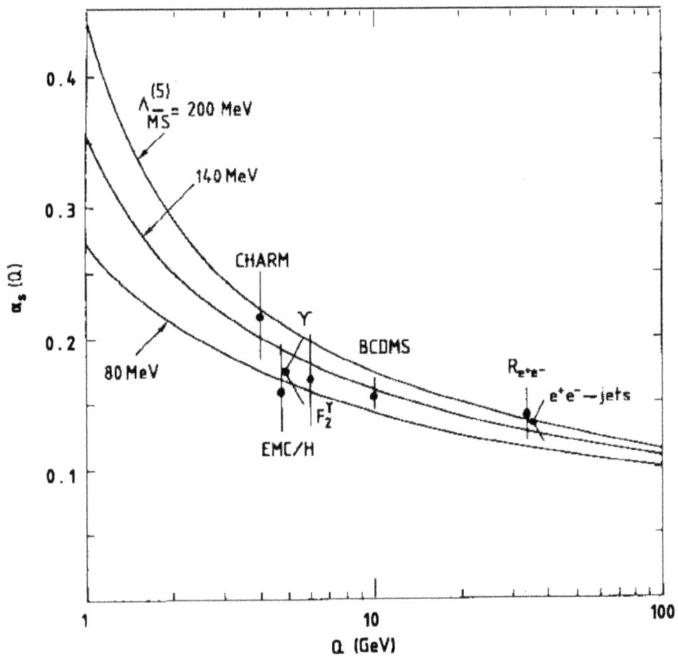

Fig. 4. Summary of the α_s determinations by Altarelli[19] in 1989.

The first extensive overview of α_s measurements was given by Altarelli[19] in 1989. In that review, he summarised measurements based on observables such as (i) the total hadronic cross section in e^+e^- annihilation (at that time known at NNLO in pQCD, i.e., up to α_s^3) from data in the range $7 < Q < 56\,\text{GeV}$; (ii) scaling violations in deep inelastic leptoproduction with structure function data from BCDMS, EMC and CHARM; (iii) quarkonium decays, especially ratios of Υ partial widths ($\Gamma_{\text{ggg}}/\Gamma_{\mu\mu}, \Gamma_{\gamma\,\text{gg}}/\Gamma_{\text{ggg}}$) measured by the CUSB, CLEO, ARGUS and Crystal Ball collaborations; and (iv) jet production, energy–energy correlations and the photon structure function from $\gamma\gamma$ reactions, using e^+e^- data collected by the PEP/PETRA experiments. A summary of these measurements is shown in Fig. 4. In an attempt to combine all those results, obtained at Q values from a few up to several tens of GeV, and using the QCD prediction for the running of $\alpha_s(Q^2)$, he arrived at a prediction for the coupling evaluated at the Z boson mass, $\alpha_s(Q \approx M_Z) \approx 0.11 \pm 0.01$; that is, a determination of the strong coupling constant at the 10% level. Interestingly, he concluded with the following statement: *Establishing that this prediction is experimentally true would be a very quantitative and accurate test of QCD, conceptually equivalent but more reasonable than trying to see the running in a given experiment.* It is impressive to note that his prediction turned out to be nicely consistent with the current WA value[2] of $\alpha_s(M_Z^2) = 0.1185 \pm 0.0006$. In addition, we see that over the past 25 years the relative uncertainty has been reduced by a factor of 18, gauging the enormous progress made during these decades.

The year 1989 also saw the start of the LEP experiments (ALEPH, DELPHI, L3, OPAL), and the following decade was characterised by great advances, both experimentally and theoretically, in the field of pQCD in general and α_s measurements in particular. Extensive overviews can be found, for instance, in Refs. 20, 21 which also discuss the application to earlier JADE data of the developments that occurred during the LEP era.

Event-shape observables were studied in great detail by the LEP experiments, showing first that pQCD at NLO does not provide an adequate accuracy in order to go well below the 10% level in relative uncertainty on $\alpha_s(M_Z^2)$. At the same time, calculations that resum large logarithms of the event-shape variable to all orders in α_s appeared and were used to improve the NLO prediction for a number of event-shape observables. This also triggered the development of a novel jet algorithm, the so-called Durham algorithm,[22] with a modified metric compared to the previously used JADE algorithm.[23] The modification of the jet metric, which defines the distance in phase-space between two particles that might or might not be combined into a new pseudo-particle, was motivated by the fact that the pQCD predictions for jet rates and the differential 3-jet rate, based on the JADE metric, did not show the exponentiation behaviour of the large logarithms,[24] whereas using the Durham metric led to exponentiation and ultimately to improved resummation predictions. Note that this Durham algorithm became the standard algorithm for jet finding at LEP, and was at the basis for later developments of iterative recombination algorithms, nowadays widely used at the LHC (cf. Ref. 2 and references therein). As a consequence of the combined NLO+resummed predictions, corrected for non-perturbative hadronization effects using phenomenological Monte Carlo (MC) models, the relative uncertainty of α_s was reduced to the $\sim 5\%$ level, still dominated by theoretical uncertainties due to missing higher orders and estimated from variations of the renormalisation scale. Attempts to replace the MC models by analytical power corrections[25] of order λ/Q did not lead to substantially different results. It became clear that only a complete NNLO calculation for jet rates and event shapes might lead to a significant reduction of uncertainties. Indeed, such a calculation[26,27] appeared after the end of LEP, and its first applications[28,29] to the 3-jet rate and to event shapes, including next-to-leading log resummations, resulted in more precise α_s measurements at the 2–3% precision level.

Most of the aforementioned determinations gave $\alpha_s(M_Z^2)$ values in a range of, very roughly speaking, 0.117–0.125. However, as summarized in Ref. 2, more recent re-analyses of the Thrust distribution, based on novel developments in soft-collinear effective field theory, resummation at next-to-next-to-next-to-leading logarithmic accuracy, and analytic calculations of non-perturbative effects, resulted in values as low as 0.1131, while at the same time claiming very small uncertainties at the 2% level. Thus, further work will be required to understand this spread of results from jet and event-shape observables, which covers a larger range than most of the individually quoted uncertainties.

In terms of inclusive observables, the LEP experiments quoted precise α_s measurements by using the total hadronic cross section (or equivalently, the hadronic width of the Z boson), as well as by analyzing hadronic τ decays. Contrary to event shapes, NNLO predictions for these observables were already available in the nineties, leading to rather small renormalisation scale uncertainties. By now, they are even known to N^{3}LO accuracy. This implies an almost negligible theoretical uncertainty in the case of the hadronic Z decay width; for instance, when included in a global fit[30] to electroweak precision data a value of $\alpha_s(M_Z^2) = 0.1193 \pm 0.0028$ is found, where the dominant part of the uncertainty is of statistical nature.

Naively speaking, similarly precise results might not have been expected from the analyses of hadronic τ decays, since here the relevant scale is the τ mass, close to the scale where pQCD is supposed to break down. Thus, non-perturbative effects and missing higher order contributions should significantly impact the attainable precision. However, it was realised that measuring different moments of the τ spectral function allows one to determine $\alpha_s(M_\tau^2)$ and to constrain non-perturbative power-suppressed contributions at the same time. Several methods were developed to estimate missing higher-order terms, beyond NNLO and N^{3}LO, such as so-called contour-improved perturbative expansions, claiming very small scale uncertainties. It is worth noting[2] that these methods are still matter of intense discussions, in particular since some of the results obtained by different groups differ by several standard deviations in terms of the quoted uncertainties. In an attempt to combine all these results and to take into account the observed spread, Ref. 2 quotes $\alpha_s(M_Z^2) = 0.1197 \pm 0.0016$. Thus, somewhat surprisingly, the study of τ decays results in one of the most precise α_s determinations, basically at the level of 1% relative uncertainty. It should be emphasised that the already precise measurements, obtained at the scale of the τ mass, turn into this even more precise result at the Z mass, because of the running of α_s as discussed at the end of Section 2.

Many of the developments of the LEP era, in the field of event shapes and jet observables, were also applied to HERA data of deep inelastic electron–proton scattering (DIS). Although here the pQCD predictions are only known up to NLO approximation so far, and the α_s extraction from jet final states is somewhat more complicated because of the additional implicit α_s dependence of the PDFs, it is impressive to see that a combination[31] based on precise HERA data of inclusive jet cross sections in neutral current DIS at high Q^2 results in $\alpha_s(M_Z^2) = 0.1198 \pm 0.0032$, which includes a theoretical uncertainty of ± 0.0026. These HERA measurements also allow covering a large range of Q^2 values and thus probing directly the running of α_s.

More inclusive DIS observables, in particular structure functions and their scaling violations as discussed in Section 3, have been used in global PDF fits based on NNLO pQCD and resulted in smaller relative uncertainties, even at the 1% level as quoted by some groups. However, quite similarly to the case of event shapes and τ observables, also here a spread of $\alpha_s(M_Z^2)$ values (roughly covering a range of 0.113[32] to 0.117[33,34]) is observed[2] that is larger than some of the individually quoted uncertainties. Two remarks are in place here: (i) these differences are still matter of intense discussions among the various groups performing global PDF fits, and (ii) it is kind of a tradition that α_s determinations from DIS and global PDF fits result in smaller values than those obtained from e^+e^- annihilations, without understanding the origin of this apparent bias.

Jet observables at hadron colliders, such as the inclusive jet cross section as a function of jet transverse momentum or invariant multi-jet masses, jet angular correlations or jet rates, are only known to NLO approximation so far. Furthermore, important systematic uncertainties due to the jet energy scale and choice of PDF set are expected to limit the attainable precision, and similarly to the case of DIS, the intrinsic α_s dependence of the PDFs has to be carefully taken into account in any α_s fit. As discussed in Ref. 2, first measurements at the Tevatron and at the LHC gave results consistent with the WA value and with relative uncertainties in the range of 4% to 8%. However, very important developments have taken place at the LHC recently, as e.g. summarised in Ref. 35. First, in both the ATLAS and CMS experiments the jet energy scale uncertainty is now known at an impressive level of about 1–2% for jets in the ~100 GeV range.[36,37] Since jet cross sections are steeply falling functions of jet momentum this has an enormous impact on the finally attainable precision. Furthermore, ratios of observables, such as the ratio of the 3-jet over the 2-jet rate, allow one to eliminate this systematic uncertainty to a large extend, as shown in Refs. 38 and 39. Finally, NNLO calculations for the inclusive jet cross section appear to be around the corner,[40] which will further boost the importance of such measurements.

In fact, recently the first α_s determination[41] at a hadron collider, using pQCD at NNLO, has been published. However, here an inclusive quantity, namely the top-pair production cross section, has been successfully exploited thanks to its strong sensitivity to both α_s and the top quark mass. Fixing the latter to its WA value allowed the CMS collaboration to extract $\alpha_s(M_Z^2)$ at an impressive relative precision of ~3%, also thanks to the remarkable experimental precision (4%) of the top cross section measurement[42] that served as input. Because of this recent progress, and because of the large Q^2 range covered by the measurements at the LHC, the running of the strong coupling constant is now being precisely studied over an unprecedented energy range.

As mentioned at the end of Section 3, a discussion of α_s determinations using lattice QCD can be found in a separate review[18] in this Book. For completeness it should be stated here that this very complementary approach, compared to the measurements described above, results in the world's most precise $\alpha_s(M_Z^2)$ determinations to date, with some of the analyses quoting relative uncertainties at the 0.5% level (cf. Ref. 2). However, the community is having intense discussions about the validity of these apparently rather optimistic estimates of systematic uncertainties. In any case, the lattice results dominate the current WA value: not including them in the averaging procedure results in $\alpha_s(M_Z^2) = 0.1183 \pm 0.0012$,[2] i.e., the uncertainty doubles.

This historical overview can not be concluded without a brief discussion of the general issue of combining α_s determinations. As discussed in Ref. 2, this is a non-trivial exercise. Since most of the individual measurements are dominated by systematic uncertainties, which cannot be expected to follow a normal distribution, and since very often the correlations among these uncertainties are not very well known, simple averaging methods as applicable to measurements with statistical errors only might not be appropriate. In 1995 Schmelling[43] proposed a method for estimating such unknown correlations, which rescales individual uncertainties according to the assumption that the normalized χ^2 of the combination should be 1. This method is also used for the current WA determination.[2] Furthermore, there is a certain arbitrariness in the choice of results included in the average. Finally, as mentioned earlier, often α_s determinations based on the same observable but using different methods give results that differ by a larger amount than would be

	Year	World average $\alpha_s(M_Z^2)$
Altarelli[19]	1989	0.11 ± 0.01
Hinchcliffe[44] (PDG)	1992	0.1134 ± 0.0035
Hinchcliffe[45] (PDG)	1995	0.118 ± 0.003
Schmelling[46]	1997	0.118 ± 0.003
Bethke[9]	2000	0.1184 ± 0.0031
Bethke[48]	2006	0.1189 ± 0.0010
Bethke, Dissertori, Salam (PDG)[2]	2013	0.1185 ± 0.0006

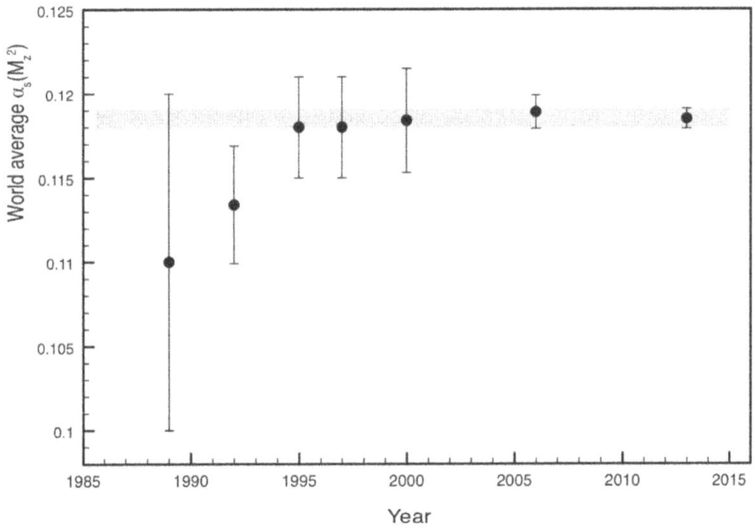

Fig. 5. A selection of world average values for $\alpha_s(M_Z^2)$ as a function of time; the band indicates the current world average value[2] of $\alpha_s(M_Z^2) = 0.1185 \pm 0.0006$.

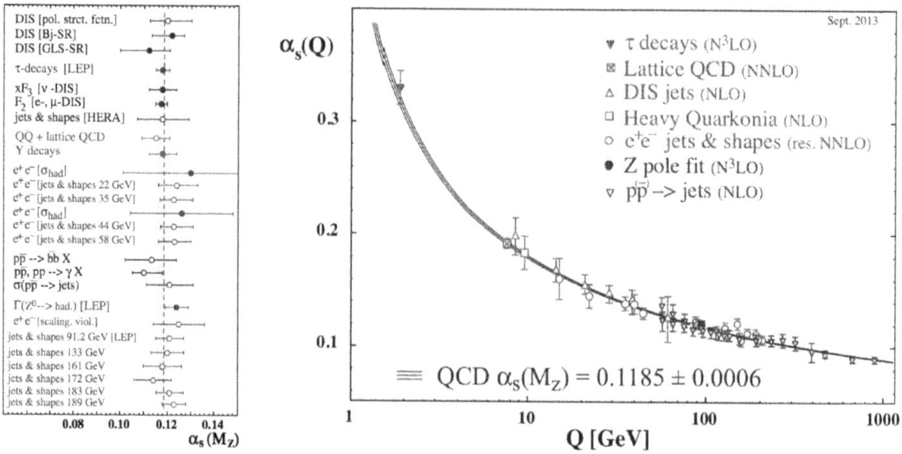

Fig. 6. Left: List of individual $\alpha_s(M_Z^2)$ measurements and their comparison to the world average from Ref. 9 in 2000; Right: current status of the running of α_s, as summarised in Ref. 2.

expected from the individually quoted uncertainties, rendering the estimate of the combined uncertainty a difficult exercise.

Throughout these years, several individuals and/or groups have compiled the available α_s measurements and combined them into a single value. The earliest attempt by Altarelli has already been discussed above. During the nineties, the reference in terms of $\alpha_s(M_Z^2)$ was established by the PDG, in particular thanks to the PDG review on QCD by Hinchcliffe (see, e.g., Refs. 44, 45). An independent estimate of the WA value was published by Schmelling[46] in 1997, based on his proposal for handling unknown correlations. Then, during the first decade of this century, Bethke[9,47,48] provided a number of comprehensive studies, that established the de-facto WA value, despite the PDG still publishing an independent combination. Since a few years this situation has been resolved, with Bethke now being co-author (together with Dissertori and Salam) of the PDG review on QCD that also contains the WA determination of α_s. Figure 5 displays this, most likely incomplete, collection of WA results as a function of time, nicely showing the impressive progress made throughout the last decades.[a] Finally, Fig. 6 presents an example[9] of inputs to the averaging procedure and the current experimental status of the running of α_s, showing excellent agreement with the theoretical expectation.

5. Conclusions

The strong coupling constant is one of the fundamental parameters of the Standard Theory of particle physics. In this article I have reviewed the theoretical and

[a]At the time of writing, an update of the PDG world average is in preparation, but not yet published. That update, based on a series of new inputs and a somewhat modified combination procedure, is anticipated to lead to a similar central value, but to an increase in the uncertainty.

experimental developments that have led to a precise knowledge of this important parameter, representing a cornerstone in our understanding of the strong interactions sector of the Standard Theory.

Acknowledgments

I would like to thank G. Rolandi and L. Maiani for inviting me to contribute to this collection of essays on the Standard Theory. I would also like to thank S. Bethke and G. Salam for many interesting discussions on the topic of α_s and for their comments on the manuscript. Finally, my thanks go to R. Miquel for providing me Fig. 3 (right).

References

1. http://physics.nist.gov/cuu/Constants/index.html
2. S. Bethke, G. Dissertori and G. Salam, *Quantum Chromodynamics*, in K. A. Olive *et al.* (Particle Data Group), *Chin. Phys. C* **38**, 090001 (2014).
3. S. Heinemeyer *et al.* [LHC Higgs Cross Section Working Group Collaboration], Handbook of LHC Higgs cross sections: 3. Higgs properties, arXiv:1307.1347 [hep-ph].
4. R. K. Ellis, Chapter 3 of this book.
5. G. Dissertori, I. G. Knowles and M. Schmelling, *Quantum Chromodynamics: High Energy Experiments and Theory*, International Series of Monographs on Physics, Vol. 115, (Oxford Univ. Press, 2009).
6. R. K. Ellis, W. J. Stirling and B. R. Webber, *QCD and Collider Physics*, Camb. Monogr. Part. Phys. Nucl. Phys. Cosmol. Vol. 8, 1 (1996).
7. H. D. Politzer, *Phys. Rev. Lett.* **30**, 1346 (1973).
8. D. Gross and F. Wilczek, *Phys. Rev. D* **8**, 3633 (1973); D. Gross and F. Wilczek, *Phys. Rev. D* **9**, 980 (1980).
9. S. Bethke, *J. Phys. G* **26**, R27 (2000) [hep-ex/0004021].
10. V. N. Gribov and L. N. Lipatov, *Sov. J. Nucl. Phys.* **15**, 438 (1972) [*Yad. Fiz.* **15**, 781 (1972)].
11. L. N. Lipatov, *Sov. J. Nucl. Phys.* **20**, 94 (1975) [*Yad. Fiz.* **20**, 181 (1974)].
12. Y. L. Dokshitzer, *Sov. Phys. JETP* **46**, 641 (1977) [*Zh. Eksp. Teor. Fiz.* **73**, 1216 (1977)].
13. G. Altarelli and G. Parisi, *Nucl. Phys. B* **126**, 298 (1977).
14. S. Brandt, C. Peyrou, R. Sosnowski and A. Wroblewski, *Phys. Lett.* **12**, 57 (1964).
15. E. Farhi, *Phys. Rev. Lett.* **39**, 1587 (1977).
16. A. Heister *et al.* [ALEPH Collaboration], *Eur. Phys. J. C* **35**, 457 (2004).
17. R. Barate *et al.* [ALEPH Collaboration], *Phys. Rept.* **294**, 1 (1998).
18. C. T. Sachrajda, Chapter 5 of this Book.
19. G. Altarelli, *Ann. Rev. Nucl. Part. Sci.* **39**, 357 (1989).
20. S. Kluth, *Rept. Prog. Phys.* **69**, 1771 (2006).
21. O. Biebel, *Phys. Rept.* **340**, 165 (2001).
22. S. Catani, Y. L. Dokshitzer, M. Olsson, G. Turnock and B. R. Webber, *Phys. Lett. B* **269**, 432 (1991).
23. W. Bartel *et al.* [JADE Collaboration], *Z. Phys. C* **33**, 23 (1986).
24. N. Brown and W. J. Stirling, *Phys. Lett. B* **252**, 657 (1990).
25. Y. L. Dokshitzer and B. R. Webber, *Phys. Lett. B* **352**, 451 (1995).

26. A. Gehrmann-De Ridder, T. Gehrmann, E. W. N. Glover and G. Heinrich, *Phys. Rev. Lett.* **99**, 132002 (2007).
27. A. Gehrmann-De Ridder, T. Gehrmann, E. W. N. Glover and G. Heinrich, *JHEP* **0712**, 094 (2007).
28. G. Dissertori, A. Gehrmann-De Ridder, T. Gehrmann, E. W. N. Glover, G. Heinrich and H. Stenzel, *JHEP* **0802**, 040 (2008).
29. G. Dissertori, A. Gehrmann-De Ridder, T. Gehrmann, E. W. N. Glover, G. Heinrich, G. Luisoni and H. Stenzel, *JHEP* **0908**, 036 (2009).
30. H. Flacher, M. Goebel, J. Haller, A. Hocker, K. Monig and J. Stelzer, *Eur. Phys. J. C* **60**, 543 (2009) [*Eur. Phys. J. C* **71**, 1718 (2011)].
31. C. Glasman [H1 and ZEUS Collaborations], *J. Phys. Conf. Ser.* **110**, 022013 (2008).
32. S. Alekhin, J. Blumlein and S. Moch, *Phys. Rev. D* **86**, 054009 (2012).
33. A. D. Martin, W. J. Stirling, R. S. Thorne and G. Watt, *Eur. Phys. J. C* **64**, 653 (2009).
34. R. D. Ball *et al.*, *Phys. Lett. B* **707**, 66 (2012).
35. J. Rojo, Constraints on parton distributions and the strong coupling from LHC jet data, arXiv:1410.7728 [hep-ph].
36. G. Aad *et al.* [ATLAS Collaboration], *Eur. Phys. J. C* **75**, 1, 17 (2015).
37. S. Chatrchyan *et al.* [CMS Collaboration], *JINST* **6**, P11002 (2011).
38. S. Chatrchyan *et al.* [CMS Collaboration], *Eur. Phys. J. C* **73**, 10, 2604 (2013).
39. ATLAS Collaboration, ATLAS-CONF-2013-041, 2013.
40. J. Pires, *EPJ Web Conf.* **90**, 07005 (2015).
41. S. Chatrchyan *et al.* [CMS Collaboration], *Phys. Lett. B* **728**, 496 (2014) [*Phys. Lett. B* **728**, 526 (2014)].
42. S. Chatrchyan *et al.* [CMS Collaboration], *JHEP* **1211**, 067 (2012).
43. M. Schmelling, *Phys. Scripta* **51**, 676 (1995).
44. K. Hikasa *et al.* [Particle Data Group Collaboration], *Phys. Rev. D* **45**, S1 (1992) [*Phys. Rev. D* **46**, 5210 (1992)].
45. L. Montanet *et al.* [Particle Data Group Collaboration], *Phys. Rev. D* **50**, 1173 (1994).
46. M. Schmelling, Status of the strong coupling constant, in *Warsaw 1996, ICHEP '96*, Vol. 1 91–102 [hep-ex/9701002].
47. S. Bethke, *Nucl. Phys. Proc. Suppl.* **135**, 345 (2004).
48. S. Bethke, *Prog. Part. Nucl. Phys.* **58**, 351 (2013).

Chapter 7

Hadron Contribution to Vacuum Polarisation

M. Davier*, A. Hoecker†, B. Malaescu‡ and Z. Zhang*

*Laboratoire de l'Accélérateur Linéaire, Univ. Paris-Sud 11 et IN2P3/CNRS, 91898 Orsay Cedex, France

†CERN, 1211 Geneva, Switzerland

‡Laboratoire de Physique Nucléaire et des Hautes Energies, IN2P3-CNRS et Univ. Pierre-et-Marie-Curie et Denis-Diderot, 75252 Paris Cedex 05, France

Precision tests of the Standard Theory require theoretical predictions taking into account higher-order quantum corrections. Among these vacuum polarisation plays a predominant role. Vacuum polarisation originates from creation and annihilation of virtual particle–antiparticle states. Leptonic vacuum polarisation can be computed from quantum electrodynamics. Hadronic vacuum polarisation cannot because of the non-perturbative nature of QCD at low energy. The problem is remedied by establishing dispersion relations involving experimental data on the cross section for e^+e^- annihilation into hadrons. This chapter sets the theoretical and experimental scene and reviews the progress achieved in the last decades thanks to more precise and complete data sets. Among the various applications of hadronic vacuum polarisation calculations, two are emphasised: the contribution to the anomalous magnetic moment of the muon, and the running of the fine structure constant α to the Z mass scale. They are fundamental ingredients to high precision tests of the Standard Theory.

1. Introduction and Historical Perspective

Vacuum polarisation (VP) originates from quantum fluctuations in the exchange of gauge bosons occurring in particle interactions. The simplest case is that of electron–positron pairs in the propagation of a virtual photon: an e^+e^- pair is emitted and re-absorbed by the photon. It is the quantum analogue of the polarisation of molecules in a dielectric material when an external electric field is applied. The field generated by the distorted molecules has the effect of reducing the field in the medium. Similarly at the quantum level the e^+e^- pairs cause a screening effect that reduces the strength of the electromagnetic force carried by the exchanged photon. It is therefore important to take into account VP when evaluating the effective interaction resulting from photon exchange.

Experimental evidence of photon VP dates back to 1947 with the precise measurement of the Lamb shift giving the energy difference between the ^2S and ^2P levels of the hydrogen atom.[1] Whereas this difference should vanish in the Dirac theory, the measurement reveals a non-zero value mostly from electron self-energy, but also including a small contribution from VP. The explanation, first proposed by Bethe, was refined in the quantum electrodynamic theory (QED) developed by R. P. Feynman, J. Schwinger and S. Tomonaga. The observation of the Lamb shift had therefore a dramatic impact on establishing QED and more generally quantum field theory.

Not only electrons and positrons, but any pair of charged particles needs to be considered when evaluating photon VP effects. In particular pairs of charged leptons heavier than the electron, $\mu^+\mu^-$ and $\tau^+\tau^-$, should occur as well. Their VP effects can be calculated with QED and are found to be suppressed with respect to e^+e^- because of their larger masses. These contributions can be felt only when the experimental observable is very precisely measured or for a large virtuality (large q^2) of the exchanged photon. An example of the first type is the anomalous magnetic moment of the electron, which is experimentally known with a precision of 0.25 ppb[2] and for which the QED prediction has to include contributions up to the α^5 level to match the experimental accuracy.

The vacuum can also be polarised by fluctuations involving strongly interacting particles, an effect called hadronic vacuum polarisation (HVP). To first order in quantum chromodynamics (QCD) these fluctuations are induced by quark-antiquark pairs. Their effect can be computed at large q^2 in perturbative QCD but not at low scales due to the non-perturbative nature of QCD at large distance. It is possible to overcome this problem by means of a dispersion relation technique involving experimental data measuring the cross section $e^+e^- \to$ hadrons. This technique was pioneered by C. Bouchiat and L. Michel[3] and by N. Cabibbo and R. Gatto,[4] and provided a first estimate of the HVP contribution to the anomalous magnetic moment of the muon.

The first evidence for HVP came from an experiment at the ACO storage ring at LAL-Orsay[5] measuring the cross section for the purely leptonic process $e^+e^- \to \mu^+\mu^-$. A characteristic interference pattern was found around the mass of the $\phi(1020)$ resonance, in agreement with the expected ϕ HVP contribution. A broader range of energies could be explored with the study of the muon magnetic moment exhibiting ever increasing sensitivity. The CERN effort to measure this quantity started in 1958 with an experiment at the synchrocyclotron,[6] followed by a more and more elaborate programme from 1962 to 1976 using storage rings fed by the proton synchrotron.[7–9] In the last phase of the programme the precision reached on the muon magnetic anomaly ($a_\mu = (g_\mu - 2)/2$, where g_μ is the muon gyromagnetic ratio) was 7 ppm, enough to detect the effect of HVP. The final experimental value achieved at CERN for a_μ was $(1\,165\,924 \pm 8.5) \cdot 10^{-9}$, yielding a deviation from pure QED of $(70.4 \pm 8.1) \cdot 10^{-9}$, to be compared to the HVP estimate[10] then of $(70.2 \pm 8.0) \cdot 10^{-9}$. This measurement established clearly the presence of HVP as a manifestation of the strong interaction in lepton properties. It ranks as one of the highlights achieved in the CERN scientific programme. The pursuit of even higher accuracies was continued at Brookhaven using the CERN technique, and it is still going on now at Fermilab, USA and J-PARC, Japan.[11] It is interesting to note that the recent measurements of the electron magnetic moment are now also sensitive[12] to HVP, which must be included in the theoretical prediction in spite of its tiny amount, approximately $m_\mu^2/m_e^2 \sim 40\,000$ time smaller than that of the muon.

Apart from predicting the lepton magnetic moments, the need for accurate calculation of HVP arose with the advent of precision tests in the electroweak sector of the Standard Theory (ST) at LEP, owing to the large statistics accumulated at the Z pole, and at SLC using the electron beam polarisation. The interpretation of these measurements required taking into account VP which affects the effective electromagnetic coupling when evolving from low energies to the Z mass scale, providing a strong impetus to improve HVP calculations and the corresponding e^+e^- cross sections. The variation of the fine structure "constant" amounts to 3.150% from leptonic VP with negligible uncertainty[13] and to about 2.7% from HVP. Matching the per mil accuracy of the electroweak measurements requires the evaluation of the HVP contribution with percent precision, pushing the state of the art for e^+e^- data.

In this review, after introducing the dispersion approach to HVP, we will discuss the sources of input data on the $e^+e^- \to$ hadrons cross section which are the necessary ingredient to compute the dispersion integrals. Over the last 50 years the quality and the accuracy of these data has improved dramatically, especially in the last decade, allowing more precise tests of the Standard Theory.

2. Dispersion Relations

Using unitarity and analyticity, spectra of hadron production from e^+e^- annihilation via a spin-one photon propagator are connected to the imaginary part of the two-point correlation (or hadronic vacuum polarisation) functions

$$\Pi_{i,V}^{\mu\nu}(q) \equiv i \int d^4x\, e^{iqx} \langle 0|T(V_i^\mu(x) V_i^\nu(0)^\dagger)|0\rangle = \left(q^\mu q^\nu - g^{\mu\nu} q^2\right) \Pi_{i,V}^{(1)}(q^2) \quad (1)$$

of vector colour-singlet quark currents involving all flavours i, and for time-like momenta-squared $q^2 > 0$. The functions $\Pi_{i,V}^{\mu\nu}(s)$ have a branch cut along the real axis in the complex $s = q^2$ plane. Their imaginary parts reproduce the hadronic annihilation spectrum and provide the basis for comparing short-distance theory with hadronic data. The analytic vacuum polarisation function $\Pi_{i,V}^{(1)}(q^2)$ obeys the dispersion relation

$$\Pi_{i,V}^{(1)}(q^2) = \frac{1}{\pi} \int_0^\infty ds\, \frac{\mathrm{Im}\Pi_{i,V}^{(1)}(s)}{s - q^2 - i\varepsilon} + \text{subtractions}, \quad (2)$$

where the unknown but in general irrelevant subtraction constants can be removed by taking the derivative of $\Pi_{i,V}(q^2)$. The dispersion relation allows one to connect the experimentally accessible hadron spectrum to the correlation functions $\Pi^{(1)}_{i,V}(q^2)$, which can be derived from theory (QCD).

As an example this formalism is applied to the running electromagnetic coupling. In quantum field theory all contributions from self-energy and vertex correction graphs to the photon vacuum polarisation function $\Pi_\gamma(q^2)$ cancel and only vacuum polarisation modifies the charge of an elementary particle. One can therefore write for the running electron charge (charge screening) at energy scale s

$$\alpha(s) = \frac{\alpha(0)}{1 - \Delta\alpha(s)}, \tag{3}$$

where $\alpha(0)$ is the fine structure constant in the long-wavelength Thomson limit and $\Delta\alpha(s) = \Delta\alpha_{\rm lep}(s) + \Delta\alpha_{\rm had}(s) = -4\pi\alpha \cdot {\rm Re}[\Pi_\gamma(s) - \Pi_\gamma(0)]$. The leptonic part $\Delta\alpha_{\rm lep}(s)$ is calculable within QED and known to high accuracy. Quark loops, however, are modified by long-distance hadronic physics that cannot be calculated within QCD. Instead, the optical theorem

$$12\pi {\rm Im}\Pi_\gamma(s) = \frac{\sigma(e^+e^- \to {\rm hadrons})}{\sigma(e^+e^- \to \mu^+\mu^-)} \equiv R(s), \tag{4}$$

and the dispersion relation

$$\Pi_\gamma(s) - \Pi_\gamma(0) = \frac{s}{\pi}\int_0^\infty ds' \frac{{\rm Im}\Pi_\gamma(s')}{s'(s'-s) - i\varepsilon}, \tag{5}$$

lead to

$$\Delta\alpha_{\rm had}(s) = -\frac{\alpha(0)s}{3\pi}{\rm Re}\int_0^\infty ds' \frac{R(s')}{s'(s'-s) - i\varepsilon}. \tag{6}$$

The hadronic vacuum polarisation contribution to the anomalous magnetic moment of the muon is derived in a similar way. The dominant leading order part is given by the dispersion integral[3,14]

$$a_\mu^{\rm had,LO} = \frac{1}{3}\left(\frac{\alpha}{\pi}\right)^2 \int_{m_\pi^2}^\infty ds \frac{K(s)}{s} R^{(0)}(s). \tag{7}$$

The integration kernel $K(s)/s \sim s^{-2}$ strongly emphasises the low-energy part of $R(s)$ with the main contribution and uncertainty stemming from the $\rho(770) \to \pi\pi$ resonance.

The computation of the dispersion integrals in Eqs. (6) and (7) requires the knowledge of $R(s)$ at any scale s, where in practice a mix of information is exploited. At low scales and just above the heavy quark thresholds e^+e^- annihilation data to hadrons (or, in some cases data from hadronic τ decays owing to isospin symmetry) are used. Narrow resonances can be described using analytical formulas with experimentally determined Breit–Wigner parameters, and the continuum region can be obtained with the use of perturbative QCD.

3. e^+e^- Data

3.1. *Experimental progress toward precision*

Exclusive low-energy $e^+e^- \to$ hadrons cross sections have been measured by experiments running at e^+e^- colliders. Fixed-energy measurements have been performed in Orsay and Novosibirsk. The most recent measurements of the $\pi^+\pi^-$ channel by the CMD2[15–18] and SND[19] experiments at the VEPP-2M collider at Novosibirsk have achieved comparable statistical uncertainties and energy-dependent systematic uncertainties down to 0.8% and 1.3%, respectively. These measurements involve running the collider at a series of different centre-of-mass energies. However, doing so, the detector conditions can also evolve between the different data-taking periods, hence the problem of evaluating precisely the systematic uncertainties and their correlations between measurements at different $\sqrt{s}$ values.

With the enhanced luminosity at colliders like the DAPHNE ϕ-factory in Frascati and PEP-II B-factory at SLAC, the use of the innovative technique of radiative return became possible. For this technique, events with a hard photon emitted from the initial state (ISR) are considered. This allows to cover, for a fixed value of the centre-of-mass energy of the e^+e^- collider and with the same detector conditions, a wide range of the spectrum of the hadronic final state. This type of measurement was first performed by the KLOE experiment,[20] where events with the ISR photon emitted along the beam-pipe were considered, using theory for the evaluation of the ISR effective luminosity. The BABAR experiment pioneered a luminosity-independent technique which consists in measuring the ratio of the cross sections $e^+e^- \to \pi^+\pi^-\gamma(\gamma)$ to $e^+e^- \to \mu^+\mu^-\gamma(\gamma)$,[21,22] where large cancellations of systematic effects occur. Considering events with a hard ISR photon reconstructed in the detector, the BABAR analysis covers the full region of interest for $g-2$, from the $\pi\pi$ threshold up to 3 GeV. In order to achieve the desired sub-percent systematic precision, additional photons emitted either from the initial or from the final state have been taken into account in this analysis. The experimental systematic uncertainties are kept to 0.5% in the ρ peak region (0.6–0.9 GeV), increasing to 1% outside. More recently, the KLOE experiment has also performed measurements with the hard photon reconstructed in the detector,[23] as well as considering the ratio between the $\pi\pi$ and $\mu\mu$ spectra.[24] Figure 1 shows a summary of the cross section measurements for $e^+e^- \to \pi^+\pi^-$, in the dominant ρ resonance region.

In addition to the $\pi\pi$ channel, which brings about 75% of the hadronic contribution to the $g-2$ of the muon, the channels with kaons and those with higher multiplicities have a non-negligible contribution. The BABAR experiment has implemented a systematic programme for a precise measurement of these cross sections, exploiting the ISR method. This allowed to complete and often to significantly improve the precision of past measurements of these channels, performed at Orsay and Novosibirsk.

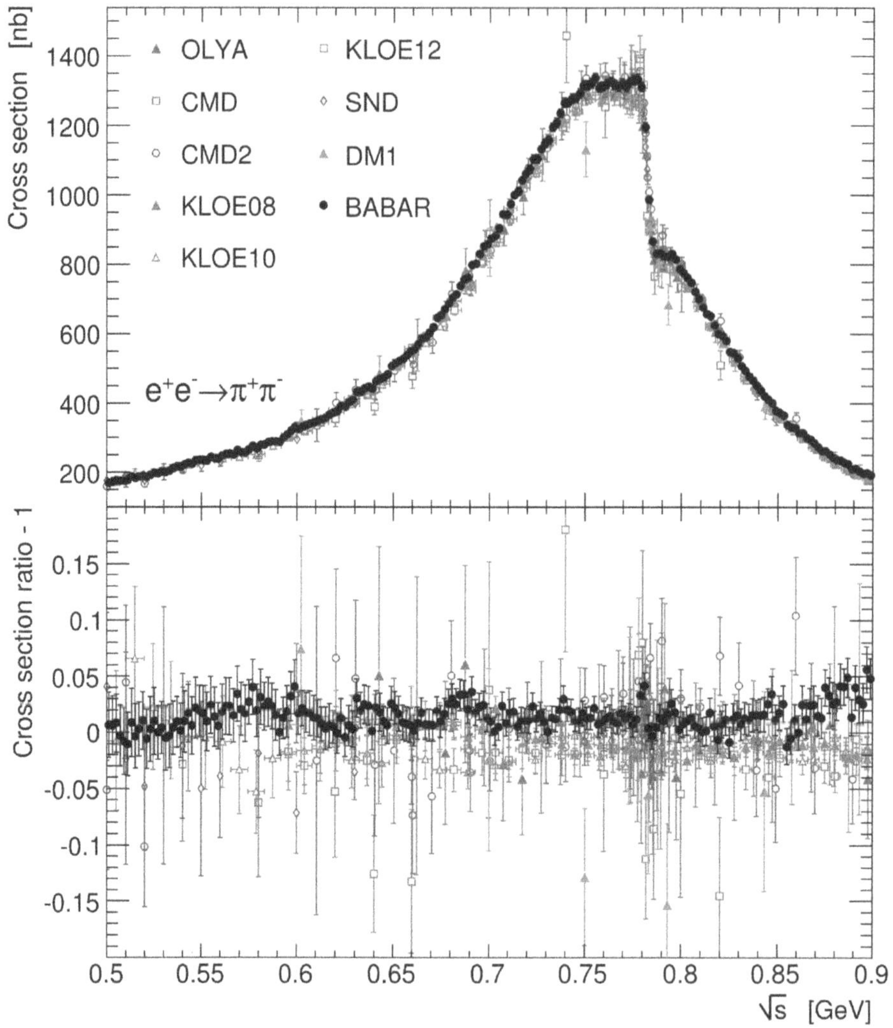

Fig. 1. Cross section for the process $e^+e^- \to \pi^+\pi^-$ measured by different experiments (top) and their relative difference with respect to the combined cross section (bottom). The uncertainty bars contain both the statistical and systematic components added in quadrature.

Several five- and six-pion modes involving π^0's, as well as $K\overline{K}[n\pi]$ final states are still unmeasured. The isospin invariance is exploited to relate their contributions to those of known channels using dynamics information obtained in the BABAR study of the final states.[25]

3.2. *Progress in combining data*

Using the input described in the previous section to compute the dispersion integral in Eq. (7) requires some special care. The requirements for averaging and integrating cross section data are: (i) properly propagate all the uncertainties in the data to the final integral uncertainty, (ii) minimise biases, i.e. reproduce the true integral as closely as possible in average and measure the remaining systematic uncertainty, and (iii) optimise the integral uncertainty after averaging while respecting the two previous requirements. Different techniques have been used in past studies. For example, a "clustering" method (see e.g. Refs. 26–28) consisted in combining weighted measurements of different experiments within a prescribed energy interval. We describe here the approach used in our recent studies.

In order to better address the three requirements above, a new methodology has been developed.[29] The first item practically requires the use of pseudo-Monte Carlo (MC) simulation, which needs to be a faithful representation of the measurement ensemble and to contain the full data treatment chain (interpolation, averaging, integration). The second item requires a flexible data interpolation method (the trapezoidal rule is not sufficient as explained below) and a realistic truth model used to test the accuracy of the integral computation with pseudo-MC experiments. Finally, the third item requires optimal data averaging taking into account all known correlations to minimise the spread in the integral measured from the pseudo-MC sample.

The combination procedure transforms the bare cross section data and associated statistical and systematic covariance matrices into fine-grained energy bins, taking into account to our best knowledge the correlations within each experiment as well as between the experiments (such as uncertainties in radiative corrections). A detailed splitting of the systematic uncertainties in sub-components allows for an improved treatment of these correlations. The covariance matrices are obtained by assuming common systematic uncertainty sources to be fully correlated. To these matrices are added statistical covariances, present for example in binned measurements as provided by KLOE, BABAR or the τ data, which are subject to bin-to-bin migration that has been unfolded by the experiments, thus introducing correlations.

The interpolation between adjacent measurements of a given experiment uses second order polynomials. This is an improvement with respect to the previously applied trapezoidal rule, corresponding to a linear interpolation, which leads to systematic biases in the integral.[29] The averaging of the interpolated measurements from different experiments contributing to a given energy bin is the most delicate step in the analysis chain. It takes into account correlations between measurements and experiments, as well as the different measurement densities or bin widths that the experiments have within a given energy interval.[29] If the χ^2 value of a bin-wise average exceeds the number of degrees of freedom ($n_{\rm dof}$), the uncertainty in this averaged bin is rescaled by $\sqrt{\chi^2/n_{\rm dof}}$ to account for inconsistencies. This rescaling is applied locally, correctly taking into account the energy dependence of the potential tension between the measurements. Such inconsistencies frequently occur because most experiments are dominated by systematic uncertainties, which are difficult to estimate. For example, in the $\pi\pi$ channel, the existing tension between the BABAR and KLOE measurements (see Fig. 1) prevents the expected precision improvement in the combination.[29]

The consistent propagation of all uncertainties into the evaluation of $a_\mu^{\rm had,LO}$ is ensured by generating large samples of pseudo experiments, representing the full list of available measurements and taking into account all known correlations. For each generated set of pseudo measurements, the identical interpolation and averaging treatment leading to the computation of Eq. (7) as for real data is performed, hence resulting in a probability density distribution for $a_\mu^{\rm had,LO}$, the mean and RMS of which define the 1σ allowed interval. Common sources of systematic uncertainties also occur between measurements of different final state channels and are taken into account when summing up the exclusive contributions. Such correlations mostly arise from luminosity uncertainties, if the data stem from the same experimental facility, and from radiative corrections. These correlations have a non-negligible impact on the evaluated uncertainty of $a_\mu^{\rm had,LO}$.

Figure 2 shows the cross section for the process $e^+e^- \to$ hadrons versus centre-of-mass energy $\sqrt{s}$. The result of the combination of the experimental measurements with its uncertainty, as well as the QCD prediction are shown.

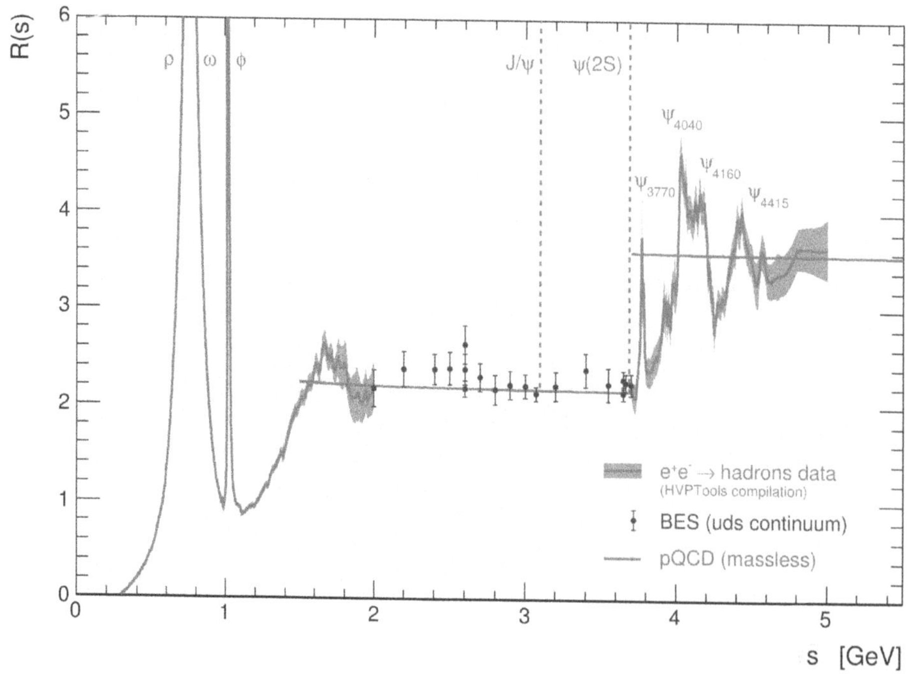

Fig. 2. Cross section for the process $e^+e^- \to$ hadrons versus centre-of-mass energy $\sqrt{s}$. The band represents the combined experimental measurements within their uncertainty.[29] The red line shows the perturbative QCD prediction, the data points show the inclusive measurements from the BES experiment.[30–33]

4. Use of tau Data

The use of tau data of semi-leptonic τ decays in the evaluation of $a_\mu^{\rm had}$ and $\Delta\alpha_{\rm had}^{(5)}(M_Z^2)$ was originally proposed in Ref. 26 based on the fact that hadronic spectral functions (or normalised invariant mass-squared distributions) from τ decays are directly related to the isovector vacuum polarisation currents when isospin invariance, or the conserved vector current (CVC), and unitarity hold. The CVC hypothesis relates the isovector, vector matrix element of the decay $\pi^- \to \pi^0 e^- \bar{\nu}_e$ to the electromagnetic form factor of the pion. At that time, the spectral functions, in particular of the $\tau^- \to \pi^-\pi^0\nu_\tau$ decay mode, and the τ-based evaluation of $a_\mu^{\rm had}$, were more precise than the corresponding e^+e^- results by a factor of two, as it is shown in Fig. 3. The precision of the τ-based evaluation for $\Delta\alpha_{\rm had}^{(5)}(M_Z^2)$, however, was comparable with the e^+e^--based one. This is due to different QED kernels in the two evaluations with the former one giving a much larger weight to the low energy data.

With more precise data from both the e^+e^- annihilation and the τ decays, significant deviations were observed between the two sets of the evaluation.[27,59] A few isospin-breaking sources are known. These include the short-distance electroweak radiative correction $S_{\rm EW}$, final state radiation corrections in the $\pi^+\pi^-$ channel FSR(s), long-distance radiative corrections $G_{\rm EM}(s)$ of order α to the photon inclusive $\tau^- \to \pi^-\pi^0\nu_\tau$ spectrum (including both virtual and real photonic corrections), the $\pi^\pm - \pi^0$ mass splitting effects $\beta_0^3(s)/\beta_-^3(s)$ (important close to the threshold), and the ratio of electromagnetic to weak form factors $|F_0(s)/F_-(s)|$ corresponding to effects of the π mass splitting $\delta m_\pi = m_{\pi^\pm} - m_{\pi^0}$, the ρ mass splitting $\delta m_\rho = m_{\rho^\pm} - m_{\rho^0}$ expected from $\gamma - \rho$ mixing, and the difference $\delta\Gamma_\rho$ in the ρ meson widths. These isospin breaking effects were carefully studied in Refs. 65–68, which improved the agreement between the two calculations, though not resolving all discrepancies.[25,65]

In Ref. 70, an additional correction originating from the γ–ρ mixing was considered for the τ data, with a claimed improvement in the compatibility of the e^+e^-- and τ-based evaluations. However, unlike for the analogous γ–Z mixing, the correction here is model-dependent because of the ρ hadronic structure.

5. Use of Theory

Sufficiently far from the quark thresholds, the inclusive hadronic cross section can be computed reliably using perturbative QCD. In a recent evaluation[25] we use four-loop perturbative QCD,[52] including $\mathcal{O}(\alpha_s^2)$ quark mass corrections, between 1.8 and 3.7 GeV, as well as above 5 GeV. Non-perturbative contributions at 1.8 GeV were determined from data[53] and found to be small. The uncertainties of the $R_{\rm QCD}$ contributions account for the uncertainty in α_s, the truncation of the perturbative series, the full difference between fixed-order perturbation theory (FOPT) and, so-called,

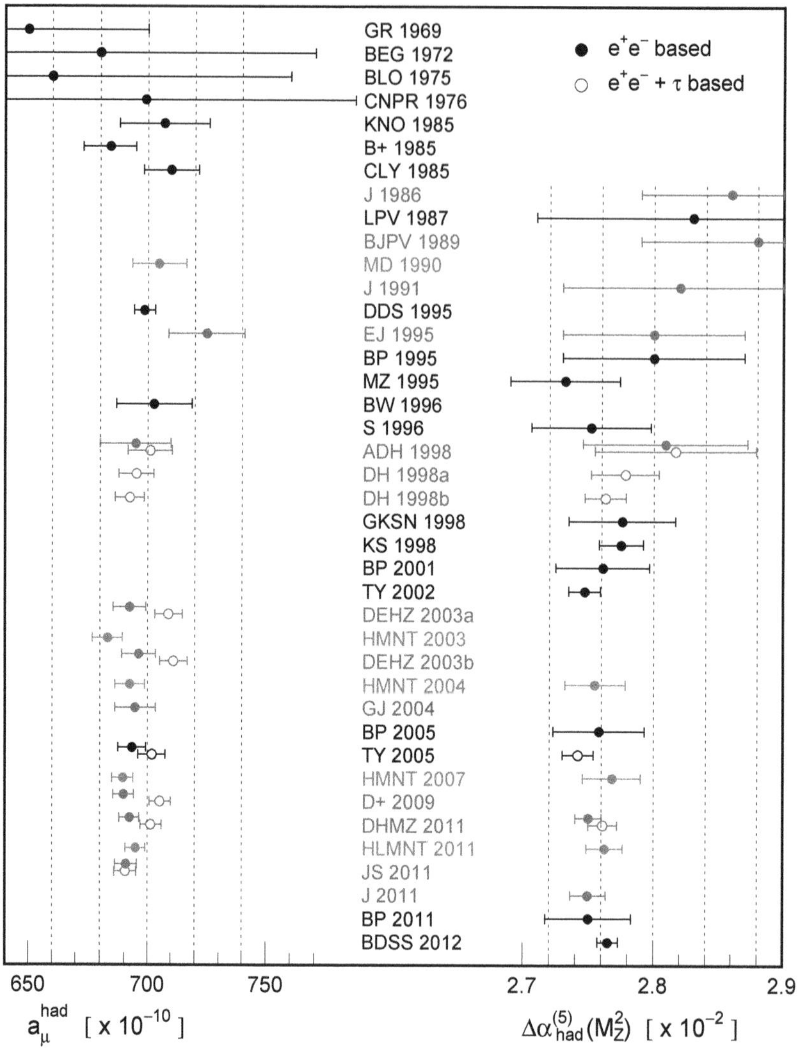

Fig. 3. Summary of the evolution as function of time for $a_\mu^{\rm had}$ and $\Delta\alpha^{(5)}_{\rm had}(M_Z^2)$. The corresponding references are Refs. 25–28,34–51,53–65,69–73. The error bars represent the total uncertainty by adding the experimental and theoretical uncertainties in quadrature.

contour-improved perturbation theory (CIPT), as well as quark mass uncertainties. The former three uncertainties are taken to be fully correlated between the various energy regions, whereas the (smaller) quark-mass uncertainties are taken to be uncorrelated.

Agreement within uncertainties is observed between the BES data and the QCD prediction below the $D\overline{D}$ threshold. Comparing the $a_\mu^{\rm had,LO}$ predictions in the energy interval 2–3.7 GeV, we find $26.5 \pm 0.2 \pm 1.7$ for BES data, and 25.2 ± 0.2 for perturbative QCD. To study the transition region between the sum of exclusive measurements and QCD, we have computed $a_\mu^{\rm had,LO}$ in two narrow energy intervals around 1.8 GeV. For the energy interval 1.75–1.8 GeV we find (in units of 10^{-10}) $2.74 \pm 0.06 \pm 0.21$ (statistical and systematic uncertainties) for the sum of the exclusive data, and 2.53 ± 0.03 for perturbative QCD. For the interval 1.8–2.0 GeV we find $8.28 \pm 0.11 \pm 0.74$ and 8.31 ± 0.09 for data and QCD, respectively. The excellent agreement represents another support for the use of QCD beyond 1.8 GeV centre-of-mass energy.

6. Applications

We recall the two historically most important applications of dispersion-relation based hadronic vacuum polarisation evaluations: the hadronic contributions to the anomalous magnetic moment of the muon and to the running of the electromagnetic coupling constant.

6.1. *The anomalous magnetic moment of the muon*

The Dirac equation predicts a muon magnetic moment, $\vec{M} = g_\mu \frac{e}{2m_\mu}\vec{S}$, with gyromagnetic ratio $g_\mu = 2$. Quantum loop effects lead to a small calculable deviation from $g_\mu = 2$, parameterised by the anomalous magnetic moment a_μ. That quantity has been accurately measured by the E821 experiment at Brookhaven National Lab (BNL) which studied the precession of μ^+ and μ^- in a constant external magnetic field as they circulated in a confining storage ring. Assuming CPT invariance, it found the combined value[74,75] $a_\mu^{\rm exp} = (11\,659\,209.1 \pm 5.4 \pm 3.3) \cdot 10^{-10}$, where the first error is statistical and the second systematic.

Comparison of experiment and theory tests the Standard Theory at its quantum loop level. A deviation in $a_\mu^{\rm exp}$ from the ST expectation would signal effects of new physics, with current sensitivity reaching up to mass scales of $\mathcal{O}(\rm TeV)$[76,77] (see Ref. 78 for a thorough review).

The ST prediction for $a_\mu^{\rm ST}$ is generally divided into three parts,

$$a_\mu^{\rm ST} = a_\mu^{\rm QED} + a_\mu^{\rm EW} + a_\mu^{\rm had} \,. \tag{8}$$

The dominant QED and much smaller electroweak (EW) contributions have been calculated to five and two-loop levels, respectively, with negligible uncertainty.

The main uncertainty stems from the hadronic contribution $a_\mu^{\rm had}$, which is further split into lowest order and higher order parts. The dominant contributions originate from two-vertex functions and are obtained via a dispersion relation approach such as Eq. (7) for the lowest order term. Accurate measurements and limited use of perturbative QCD lead to the results $a_\mu^{\rm had,LO} = (692.3 \pm 4.2) \cdot 10^{-10}$ and $(701.5 \pm 4.6) \cdot 10^{-10}$ for the e^+e^- based and $e^+e^- + \tau$ based evaluations, respectively. The higher order hadronic contribution,[79] $a_\mu^{\rm had,NLO} = (7 \pm 26) \cdot 10^{-11}$, is dominated by the uncertainty in a term involving four photon vertices that is theoretically estimated and model dependent.

Adding all terms together gives the representative e^+e^- data based ST prediction[75]

$$a_\mu^{\rm ST} = (11\,659\,180.3 \pm 0.1 \pm 4.2 \pm 2.6) \cdot 10^{-10}\,, \tag{9}$$

where the errors are due to the electroweak, lowest-order hadronic, and higher-order hadronic contributions, respectively. The difference to experiment $a_\mu^{\rm exp} - a_\mu^{\rm ST} = (28.8 \pm 6.3 \pm 4.9) \cdot 10^{-11}$ (with all errors combined in quadrature) represents an interesting but not conclusive discrepancy of 3.6 times the estimated uncertainty. Using τ data in the relevant channels reduces that number to 2.4. The left column of Fig. 3 gives a historical view of the evaluations of $a_\mu^{\rm had,LO}$. The individual results differ in the experimental datasets used, the averaging procedures, and the amount of theory injected into the calculation.

6.2. *Running electromagnetic fine structure constant at M_Z^2*

The running electromagnetic fine structure constant, $\alpha(s) = \alpha(0)/(1 - \Delta\alpha_{\rm lep}(s) - \Delta\alpha_{\rm had}(s))$, at the scale of the Z mass-squared, $s = M_Z^2$, is an important ingredient of the ST fit to electroweak precision data at the Z pole. Similar to a_μ, the leptonic contribution has a negligible uncertainty so that $\alpha(M_Z^2)$ is limited by the hadronic vacuum polarisation part.

Summing all the contributions to the integral (6) gives for the e^+e^- based hadronic term[80]

$$\Delta\alpha_{\rm had}(M_Z^2) = (275.0 \pm 1.0) \cdot 10^{-4}\,, \tag{10}$$

which is, contrary to the evaluation of $a_\mu^{\rm had,LO}$, not dominated by the uncertainty in the low-energy data, but by contributions from all energy regions, where both experimental and theoretical errors are of similar magnitude. The corresponding τ-based result is $\Delta\alpha_{\rm had}(M_Z^2) = (276.1 \pm 1.1) \cdot 10^{-4}$. The use of perturbative QCD instead of experimental data in the energy interval between 1.8 and 3.7 GeV (cf. Fig. 2) significantly improves the accuracy in the evaluation of $\Delta\alpha_{\rm had}(M_Z^2)$. Excluding the analytical top-quark contribution from the result (10) gives $\Delta\alpha_{\rm had}^{(5)}(M_Z^2) = (275.7 \pm 1.0) \cdot 10^{-4}$. A historical compilation of $\Delta\alpha_{\rm had}^{(5)}(M_Z^2)$ evaluations is given on the right column of Fig. 3.

Adding the three-loop leptonic contribution,[81] $\Delta\alpha_{\rm lep}(M_Z^2) = 314.97686 \cdot 10^{-4}$, with negligible uncertainty, one finds

$$\alpha^{-1}(M_Z^2) = 128.962 \pm 0.014. \tag{11}$$

The running electromagnetic coupling at M_Z enters at various levels the global ST fit to electroweak precision data. It contributes to the radiator functions that modify the vector and axial-vector couplings in the partial Z boson widths to fermions, and also to the ST prediction of the W mass and the effective weak mixing angle. Overall, the fit exhibits a -39% correlation between the Higgs boson mass and $\Delta\alpha_{\rm had}(M_Z^2)$.[82] The current precision of $\Delta\alpha_{\rm had}(M_Z^2)$ is sufficient to not deteriorate the constraints obtained by the present electroweak fit.

7. Perspectives

Recent measurements of the cross section for $e^+e^- \to$ hadrons at energies less than 5–10 GeV have played a major role in particle physics. The increase in precision and completeness of these data have been essential for progressing in important areas of the electroweak and strong sectors. However even better measurements are necessary to match future needs. Fortunately, the BABAR ISR programme is near completion with results on the few last significant processes to be published soon. Also new experiments are underway: CMD-3 and SND-2 at VEPP-2000, BES3 at BEPC2, and in the future Belle-2 at SuperKEKB. Experimental progress requires an even better control of systematic uncertainties and also resolving inconsistencies in the present data which limit the accuracy of the combined results. It is not unreasonable to gain a factor of two in the course of the next few years over the energy range up to 5 GeV.

A short-term goal is to improve the theoretical prediction for the muon magnetic anomaly as a new generation of experiments are aiming at an increase in precision by a factor of four.[83,84] Assuming the expected improvement in the leading order hadronic vacuum polarisation part, and even if there is no improvement in the current uncertainty assigned to the light-by-light contribution, the discrepancy would reach about 8 standard deviations if measured at the same central value. This would represent a compelling signal for new physics beyond the Standard Model.

An independent approach being explored by several groups is a first-principle calculation of HVP using lattice QCD.[85] This method, which allows one to cover the low-energy region in a purely theoretical way, involves the computation of the two-point quark-connected correlator for Euclidean q^2. The major problem at present is to get precise enough lattice data at small q^2 and extrapolation at $q^2 = 0$ introduces systematic uncertainties which are difficult to estimate.[86] Currently the precision claimed is ranging from a few times to an order of magnitude worse than for the dispersive approach based on experimental data, but further progress is expected.

Improving the determination of $\alpha(M_Z^2)$ would also be required for future precision electroweak tests for example with the very large statistics available at a Z factory, as foreseen in the GigaZ option of the International Linear Collider (ILC).[87,88]

Again with the same improvement of a factor of two in the accuracy of $\Delta\alpha_{\rm had}(M_Z^2)$ and the precision expected for the electroweak observables at ILC with GigaZ, an indirect determination of M_H with an uncertainty of 7 GeV could be obtained[88] and compared to the known value of M_H as a precise test of the Standard Theory.

References

1. W. E. Lamb and R. C. Retherford, *Phys. Rev.* **72**, 241–243 (1947).
2. D. Hanneke, S. Fogwell, and G. Gabrielse, *Phys. Rev. Lett.* **100**, 120801 (2008).
3. C. Bouchiat and L. Michel, *J. Phys. Radium.* **22**, 121 (1961).
4. N. Cabibbo and R. Gatto, *Phys. Rev.* **124**, 1577–1595 (1961).
5. J. Augustin *et al.*, *Phys. Rev. Lett.* **30**, 462–464 (1973).
6. G. Charpak *et al.*, *Phys. Rev. Lett.* **6**, 128–132 (1961).
7. J. Bailey *et al.*, *Phys. Lett. B* **28**, 287–290 (1968).
8. J. Bailey *et al.*, *Nuovo Cim. A* **9**, 369–432 (1972).
9. J. Bailey *et al.*, *Phys. Lett. B* **67**, 225 (1977).
10. J. Calmet, S. Narison, M. Perrottet, E. de Rafael, *Rev. Mod. Phys.* **49**, 21 (1977).
11. B. L. Roberts. *Nucl. Phys. Proc. Suppl.* **218**, 237 (2011).
12. Mibe T. *Nucl. Phys. Proc. Suppl.* **218**, 242 (2011).
13. M. Steinhauser, *Phys. Lett. B* **429**, 158–161 (1998).
14. M. Gourdin and E. De Rafael, *Nucl. Phys. B* **10**, 667–674 (1969).
15. R. Akhmetshin *et al.*, *JETP Lett.* **84**, 413–417 (2006).
16. R. Akhmetshin *et al.*, *Phys. Lett. B* **578**, 285–289 (2004).
17. R. Akhmetshin *et al.*, *Phys. Lett. B* **648**, 28–38 (2007).
18. V. Aulchenko *et al.*, *JETP Lett.* **82**, 743–747 (2005).
19. M. Achasov *et al.*, *J. Exp. Theor. Phys.* **103**, 380–384, (2006).
20. F. Ambrosino *et al.*, *Phys. Lett. B* **670**, 285–291 (2009).
21. B. Aubert *et al.*, *Phys. Rev. Lett.* **103**, 231801 (2009).
22. J. Lees *et al.*, *Phys. Rev. D* **86**, 032013 (2012).
23. F. Ambrosino *et al.*, *Phys. Lett. B* **700**, 102–110, (2011).
24. D. Babusci *et al.*, *Phys. Lett. B* **720**, 336–343 (2013).
25. M. Davier, A. Hoecker, B. Malaescu, and Z. Zhang, *Eur. Phys. J. C* **71**, 1515 (2011).
26. R. Alemany, M. Davier, and A. Hoecker, *Eur. Phys. J. C* **2**, 123–135, (1998).
27. M. Davier, S. Eidelman, A. Hoecker, and Z. Zhang, *Eur. Phys. J. C* **27**, 497–521, (2003).
28. K. Hagiwara, A. Martin, D. Nomura, and T. Teubner, *Phys. Lett. B* **557**, 69–75 (2003).
29. M. Davier, A. Hoecker, B. Malaescu, C. Yuan, and Z. Zhang, *Eur. Phys. J. C* **66**, 1–9 (2010).
30. J. Bai *et al.*, *Phys. Rev. Lett.* **84**, 594–597 (2000).
31. J. Bai *et al.*, *Phys. Rev. Lett.* **88**, 101802 (2002).
32. M. Ablikim *et al.*, *Phys. Lett. B* **641**, 145–155 (2006).
33. M. Ablikim *et al.*, *Phys. Lett. B* **677**, 239–245 (2009).
34. M. Gourdin and E. De Rafael, *Nucl. Phys. B* **10**, 667–674 (1969).
35. A. Bramon, E. Etim, and M. Greco, *Phys. Lett. B* **39**, 514–516 (1972).
36. V. D. Barger, W. Long, and M. Olsson, *Phys. Lett. B* **60**, 89 (1975).
37. J. Calmet, S. Narison, M. Perrottet, and E. de Rafael, *Phys. Lett. B* **61**, 283 (1976).
38. T. Kinoshita, B. Nizic, and Y. Okamoto, *Phys. Rev. D* **31**, 2108 (1985).
39. L. Barkov *et al.*, *Nucl. Phys. B* **256**, 365–384 (1985).
40. J. Casas, C. Lopez, and F. Yndurain, *Phys. Rev. D* **32**, 736–742 (1985).
41. F. Jegerlehner, *Z. Phys. C* **32**, 195 (1986).
42. B. Lynn, G. Penso, and C. Verzegnassi, *Phys. Rev. D* **35**, 42 (1987).
43. H. Burkhardt, F. Jegerlehner, G. Penso, and C. Verzegnassi, *Z. Phys. C* **43**, 497–501 (1989).
44. L. Martinovic and S. Dubnicka, *Phys. Rev. D* **42**, 884–892 (1990).
45. F. Jegerlehner, *Prog. Part. Nucl. Phys.* **27**, 1–76 (1991).

46. A. Dubnickova, S. Dubnicka, and P. Strizenec, *Acta Phys. Slov.* **45**, 467 (1995).
47. S. Eidelman and F. Jegerlehner, *Z. Phys. C* **67**, 585–602 (1995).
48. H. Burkhardt and B. Pietrzyk, *Phys. Lett. B* **356**, 398–403, (1995).
49. A. D. Martin and D. Zeppenfeld, *Phys. Lett. B* **345**, 558–563 (1995).
50. D. Brown and W. Worstell, *Phys. Rev. D* **54**, 3237–3249 (1996).
51. M. L. Swartz, *Phys. Rev. D* **53**, 5268–5282 (1996).
52. P. A. Baikov, K. G. Chetyrkin, J. H. Khn, *Nucl. Phys. B* **482**, 213 (1996).
53. M. Davier and A. Hoecker, *Phys. Lett. B* **419**, 419–431 (1998).
54. M. Davier and A. Hoecker, *Phys. Lett. B* **435**, 427–440 (1998).
55. S. Groote, J. Korner, K. Schilcher, and N. Nasrallah, *Phys. Lett. B* **440**, 375–385 (1998).
56. J. H. Kuhn and M. Steinhauser, *Phys. Lett. B* **437**, 425–431 (1998).
57. H. Burkhardt and B. Pietrzyk, *Phys. Lett. B* **513**, 46–52 (2001).
58. J. De Troconiz and F. Yndurain, *Phys. Rev. D* **65**, 093001 (2002).
59. M. Davier, S. Eidelman, A. Hoecker, and Z. Zhang, *Eur. Phys. J. C* **31**, 503–510 (2003).
60. K. Hagiwara, A. Martin, D. Nomura, and T. Teubner, *Phys. Rev. D* **69**, 093003 (2004).
61. S. Ghozzi and F. Jegerlehner, *Phys. Lett. B* **583**, 222–230 (2004).
62. H. Burkhardt and B. Pietrzyk, *Phys. Rev. D* **72**, 057501 (2005).
63. J. de Troconiz and F. Yndurain, *Phys. Rev. D* **71**, 073008 (2005).
64. K. Hagiwara, A. Martin, D. Nomura, and T. Teubner, *Phys. Lett. B* **649**, 173–179 (2007).
65. M. Davier *et al.*, *Eur. Phys. J. C* **66**, 127–136 (2010).
66. V. Cirigliano, G. Ecker, H. Neufeld, *Phys. Lett. B* **513**, 361 (2001); V. Cirigliano, G. Ecker, H. Neufeld, *JHEP* **0208**, 2 (2002).
67. A. Flores-Tlalpa, F. Flores-Baez, G. Lopez Castro, G. Toledo Sanchez, *Phys. Rev. D* **74**, 071301 (2006).
68. F. Flores-Baez, G. Lopez Castro, G. Toledo Sanchez, *Phys. Rev. D* **76**, 096010 (2007).
69. K. Hagiwara, R. Liao, A. D. Martin, D. Nomura, and T. Teubner, *J. Phys. G* **38**, 085003 (2011).
70. F. Jegerlehner and R. Szafron, *Eur. Phys. J. C* **71**, 1632 (2011).
71. F. Jegerlehner, *Nuovo Cim. C* **034S1**, 31–40 (2011).
72. H. Burkhardt and B. Pietrzyk, *Phys. Rev. D* **84**, 037502 (2011).
73. S. Bodenstein, C. Dominguez, K. Schilcher, and H. Spiesberger, *Phys. Rev. D* **86**, 093013 (2012).
74. G. Bennett *et al.*, *Phys. Rev. Lett.* **89**, 101804 (2002).
75. K. Olive *et al.*, Review of Particle Physics, The Muon Anomalous Magnetic Moment by A. Hoecker and B. Marciano, *Chin. Phys. C* **38**, 090001 (2014).
76. A. Czarnecki and W. J. Marciano, *Phys. Rev. D* **64**, 013014, (2001).
77. M. Davier and W. Marciano, *Ann. Rev. Nucl. Part. Sci.* **54**, 115–140 (2004).
78. J. P. Miller, E. de Rafael, and B. L. Roberts, *Rept. Prog. Phys.* **70**, 795 (2007).
79. J. Prades, E. de Rafael, and A. Vainshtein, *Adv. Ser. Direct. High Energy Phys.* **20**, 303 (2009).
80. M. Davier, A. Hoecker, B. Malaescu, and Z. Zhang, *Eur. Phys. J. C* **71**, 1515 (2011).
81. M. Steinhauser, *Phys. Lett. B* **429**, 158–161 (1998).
82. H. Flacher *et al.*, *Eur. Phys. J. C* **60**, 543–583 (2009).
83. J. Kaspar, *Nucl. Part. Phys. Proc.* **260**, 243 (2015).
84. N. Saito, *AIP Conf. Proc.* **1467**, 45–56 (2012).

85. T. Blum, *Phys. Rev. Lett.* **91**, 052001 (2003).
86. C. Aubin, T. Blum, M. Golterman, and S. Peris, *Phys. Rev. D* **86**, 054509 (2012).
87. H. Baer *et al.*, The International Linear Collider Technical Design Report — Volume 2: Physics arXiv:1306.6352 [hep-ph] (2013).
88. M. Baak *et al.*, *Eur. Phys. J. C* **74**, 3046 (2014).

Chapter 8

The Number of Neutrinos and the Z Line Shape

Alain Blondel

Université de Genève, D.P.N.C., Faculté des Sciences, CH1211 Genève 4, Switzerland

The Standard Theory can fit any number of fermion families, as long as the number of leptons and quark families are the same. At the time of the conception of LEP, the number of such families was unknown, and it was feared that the Z resonance would be washed out by decaying into so many families of neutrinos! It took only a few weeks in the fall of 1989 to determine that the number is three. The next six years (from 1990 to 1995) were largely devoted to the accurate determination of the Z line shape, with a precision that outperformed the most optimistic expectations by a factor of 10. The tale of these measurements is a bona fide mystery novel, the precession of electrons being strangely perturbed by natural phenomena, such as tides, rain, hydroelectric power, fast trains, not to mention vertical electrostatic separators. The number hidden in the loops of this treasure hunt was 179, the first estimate of the mass of the top quark; then, once that was found, where predicted, the next number was close to zero: the logarithm of Higgs mass divided by that of the Z. Twenty years later, the quality of these measurements remains, but what they tell us is different: it is no longer about unknown parameters of the Standard Theory, it is about what lies beyond it. This is so acutely relevant, that CERN has launched the design study of a powerful Z, W, H and top factory.

1. Introduction: What is the Number of Families of Fermions?

At the time LEP started, the basic properties of the weak interactions were already well known. One pressing question, however, could not be answered either by theoretical arguments or by direct experiments: what is the number of families of fermions? LEP answered this fundamental question in a few weeks by measuring the Z resonance. With six years of data and meticulous measurements of luminosity and energy the LEP experimentalists determined the Z boson mass and width, as well as the Z decay rates, with a precision which is unlikely to be surpassed soon.

All elementary quarks and leptons that have been observed are organised in *exactly* three families: (or generations).

$$\begin{pmatrix} u \\ d' \end{pmatrix} \begin{pmatrix} c \\ s' \end{pmatrix} \begin{pmatrix} t \\ b' \end{pmatrix} \quad \text{doublets of left-handed quarks}$$

$$\begin{matrix} (u) & (c) & (t) \\ (d) & (s) & (b) \end{matrix} \quad \text{singlets of right-handed quarks}$$

$$\begin{pmatrix} \nu_e \\ e \end{pmatrix} \begin{pmatrix} \nu_\mu \\ \mu \end{pmatrix} \begin{pmatrix} \nu_\tau \\ \tau \end{pmatrix} \quad \text{doublets of left-handed leptons}$$

$$\begin{matrix} (?\nu_e) & (?\nu_\mu) & (?\nu_\tau) \\ (e) & (\mu) & (\tau) \end{matrix} \quad \text{singlets of right-handed leptons}$$

As far as we can tell the Electroweak theory could not easily accomodate further isolated fermions, but it could accomodate any number of families of the same type. One could easily envisage a situation where many families including heavy charged quarks and leptons could exist, without these heavy leptons being ever produced in accessible experiments, by lack of available energy. Nevertheless, since the known neutrinos are very light, it is natural to expect that these additional families would include light neutrinos as well, leading to the possibility that many families of light neutrinos would exist.

The existence of many light neutrinos would have considerable cosmological consequences. In particular, the evolution of the universe within the first second after the Big Bang would be profoundly affected. The argument, developed in Ref. 1, is the following. At the time where energies are large enough, reactions such as $e^+e^- \to \nu\bar{\nu}$ transform a fraction of available energy into neutrinos in a democratic way. The creation of neutrons and protons however is controlled by reactions involving the electron neutrino, such as $\nu_e + n \to p + e^-$, and is consequently very sensitive to the number of light neutrino families, N_ν, which compete with electron neutrinos. The relative abundance of Hydrogen, Deuterium and Helium, and therefore the entire chemical constitution of our universe is a sensitive function of this number.

Before SLC and LEP started, limits on the number of light neutrinos were given from the above cosmological considerations, since there are data on the relative abundance of various nuclei in the universe, in particular the ratio of helium to hydrogen, or, with similar arguments, from the time development of the supernova 1987A. There were also indications from the direct search for the process $e^+e^- \to \nu\bar{\nu}\gamma$ (single-photon experiments), or from the early measurements of the Z and W boson properties in the CERN and FERMILAB $\mathrm{p\bar{p}}$ experiments. A review[2] of these constraints published in 1989 evaluated the best estimate of N_ν to be $\mathrm{N}_\nu = 2.1^{+0.6}_{-0.4}$, and stated that "$\mathrm{N}_\nu = 3$ is perfectly compatible with all data, but four families still provide a reasonable fit".

In searching for further families of neutrinos, it will be assumed that their couplings are the same as those of ν_e, ν_μ and ν_τ. Universality is deeply embedded in the Standard Theory, identical multiplets having the same coupling constant. It is very well verified for Charged Current interactions of e-μ-τ leptons, including the neutrinos, and for Neutral Current interactions of charged leptons.

2. Determination of the Number of Light Neutrino Species at LEP and SLC

The most precise determination of the number of light neutrino species is obtained from measurements of the visible cross-sections of e^+e^- annihilation at and around the Z resonance, as is made explicit in Fig. 1. If the Z is allowed to decay into more types of light neutrinos which will lead to an invisible final state, it will decay less often into the visible ones.

The realization that the visible cross-sections might be sensitive to the number of light neutrinos is rather ancient, one finds the question asked in John Ellis "Zedology",[3] where one can find this anguished question: *The Z peak is large and*

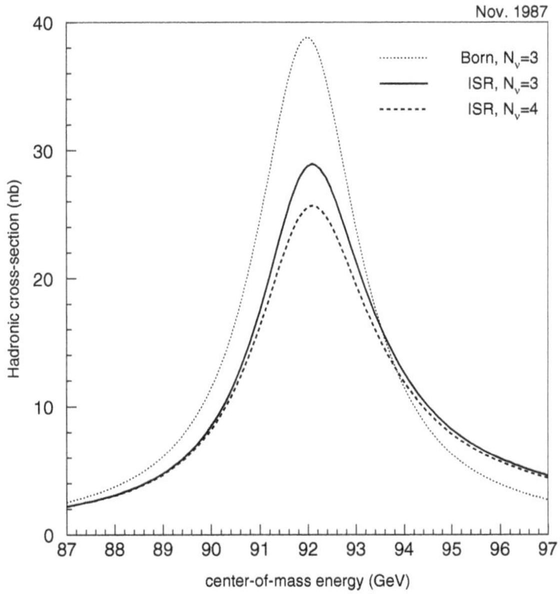

Fig. 1. The $e^+e^- \to hadrons$ cross section as a function of center-of-mass energy. This curve was drawn before LEP start-up in 1987. At that time the Z mass was measured to be around 92 GeV with an error larger than 1.5 GeV. The dotted line represents the born approximation prediction for three species of light neutrinos. The full line includes the effect of initial state radiation. The dashed line represents the effect of adding one more type of light neutrino with the same couplings as the first three. It is clear from this picture that the cross-section at the peak of the resonance contains most of the information on the number of light neutrino species.

dramatic, as long as there are not too many generations of fermions. Is it conceivable that there might be so many generations as to wash out the Z peak?. Since at that time the bound on the number of light neutrinos was very weak (about 6000), this certainly was a frightening possibility for those planning to build LEP! Dramatic also were the few first weeks of SLC and LEP operation where it was quickly realized that the Z peak was there indeed, large and dramatic, and that, alas, the number of light neutrinos was three.

There was intense competition between the SLC at SLAC (California, USA) and LEP at CERN (Switzerland). The two projects were with rather different in concept. LEP was build as the largest possible conventional e^+e^- storage ring, with a circumference of 27 km. This standard technique would ensure few surprises and and reliable high luminosity. SLC on the other hand, was the prototype of a new concept of accelerator, the linear collider; it was re-using the old Stanford linac, with improvements in the acceleration technique (SLED) and addition of arcs to bring e^+ and e^- in collisions, as well as of challenging positron source and damping rings.

The commissioning of SLC started in early 1987, and lead to a number of technical difficulties, not surprising in retrospect for such a new project. The first Z hadronic decay was produced on April 11 1989, and recorded in the MarkII detector. Luminosity was very low, a few $10^{-27}/\text{cm}^2/\text{s}$, leading to a few Z hadronic decays per day. With the LEP start-up advertised for the 14 of July, the time where SLC would hold the lead was going to be short, and intense. Nevertheless the SLC collaboration was able to collect a total of 106 Z decays by 24 July and submit a publication,[4] where the Z mass was determined to be $M_Z = 91.11 \pm 0.23$ GeV, and the number of light neutrinos species $N_\nu = 3.8 \pm 1.4$.

LEP did not start collisions on 14 of July but on 13 August for one week. The high luminosity optics were however not yet commissioned and events came at a rate of 1 a day for the four experiments; this was not enough to make a measurement. Running resumed on 20 September with superconducting quadrupoles and in just three weeks, until 9 October, 3000 Zs were collected in each of the experiments. By 13 October, a seminar was organized at CERN where the four collaborations presented their first results,[5–9] shown in Table 1. The day before, SLC had organized a public conference where updated results had been presented,[9] based on 480 events. The "online average" of these results is also shown in Table 1, $N_\nu = 3.12 \pm 0.19$. The number of light neutrinos was three.

Following this important contribution, SLC was shaken by an earthquake on 24 October 1989, from which it took more than a year to recover, and then concentrated on polarized beam physics. LEP went on, to the end of 1989 and for 6 more years (1989 to 1995), each experiment collecting 4 million hadronic Z decays. With the final results now available, the number of light neutrinos determined to be $N_\nu = 2.9841 \pm 0.0083$,[10] as will be illustrated on Fig. 4.

The early results were unexpectedly precise, and the precision of final ones exceed by a factor 100 the expectations that could be found in the studies preceding the

Table 1. First results from LEP and SLC on the Z mass and the number of light neutrino species, as published around 12 October 1989 (in order of submission to the journal).

Experiment	Hadronic Zs	Z mass (GeV)			N_ν		
MARKII	450	91.14	±	0.12	2.8	±	0.60
L3	2538	91.13	±	0.06	3.42	±	0.48
ALEPH	3112	91.17	±	0.05	3.27	±	0.30
OPAL	4350	91.01	±	0.05	3.10	±	0.40
DELPHI	1066	91.06	±	0.05	2.4	±	0.64
Average		91.10	±	0.05	3.12	±	0.19

Table 2. Numerical values of quantum numbers, Neutral Current couplings, and Z decay partial widths, for the four types of fermions, for hadrons and total width. The value of $\sin^2\theta_W^{\rm eff}$ is 0.2315.

f	I_{3f}	Q_f	g_{Af}	g_{Vf}	Γ_f (MeV)
ν	1/2	0	1/2	1/2	167
e	−1/2	−1	−1/2	−0.04	84
u	1/2	2/3	1/2	0.19	300
d	−1/2	−1/3	−1/2	−0.35	383
b	−1/2	−1/3	−1/2	−0.35	376
Hadrons $= u + d + c + s + b$	–	–	–	–	1740
Total for three neutrinos	–	–	–	–	2500
Total for four neutrinos	–	–	–	–	2670

start of LEP. Once the method is explained in more detail, it will become clear that the unexpected capacity of the experiments to perform precise measurement of hadronic cross-sections is the reason for this success.

3. Determination of the Z Line Shape Parameters

Around the Z pole, the $e^+e^- \to Z \to f\bar{f}$ annihilation cross-section is given by

$$\sigma_f = \frac{12\pi(\hbar c)^2}{\mathrm{M}_\mathrm{Z}^2}\frac{s\Gamma_e\Gamma_f}{(s-\mathrm{M}_\mathrm{Z}^2)^2 + s^2\frac{\Gamma_\mathrm{Z}^2}{\mathrm{M}_\mathrm{Z}^2}}, \tag{1}$$

which is a general formula or a spin one particle produced in e^+e^- annihilation into a visible channel f. This typical resonance shape peaks around the Z mass, $\sqrt{s} = \mathrm{M_Z}$, and has a width Γ_Z. If the Z decays a fraction B_f of the time into a final state f the corresponding partial width is defined as $\Gamma_f = B_f\Gamma_\mathrm{Z}$.

The Standard Theory predicts the numerical values for the Z partial widths, as displayed in Table 2. The main decay mode of the Z is into hadrons (70%), each of the leptons representing only 3% and three neutrinos would contribute 20%. There are no other substantial decay modes unless there are new particles. In particular the branching ratio into an hypothetic Higgs boson lighter than the Z is expected to be very small. In this expression the number of neutrinos intervenes through the total Z width Γ_Z:

$$\Gamma_Z = 3\Gamma_\ell + \Gamma_{\rm had} + N_\nu \Gamma_\nu. \tag{2}$$

If N_ν increases, the total width which is in the denominator of Eq. (1) increases, and the cross-section is decreased.

Equation (1) receives a number of modifications to account for the contribution of the photon exchange process (this is less than one percent), and more importantly what is called 'initial state radiation' (ISR), in which one or both of the initial state electrons lose energy into photons. This phenomenon reduces the initial state energy and smears out the resonance significantly as can be seen in Fig. 1. Due to the availability of calculations up to second order in perturbation theory, this large (30% at the peak) effect can be corrected for with a relative precision of 5×10^{-4}.

The principle of the analysis is then as follows: all visible channels are detected by large acceptance detectors and classified according to four categories: (i) hadrons, (ii) electron pairs, (iii) muon pairs, (iv) tau pairs. Examples of such events are shown in Fig. 2. These events are easy to detect, with high and well known efficiencies (as high as $99 \pm 0.05\%$ for hadronic decays), and to separate from each other.

In order to extract a cross-section from the number of events, the luminosity of the accelerator needs to be determined ($N = \mathcal{L} \cdot \sigma$). This is done by measuring at the same time another process with known cross-section, the elastic scattering $e^+e^- \to e^+e^-$, known as Bhabha scattering, which results in two low angle electron and positron. To this effect, the LEP experiments were equipped ab initio with low angle detectors, to detect these scattered electrons. The precision with which these detectors can measure this process is determined by the knowledge of the solid angle they cover and by the accuracy with which they measure the angle of the scattered electrons. The initial detectors were able to reach a precision of about one percent, but were progressively replaced with extremely precisely machined silicon tungsten calorimeters or silicon trackers (as in Fig. 3) which allowed a determination of the luminosity with an accuracy of 5×10^{-4} or better. It took many years of detailed higher order calculations to achieve a similar precision on the theoretical estimate of the cross-section within this well defined acceptance.

Measurements of cross-section for a given final state $f\bar{f}$ around the Z pole allow the extraction of three parameters: the position of the peak, the width of the resonance and an overall normalization, that is best obtained from the peak cross-section,

$$\sigma_f^0 = \frac{12\pi(\hbar c)^2}{M_Z^2}\frac{\Gamma_e\Gamma_f}{\Gamma_Z^2} = \frac{12\pi(\hbar c)^2}{M_Z^2} B_e.B_f\ . \tag{3}$$

By measuring cross-sections for hadrons, electron pairs, muon pairs and tau pairs, one can obtain six numbers: the mass, the total width, and four other parameters which could be four branching ratios, or four partial widths. A better choice is to use the peak cross-section for hadrons $\sigma_{\rm had}^{\rm peak,0}$, and the ratios of hadrons to the

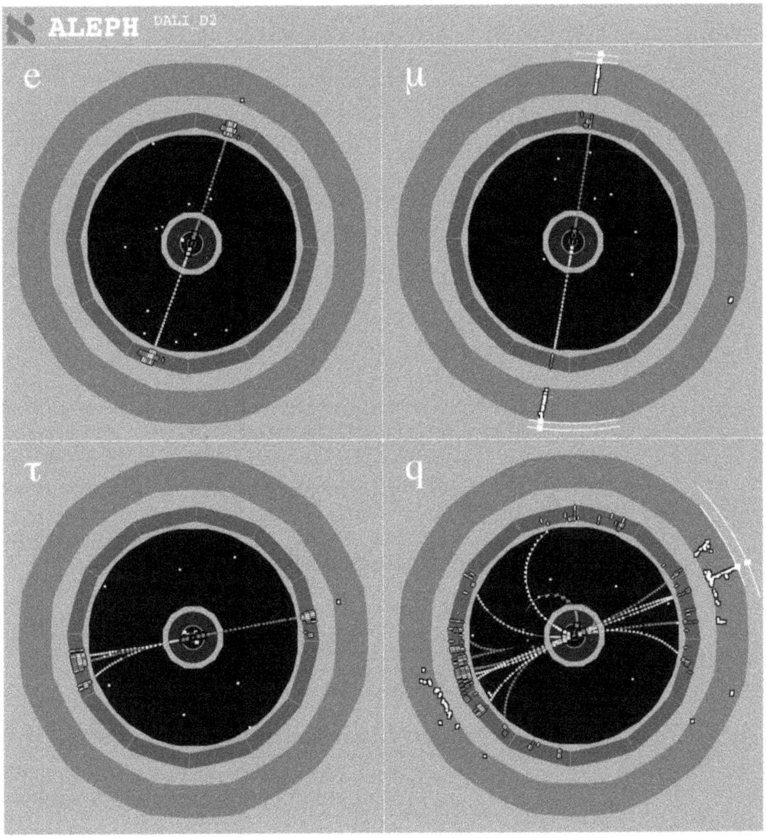

Fig. 2. The four types of Z boson decays. Up left a decay into a pair of electrons, up right into a pair of muons, bottom left a pair of tau leptons, one decaying into an electron and the other into three particles, bottom right a pair of quarks that fragments into a number of hadrons.

various leptonic partial widths, $R_\ell \equiv \Gamma_{had}/\Gamma_\ell$. The Standard Theory implies lepton universality, and if this is assumed, the number of parameters can be reduced to four, $M_Z, \Gamma_Z, \sigma_{had}^{peak,0}, R_\ell$. The choice of these observables to fit the line-shape measurements is dictated by the fact that they are experimentally uncorrelated, both from the point of view of statistical and systematic errors.

By reporting the expression for Γ_Z of Eq. (2) into the peak cross-section for hadrons, Eq. (3), the number of neutrinos can be extracted from quantities that are measured at the peak only:

$$N_\nu = \frac{\Gamma_\ell}{\Gamma_\nu} \cdot \left(\sqrt{\frac{12\pi R_\ell}{M_Z^2 \sigma_{had}^{peak,0}}} - R_\ell - 3 \right). \tag{4}$$

The sensitivity of N_ν to R_ℓ is small, as there is a cancellation between the two terms containing this quantity. As a result the experimental measurement that enters most in the determination of N_ν is the peak cross-section, as already guessed intuitively

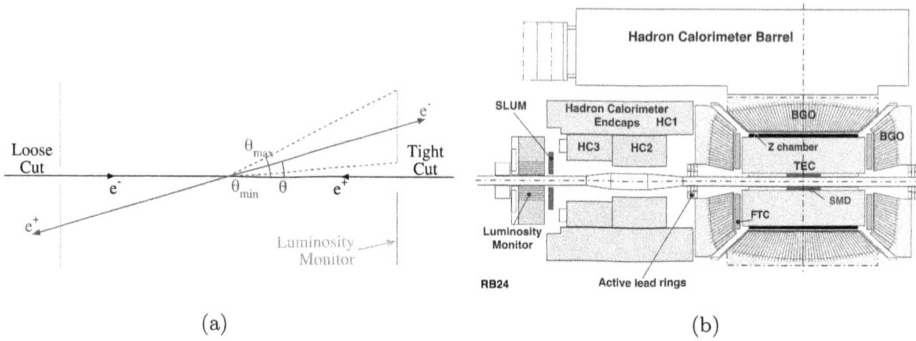

Fig. 3. Example of a LEP luminosity monitor (the L3 experiment). The process to be detected is elastic $e^+e^- \to e^+e^-$ scattering, shown on the left, which is seen by a coincidence of an electron in each of the detectors placed in the forward regions of the detector as shown on the right. In order to determine the solid angle seen by this coincidence in a way which is independent of the exact location of the collision point (which is difficult to determine) a set of two different acceptances is defined for the two arms. The precise definition of the acceptance is obtained by a precisely machined silicon telescope (SLUM) positioned in front of the electron calorimeters (luminosity monitor); there is a symmetric device on the other side of the detector.

from Fig. 1. This explains how quickly the number of neutrinos was obtained, a few weeks after the start-up of LEP and SLC.

4. Precision Measurements of the Mass and Width of the Z

The years 1990–1995, following this first success, were devoted to high-statistics, high-precision running of LEP at and around the Z pole. The LEP line shape results have been extensively documented in Ref. 10, in which the theoretical framework and all experimental results are described and combined.

The interest of precise measurements of the Z line shape parameters is evident when considering the Standard Theory expressions for the Z partial widths:

$$\Gamma_f = \frac{\alpha}{6\sin^2\theta_{\rm w}\cos^2\theta_{\rm w}} \mathrm{M_Z}\left({g_{Lf}}^2 + {g_{Rf}}^2\right)\cdot N_c\cdot\left(1+\frac{3Q_f^2}{4}\frac{\alpha}{\pi}\right)(1+\alpha_s/\pi+\cdots), \quad (5)$$

where the couplings are given in a universal way by: $g_{L,R\ f} = I^3_{L,R\ f} - Q_f\sin^2\theta_{\rm w}$, with $I^3_{R\ f} = 0$ for all known fermions. Equivalently one can define the coupling of the Z to the vector or axial vector fermion current:

$$\begin{aligned} g_{Vf} &= (g_{Lf}+g_{Rf}) = I^3_{Lf} - 2Q_f\sin^2\theta_{\rm w}, \\ g_{Af} &= (g_{Lf}-g_{Rf}) = I^3_{Lf}. \end{aligned} \quad (6)$$

Electroweak corrections to these formulae are largely accounted for by using universal effective couplings at the Z energy scale, both for $\alpha \to \alpha(\mathrm{M_Z^2})$ for the QED coupling constant, $\alpha_s \to \alpha_s(\mathrm{M_Z^2})$ for the strong coupling constant and for the weak mixing angle $\sin^2\theta_{\rm w} \to \sin^2\theta_{\rm w}^{\rm eff}$; the so-called ρ parameter is introduced to account for the change in the relative strength of the neutral current and charged

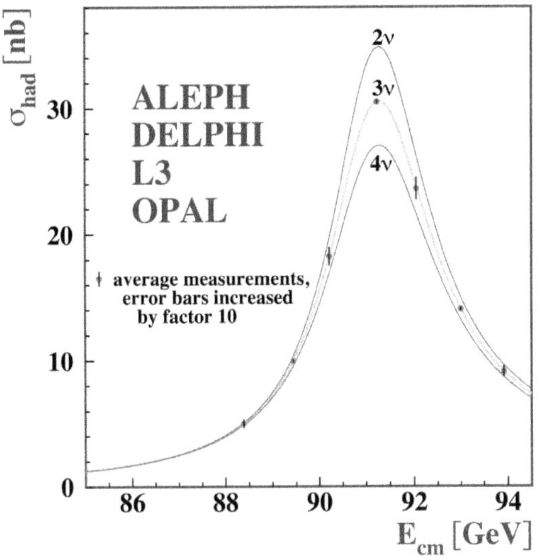

Fig. 4. The $e^+e^- \to$ hadrons cross section as a function of center-of-mass energy, as measured by the LEP experiments. The curves represent the Standard Theory predictions for two, three and four species of light neutrinos. It is clear from this picture that there is no further light neutrino species with couplings identical to the first three.

current weak interactions, and normalize by a factor $\sqrt{\rho}$ all the couplings of eq. 6, and consequently all the Z partial widths.

These corrections amount to 6% for $\alpha(\mathrm{M}_\mathrm{Z}^2)$, and to 1–2% for $\sin^2\theta_\mathrm{w}^\mathrm{eff}$ and ρ due to the large mass of the top quark. Small additional non-universal corrections (vertex corrections) amount to a few 10^{-3}; they are insensitive to such effects as the top quark or Higgs boson masses. The b partial width constitutes the well known exception, the vertex correction involving the top quark amounting to 2%.

The interest of the various parameters extracted from the line shape is then schematically summarized as follows:

— The Z mass is one of the precise inputs to the electroweak theory calculations.
— The Z width is sensitive to the strong coupling constant and to the electroweak corrections involving the top quark mass. In two different ways: first, because of the strong top mass dependence of the ρ parameter; and in addition because 21.5% of hadronic Z decays are into b-quark pairs, it also acquires a sensitivity to the top mass from the b vertex correction.
— The peak cross-section for hadrons is very sensitive to the partial width of the Z into invisible modes, in particular N_ν and, in effect, to hardly anything else.
— The ratio of hadrons to leptons is very sensitive to the strong coupling constant $\alpha_s(\mathrm{M}_\mathrm{Z}^2)$, and to a lesser extent to the b vertex correction; the equality of the ratios for different lepton species constitute an essential test of the universality of the couplings of the Z boson.

By combination of these measurements one can obtain the most precise measurement of $\alpha_s(M_Z^2)$, and, more importantly at the time, a prediction for the mass of the top quark from radiative corrections. For instance the Z width varies by 2 MeV of the top quark mass varies by 10 GeV.

In 1992, searches for the top quark mass in high energy hadron colliders had not been successful yet. For this reason, there was great interest in performing precise measurement of the Z width. This can be done by measuring e^+e^- cross-sections across the Z resonance. The choice of points and the amount of running necessary at each point was subject to careful studies:[11]

— The best scan would correspond to taking data at the peak, and at two points situated at +2 Gev and −2 GeV from the peak, with nearly equal amounts of data taken at each point.
— The beam energies were chosen so as to allow precise measurement of the LEP beam energies by the technique of resonant depolarization.

The resonant depolarization is a powerful technique which allows the determination of the beam energy in electron (and positron) storage rings, with precision of typically one part per million. So was measured in Novosibisrk the J/ψ mass $m = 3096.916 \pm 0.011$ MeV[12] and several other particles produced in e^+e^- annihilation. Beam polarization builds up slowly in the large storage ring (typically 5 hours rise time in the Z region), and requires exquisite beam orbit corrections. First observation of spin polarization was realized in 1990, by a small collaboration of accelerator and particle physicists[13] and a first precise determination of the beam energy was performed by resonant depolarization in 1991.[14] The method was determined to have an instantaneous precision of 100 KeV on the beam energy, as shown on Fig. 5(a).

To allow build up of the transverse polarization by Sokolov–Ternov effect, the energy had to be such that the spin tune was near a half-integer. By a lucky coincidence, the Z peak happens to correspond to $\nu = 103.5$.

It was thus decided to scan the Z peak at energies corresponding to spin tunes of 101.5, 103.5, 105.5. One unit of spin tune corresponding to 0.44065 GeV beam energy, or 0.8813 in center-of-mass energy, this is close enough to ±2 GeV.

A very systematic scan of the Z resonance was performed in 1993, with regular energy calibrations of the LEP energies.

It was soon realized that the measurement of beam energy showed time variations at short time scale (a few hours) well beyond the intrinsic precision of the method. There began one of the most exciting sagas of the LEP era. The most evident effect was the effect of ground motion due to various effects, the most popular one being the earth tides.[15] That a laboratory measurement of such scale could be associated with something as visible as the phase of the moon made great impact on the public! A precise measurement of the effect was undertaken at full moon on

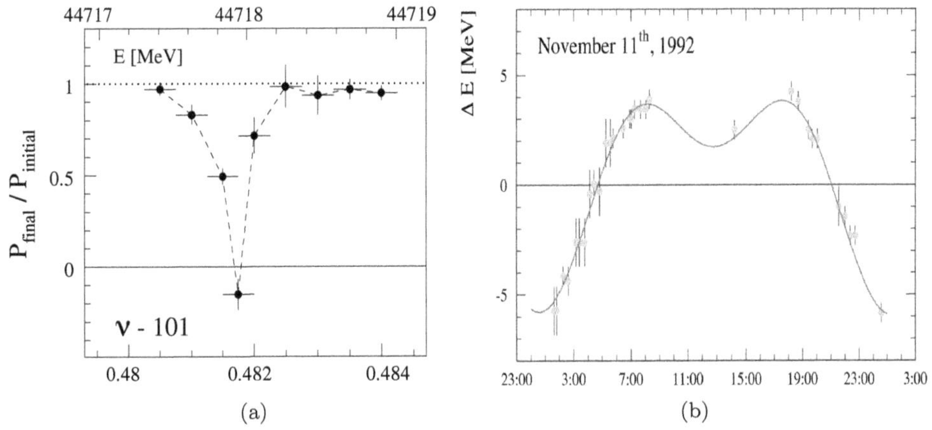

Fig. 5. (a) Measurement of the LEP beam energy by resonant depolarization. The graph shows the ratio of polarization level before and after excitation by a small RF kicker oscillating at a frequency in the vicinity of the spin precession frequency. The lower horizontal axis shows the oscillation frequency reported in unit of the spin tune, while the top axis indicates the corresponding beam energy in MeV. (b) The LEP tide experiment. Measurement of the LEP beam energy by resonant depolarization were performed on a regular basis over 24 hours at full moon and compared with the calculation from the earth tides assuming elastic deformation.

11 November 1992 and the result is shown in Fig. 5(b) showing a rather significant swing of more than 10 MeV.

The Z width measurement is sensitive to relative point-to-point errors, and relatively safe against short term energy variations, given that the scan pattern of -2, peak and $+2$ energy points was very well randomized by the running incidents of the accelerator. The Z mass, however, is sensitive to the absolute calibration of the beam energy.

It was realized in 1994 that the time evolution of the LEP energy was also disturbed by human activity. Jumps were observed during a long-term stability experiment, with a perplexing daily pattern. This may have biased the Z mass measurement, since, during the 1993 scan, the energy calibrations were always performed at the end of data taking periods of 8–10 hours. To monitor this effect, 16 NMR probes were inserted in a sample of the LEP magnets, to observe possible time variations. This should allow a complete study of these jumps, and a new scan was decided for 1995, in conditions similar to those of 1993, and additional monitoring. In 1994 data had been taken at the Z peak to accumulate large statistics.

As it turned out, the cause of the constant rise in the magnetic field was attributed to electric perturbations due to the passage of the french trains (TGV) in the area. This is shown in a spectacular way in Fig. 6. Due to hysteresis, this led to a positive drift of the beam energy during data taking and to a shift of about 3 MeV in the determination of the average beam energy. The corresponding correction to the Z mass was somewhat larger than the quoted systematic error from the 1993 scan.

The LEP energy determination revealed many such effects and resulted in overall precision of 1.7 MeV systematic error on the Z mass and 1.1 MeV on the Z width, as described in several papers.[16,17]

In addition to this careful determination of the LEP energies, it was necessary to ensure high precision for the experimental determination of the cross-sections, with particular attention to possible energy-dependent errors. High statistics were by then available, each experiment collecting around one million events. Major sources of systematic errors on the determination of the Z mass and width are:

— The contamination of the Z decays by non-resonant backgrounds, which tend to widen the apparent shape of the peak.
— The fact that the effect of initial-state radiation is not the same above the peak, where it is dominated by emission of a photon back to the Z peak, and below, where its effect is essentially to reduce the visible cross-section.

Figure 4 highlights the precision of the final LEP data.

5. The Discovery of the Top Quark, the Higgs Boson Mass

After the 1993 Z peak scan, the LEP experiments were able to produce for the 1994 winter conferences a combined measurement of the Z width with a precision of 3 MeV, which was the main ingredient to a prediction[18] of the top quark mass of 172^{+13+18}_{-14-20} (the first error corresponds to the experimental error, the second one to the unknown mass of the Higgs boson). A month later, the first indicative observations of the top quark with a mass of 176 ± 16 GeV were announced by the CDF collaboration[19] at the Fermilab Tevatron proton–antiproton collider! The discovery of the top quark was made officially[20] one year later in 1995.

As soon as the top quark was discovered, a large unknown was removed from the interpretation of the precision measurements. As a consequence, these measurements could be turned into predictions of the Higgs boson mass. The Higgs boson sensitivity of the Electroweak Radiative Corrections scales as $\frac{\alpha}{\pi} \cdot \log(\frac{m_h}{m_Z})^2$, and a precise determination of the Higgs boson mass (should it exist) is difficult, however the last result from the Z line shape as given in[10] was $m_h = 129 + 74 - 49$ GeV.

6. Discussion and Outlook

It took several years of analysis by the LEP experimenters to extract the final values of the Z line shape parameters documented in Ref. 10 from which all precise numbers, statistical correlations, etc. should be taken. In the end, the Z mass was measured with a precision of 2×10^{-5}, the Z width, peak cross-section and ratio of hadrons to leptons to 10^{-3}, and the relative couplings of the various leptons

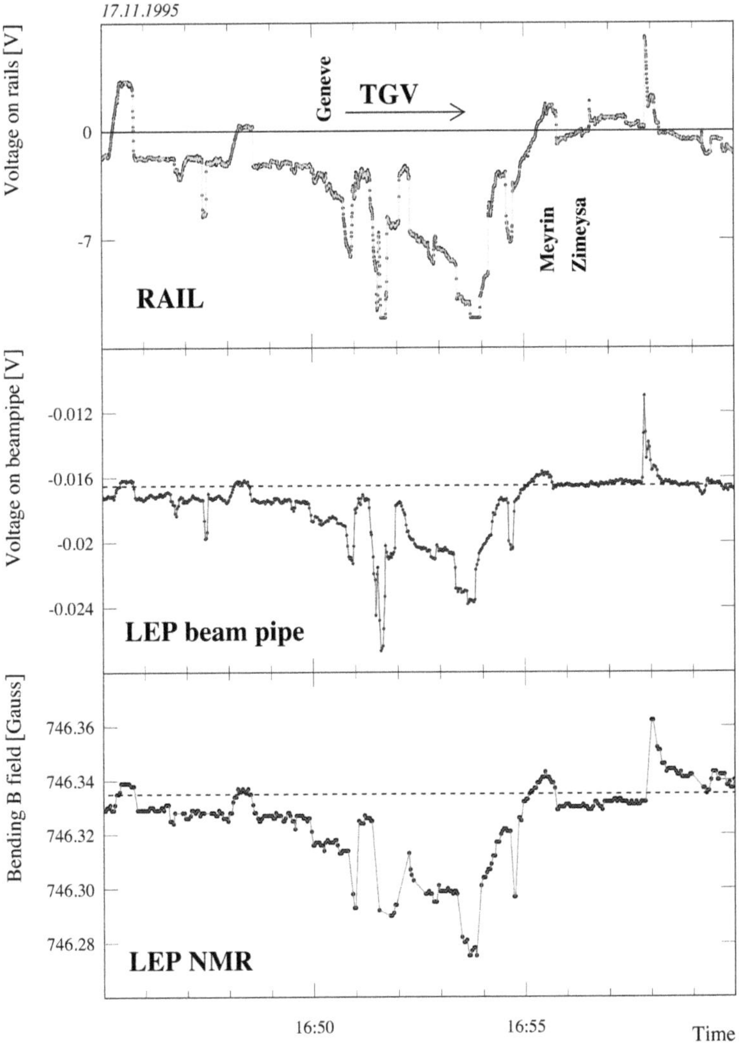

Fig. 6. The effect of trains on the LEP energy. The top plot shows the time variation of the voltage on the Geneva-Paris train track, the middle plot shows the voltage measured on the LEP beam pipe and the bottom plot shows the recorded NMR probe signal in the sampled LEP magnets.

to a precision of 2×10^{-3}. This precision exceeds the pre-LEP expectations by one order of magnitude. Such a precision, obtained after many years of hard work and clever tricks, was possible at LEP due to the extremely clean conditions, and the availability of resonant depolarization for the energy calibration in a circular machine. When LEP was closed in 2000, it was thought unlikely that, at least for the Z mass and width, these measurements would be improved in any foreseeable future.

Times have changed. The Higgs boson has been discovered at a mass of 125 GeV, and with it the Standard Theory in its minimal form is beautifully verified and supposedly complete. As far as Radiative corrections are concerned, it can be said that "The Standard Theory has nowhere to go", the predictions for the relations between e.g. the Z and W boson masses being now fixed, up to the precision of the calculations. This places particle physics in a new situation: the Standard Theory does not explain a number of well established experimental facts among which:

- the neutrino masses — discovery of the neutrino oscillations in 1998 implies that neutrinos have mass. This in turn almost certainly leads to the existence of the right-handed neutrinos.
- dark matter — the long standing astronomical observation of velocity of stars in galaxies requires matter that we cannot see, either weakly interacting (such as the popular Lightest Supersymmetric Particle should Supersymmetry be realized at relatively low energy (100 TeV or less) scale) or interacting even less, like right-handed neutrinos, or axions.
- the existence of a complete baryon-antibaryon asymmetry in the Universe — albeit with a small matter content. This points to a source of CP violation in excess of that present in quarks, and to the existence of matter-antimatter transitions.

The precision measurements of electroweak quantities were able to predict successfully the top mass, then the Higgs boson mass. Now that these particles are well known they could be used to test the existence of further weakly interacting particles or the existence of non-standard mechanisms for the Electroweak symmetry breaking. In addition the question of the existence of the right handed neutrinos, who can acquire mass not only from Yukawa couplings (Dirac terms) like all other fermions, but also from a Majorana mass term triggering the 'see-saw' mechanism, is a very open problem. Their mass can extend from a few eV up to the GUT scale.

In front of this very open field the experimental means of investigation should be as broad as possible. One of the tools of choice in this endeavor is the Future Circular Collider. In 2011, while the first evidence for the Higgs boson at the LHC[22,23] was reported by ATLAS and CMS, with a mass of 125 GeV, shown on 13 December 2011 at CERN, it was pointed out that a circular e^+e^- collider of the new generation with the improvements developed for the b-factories could be built in the LEP/LHC tunnel to measure precisely the properties of this new particle.[24] There are serious issues with installing a new accelerator in the LHC, however, and the proposal was quickly modified to a grand scenario involving the construction of a new 100 km tunnel around Geneva, able to host in succession a high luminosity e^+e^- collider (dubbed TLEP then FCC-ee) and a 100 TeV superconducting collider (dubbed FCC-hh).

A first look a the physics that can be achieved in the FCC-ee has been published in Ref. 25. Considerable improvements can still be made on the Z line shape measurements given the huge luminosity of the future accelerator, and are discussed in that paper. To give an idea, Z mass and width measurement with a precision of better than 100 KeV (yes, keV), can be envisaged, a measurement of the W mass to 500 keV, $\alpha_s(m_Z)$ with a precision of ± 0.0001 and $\alpha_{\mathrm{QED}}(m_Z)$ with a precision of $\pm 210^{-5}$. More recent updates concerning the search for right-handed neutrinos can be found in[26] and for the top quark measurements in Ref. 27.

The combination of the FCC-ee and the FCC-hh, which are highly complementary, constitutes the most powerful means of investigation into the unknown that seems conceivable today. CERN has launched the FCC study,[28] with the aim of producing a conceptual design report in 2018, in view of offering an ambitious post-LHC accelerator project.

References

1. S. Weinberg, *Gravitation and Cosmology* (John Wiley, 1972).
2. D. Denegri, B. Sadoulet, M. Spiro, *Rev. Mod. Phys.* **62**, 1 (1990).
3. J. R. Ellis, Zedology, in *LEP summer study*, CERN Yellow report 79-01 (1979) 618.
4. G. S. Abrams *et al.* (MarkII Collab.) *Phys. Rev. Lett.* **63**, 724 (1989).
5. D. Decamp *et al.* (ALEPH Collab.) *Phys. Lett. B* **231**, 519 (1989).
6. P. Aarnio *et al.* (DELPHI Collab.) *Phys. Lett. B* **231**, 539 (1989).
7. B. Adeva *et al.* (L3 Collab.) *Phys. Lett. B* **231**, 509 (1989).
8. M. Z. Akrawy *et al.* (OPAL Collab.) *Phys. Lett. B* **231**, 530 (1989).
9. G. S. Abrams *et al.* (MarkII Collab.) *Phys. Rev. Lett.* **63**, 2173 (1989).
10. The LEP collaborations Electroweak working group has produced regular updates since 1991. The last word was reported in *Precision Electroweak Measurements on the Z Resonance*, The ALEPH, DELPHI, L3, OPAL, SLD Collaborations, the LEP Electroweak Working Group, the SLD Electroweak and Heavy Flavour Groups http://arxiv.org/abs/hep-ex/0509008; *Phys. Rep.* **427**, 257–454 (2006). This contains references to the latest experimental and theoretical papers.
11. A. Blondel, LEP energy calibration in 1992, in *Proceedings of the second workshop on LEP performance*, Chamonix (Jan. 1992), 339 sq., ed., J. Poole, CERN SL/92-29(DI). CERN document server http://cds.cern.ch/record/260389/files/.
12. K. A. Olive *et al.* (Particle Data Group), *Chin. Phys. C* **38**, 090001 (2014).
13. L. Knudsen *et al.*, *Phys. Lett. B* **270**, 97 (1991).
14. L. Arnaudon *et al.*, *Phys. Lett. B* **284**, 431 (1992).
15. L. Arnaudon *et al.*, *Nucl. Inst. Meth. A* **357**, 249 (1995).
16. R. Assmann *et al.*, *Polarization Studies at LEP in 1993*, CERN SL/94-08(BI). R. Assmann *et al.*, *Accurate determination of the LEP beam Energy by resonant Depolarization*, CERN SL/94-71(BI), *Zeit. Phys. C* **66**, 45 (1995).
17. R. Assmann *et al.*, *Zeit. Phys. C* **66**, 567 (1995). R. Assmann *et al.*, *Eur. Phys. J. C* **6**, 187–223 (1999).
18. B. Pietrzyk, in *XXIX Rencontres de Moriond, Meribel, March 1994, 'Electroweak Interactions and Unified Theories'*, ed. J. Tran Than Van (Frontières), p. 137; LAPP-EXP-94.07, 1994.
19. F. Abe *et al.* (CDF Collab.) *Phys. Rev. D* **50**, 2966 (1994). (This paper gentlemanly refers to Ref. 18).
20. F. Abe *et al.* (CDF Collab.) *Phys. Rev. Lett.* **74**, 2626 (1995); D0 Collab., *Phys. Rev. Lett.* **74**, 2632 (1995).
21. *Physics at LEP*, CERN 86-02, Geneva (1986).
22. G. Aad *et al.* (ATLAS Collab.), *Phys. Lett. B* **716**, 1 (2012).

23. S. Chatrchyan *et al.* (CMS Collab.), *Phys. Lett. B* **716**, 30 (2012).
24. A. Blondel and F. Zimmermann, *A High Luminosity* e^+e^- *Collider in the LHC tunnel to study the Higgs Boson*, arXiv:1112.2518.
25. M. Bicer *et al.*, *JHEP* **1401**, 164 (2014), arXiv:1308.6176.
26. A. Blondel *et al.*, Search for heavy right handed neutrinos at the FCC-ee, arXiv:1411.5230, to appear in ICHEP2014 proceedings.
27. P. Janot, *JHEP* **1504**, 182 (2015).
28. The FCC design study, http://cern.ch/fcc. The FCC-ee design study, http://cern.ch/fcc-ee.

Chapter 9

Asymmetries at the Z pole: The Quark and Lepton Quantum Numbers

R. Tenchini

INFN, Sezione di Pisa,
Largo B. Pontecorvo 3, I-56127 Pisa, Italy

The impressive progress on the knowledge of lepton and quark electroweak couplings over the LEP and SLC decade is reviewed. The experimental methods for measuring the forward–backward asymmetry of charged-fermion pair-production are described, for different fermion species. The precise measurements of the left–right asymmetry and of tau polarisation at the Z resonance are also reminded. After discussing the determination of the Weinberg electroweak mixing angle, lepton and quark couplings are extracted by combining asymmetry and polarisation measurements with measurements of partial decay widths of the Z boson, performed at LEP in the same years.

1. Introduction

The present knowledge of quark and lepton electroweak neutral couplings is largely based on the data collected in e^+e^- collisions at the Z pole, between 1989 and 1998, by LEP and SLC. In that years, measurements of lepton couplings improved by two order of magnitudes with respect to previous experiments, based on neutrino scattering, and individual quark couplings were measured for the first time. The jump in precision was possible because parity violation in Z boson production and decay has a direct consequence on experimental data, yielding measurable asymmetries that can be used to determine the the Weinberg electroweak mixing angle $\sin^2\theta_W$ and, by including measurements of the Z partial widths, the couplings themselves.

Before describing the legacy measurements performed by ALEPH,[1] DELPHI,[2] L3,[3] OPAL,[4] SLD[5] and their interpretations, in next Section definitions and properties of the basic asymmetries sensitive to the electroweak mixing angle and neutral fermion couplings are briefly recalled.

2. Asymmetries and Polarisations at the Z pole

The differential cross section for the process $e^+e^- \to f^+f^-$ via exchange of a Z boson can be written as

$$\frac{d\sigma}{d\Omega} = B(s) \times \{[1+\cos^2\theta][(g_{Le}^2 + g_{Re}^2)(g_{Lf}^2 + g_{Rf}^2)] + 2\cos\theta[(g_{Le}^2 - g_{Re}^2)(g_{Lf}^2 - g_{Rf}^2)]\}, \tag{1}$$

where $g_{L(R)e}$ and $g_{L(R)f}$ are the left (right) couplings for the initial and final state, respectively, θ is the scattering angle of the final-state fermion with respect to the initial-state fermion direction (the electron in this case) and $B(s)$ is a *Breit–Wigner* term angular-independent. The coefficient of the cosine in Eq. (1) would vanish if the initial or final-state left and right couplings were identical, as in the electromagnetic case. Parity violation in Z production and decay leads to a non-zero forward–backward asymmetry, A_{FB}, defined as:

$$A_{FB} = \frac{\int_{\theta=0}^{\theta=\pi/2} \frac{d\sigma}{d\cos\theta} - \int_{\theta=\pi/2}^{\theta=\pi} \frac{d\sigma}{d\cos\theta}}{\int_{\theta=0}^{\theta=\pi/2} \frac{d\sigma}{d\cos\theta} + \int_{\theta=\pi/2}^{\theta=\pi} \frac{d\sigma}{d\cos\theta}}. \tag{2}$$

Integration over the scattering angle gives

$$A_{FB} = \frac{3}{4}\mathcal{A}_e\mathcal{A}_f \tag{3}$$

where

$$\mathcal{A}_e = \frac{g_{Le}^2 - g_{Re}^2}{g_{Le}^2 + g_{Re}^2} = \frac{2\, g_{V\mathrm{e}}/g_{A\mathrm{e}}}{1+(g_{V\mathrm{e}}/g_{A\mathrm{e}})^2}, \tag{4}$$

and similarly

$$\mathcal{A}_f = \frac{g_{Lf}^2 - g_{Rf}^2}{g_{Lf}^2 + g_{Rf}^2} = \frac{2\, g_{Vf}/g_{Af}}{1+(g_{Vf}/g_{Af})^2}. \tag{5}$$

In Eqs. (4) and (5) vector and axial couplings are introduced ($g_{V\mathrm{e}} = g_{Le} + g_{Re}$, $g_{A\mathrm{e}} = g_{Le} - g_{Re}$ and similarly for f). One can see that $\mathcal{A}_\mathrm{e}$ depends on the ratio between vector ($g_{V\mathrm{e}}$) and axial vector($g_{A\mathrm{e}}$) coupling constants of the electron, which are in turn related to the *effective* electroweak mixing angle, defined as[6,7]

$$\sin^2\theta_{\text{eff}}^{\ell} \equiv \frac{1}{4}\left(1 - \frac{g_{V\ell}}{g_{A\ell}}\right). \tag{6}$$

If polarised beams are available, or if polarisation can be measured in the final state, other useful asymmetries can be defined. As an example, if a polarised electron beam collides with unpolarised positrons at a centre-of-mass energy equal to m_Z, the total cross section is much bigger if left-handed polarisation is used. The relative difference between the two cross sections (σ_L and σ_R) is the left–right asymmetry

(A_{LR}), which is related to the right-handed (g_{Le}) and left-handed (g_{Re}) electron couplings by

$$A_{LR} = \frac{\sigma_{e_L^-} - \sigma_{e_R^-}}{\sigma_{e_L^-} + \sigma_{e_R^-}} = \frac{g_{Le}^2 - g_{Re}^2}{g_{Le}^2 + g_{Re}^2} \equiv \mathcal{A}_e. \tag{7}$$

A measurement of A_{LR} provides direct access to the asymmetry between the initial-state couplings, $\mathcal{A}_e$, which represents also the net polarisation acquired by the Z boson along the direction of the electron beam in case of collisions of unpolarised beams. In this case the Z polarisation, and therefore $\mathcal{A}_e$, can be measured by analysing the polarisation of the fermion emitted by the Z boson since angular momentum conservation relates the two quantities. In practice at the Z pole this is possible only if the emitted fermion is a tau lepton, by measuring the tau polarisation in $e^+e^- \to Z \to \tau\bar{\tau}$.

Polarised beams are also useful to provide a direct measurement of the asymmetry between final state couplings, $\mathcal{A}_f$, by measuring the polarised forward–backward asymmetry, defined as

$$A_{FB}^{pol}(f) = \frac{(\sigma_{e_L^- f_F} - \sigma_{e_R^- f_F}) - (\sigma_{e_L^- f_B} - \sigma_{e_R^- f_B})}{\sigma_{e_L^- f_F} + \sigma_{e_R^- f_F} + \sigma_{e_L^- f_B} + \sigma_{e_R^- f_B}}, \tag{8}$$

where f_F and f_B indicate forward and backward outgoing fermions, respectively. At the Z peak

$$A_{FB}^{pol}(f) = \frac{3}{4}\frac{g_{Lf}^2 - g_{Rf}^2}{g_{Lf}^2 + g_{Rf}^2} \equiv \frac{3}{4}\mathcal{A}_f \tag{9}$$

showing that Eq. (8) is only dependent on the final state couplings, as anticipated.

3. Forward–Backward Asymmetries

The forward–backward asymmetry at the Z pole (Eq. (3)) has been measured at LEP for individual lepton species (e, μ, τ) and for heavy quarks (c and b). Measurements were also performed inclusively for hadrons and, though with much lower precision, for s and u, d quarks. The data were collected by ALEPH, DELPHI, L3 and OPAL in the years 1986–1995 (LEP1 phase) and correspond to an integrated luminosity of 150 pb^{-1} per experiment, yielding a total of more than 15 millions hadronic Z decays and 1.7 millions leptonic Z decays.

Lepton and quark forward–backward asymmetries have several conceptual and experimental differences. Assuming lepton universality the ratios between vector ($g_{V\ell}$) and axial vector($g_{A\ell}$) coupling of the Z to charged leptons are equal, therefore asymmetries involving leptons provide a direct determination of the effective mixing angle (Eq. (6)) using the relation $A_{FB} = \frac{3}{4}\mathcal{A}_e\mathcal{A}_f = \frac{3}{4}{\mathcal{A}_\ell}^2$. The quark forward–backward asymmetries depends on $\mathcal{A}_f = \frac{2g_{Vq}/g_{Aq}}{1+(g_{Vq}/g_{Aq})^2}$ where the subscript q indicates the quark flavour. The ratio of quark couplings can be expressed in terms of

$\sin^2\theta^{\ell}_{\rm eff}$ and non-universal corrections as[8]

$$\frac{g_{Vq}}{g_{Aq}} = 1 - \frac{2Q_q}{I_{3L,q}}(\sin^2\theta^{\ell}_{\rm eff} + C_q), \tag{10}$$

where Q_q is the electric charge and $I_{3L,q}$ is the third component of the weak isospin. The residual vertex correction C_q can be computed assuming the Standard Theory. For *udsc* quarks it is small and has very little dependence on the parameters of the model, while for b it depends on the top mass because of additional $\mathrm{Z} \to \mathrm{b\bar{b}}$ vertex corrections ($C_b = +0.0014$). In case of quarks $\mathcal{A}_f$ is large and weakly dependent on $\sin^2\theta^{\ell}_{\rm eff}$ leaving most of the dependence on the weak mixing angle to $\mathcal{A}_{\rm e}$. Therefore for quarks A_{FB} is essentially linearly dependent on $\sin^2\theta^{\ell}_{\rm eff}$, while for leptons it shows a quadratic dependence. The consequence of this behaviour is shown in Table 1 where the magnitude of A_{FB} and its sensitivity to $\sin^2\theta^{\ell}_{\rm eff}$ is given for leptons, for u-type and d-type quarks, showing that the latter can provide the most precise measurements of the electroweak mixing angle, within the Standard Theory.

Most measurements of forward–backward asymmetries have been determined by fitting the data to the differential angular distribution

$$\frac{dN}{d\cos\theta} = C(\cos\theta)\cdot\left(1 + \cos^2\theta + \frac{8}{3}A_{FB}\cos\theta\right) \tag{11}$$

where θ is the scattering angle of the fermion in the centre-of-mass system and $C(\cos\theta)$ is an acceptance function modifying the differential cross section (Eq. (1)). Experimentally the A_{FB} measurement requires the identification of the fermion in the final state, i.e. a measurement of its charge. Again, lepton and hadron asymmetries are different in this respect:

- in leptonic Z decays typically the selection of two fermions of opposite charge is required, automatically defining the final-state fermion.
- in hadronic Z decays at least one fermion must be tagged and its charge measured, as for the measurement of heavy quark asymmetries described in Section 3.2.

Table 1. Magnitude of the forward–backward asymmetry and its sensitivity on $\sin^2\theta^{\ell}_{\rm eff}$ for various fermion species at the pole of the Z. The value of 0.2316 is used for $\sin^2\theta^{\ell}_{\rm eff}$. For comparison the last line gives the magnitude and the sensitivity for A_{LR}.

	A_{FB}	$\frac{\partial A_{FB}}{\partial \sin^2\theta^{\ell}_{\rm eff}}$
Leptons	.02	−1.7
u and c quarks	.07	−4.0
d, s and b quarks	.10	−5.6
A_{LR}	.15	−7.8

In both cases it can be shown that the acceptance function is symmetric, provided the selection efficiency does not depend on the fermion charge.

Forward–backward asymmetries depend on the centre-of-mass energy, because of the interference between photon and Z exchange diagrams. Near the Z peak the asymmetry depends on the electric charge of the final fermion and on its axial coupling and has very little dependence on other electroweak parameters. The functional dependence can be approximated as

$$A_{FB}^{f}(s) \simeq A_{FB}^{f}(m_Z^2) + \frac{(s-m_Z^2)}{s}\frac{3\pi\alpha(s)}{\sqrt{2}\mathrm{G_F}m_Z^2}\frac{2Q_eQ_fg_{Ae}g_{Af}}{(g_{Ve}^2+g_{Ae}^2)(g_{Vf}^2+g_{Af}^2)}, \tag{12}$$

where the running fine structure constant ($\alpha(s)$), the Fermi constant ($\mathrm{G_F}$) and the electric charges (Q_e and Q_f) have been introduced. The dependence is maximal for leptons ($\Delta A_{FB}^{\ell}/\Delta E_{CM} \simeq 0.00009/\mathrm{MeV}$), while down-type quarks show the smallest energy dependence. This effect is corrected for using the precise determination of the LEP beam energy,[9] by extrapolating the measured asymmetry to m_Z. All forward–backward asymmetries have also to be corrected for the effect of initial state radiation, for imaginary parts of the couplings (in particular for $\mathrm{Im}(\Delta\alpha)$), for the effect of pure photon exchange and the presence of box diagrams. Specific corrections are also applied for final state photon radiation (leptons) and gluon emission (hadrons). The uncertainty of these corrections is, in all cases, much smaller than the total error, dominated by the statistical uncertainty for all measurements. The corrected asymmetry for a fermion f is indicated as $A_{FB}^{0}(f)$ in next Sections.

3.1. *Lepton forward–backward asymmetries*

At LEP the forward–backward asymmetry of $\mathrm{e^+e^-} \to \ell^+\ell^-(\gamma)$ events was determined by fitting the data to Eq. (11); the angle θ was defined by the scattering angle of the final-state negative lepton. For tau leptons the direction was given by the sum of the momenta of charge particles associated to the tau decays; the tau charge was measured in the same way.

In the case of $\mathrm{e^+e^-}$ final state, the t-channel photon-exchange process induces an important asymmetric correction and requires a careful treatment. The contribution of this process was taken into account by subtracting it from the measured angular distribution. Semi-analytical calculations incorporating leading-log photonic corrections, first-order non-log terms and first-order weak corrections[10] were used for this correction. The t-channel influence was reduced by analysing the data in a restricted angular region, typically in the $-0.9 \leq \cos\theta \leq +0.7$ range. (Within this range the t-channel contributes 12% to the total cross section, therefore calculations with 1% precision yielded an uncertainty of 0.1%.)

The asymmetries $A_{FB}(\ell)$ ($\ell = \mathrm{e}, \mu$ and τ) measured at LEP[11–14] were extracted with a fit to the measured $A_{FB}(s)$ using data collected near the Z peak and at the off-peak points used to measure the Z lineshape. The fitting formula took into account the energy dependence of the asymmetry and the fit was done simultaneously with the lineshape data to account for the effect of the energy uncertainty. In the simultaneous fit of the Z lineshape and $A^{\ell}_{FB}(s)$, the axial couplings were essentially determined by the lineshape and then used to transport the off-peak measurements of $A^{\ell}_{FB}(s)$ to $\sqrt{s} = m_Z$.

The measurement of $A^0_{FB}(\ell)$ was a rather straightforward measurement with low systematic uncertainties. For the μ and τ channels the systematic uncertainties were related to the applied corrections, to the presence of background and to possible detector asymmetries. Typical systematic errors quoted by the LEP experiments were of the order of $\Delta A_{FB} = 0.0005$–0.001 for muons and $\Delta A_{FB} = 0.001$–0.003 for taus, depending on the experiment. For electrons, the theoretical uncertainty introduced in the treatment of the t-channel terms (≈ 0.0014) had to be taken into account increasing the typical error to $\Delta A_{FB} \approx 0.002$. The uncertainty on the centre-of-mass energies gives a contribution of $\Delta A^0_{FB}(\ell) = 0.0004$, comparable to the experimental systematics. These last two uncertainties were common to the four experiments and had to be treated in a correlated way when averaging the measurements.

The combination of the results of the four LEP experiments gave

$$A^0_{FB}(e) = 0.0145 \pm 0.0025, \tag{13}$$

$$A^0_{FB}(\mu) = 0.0169 \pm 0.0013, \tag{14}$$

$$A^0_{FB}(\tau) = 0.0188 \pm 0.0017. \tag{15}$$

The three measurement can be combined assuming lepton universality, giving

$$A^0_{FB}(\ell) = 0.0171 \pm 0.0010, \tag{16}$$

$$\sin^2\theta^{\ell}_{\text{eff}} = 0.23099 \pm 0.00053. \tag{17}$$

The dependence of the asymmetries on the centre-of-mass energy, $A^{\ell}_{FB}(s)$, is consistent with the expected value and sign of the lepton axial couplings and it is shown in Fig. 1.

3.2. *Heavy quark asymmetries*

Forward–backward asymmetries for b and c quarks were determined at LEP with three different techniques. The first method (*lepton tagging*) was based on the presence of a lepton in a jet as a tag for $Z \to b\bar{b}$ or $Z \to c\bar{c}$ events. Lepton kinematics was used to discriminate the different lepton sources, on a statistical basis. With this method the charge of the lepton can be used to define the scattered fermion. The second method (*lifetime tagging*) relied on the selection of $Z \to b\bar{b}$ events using the properties of long-lived B hadrons, followed by the use of a jet-charge measurement. The third method was conceptually similar to the first one, with lepton

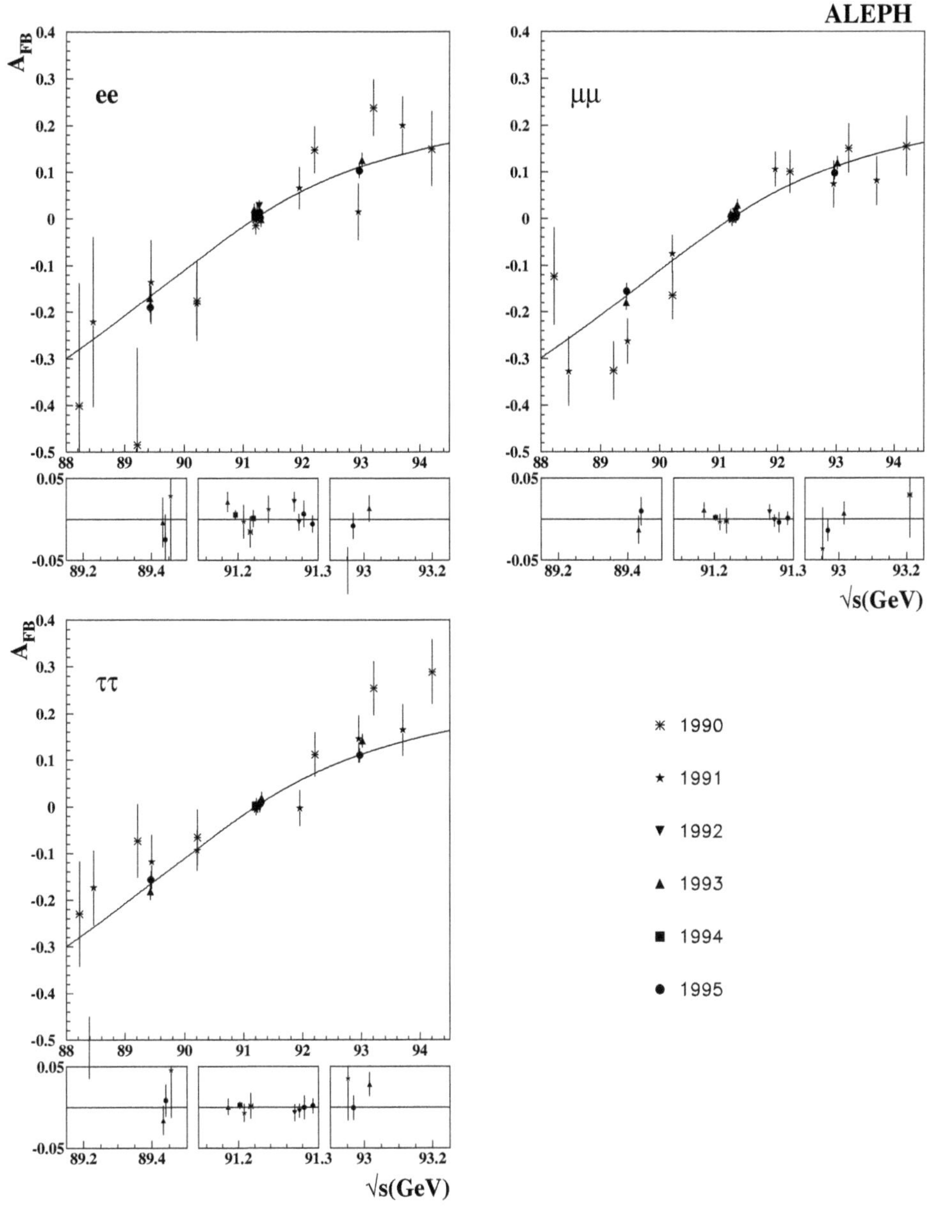

Fig. 1. Measurement of the forward–backward asymmetries for the three lepton species at various centre-of-mass energies, as measured by the ALEPH experiment.[11]

tagging replaced by *D meson tagging* and was solely used for the measurement of the c asymmetry. Here only the first two methods are briefly described, but results from D meson tagging are used in the LEP heavy quark asymmetry combination mentioned at the end of this Section.

3.2.1. *Lepton tagging*

In order to illustrate the main aspects of the lepton tagging method it is useful to consider hadronic Z decays selected by requiring leptons at high-$p_\perp$ with respect to the axis of their associated jet. A proper $p_\perp$ cut can provide a high-purity sample of $\mathrm{Z} \to \mathrm{b\bar{b}}$ events, with enhanced $\mathrm{b} \to \ell^-$ content. Such a sample can be used to measure the b asymmetry by fitting the polar angle distribution of the thrust axis to the functional form provided by Eq. (11). The thrust axis is conventionally defined as oriented towards the hemisphere containing the lepton if this is negatively charged, towards the other hemisphere otherwise. Semileptonic $\mathrm{b} \to \ell^-$ decays carry the correct charge correlation between quark and tagging lepton, yielding a forward–backward asymmetry in the oriented thrust axis polar angle distribution. The observed asymmetry (A_{FB}^{obs}) can be related to the $\mathrm{Z} \to \mathrm{b\bar{b}}$ forward–backward asymmetry (A_{FB}^{b}) in terms of the contributions of the different components of the selected sample:

$$A_{FB}^{obs} = (1 - 2\bar{\chi})(f_{\mathrm{r.s.}}^{\mathrm{b}} - f_{\mathrm{w.s.}}^{\mathrm{b}})A_{FB}^{b} + f_{\mathrm{bkg}}^{\mathrm{b}}\eta_{\mathrm{bkg}}^{\mathrm{b}}A_{FB}^{b}$$
$$-f^{\mathrm{c}\to\ell^+}A_{FB}^{c} - f_{\mathrm{bkg}}^{\mathrm{c}}\eta_{\mathrm{bkg}}^{\mathrm{c}}A_{FB}^{c} + f^{\mathrm{uds}}\overline{A_{FB}^{uds}}, \tag{18}$$

where $f_{\mathrm{r.s.}}^{\mathrm{b}}$ and $f_{\mathrm{w.s.}}^{\mathrm{b}}$ are the fractions of prompt leptons from b decays with right/wrong charge correlation between the lepton charge and the b quark charge, dominated by $\mathrm{b} \to \ell^-$ an $\mathrm{b} \to \mathrm{c} \to \ell^+$ decays, respectively, and the term $(1 - 2\bar{\chi})$ is introduced to correct for neutral b meson mixing ($\bar{\chi} \approx 0.13$). The charm asymmetry is indicated by A_{FB}^{c} and the fraction of prompt leptons from charm decays by $f^{\mathrm{c}\to\ell^+}$. The factors $\eta_{\mathrm{bkg}}^{\mathrm{b}}$ and $\eta_{\mathrm{bkg}}^{\mathrm{c}}$ describe the (small) correlation between the charge of fake and non-prompt leptons and the charge of the primary quark in b and c events, and the last term accounts for the contribution of light quark background.

High-$p_\perp$ leptons can be selected in both hemispheres, and the counting of same sign and opposite sign pairs ($N_{\mathrm{o.s.}}$, $N_{\mathrm{s.s.}}$) gives the possibility to measure from the data the charge correlation in b events $\mathcal{P}_{\mathrm{b}} = (1 - 2\bar{\chi})(f_{\mathrm{r.s.}}^{\mathrm{b}} - f_{\mathrm{w.s.}}^{\mathrm{b}})$, once the small contribution from charm and light quark events has been subtracted:

$$f_{\mathrm{o.s.}} = \frac{N_{\mathrm{o.s.}}}{N_{\mathrm{o.s.}} + N_{\mathrm{s.s.}}} = \mathcal{P}_{\mathrm{b}}^2 + (1 - \mathcal{P}_{\mathrm{b}})^2 + \mathrm{c\ and\ uds\ corrections.} \tag{19}$$

The analysis of the dilepton sample to evaluate $\mathcal{P}_{\mathrm{b}}$ from the data, lowers considerably the dependence of the measurement upon the knowledge of semileptonic b decays (rates and kinematic properties), as well as b meson mixing. It should be also noted that since the measurement makes use of both forward negative leptons and backward positive leptons to tag forward b quarks, the detector acceptance has nearly no effect on the extracted value of the asymmetry: a sizeable effect could arise only in case of inefficiencies that are both forward–backward asymmetric and different for positive and negative leptons, which was very unlikely for LEP detectors.

The extraction of A^b_{FB} from high-$p_\perp$ leptons, using Eq. (18), depends on the evaluation and on the modelling of the charm component, and on the assumed value of A^c_{FB}. At LEP[15–18] the whole $p_\perp$ spectrum of lepton candidates was studied, providing a simultaneous measurement of A^b_{FB} and A^c_{FB}.

3.2.2. *Inclusive measurements*

Heavy quark tagging methods based on lifetime have high performance, but they do not provide information about the quark charge. Inclusive methods have been developed to estimate the charge of the b quark, to complement lifetime tags for the measurement of forward–backward asymmetries.

The jet charge is usually defined as

$$Q_{\mathrm{h}} = \frac{\sum_i q_i p^k_{\parallel i}}{\sum_i p^k_{\parallel i}}, \tag{20}$$

where $p_{\parallel i}$ is the momentum of a particle parallel to the thrust axis, and the sum runs over all charged particles in a hemisphere. The parameter k can be tuned to obtain high sensitivity to the quark charge, while keeping low correlation between the charge of the two hemispheres (values used at LEP were between 0.3 and 1).

In a pure sample of b events, the forward–backward asymmetry is proportional to the mean charge flow between the two hemispheres

$$Q^{\mathrm{b}}_{\mathrm{FB}} \equiv \langle Q^{\mathrm{b}}_{\mathrm{F}} - Q^{\mathrm{b}}_{\mathrm{B}} \rangle = \delta_{\mathrm{b}} A^b_{FB} \tag{21}$$

where δ_{b} is a parameter called charge separation. At parton level δ_q (the charge separation for a generic quark q) is equal to twice the quark charge, but hadronization and decays lower its value, diluting the measured charge flow. A precise determination of the forward–backward asymmetry requires an evaluation of δ_q with the lowest possible uncertainty. The advantage of high-purity single-flavour samples, that in practice can be obtained for b quarks only, lies on the possibility of measuring δ_q from the data, lowering considerably the use of theoretical assumptions in the evaluation of this parameter, and therefore lowering its uncertainty. A sketch illustrating the technique used at LEP to measure the charge separation is presented in Fig. 2.

In an asymmetry analysis, pure b samples cannot be selected and contributions of the other flavours have to be taken into account; for instance, the charge flow can be written as

$$Q_{\mathrm{FB}} = f_{\mathrm{b}} \delta_{\mathrm{b}} A^b_{FB} + f_{\mathrm{c}} \delta_{\mathrm{c}} A^c_{FB} + f_{\mathrm{uds}} \delta_{\mathrm{uds}} A^{uds}_{FB}, \tag{22}$$

where the f_{b}, f_{c}, f_{uds} are the fraction of b, c and light quark events in the selected sample, and light quark have been treated as a single class.

In the initial LEP measurements, the sample composition as well as the charm and light quark charge separations was estimated with the simulation, δ_{b} was extracted from the data and the b asymmetry derived from the observed charge

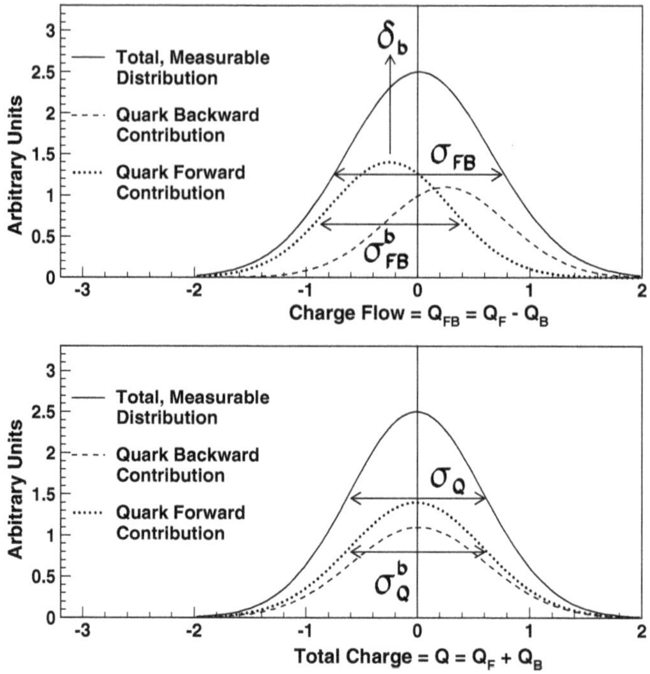

Fig. 2. Sketch showing the distributions of charge flow and total charge: the difference in width between the two distribution is related to the charge separation.

flow. In the final measurements[19–22] sophisticated methods using double tagging techniques were employed and most parameters were determined in situ from data.

3.2.3. *Heavy quark asymmetries: Combined results and QCD corrections*

The LEP measurements of b and c forward–backward asymmetries using lepton tagging, lifetime tagging and D mesons tagging[15–25] were combined to merge the experimental information in an optimal way.[26] As mentioned earlier, the extraction of the effective electroweak mixing angle and of the quark couplings requires the evaluation of the corrected b and c asymmetries $A^0_{FB}(b)$ and $A^0_{FB}(c)$, from the measured asymmetries. Heavy quark asymmetries are affected by radiative effects due to strong interactions related to virtual vertex and gluon bremsstrahlung diagrams, which modify the angular distribution of the fermions emitted in the final state. The emission of an hard gluon, for example, may scatter both b and $\bar{\text{b}}$ in the same hemisphere (forward or backward): in such events the original electroweak asymmetry is destroyed. The effect of such radiative effects is to lower the experimentally observed asymmetry by a few percent. Detailed calculation based on perturbative QCD, including second-order corrections for massless quarks and quark mass effects at first-order, were used.[27] In practice experimental cuts reduce considerably the QCD corrections.[28] For instance momentum cuts, which are applied in lepton tagging, select events with reduced gluon radiation. Furthermore in some cases the effect of hard gluon radiation is automatically incorporated by analysis procedure. This is the case for the inclusive measurements based on jet charge techniques, because the b charge separation, measured with data, is an effective parameter that includes the QCD smearing.

The final LEP averages for the b and c forward–backward asymmetries at the Z pole are[29]

$$A^0_{FB}(b) = 0.0992 \pm 0.0016,$$
$$A^0_{FB}(c) = 0.0707 \pm 0.0035.$$

There is a +15% correlation between the two results. Both results are dominated by the statistical uncertainties. In particular, for the b asymmetry, the systematic uncertainties related to the QCD corrections is a factor three lower than the statistical error.

The dependence of the b and c asymmetries on the centre-of-mass energy, $A^b_{FB}(s)$ and $A^c_{FB}(s)$, is regulated by the quark electric charge and its axial coupling (Eq. (12)). Their observed energy dependence is shown in Fig. 3 and compared with the ST prediction. The different slope for b and c quarks is due to the absolute value of their electric charge, that is twice larger for up-type quarks. The asymmetry is increasing in both cases because the two quark types have opposite sign (and same absolute value) for the axial couplings.

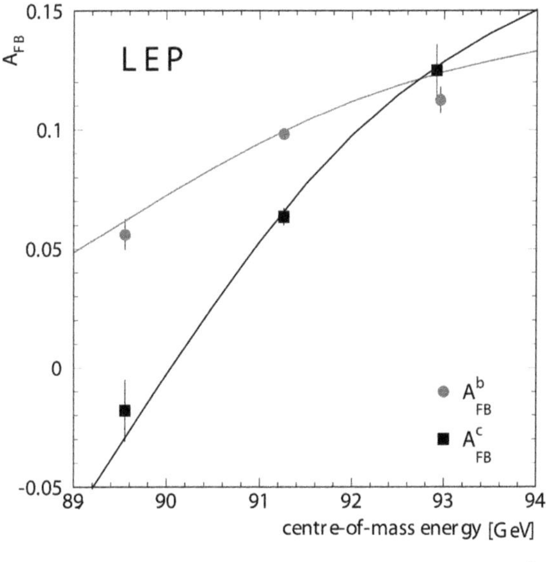

Fig. 3. Measurement of the b and c forward–backward asymmetries as a function of the centre-of-mass energy.[29] The ST expectation for the two quark types is shown. The value of $\sin^2\theta^\ell_{\rm eff}$ given in Section 6 is used to normalise the vertical scale for the ST prediction.

4. Asymmetries with Polarised Beams

4.1. *Measurement of the left–right asymmetry* (A_{LR})

The measurement of A_{LR} requires the availability of longitudinally polarised beams. At SLC longitudinal polarisation of the electron beam was achieved by a circularly polarised laser source hitting a GaAs photocathode,[30] allowing SLC to be operated with an electron beam polarisation of about 75%. Equation (7), modified to take into account the average beam polarisation ($\mathcal{P}_e$), reads

$$A_{LR} = \frac{1}{\mathcal{P}_e} \frac{\sigma_L - \sigma_R}{\sigma_L + \sigma_R}. \tag{23}$$

The main experimental issue, for a precise measurement of A_{LR}, is an accurate determination of the beam polarisation. This need could be overcome if both electron and positron beams were independently polarised, a scheme originally proposed for LEP[31] that never went into operation, because it would have required the installation of spin rotators. The standard SLC operating cycle consisted of two close electron bunches, the first of which was polarised, while the other was used to produce unpolarised positrons. The sign of the electron polarisation was randomly chosen, so that the measurement was not affected by time variations of the apparatus efficiency.

The SLD experiment monitored the longitudinal SLC-beam polarisation with a polarimeter based on Compton scattering of electrons by circularly polarised laser light taking place after the interaction point. The Compton cross sections for spin-parallel (j=3/2) and spin anti-parallel (j=1/2) interactions are different, and this difference is a function of the normalized scattered-photon energy fraction (x). The difference can be written, in terms of x, as

$$\frac{d\sigma^{3/2}}{dx} - \frac{d\sigma^{1/2}}{dx} = \frac{d\sigma}{dx}(1 - \mathcal{P}_\gamma \mathcal{P}_e A(x)) \tag{24}$$

where $A(x)$ is the Compton asymmetry function.[32] The laser beam polarisation ($\mathcal{P}_\gamma$, typically 99.8%) was continuously monitored. The statistical accuracy on $\mathcal{P}_e$ was of ±1% every three minutes. The relative systematic uncertainties[33,34] in the polarisation measurement are summarized in Table 2. The depolarisation of the electron beam during the e^+e^- collision was checked by measuring the polarisation with and without beam collisions and was found to be negligible. The total contribution of the systematic uncertainty on the beam polarisation to the measurement of A_{LR} was 0.52 % (high-statistics 1997/8 run); this is the main source of systematic uncertainty in the left–right asymmetry measurement.

The asymmetry of the left–right rates was measured with a simple event selection, since A_{LR} does not depend on the final state as long as this is an s-channel process. Care must be taken in rejecting Bhabha scattering events, because of the t-channel contribution to $e^+e^- \to e^+e^-$. The total sample comprised approximately 537000 Z decays and was mostly made of hadronic events,

Table 2. Relative (%) systematic uncertainties on the electron beam polarisation at SLD in two data-taking periods.[33,34] The last entry represents the uncertainty on the difference between the measured polarisation and the polarisation at the interaction point (IP).

Source of uncertainty	1994/5 (%)	1997/8 (%)
Laser polarisation	0.2	0.1
Detector Linearity	0.5	0.2
Detector Calibration	0.29	0.40
Electronic Noise	0.20	0.20
Transport from polarimeter to SLD IP	0.17	0.15

with a small tau contributions ($\sim 0.3\%$). The events produced with left-handed (N_L) and right-handed (N_R) polarisation were counted and their asymmetry $A_m = (N_L - N_R)/(N_L + N_R) \sim 0.12$ was measured. The measured asymmetry is related to A_{LR} by the following expression

$$A_{LR} = \frac{A_m}{\langle \mathcal{P}_{\rm e} \rangle} + \frac{1}{\langle \mathcal{P}_{\rm e} \rangle} \left[f_b(A_m - A_b) - A_L + A_m^2 A_{\mathcal{P}} \right.$$

$$\left. - E_{cm} \frac{\sigma'(E_{cm})}{\sigma(E_{cm})} A_E - A_\epsilon + \langle \mathcal{P}_{\rm e} \rangle \mathcal{P}_{\rm p} \right] \tag{25}$$

where a number of small corrections, listed below, are incorporated. In Eq. (25) A_X indicates the left–right asymmetry of X, defined as $A_X \equiv \frac{X_L - X_R}{X_L + X_R}$. The first term in the square bracket represents the correction for the background: f_b is the background fraction and A_b the background left–right asymmetry. The second term represents the asymmetry of the integrated luminosity, while the third term takes into account the asymmetry in the beam left and right absolute polarisations. The fourth term corrects for the different centre-of-mass energies when left or right beams are used: $\sigma'(E_{cm})$ is the derivative of the cross section with respect to E_{cm}. The fifth term represents the left–right asymmetry in selection efficiency: it is totally negligible for an apparatus with symmetric acceptance in polar angle. Finally, the last term corrects for possible longitudinal polarisation of the positron beam. This was measured with a dedicated experiment based on Möller scattering and found negligible. The sum of the corrections in the square brackets of Eq. 25 gives $[+0.16 \pm 0.07]\%$ for the 1997/8 high luminosity run.

Also for the left–right asymmetry there is a dependence on the centre-of-mass energy because of the Z–γ s-channel interference:

$$A_{LR}(s) = \mathcal{A}_{\rm e} + 0.00002 \Delta E({\rm MeV}) + 0.00005 \tag{26}$$

where ΔE is the difference between m_Z and the actual centre-of-mass energy, while the constant term accounts for the correction due to the imaginary part of $\Delta\alpha$. In order to apply the correction and compute the asymmetry at the Z pole the centre-of-mass energy of the experiment must be precisely known. SLC employed two energy spectrometers (one for the electron and one for the positron beam) calibrated, through an energy scan, to the precise measurement of m_Z at LEP. The measured average offset was $-46\,\mathrm{MeV}$ and the total centre-of-mass energy uncertainty 29 MeV. The measured left–right asymmetry was also corrected for the effect of initial state radiation (the most sizeable QED correction, which lowers the asymmetry as expected from Eq. (26)), for the effect of pure photon exchange (which slightly dilutes the asymmetry) and for other higher order QED/electroweak effects (as the already mentioned imaginary part of $\Delta\alpha$). The total correction (including the centre-of-mass energy offset) was 0.00358 ± 0.00058, the error being essentially due to the uncertainty on the beam energy. When this uncertainty is added in quadrature to the uncertainty on the electron beam polarisation and the uncertainty on the corrections of Eq. (25) a total systematic error on A_{LR} of 0.64 % is obtained.

The final result,[34] including the statistical errors, was

$$A_{LR} = 0.15138 \pm 0.00216, \tag{27}$$

$$\sin^2\theta^{\ell}_{\mathrm{eff}} = 0.23097 \pm .00027 \tag{28}$$

giving the most precise measurement of the electroweak mixing angle.

4.2. *Heavy quark asymmetries with polarised beams*

The quark asymmetries discussed in Section 3.2, based on measurements employing unpolarised beams, are probing the product of initial and final state couplings, $\mathcal{A}_e\mathcal{A}_q$. On the other hand the polarised forward–backward asymmetry ($A^{pol}_{FB}(q)$), defined by Eq. (8), is solely dependent on $\mathcal{A}_q$. The polarised forward–backward asymmetry of b and c quarks have been measured by SLD[35] using flavour tagging methods very similar to the ones used for the unpolarised case. Inclusive samples of $Z \to b\bar{b}$ events selected thanks to the long b lifetime provided a precise determination of $\mathcal{A}_b$ using jet-charge techniques. Semileptonic b and c decays gave a simultaneous determination of $\mathcal{A}_b, \mathcal{A}_c$ through the analysis of inclusive lepton spectra; D mesons were used to measure $\mathcal{A}_c$. Measurements were corrected for QCD effects, which are similar to the unpolarised case.

The combination of SLD results gave[29]

$$\mathcal{A}_b = 0.923 \pm 0.020,$$
$$\mathcal{A}_c = 0.670 \pm 0.027$$

with a small correlation between $\mathcal{A}_b$ and $\mathcal{A}_c$ (11%). The measurements are consistent with the ST predictions of $\mathcal{A}_b = 0.935$ and $\mathcal{A}_c = 0.668$, respectively. These predictions have a small uncertainty (≈ 0.001 for b quarks) because the quark $\mathcal{A}_q$ parameters are only weakly dependent on $\sin^2\theta^{\ell}_{\mathrm{eff}}$.

5. Measurement of the tau Polarisation in Z Decays

As mentioned in Section 2, in e^+e^- collisions the Z boson acquires a net polarisation equivalent to $\mathcal{A}_e$, which can be measured if the polarisation of the outgoing fermions is analysed. Tau leptons in $Z \to \tau^+\tau^-$ events can be used as polarimeters by measuring statistically the properties of their decay products.

The helicity of the two taus from Z decay are nearly 100% anti-correlated, except for very small $\mathcal{O}(\frac{m_f^2}{m_Z^2})$ corrections. It is convenient to measure the τ polarisation as a function of the angle (θ) between the τ^- and the electron beam. The definition of the tau polarisation for any $\cos\theta$ bin is given by

$$P_\tau = \frac{\sigma_R - \sigma_L}{\sigma_R + \sigma_L}, \tag{29}$$

where σ_R is the cross section to produce a right-handed τ^- and σ_L is the cross section to produce a left-handed τ^-. It can be shown that

$$P_\tau(\cos\theta) = \frac{A_{pol}(1+\cos^2\theta) + \frac{8}{3}A_{pol}^{FB}\cos\theta}{(1+\cos^2\theta) + \frac{8}{3}A_{FB}\cos\theta} \tag{30}$$

where A_{pol} is the total τ^- polarisation (integrated over $\cos\theta$), A_{FB} indicates the forward–backward asymmetry of the tau pairs and $A_{pol}^{FB} = \frac{\sigma_{F,R}-\sigma_{F,L}+\sigma_{B,L}-\sigma_{B,R}}{\sigma_{tot}}$ is the forward–backward polarisation asymmetry. The total polarisation and the polarisation asymmetry are related to the couplings as

$$A_{pol}^{FB} = -\tfrac{3}{4}\mathcal{A}_e, \tag{31}$$

$$A_{pol} = -\mathcal{A}_f. \tag{32}$$

The LEP experiments determined $\mathcal{A}_e$ and $\mathcal{A}_\tau$ simultaneously by measuring the $P_\tau(\cos\theta)$ distribution and fitting it to the functional dependence given by Eq. (30). This procedure gives better total error than measurements integrated over the hemispheres by giving more weight to $\cos\theta$ bins with higher sensitivity.

The polarisation of the τ is measured exploiting the parity violation of its weak decay,[36] that is mediated by a pure V-A current. Five tau decay channels, amounting to a branching ratio of about 90% have been used ($\tau \to \pi\nu$, $\tau \to \rho\nu$, $\tau \to a_1\nu$, $\tau \to e\nu\bar{\nu}$, $\tau \to \mu\nu\bar{\nu}$). Tau decays to charged kaons, having relatively low branching ratio, were included in the corresponding pion channels.

The principle of the polarisation analysis is easily understood by taking the simplest channel, $\tau \to \pi\nu$. The tau decay in this channel, for the two helicity cases, is sketched in Fig. 4. Because of the left-handedness of the neutrino, in case of decays of right-handed taus, the pion is boosted in the direction of the tau. The opposite is true for decays of left-handed taus. It follows that the energy of the pion discriminates between the two parent-tau helicity states. The tau differential decay

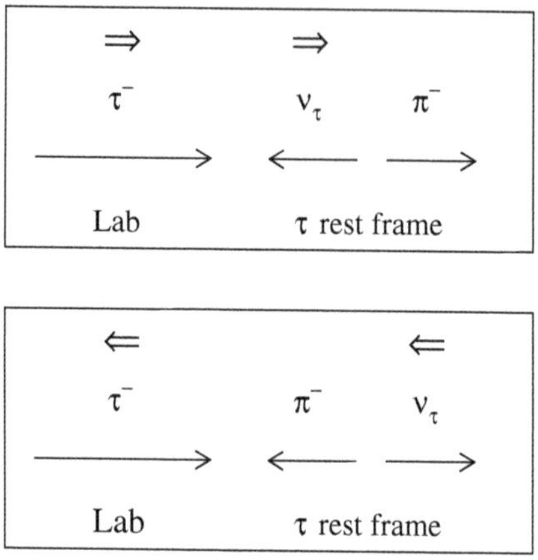

Fig. 4. The principle of the tau polarisation analysis taking the $\tau \to \pi\nu$ channel as an example.

width, given in term of the scaled pion energy $x_\pi = \frac{E_\pi}{E_{\rm beam}}$ is

$$\frac{1}{\Gamma}\frac{d\Gamma}{dx_\pi} = 1 + P_\tau(2x_\pi - 1) \tag{33}$$

as can be shown by boosting into the laboratory frame the rest-frame decay angular distribution of a spin 1/2 particle decaying into two particles of spin 1/2 and spin 0, respectively.

The measurement of the polarisation used two sets of reference decay distributions, one for $P_\tau = -1$ and one for $P_\tau = 1$, obtained applying the $\tau \to \pi\nu$ selection cuts to simulated data. The tau polarisation could be extracted by performing a binned maximum likelihood fit of the measured distributions to the sum of the corresponding simulated distributions normalized by the coefficients $\frac{N}{2}(1 + P_\tau)$ and $\frac{N}{2}(1 - P_\tau)$, where N is the number of events. Background events, mostly coming from cross-contamination from other τ decays passing the $\tau \to \pi\nu$ selection, were included in the simulated data. Similar procedures were used for the other channels.

The angular dependence of the tau polarisation measured at LEP[37–40] can be seen in Fig. 5. The LEP combined fit[29] gives

$$\mathcal{A}_{\rm e} = -\frac{4}{3}A^{FB}_{pol} = 0.1498 \pm 0.0049, \tag{34}$$

$$\mathcal{A}_\tau = -A_{pol} = 0.1439 \pm 0.0043. \tag{35}$$

The correlation between $\mathcal{A}_{\rm e}$ and $\mathcal{A}_\tau$ is small (+1.2%). To the measurements of A_{pol} and $\frac{4}{3}A^{FB}_{pol}$ a small correction is applied to take into account the difference between the centre-of-mass energy and the Z pole, the effects of the photon exchange, the $Z{-}\gamma$ interference and initial and final state radiation. The correction amounts to $\sim +0.005$ in both cases and its uncertainty ($\sim$0.0002) is small because of the

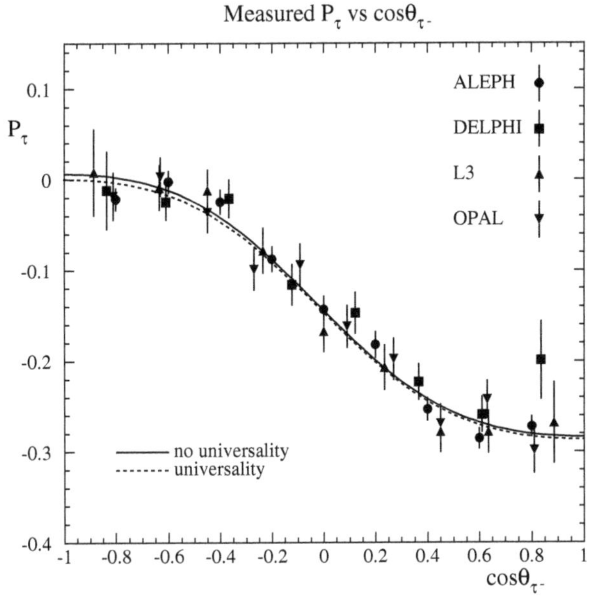

Fig. 5. Angular distribution of the tau polarisation measured by the four LEP experiments.[29] The solid and dashed lines represents the result of fits without and with the assumption of lepton universality, respectively.

precise knowledge of the beam energy at LEP. Assuming lepton universality the two measurements can be combined, giving

$$\mathcal{A}_\ell = 0.1465 \pm 0.0033.$$

The uncertainty on $\mathcal{A}_\ell$ was statistically dominated, the systematic component is equal to 0.0015. The corresponding value of the effective weak mixing angle is

$$\sin^2 \theta^\ell_{\rm eff} = 0.23159 \pm 0.00041.$$

The $\mathcal{A}_{\rm e}$ and $\mathcal{A}_\tau$ measurements from the four LEP Collaborations are shown in Fig. 6 and compared to the $\mathcal{A}_{\rm e}$ measurement of SLD.

6. Interpretations

6.1. *The determinations of* $\sin^2 \theta^\ell_{\rm eff}$

The measurements of the asymmetries presented in the previous Sections can be interpreted as a measurement of $\sin^2 \theta^\ell_{\rm eff}$. For the leptonic forward–backward asymmetries, for the measurements of $\mathcal{A}_{\rm e}$ and $\mathcal{A}_\tau$ from tau polarisation, and for the measurement of A_{LR} the interpretation requires the only assumption of lepton universality. The derivation of $\sin^2 \theta^\ell_{\rm eff}$ from hadronic measurements requires the

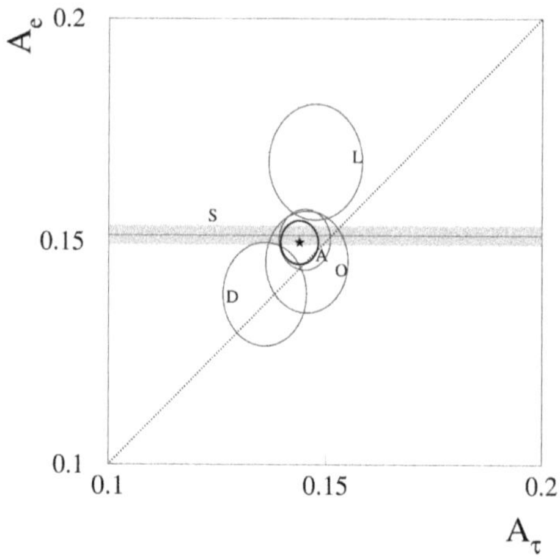

Fig. 6. Comparison of the $\mathcal{A}_e$ and $\mathcal{A}_\tau$ measurements by the ALEPH (A), DELPHI (D), L3 (L) and OPAL (O) experiments. The elipses give the standard error countours (corresponding to 39% CL for a two-dimension gaussian). The combination of the four experiments is represented by a star (central value) and a thicker elipse. The horizontal band indicates the $\mathcal{A}_e$ measurement by SLD (S) from the left–right asymmetry; the allowed range is given by plus and minus one standard deviation with respect to the central value. (Courtesy of André Rougé.)

knowledge of the $\mathcal{A}_q$ terms that, as already discussed, have only a mild dependence on $\sin^2\theta^\ell_{\text{eff}}$ in the Standard Theory (ST). For this class of measurements the validity of the ST for the $\mathcal{A}_q$ terms is assumed; this assumption is corroborated by the direct measurements of $\mathcal{A}_b, \mathcal{A}_c$ using polarised beams, which agree with the ST.

A compilation of the various results is shown in Fig. 7, where the dependence of $\sin^2\theta^\ell_{\text{eff}}$ on the Higgs boson mass is also indicated. The six results shown in the figure are obtained, respectively, from the lepton forward–backward asymmetry, the tau polarisation, the left–right asymmetry, the b forward–backward asymmetry, the c forward–backward asymmetry and measurements of charge asymmetry using all quark flavours. The average of the six measurements gives:

$$\sin^2\theta^\ell_{\text{eff}} = 0.23153 \pm 0.00016$$

with a χ^2 of 11.8 for five degrees of freedom corresponding to a confidence level of 3.7%. This confidence level is relatively low, because the most precise determinations, based on A_{LR} and on the b asymmetry are about 3σ apart. From the experimental point of view A_{LR} and the combination of b asymmetry measurements are both dominated by statistical errors, with accurate studies of the much lower systematic uncertainties. In particular the b asymmetry was measured by the four LEP experiments with two very different methods (Section 3.2) yielding largely uncorrelated results; the χ^2/dof of the eight-measurement combination was 0.55 showing agreement among experiments and methods. On the other hand a

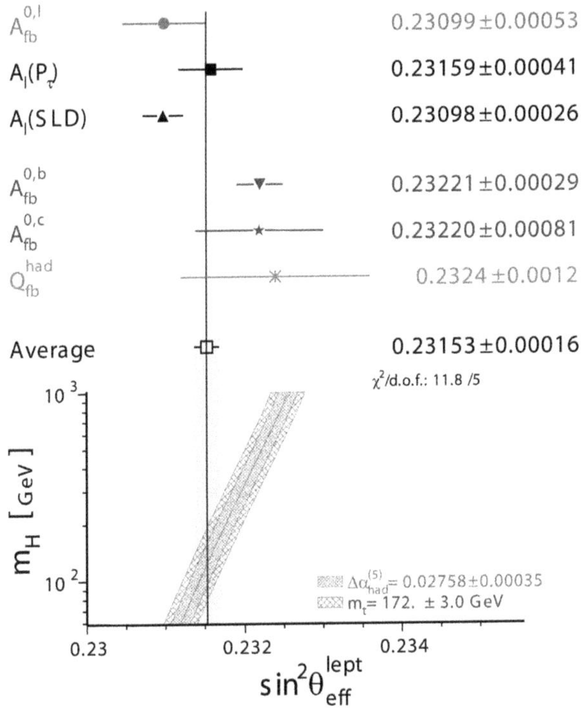

Fig. 7. The determinations of $\sin^2\theta^\ell_{\rm eff}$ from the measurements described in this chapter and their average.[29] The measurements are, starting from the top, the lepton forward–backward asymmetry, the tau polarisation, the left–right asymmetry, the b forward–backward asymmetry, the c forward–backward asymmetry and the the jet charge asymmetry using all quark flavours. The results are compared to the ST prediction, as a function of the Higgs boson mass. The uncertainty due to $\alpha(m_Z^2)$ on the ST predictions is indicate by a band. The effect of varying the top mass within the range indicated in the figure is added as two extra side bands.

departure of the b couplings from their ST expectation seems to be excluded by the precise measurements of $\mathcal{A}_b$ and R_b. Therefore in the $\sin^2\theta^\ell_{\rm eff}$ average and in the subsequent extraction of the couplings this discrepancy is assumed to be related to a statistical fluctuation.

Recently new determinations of $\sin^2\theta^\ell_{\rm eff}$ have been published by experiments at hadron colliders, Tevatron and LHC, based on lepton-pair production in $p\bar{p}/pp$ collisions (Drell–Yan events).[41] The electroweak mixing angle is determined by the forward–backward asymmetry in the centre-of-mass, approximated by the Collins-Soper frame,[42] of dileptons with invariant mass close to the Z mass. The measurement requires an assumption on the direction of the incoming quark/antiquark, therefore it depends on the knowledge of the parton distribution functions (PDFs). Precisions comparable to the less precise measurements in Fig. 7 have been obtained.

6.2. *Extraction of neutral current couplings*

The couplings of the neutral current to leptons ($\ell = e, \mu, \tau$) can be determined using three ingredients:

- the Z leptonic partial widths,[11–14]
- the $\mathcal{A}_\ell$ parameters as determined by the leptonic asymmetries,
- the energy dependence of the leptonic forward–backward asymmetries.

The partial width of the decay $Z \to \ell^+\ell^-$, gives the sum of the squares of the couplings, following the relationship

$$\Gamma_{Z\to\ell^+\ell^-} = \frac{\sqrt{2}G_F m_Z^3}{12\pi}[g_{V\ell}^2 + g_{A\ell}^2], \tag{36}$$

while the ratio of the vector and axial couplings is given by the leptonic measurements of $\mathcal{A}_\ell$ (Eq. (4)), i.e. by the measurement of A_{LR}, of the tau polarisation and of the leptonic forward–backward asymmetries. The energy dependence of the asymmetries (Eq. (12)) fixes the value of the axial couplings, up to a common sign. This last ingredient is required, since the widths and asymmetries do not change if $g_{V\ell}$ and $g_{A\ell}$ replace each other, as can be seen from Eqs. (36) and (4).

The measured vector and axial couplings to electron, muon and tau are compared in Fig. 8 to test the hypothesis of lepton universality. The measurements are in agreement and lepton universality is tested to less than 0.1% for axial couplings

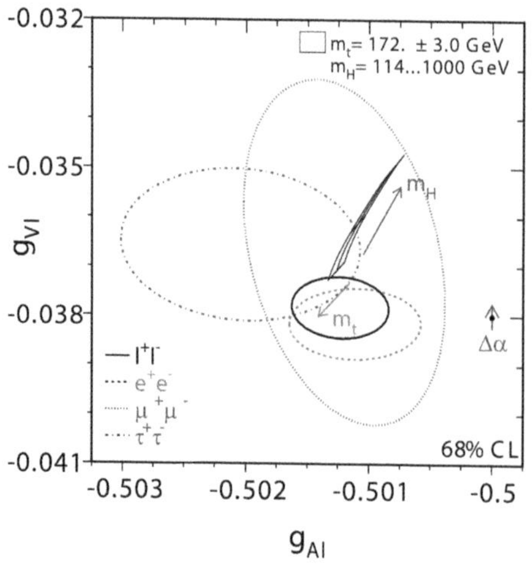

Fig. 8. The vector and axial couplings of the neutral current to electrons, muons and taus.[29] The 68% CL allowed regions are shown with dashed, dotted and dash-dotted lines, respectively. The combination of the measurements from the three lepton species, assuming lepton universality, is also shown (full line). The shaded area shows the ST prediction, within the allowed values for the top and Higgs boson masses. The uncertainty on $\alpha(m_Z)$ is indicated by the small arrow.

and to a few percent for the smaller vector couplings:

$$\frac{g_{V\mu}}{g_{Ve}} = 0.961 \pm 0.063 \quad \frac{g_{A\mu}}{g_{Ae}} = 1.0002 \pm 0.00064$$
$$\frac{g_{V\tau}}{g_{Ve}} = 0.958 \pm 0.030 \quad \frac{g_{A\tau}}{g_{Ae}} = 1.0019 \pm 0.00073.$$

The b and c quark couplings can be extracted with the same procedure adopted for the lepton case, by using the measurements of R_b[43] and R_c,[44] the values of $\mathcal{A}_b$ and $\mathcal{A}_c$ determined by the polarised heavy quark asymmetries, and the energy dependence of the b and c forward–backward asymmetries. With this method the axial (vector) b couplings can be tested to a precision of approximately 2% (3%). Similar precisions are obtained with the tests of the c couplings (the bounds in this case are somewhat weaker mainly because of the larger uncertainty on the measured value of R_c). All couplings are found to agree with the ST.

The $\mathcal{A}_q$ parameters for b and c quarks can also be evaluated from the unpolarised b and c asymmetries using Eq. (2) and the value of $\mathcal{A}_e$ derived from $A^0_{FB}(\ell)$, from the tau polarisation and from A^0_{LR}. This interpretation of the heavy quark unpolarised asymmetries is bound to lead, however, to a rather low value of $\mathcal{A}_b$ (0.881 ± 0.017, compared to the ST expectation of 0.935) because of the 3σ discrepancy between $A^0_{FB}(b)$ and A^0_{LR} already mentioned in the discussion concerning the determination of $\sin^2 \theta^{\ell}_{\rm eff}$. As a consequence a rather high (low) value of the axial (vector) b couplings is obtained and the agreement with the ST is marginal for both couplings.[29] It must be stressed, however, that this discrepancy is totally correlated with the one seen in Fig. 7.

The measurements of the vector and axial couplings for various fermion species are depicted in Fig. 9. The regions allowed by the experimental measurements at 68% CL are shown. The precision obtained for the lepton measurements is

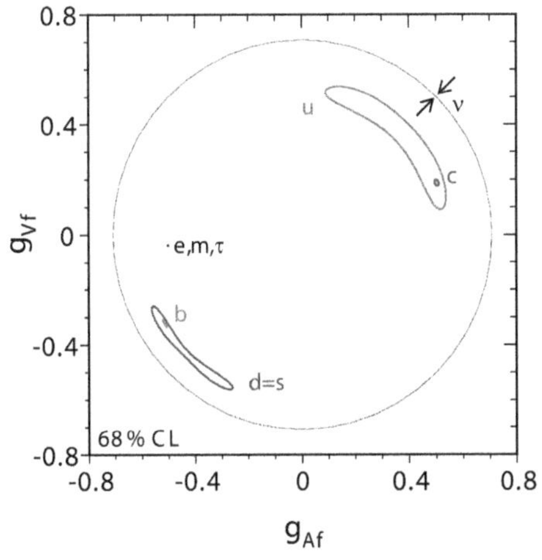

Fig. 9. The vector and axial couplings of the neutral current to various fermion species.[29] The regions allowed by the experimental measurements (68% CL) are shown.

impressive: with the scale used in this figure the three measurements are represented by three superimposed dots. Considerable precision is obtained also for the heavy quark couplings. Constraints are obtained on the couplings of lighter quarks by measurements of forward–backward asymmetries using kaons[45] and high-momentum stable particles.[46] As the large uncertainties of these measurements do not allow the study of the energy dependence, the contours indicating the allowed regions are symmetric with respect to the line $g_{Vf} = g_{Af}$. The constraints on neutrino couplings are computed from the measurement of the Z invisible width,[11–14] assuming three neutrino families with identical neutral couplings. In this case the experimentally allowed region is represented by a very thin ring.

7. Summary and Outlook

In the years 1989–1998 the precision on the Weinberg electroweak mixing angle from lepton and quark asymmetries reached an impressive relative uncertainty of 0.07%. Such a precision is not easily attainable at hadron colliders, because it requires an accurate knowledge of the initial state. Eventually it could be reached at LHC depending on the progress in understanding parton distribution functions.

The measurements of lepton neutral couplings reminded in these pages improved by two order of magnitudes the tests of neutral-current lepton-universality available before the start of LEP and SLC, based on νe and $\nu\mu$ scattering. The values of $g_{V\ell}$ and $g_{A\ell}$ compared with ST predictions (Fig. 9) clearly indicated a low mass for the Higgs boson, as confirmed in 2012 by the direct observation at LHC.[47,48]

Quark neutral couplings were also measured in the same decade. Heavy quark couplings (c and b) were determined at the level of a few percent, however the ratio of vector to axial b couplings, when extracted by comparing the measurement of the b forward–backward asymmetry to leptonic asymmetries measurements, shows a deviation of about 3σ with respect to the ST expectation. Whether this discrepancy can be ascribed to a statistical fluctuation, or to other effects, will be determined by future experiments.

References

1. ALEPH Collaboration, *Nucl. Instrum. Methods A* **294**, 121 (1990).
2. DELPHI Collaboration, *Nucl. Instrum. Methods A* **303**, 233 (1991).
3. L3 Collaboration, *Nucl. Instrum. Methods A* **289**, 35 (1990).
4. OPAL Collaboration, *Nucl. Instrum. Methods A* **305**, 275 (1991).
5. SLD Design Report, SLAC-273 (1984).
6. W. Hollik, *Fortschr. Phys.* **38**, 165 (1990).
7. R. Tenchini and C. Verzegnassi, *The Physics of the Z and W Bosons* (World Scientific, 2008), p. 161.
8. W. Hollik, "Electroweak Theory", hep-ph/9602380.
9. The LEP Energy Working Group, The Aleph, Delphi, L3 and Opal Collaborations, *Phys. Lett. B* **307**, 187 (1993).

10. W. Beenakker, F. A. Berends, S. C. Van Der Marck, *Nucl. Phys. B* **349**, 323 (1991).
11. ALEPH Collaboration, *Eur. Phys. J. C* **14**, 1 (2000).
12. DELPHI Collaboration, *Eur. Phys. J. C* **16**, 371 (2000).
13. L3 Collaboration, *Eur. Phys. J. C* **16**, 1 (2000).
14. OPAL Collaboration, *Eur. Phys. J. C* **19**, 587 (2001).
15. ALEPH Collaboration, *Eur. Phys. J. C* **24**, 177 (2002).
16. DELPHI Collaboration, *Eur. Phys. J. C* **34**, 109 (2004).
17. L3 Collaboration, *Phys. Lett. B* **448**, 152 (1999).
18. OPAL Collaboration, *Phys. Lett. B* **577**, 18 (2003).
19. ALEPH Collaboration, *Eur. Phys. J. C* **22**, 201 (2001).
20. DELPHI Collaboration, *Eur. Phys. J. C* **40**, 1 (2005).
21. L3 Collaboration, *Phys. Lett. B* **439**, 225 (1998).
22. OPAL Collaboration, *Phys. Lett. B* **546**, 29 (2002).
23. ALEPH Collaboration, *Phys. Lett. B* **434**, 415 (1998).
24. DELPHI Collaboration, *Eur. Phys. J.* **C10** 219 (1999).
25. OPAL Collaboration, *Z. Phys. C* **73**, 379 (1997).
26. ALEPH, DELPHI, L3, OPAL Collaborations, *Nucl. Instr. Meth. A* **378** 101 (1996).
27. G. Altarelli and B. Lampe, *Nucl. Phys. B* **391**, 3 (1993); V. Ravidran and W. L. van Neerven, *Phys. Lett. B* **445**, 214 (1998); S. Catani and M. Seymour, *JHEP* **9907**, 023 (1999).
28. D. Abbaneo *et al.*, *Eur. Phys. J. C* **4**, 185 (1998).
29. The ALEPH, DELPHI, L3, OPAL, SLD Collaborations, the LEP Electroweak Working Group, the SLD Electroweak and Heavy Flavour Groups, Physics Reports **427**, 257 (2006).
30. N. Phinney, *Int. J. Mod. Phys. A* **45**, 2 (1993); M. Woods, in *Proceedings of the 12th International Symposium on High Energy Spin Physics* eds. C. W. Jager *et al.* (World Scientific, Signapore, 1997), p. 621.
31. A. Blondel, *Phys. Lett. B* **202**, 145 (1988).
32. F. W. Lipps and H. A. Tolhoek, *Physica* **20**, 85 (1954).
33. B. Schumm, "Electroweak results from SLD", SLAC-PUB-7697.
34. SLD Collaboration, *Phys. Rev. Lett.* **84**, 5945 (2000).
35. SLD Collaboration, *Phys. Rev. Lett.* **88**, 151801 (2002); SLD Collaboration, *Phys. Rev. Lett.* **90**, 141804, (2003).
36. Y. S. Tsai, *Phys. Rev. D* **4**, 2821 (1971).
37. ALEPH Collaboration, *Eur. Phys. J. C* **20**, 401 (2001).
38. DELPHI Collaboration, *Eur. Phys. J. C* **14**, 585 (2000).
39. L3 Collaboration, *Phys. Lett. B* **429**, 387 (1998).
40. OPAL Collaboration, *Eur. Phys. J. C* **21**, 1 (2001).
41. CDF Collaboration, *Phys. Rev. D* **89**, 072005 (2014); D0 Collaboration, *Phys. Rev. Lett.* **115**, 041801 (2015); ATLAS Collaboration, *JHEP* **09**, 049 (2015); CMS Collaboration, *Phys. Rev. D* **84**, 112002 (2011).
42. J. C. Collins and D. E. Soper, *Phys. Rev. D* **16**, 2219 (1977).
43. ALEPH Collaboration, *Phys. Lett. B* **401**, 150 (1997); ALEPH Collaboration, *Phys. Lett. B* **401**, 163 (1997); DELPHI Collaboration, *Eur. Phys. J. C* **10**, 415 (1999); L3 Collaboration, *Eur. Phys. J. C* **13**, 47 (2000); OPAL Collaboration, *Eur. Phys. J. C* **8**, 217 (1999); SLD Collaboration, *Phys. Rev. D* **71**, 112004 (2005).
44. ALEPH Collaboration, *Eur. Phys. J. C* **4**, 557 (1998); ALEPH Collaboration, *Eur. Phys. J. C* **16**, 597 (2000); DELPHI Collaboration, *Eur. Phys. J. C* **12**, 225 (2000); OPAL Collaboration, *Eur. Phys. J. C* **1**, 439 (1998); OPAL Collaboration, *Z. Phys. C* **72**, 1 (1996); SLD Collaboration, hep-ex/0503005, (2005).

45. DELPHI Collaboration, *Eur. Phys. J. C* **14**, 613 (2000); SLD Collaboration, *Phys. Rev. Lett.* **85**, 5059 (2000).
46. OPAL Collaboration, *Z. Phys. C* **76**, 387 (1997).
47. ATLAS Collaboration, *Phys. Lett. B* **716**, 1 (2012).
48. CMS Collaboration, *Phys. Lett. B* **716**, 30 (2012).

Chapter 10

The *W* Boson Mass Measurement

Ashutosh V. Kotwal
Department of Physics, Duke University,
Durham, NC 27708, USA
ashutosh.kotwal@duke.edu

The measurement of the W boson mass has been growing in importance as its precision has improved, along with the precision of other electroweak observables and the top quark mass. Over the last decade, the measurement of the W boson mass has been led at hadron colliders. Combined with the precise measurement of the top quark mass at hadron colliders, the W boson mass helped to pin down the mass of the Standard Model Higgs boson through its induced radiative correction on the W boson mass. With the discovery of the Higgs boson and the measurement of its mass, the electroweak sector of the Standard Model is over-constrained. Increasing the precision of the W boson mass probes new physics at the TeV-scale. We summarize an extensive Tevatron (1984–2011) program to measure the W boson mass at the CDF and DØ experiments. We highlight the recent Tevatron measurements and prospects for the final Tevatron measurements.

1. Introduction

A series of fixed target and collider experiments have motivated and confirmed both the matter content of the Standard Theory (ST) in terms of fermion multiplets[1] as well as the gauge transformations under the $SU(3)_c \times SU(2)_L \times U(1)_Y$ gauge group.[1] In the area of electroweak-symmetry breaking, the predicted Higgs mechanism[3] has been spectacularly confirmed recently by the observation of the Higgs boson at the LHC.[2] In weak interaction physics, the direct measurements of the W boson mass and width have steadily improved in precision. Measurements of the effective Weinberg angle $\sin^2 \theta^\ell_{\rm eff}$ at hadron colliders are approaching the precision formerly achieved at LEP and SLD.

The direct measurement of the Higgs boson mass[2] has provided the last missing parameter defining the electroweak sector in the ST. As a result, M_W and $\sin^2 \theta^\ell_{\rm eff}$ can now be predicted at loop-level in terms of other known quantities in the ST, as well as in extensions of the ST.[4] Therefore, the ST and its extensions can be stringently tested by precise measurements of M_W and $\sin^2\theta^{\ell}_{\text{eff}}$. New physics can cause observable departures from their Standard Theory values.

2. History of the W Mass Measurement

After the W and Z bosons were discovered[5] by the UA1 and UA2 experiments at the $Sp\bar{p}S$ at CERN, these experiments made the initial measurements of their mass. The CDF experiment used the Tevatron Run 0 data, and both the CDF and DØ experiments used the Tevatron Run 1 data[6,7] to perform more precise measurements of the W boson mass. The electron-positron collider LEP II increased its energy above the Z-boson pole and started producing W boson pairs concurrently with the Tevatron Run 1. The scan of cross section as a function of collider center-of-mass energy yielded the first LEP II measurements of M_W at threshold. With further increase in collider energy, higher statistics and precision were obtained when the semi-leptonic and all-hadronic decay channels of the WW system were reconstructed for the M_W measurement.

The CDF and DØ measurements of M_W from the Tevatron Run 1 were combined to yield[6–8]

$$M_W = 80454 \pm 59\,\text{MeV}. \tag{1}$$

The final combined result[9] from ALEPH,[10] DELPHI,[11] L3[12] and OPAL[13] experiments at LEP II was

$$M_W = 80376 \pm 33\,\text{MeV}. \tag{2}$$

A recent review[14] of the Tevatron Physics program provides a summary of other electroweak physics measurements.

3. Theoretical Considerations of M_W

The loop-level expression for the predicted M_W in terms of other known quantities is[15]

$$M_W^2\left(1 - \frac{M_W^2}{M_Z^2}\right) = \frac{\pi\alpha}{\sqrt{2}G_F}\frac{1}{1-\Delta r}, \tag{3}$$

where Δr represents the loop-induced radiative corrections. The tree-level relation in the ST is obtained by setting $\Delta r = 0$. α is the electromagnetic coupling and G_F is the Fermi constant extracted from the muon decay lifetime. The coupling of the W and Z bosons to the Higgs field's vacuum expectation value v determines their tree-level masses. In the ST, the contributions to Δr are dominated by (i) the running of the electromagnetic coupling due to light-quark loops, (ii) the contribution from the top (t) and bottom (b) quark loop in the W boson propagator, and (iii) the loops in the W boson propagator involving Higgs bosons. The top and bottom quarks have large Yukawa couplings to the Higgs field, and the difference between their couplings is also large. This difference severely breaks the "custodial" $SU(2)$ symmetry which connects the tree-level relationship between the W and Z boson masses to the weak mixing angle. As a result, the $t\bar{b}$ loop makes an additional contribution to splitting between the W and Z boson masses. Furthermore, the mixing between the T_3 generator of $SU(2)_L$ and the $U(1)_Y$ generator caused by $\sin^2\theta^\ell_{\rm eff} \neq 0$ causes a difference between the WWh and ZZh couplings. Because of this difference, Higgs boson loops also contribute to the splitting between the W and Z boson masses. If a smaller value of $\sin^2\theta^\ell_{\rm eff}$ had occurred in nature, an interesting consequence would have been that the M_W measurement would be less sensitive to the Higgs boson mass m_H.

New physics that contributes to the precision electroweak observables primarily through loop corrections to the gauge-boson self-energies, i.e. through propagator corrections, can be generally described by the S, T, U *oblique* parameters.[16] These parameters relate to the gauge-boson self-energies $\Pi(Q^2)_{VV'}$ (where Q^2 is the renormalization scale) as follows: the slope $(\Pi'_{VV'})$ of $\Pi_{VV'}$ with respect to Q^2 is given by S, the difference of $\Pi(0)_{WW}$ and $\Pi(0)_{ZZ}$ is given by T, and the difference of slopes Π'_{WW} and Π'_{ZZ} is given by U. Propagator effects of new physics that violate the custodial $SU(2)$ symmetry are captured by T and U. The intercept and/or the slope of Π are more easilly affected by new physics contributions than a *difference* in the slopes for the W and Z boson propagators. Since new physics contributions to U tend to be of higher order than contributions to S and T, it is common to work in the $U = 0$ approximation for simplicity.

These oblique parameters are defined to be zero in the ST. The radiative corrections to M_W and $\sin^2\theta^\ell_{\rm eff}$ due to new physics can be written as

$$\Delta r \approx \Delta r^{SM} + \frac{\alpha}{2s_W^2}S + \frac{\alpha c_W^2}{s_W^2}T + \frac{s_W^2 - c_W^2}{4s_W^4}U,$$

$$\Delta\sin^2\theta^\ell_{\rm eff} = \Delta\sin^2\theta^{\ell,SM}_{\rm eff} + \frac{\alpha}{4(c_W^2 - s_W^2)}S + \frac{\alpha s_W^2 c_W^2}{c_W^2 - s_W^2}T. \tag{4}$$

Measurements of Δr and $\Delta\sin^2\theta^\ell_{\rm eff}$ impose a two-dimensional constraint on new physics in the ST plane, because the coefficients of S and T are different in Eqn. 4. Constraints in the ST plane from the data are shown in Fig. 1. Improved electroweak measurements can guide the search for new physics and complement direct searches, as illustrated by the ST variation from two models of new physics shown in Figs. 1 and 2 respectively.

4. Tevatron M_W Measurements from Run 2

The CDF and DØ experiments at Fermilab's Tevatron $p\bar{p}$ collider have produced four measurements of M_W from the Run 2 ($\sqrt{s}$ = 1.96 TeV) data. The two

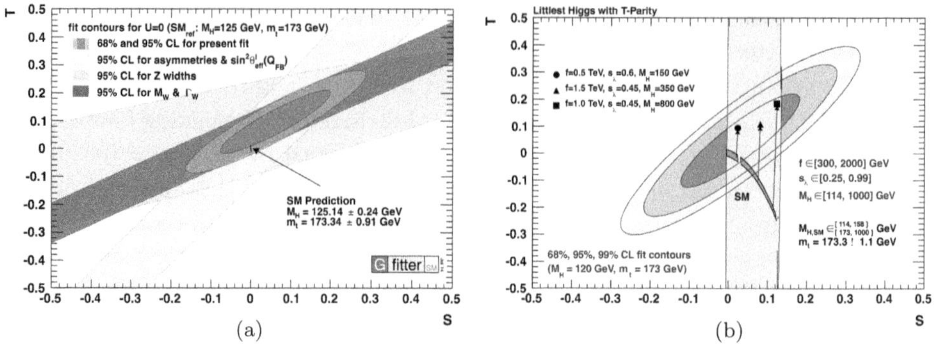

Fig. 1. (a) Illustration of the constrained region of ST parameter space from measurements, reproduced with permission from Ref. 17, and (b) illustration of the constrained region of ST parameter space from measurements, compared to a range of predictions from Littlest Higgs models. Figure (b) is reproduced with permission from Ref. 18.

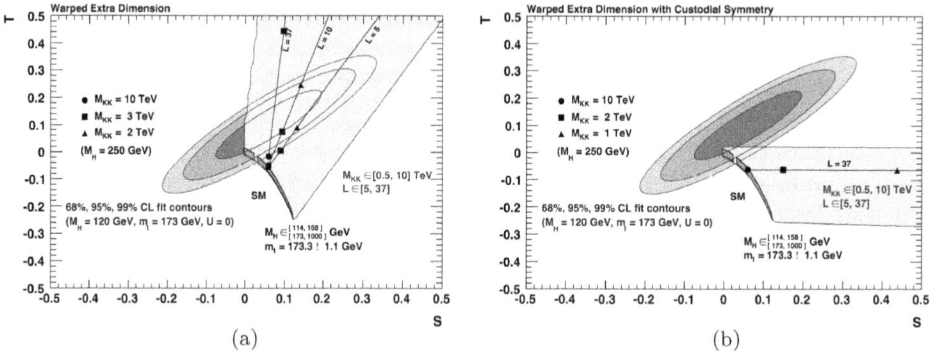

Fig. 2. Illustration of the constrained region of ST parameter space from measurements, compared to a range of predictions from warped extra-dimensional models without (a) and with (b) an additional custodial symmetry introduced. Figures reproduced with permission from Ref. 18.

measurements from CDF,[19,20] using 200 pb^{-1} and 2.2 fb^{-1}, respectively, of integrated luminosity, are $M_W = 80413 \pm 48$ MeV and $M_W = 80387 \pm 19$ MeV. The second measurement subsumed the former result as it was inclusive of the data used for the first measurement. The two measurements[21,22] from the DØ experiment used consecutive, independent datasets corresponding to 1 fb^{-1} and 4.3 fb^{-1}, respectively, of integrated luminosity to measure $M_W = 80401 \pm 43$ MeV and $M_W = 80367 \pm 26$ MeV. The combined Tevatron result[23] is

$$M_W = 80387 \pm 16\,\text{MeV}. \tag{5}$$

Hadron colliders are now the source of the most precise measurements of M_W, having significantly surpassed the precision achieved by LEP II.

5. Techniques for M_W Measurement at Hadron Colliders

The key elements of the experimental technique to measure M_W have not changed over the last two decades.[24] During this time, the simulation of W boson production and decay, the detector response and resolution, and the detector calibrations have become increasingly more accurate. Subtle effects have steadily been understood and included in the calibration of the data or in the simulation.

Inclusive production of W bosons, describable by quark-antiquark annihilation at leading order (LO) in QCD, is by far the largest source of W bosons at the Tevatron. This sample provides the basis for the M_W measurement. The two-body decay of W bosons to quark–antiquark pairs, which have the largest branching ratio, is fully reconstructible as an invariant mass peak. However, the hadronic decay of the W boson is not usable for the mass measurement for two reasons. Firstly, the systematic uncertainties on jet energy calibration are simply too large to allow competitive measurements of M_W. Secondly, the QCD dijet background is very large and the dijet mass peak from W boson decay sits above an enormous QCD background. It is difficult to trigger on this mode with high acceptance due to the background rates. Hence the use of the hadronic decay mode for the M_W measurement is practically impossible. However, electron and muon momenta can be precisely measured and accurately calibrated, and these leptonic decays can be triggered and identified with high efficiency and small backgrounds. Thus, the M_W measurement has always been performed with the leptonic decay modes.

The invariant mass cannot be reconstructed due to the undetectable neutrino in the two-body leptonic decay of the W boson. Many of the complications and systematic uncertainties associated with the M_W measurement are related to the presence of the neutrino. The observable momenta in these events are the 3-momentum of the lepton and the transverse momentum (p_T) of the hadronic "recoil" particles which balance the p_T of the boson. The longitudinal component of the momentum of these recoiling particles is not discernable because of the unknown energy-momentum flowing down the beampipe. Compared to electon-positron colliders where the final-state 4-momentum is fully reconstructible, it is a disadvantage at hadron colliders that the most of the energy associated with the interactions of the spectator partons is flowing at very small angles to the beam, outside the detector acceptance.

Hadron collider experiments exploit the characteristic feature called Jacobian edge in the p_T distribution of the charged lepton. This feature is present in any two-body decay, where the p_T distribution rises up to $p_T \sim M_W/2$ and falls rapidly past this value. The Jacobian edge can be understood in the rest frame of the W boson, where in the limit of zero intrinsic width, a rest-frame decay angle of 90° with respect to the beam axis leads to the largest possible p_T of the leptons. Since the edge occurs at half the mass of the decaying particle, its location provides sensitivity to the W boson mass.

The lepton p_T distribution is affected by the angular distribution of the boson decay in its rest frame, and by the p_T-boost of the boson. The effect of $p_T(W)$ has been taken into account using two approaches. In one approach, the lepton p_T distribution is fitted using simulated templates and the simulation includes a theoretical calculation of the boson kinematics, including the boson p_T spectrum. The theoretical model is constrained using the $p_T(Z)$ measurement where the $Z \to \ell\ell$ kinematics can be measured well. Thus, in this approach the theoretical model is only used to relate the p_T spectra of the W and Z bosons.

In the second approach, $\vec{p}_T(W)$ is measured in each event and used in conjunction with $\vec{p}_T(\ell)$. Typically, the hadronic activity recoiling against the W boson is small and diffuse without the presence of hard jets. The quantity $\vec{u}_T$ is defined as the inclusive vector sum of transverse energies over all calorimeter towers (excluding towers containing energy deposits from the charged lepton), and $\vec{p}_T(W) \equiv -\vec{u}_T$. The $\vec{u}_T$ measurement is biased by the non-linear response and resolution of the detector, as well as by underlying event energy (from spectator parton interactions) and additional $p\bar{p}$ collisions. Since $\vec{u}_T$ cannot be reliably corrected for these detector effects, they have to be carefully included in the simulation. The neutrino p_T vector $\vec{p}_T(\nu) \equiv -\vec{p}_T(\ell) - \vec{u}_T$ is inferred by imposing transverse momentum balance.

A quantity that combines the information in $\vec{p}_T(\ell)$ and $\vec{u}_T$ is the transverse mass m_T, which is analogous to the invariant mass but computed using only the transverse quantities; $m_T = \sqrt{2p_T^\ell p_T^\nu(1 - \cos\Delta\phi)}$, where $\Delta\phi$ is the azimuthal opening angle between the two leptons. The Jacobian edge is also present in the m_T distribution. In fact, while the Jacobian edge in the $p_T(\ell)$ distribution is smeared by $p_T(W)$, the edge in the m_T distribution is not affected to first order in $p_T(W)/M_W$ because the measurement of $p_T(W)$ is incorporated into m_T. The disadvantage of this approach is the systematic uncertainty associated with the $\vec{u}_T$ measurement. The distributions of m_T, $p_T(\ell)$ and $p_T(\nu)$ are all used to extract (correlated) measurements of M_W, providing cross-checks since they have different systematic uncertainties.

5.1. *Lepton momentum and energy calibration*

The precision on the charged lepton's energy/momentum calibration is the single most important aspect of the M_W measurement. The electron energy is measured using the uranium-liquid argon (U-LAr) sampling calorimeter of the DØ experiment,[22] and its direction is obtained from the scintillator fiber tracker. The track momentum resolution was considered to be inadequate for this measurement and therefore the muon channel was not used.

The absolute energy scale is set by calibrating the measured $Z \to ee$ boson mass to the world-average value.[25] In order to extrapolate the calibration from the Z boson mass to the W boson mass, non-linearity in the response has to be accounted for. One of the issues in the calibration of the energy response is understanding the sources of non-proportional response. The energy lost by the electron in the passive material upstream of the calorimeter is not proportional to its energy, hence it causes the calorimeter response to be non-proportional to the original electron's energy. This has been studied in detail in the DØ analysis using the longitudinal segmentation of the calorimeter. The U-LAr electromagnetic calorimeter provides readout in four longitudinal segments, with the first two samples corresponding to ≈ 2 radiation lengths (X_0) each, and the third (fourth) sample corresponding to 7 (10) X_0, all at normal incidence. The amount of material traversed upstream of the calorimeter affects the electron shower development, which is reflected in the fractional energy deposition in the first three longitudinal samples of the electromagnetic calorimeter. These longitudinal energy fractions are measured in $Z \to ee$ data (shown in Fig. 3) and reproduced in the GEANT-based[26] detector simulation by incorporating an additional passive layer. These studies are performed in bins of electron pseudo-rapidity, and cross-checked with the $W \to e\nu$ data. Non-linear effects are further constrained by studying the variation of the measured Z boson mass with electron energy. The energy response for electrons is characterized by a scale factor α and an offset β, with results shown in Fig. 3.

The dependence of the calibration on pseudo-rapidity and instantaneous luminosity is also studied carefully. The electron energy response depends on the instantaneous luminosity for two reasons. Firstly, it affects the underlying event energy deposited in the electron cone. Secondly, the particle flux through the calorimeter depends on the instantaneous luminosity, which in turn affects the average current through the signal boards and the high-voltage lost across them. The high voltage affects the response in the liquid Argon gaps. The electron energy resolution is also studied carefully and its dependence on pseudo-rapidity and other factors is simulated.

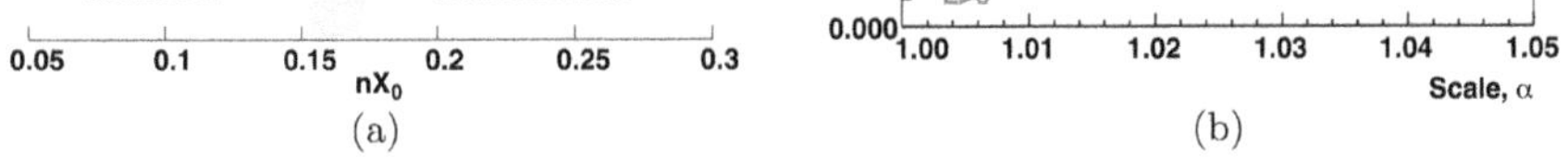

(a) (b)

Fig. 3. The amount of uninstrumented material in front of the EM calorimeter of the DØ experiment, in units of radiation lengths of copper (a), and the fit results for the electron energy scale α and offset β (b). Both results are extracted from fits to the DØ data. Different bins of instantaneous luminosity are shown as the different ellipses in (b) with consistent results. Instantaneous luminosity is shown in units of 36×10^{30} cm^{-2} s^{-1}. Figures reproduced with permission from Ref. 22.

The strategy[19,27] for the lepton momentum calibration on the CDF experiment is to understand the central drift chamber[28] and the solenoid magnetic field[29] from first principles. One advantage of this strategy is that both electron and muon tracks can be used, roughly doubling the available statistics. More importantly, a number of additional systematic cross-checks become available to verify the robustness of the result. The CDF electromagnetic (EM) sampling calorimeter[30,31] uses lead absorber and plastic scintillator with relatively coarse transverse granularity compared to DØ, and with no longitudinal segmentation. There are $\approx 19\%$ radiation lengths of material upstream of the drift chamber, in which bremsstrahlung radiation occurs. Since bremsstrahlung photons are coalesced with the electron shower in the calorimeter, the energy resolution of the calorimeter cluster is better than the track-based measurement. The W and Z boson kinematics in the electron channel are reconstructed using the calorimeter cluster energy and the direction of the track. The tracker momentum calibration is transferred to the EM calorimeter by fitting the peak of $E_{\mathrm{cal}}/p_{\mathrm{track}}$ near unity. Electrons from $W \to e\nu$ and $Z \to ee$ decays are used for this purpose.

The CDF strategy provides the opportunity to make independent measurements of the $Z \to ee$ and $Z \to \mu\mu$ masses based on the calibrations of the tracker and the EM calorimeter. Comparing these Z boson mass measurements to the world-average value provides confirmation of the calibration strategy. The Z boson mass measurements are subsequently used as additional calibration points in order to exploit the full power of the data. Though not competitive with the LEP measurements, the CDF measurements of the Z boson mass are the most precise at hadron colliders.

The first step in the calibration of the tracker is to derive precise wire-by-wire alignment constants for the drift chamber (which has $\approx 30,000$ wires). This is done using cosmic ray tracks recorded *in-situ* with collider data. CDF developed the technique to fit both sides of the cosmic ray trajectory to a single helix by creating a special reconstruction algorithm[32] for this purpose. This two-sided helix fit brings in unique constraints which are not available from collider tracks originating at the beamline. Due to these constraints, information[33] on internal deformations of the drift chamber (relative rotations of radial layers, and relative twists of the cylinder end plates) can be extracted from the hit residuals with respect to this fit.

Additional information on the gravitational and electrostatic deflections of the wires between the end plates is obtained by comparing the track parameters of the diametrically opposite segments of the same cosmic ray track, because in the absence of any biases these parameters should match. The goal is to minimize the biases in curvature and polar angle measurements and provide a highly linear tracker response in curvature by studying these effects in detail. After using these alignment constants for track reconstruction, the $\langle E_{\mathrm{cal}}/p_{\mathrm{track}} \rangle$ are compared between positrons and electrons and final tweaks to track parameters are applied to equalize them.

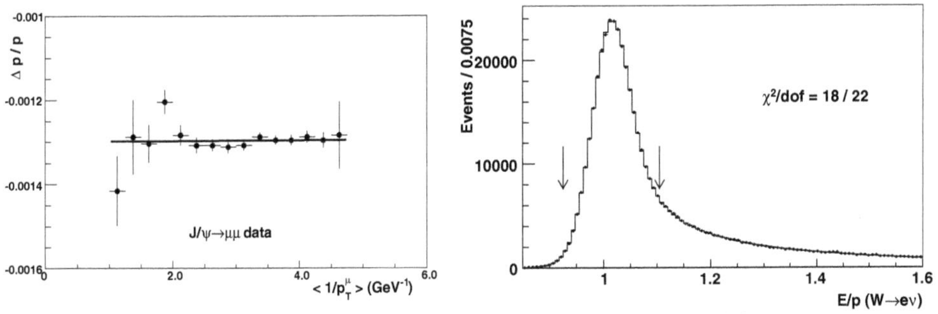

(a) (b)

Fig. 4. The fractional momentum scale correction extracted from $J/\psi \to \mu\mu$ data on CDF, shown as a function of the average p_T^{-1} of the muons (a), and the distribution of $E_{\text{cal}}/p_{\text{track}}$ used to calibrate the EM calorimeter on CDF, overlaid with the best-fit simulation (b). Figures reproduced with permission from Ref. 27.

In the CDF simulation, which is customized for the M_W measurement, particles are propagated through a high-granularity spatial grid of passive material towards the EM calorimeter. The grid is extracted from a detailed GEANT geometry which incorporates the silicon sensors, support and readout structures, the beampipe and the drift chamber's internal construction. During the particle propagation, the Landau-distributed ionization energy loss, bremsstrahlung (including detailed estimation of the Landau–Pomeranchuk–Migdal[34] suppression of soft photon bremsstrahlung), Compton scattering and e^+e^- conversion of photons, as well as multiple scattering[35–37] are simulated. Hits are deposited on the drift chamber wires according to measured efficiencies and resolutions, and used to fit tracks in the same manner as the collider data. The absolute momentum scale, the total amount of passive material and magnetic field non-uniformity are measured using fits to the $J/\psi \to \mu\mu$ and $\Upsilon \to \mu\mu$ mass peaks, including the variation as a function of muon momentum (see Fig. 4) and polar angle.

A detailed GEANT4 simulation[38] of the lead-scintillator sampling calorimeter geometry is performed to understand the longitudinal shower development of electrons and photons. Low-energy nonlinearity due to absorption of soft shower particles, and high-energy non-linearity and non-gaussian resolution due to longitudinal shower leakage, are calculated. The sampling term in EM calorimeter resolution is also estimated from this simulation. These predictions are parameterized[38] and incorporated into the custom M_W simulation. The calorimeter thickness is tuned in pseudo-rapidity bins using the rate of events with low values of $E_{\text{cal}}/p_{\text{track}}$, while the radiative material upstream of the calorimeter is tuned using rate of events with high values of $E_{\text{cal}}/p_{\text{track}}$. The $E_{\text{cal}}/p_{\text{track}}$-based energy calibration (see Fig. 4) is repeated in bins of E_{cal} to measure the residual non-linearity.

Validation of these procedures is obtained by fitting for the $Z \to ee$ mass using sub-samples of radiative and non-radiative electrons. The fits are performed separately using either the calorimeter energies or the track momenta for reconstructing the invariant mass. The consistency of all these fits gives confidence in the analysis procedures.

5.2. *Hadronic recoil simulation*

A first-principles simulation of the hadronic recoil vector $\vec{u}_T$ is hindered by its soft and diffuse nature, which also causes the calorimeter response to the boson $\vec{p}_T$ to be significantly less than one. Some of the causes are that soft particles with $p_T < 400\,\text{MeV}$ curl up in the magnetic field, soft photons may be absorbed in the upstream material, and neutron interactions and hadronic showers are more difficult to simulate than EM showers. Furthermore, the underlying event and additional $p\bar{p}$ collisions produce hadronic energy flow which is uncorrelated with the hard scatter and which degrade the resolution of the $\vec{u}_T$ measurement. The conclusion is that applying corrections to the measured $\vec{u}_T$ to account for these effects is not an effective method. Therefore, the response and resolution effects are modeled in the custom simulation.

A parametric description of the hadronic recoil response and resolution is extracted from the $Z \to \ell\ell$ data and beam-crossing triggers. The p_T-balance between $\vec{p}_T(\ell\ell)$ and $\vec{u}_T$ in Z boson events as a function of $p_T(\ell\ell)$ is used to tune the parameterization of the boson p_T response and resolution in the custom simulation. Information for the modeling of the underlying event and additional $p\bar{p}$ collisions is extracted from data events triggered randomly on beam crossings and on inelastic $p\bar{p}$ collisions (minimum-bias events).

Another important effect is that certain calorimeter towers receive large energy deposits from the charged lepton(s). This energy contamination masks the hadronic energy deposition in the same towers, therefore these towers have to be omitted from the calculation of $\vec{u}_T$. The hadronic energy that is lost in these towers creates a bias in the measured $\vec{u}_T$, and this bias $\Delta u_{||}$ is parallel to the direction of the lepton. In the calculation of m_T, the component $u_{||}$ of $\vec{u}_T$ that is parallel to the lepton enters linearly, while the perpendicular component $u_\perp$ enters at higher order, thus any bias in $u_{||}$ directly biases m_T. Therefore $\Delta u_{||}$ and its dependence on event kinematics are measured in $W \to \ell\nu$ data and parameterized with care. Figure 5 shows the $u_{||}$ distribution from CDF and the W boson mass measurements in sub-samples separated by $u_{||}$ from DØ.

5.3. *Backgrounds*

Backgrounds in the $W \to \ell\nu$ sample arise from three sources; (i) Z boson events in which a lepton is lost and most of its energy is undetected or misreconstructed, causing large missing E_T to be inferred, (ii) irreducible background from $W \to \tau\nu \to \ell\nu\bar{\nu}\nu$, and (iii) QCD jets being misidentified as leptons. Most of the backgrounds are small (of $\mathcal{O}(1\%)$ or less) except for the $Z \to \mu\mu$ background in the CDF analysis. The background fractions and kinematic shapes have to be

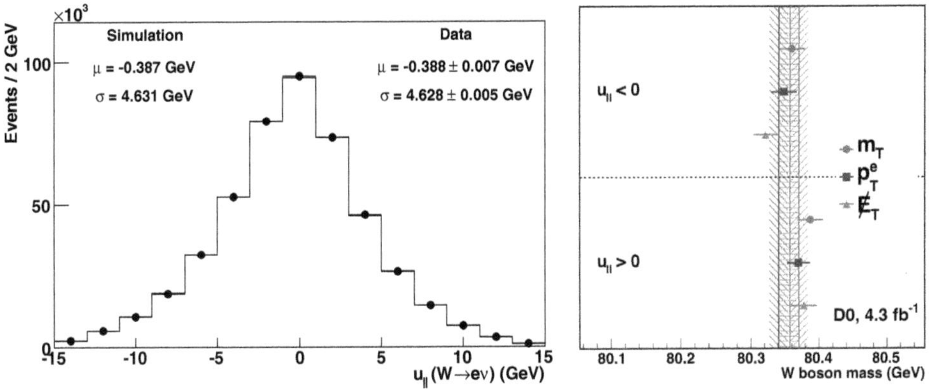

(a) (b)

Fig. 5. The distribution of $u_{||}$ from CDF overlaying data and simulation (a), and W boson mass fits from two sub-samples of DØ data binned in $u_{||}$ (b). Figures reproduced with permission from Refs. 27 and 22 respectively.

determined to a precision commensurate with the other systematic uncertainties in the analysis.

An electron can be lost in the gap between the central and endcap calorimeters or between the azimuthal modules in CDF central EM calorimeter, where the calorimeter is poorly instrumented. If the associated track is also not reconstructed, a $Z \to ee$ event mimics a $W \to e\nu$ event. This background is determined from the data by DØ and from a combination of simulation and data by CDF. If a muon from a $Z \to \mu\mu$ decay is emitted outside the pseudo-rapidity acceptance of the CDF central drift chamber (which extends up to $|\eta| \approx 1$), the event will mimic a $W \to \mu\nu$ candidate event. This background is estimated from the CDF simulation as it is geometrical in origin. The Z boson background has a bigger effect on the W mass fit than a monotonically falling background distribution because of the Jacobian edge in the Z boson distribution, which biases the fitted M_W upwards. The $W \to \tau\nu$ background is estimated from simulation including the effect of the τ polarization.

A combination of fragmentation and reconstruction effects can cause jets to be mis-identified as electrons. As these effects are rare and difficult to simulate, data-driven techniques are used to estimate the rates and distributions of such backgrounds. Multijet production where a jet fragments to an isolated, high-p_T $\pi^0 \to \gamma\gamma$, followed by an asymmetric $\gamma \to ee$ conversion in the detector material, is the typical source of mis-identified electrons. Mismeasurement of other jets can lead to the inference of sufficient missing E_T to satisfy the $W \to e\nu$ selection. A background-enriched sample, obtained by inverting or loosening the electron identification criteria or by relaxing the missing E_T requirement, is used to obtain the shapes of the background kinematic distributions. The background normalization is obtained by fitting the distribution of electron identification variables, or the distribution of missing E_T, using corresponding templates of pure signal (from simulation) and background.

The rate for misidentification of jets as muons due to punch-through is small at CDF. However, another source of background arises due to $\pi/K \to \mu$ decays-in-flight (DIF). A low p_T meson can be mis-reconstructed as a high-p_T muon if it decays within the tracking volume of the drift chamber, due to the helical fit of a kinked trajectory. This background source of $W \to \mu\nu$ candidates arises from DIF in minimum-bias events. Poor track-fit χ^2, large impact parameter, and the "seagull" — like pattern of hits are characteristics of DIF tracks that differentiate them from prompt muons. The DIF background has a hard spectrum of reconstructed p_T since the fitted track-curvature is approximately random and uniformly distributed.

5.4. *Production and decay model*

Fitting the data distributions to extract M_W requires a good theoretical understanding of W boson production and decay. The relevant aspects are (i) the parton distribution functions (PDFs) which determine the longitudinal momentum distribution of the W boson, (ii) the effect on the lepton p_T spectra from the transverse momentum of the W boson, (iii) the effect on the lepton p_T spectra and the correlation between the lepton and boson p_T due to the decay angular distribution of the W boson, and (iv) the kinematic distributions of the radiative photons which share energy with the charged lepton.

For a detector with full acceptance, the transverse momentum distributions of the decay leptons would be insensitive to the longitudinal momentum of the W boson. However, if only central leptons are used in the analysis, the acceptance for a given lepton p_T depends on the longitudinal boost. The fitted M_W therefore depends on the modeling of the longitudinal momentum distribution of the W boson, which is determined by the PDF choice. The PDFs provided by the global fitting groups have associated parametrizations of uncertainties. The uncertainty in the extracted M_W is obtained by reweighting the simulated events by the ratio of varied PDFs. The PDF error sets provided by the CTEQ[39] and MSTW[40] groups are used to quote the uncertainty in M_W. For the central value, DØ has used the CTEQ6.1 PDF set while CDF has used the CTEQ6.6 set. CDF has cross-checked that the central values obtained from the CTEQ6.6 and MSTW2008 PDF sets are consistent within the quoted uncertainty.

The relevant range of the boson p_T is $P_T(W) < 25$ GeV, where the spectrum is strongly affected by QCD parton showering and non-perturbative physics. The RESBOS[43] program includes NLO and the dominant NNLO amplitudes, resummation of parton showers and a non-perturbative form factor that depends on the beam-energy and Q^2. The dilepton p_T spectrum (in the case of CDF) and the distribution of the azimuthal opening angle in $Z \to \ell\ell$ events (in the case of DØ) have been used to tune the parameters of the non-perturbative form factor. The perturbative effects have a small impact on the M_W analysis since they are more important at high $p_T(W)$ and those events are not used for the measurement. The Q^2-dependence has a small effect because it controls the difference between the $p_T(Z)$ and $p_T(W)$ spectra. Both the perturbative effects and the Q^2-dependence have been obtained from global fits[41] to data. Recently, the CDF analysis has fitted the $p_T(Z)$ spectrum with both the non-perturbative parameter and α_S as a second parameter.

The RESBOS program also calculates the decay angular distribution at NLO and partial-NNLO level. Higher-order effects have negligible impact at the current levels of precision. As the precision of future measurements improves, a complete NNLO calculation would be desirable.

CDF and DØ incorporate the rates and distributions of final-state radiative (FSR) photons using PHOTOS[42] interfaced to RESBOS. CDF has calibrated PHOTOS-FSR against the more complete HORACE[44] calculation, which is an exact NLO electroweak calculation interfaced to a leading-logarithm photon shower. HORACE also corrects photons in the multi-photon shower for the difference between the leading-logarithm calculation and the NLO calculation. In the future, an exact NNLO electroweak calculation may be required.

5.5. *Results*

The measurements from CDF and DØ were summarized in Sec. 2. The uncertainties in these measurements are shown in Table 1. The Tevatron (world) average of $M_W = 80387(80385) \pm 16(15)$ MeV[23] is about 1.6σ above the ST prediction: $M_W = 80358 \pm 8$ MeV.[17] The comparison puts stringent bounds on new physics. The inputs for the ST prediction are the Z-pole measurements from LEP and SLD, the top quark mass from Tevatron and LHC experiments, the Higgs boson mass from LHC, and a recent determination of the hadronic vacuum polarization contribution to $\alpha_{EM}(M_Z^2)$.

Table 1. Uncertainties in units of MeV on the combined result (m_T fit) on M_W from CDF (DØ) using 2.2 (4.3) fb^{-1} of integrated luminosity. "na" denotes the uncertainty is not individually tabulated.

Source	CDF Uncertainty	DØ Uncertainty
Lepton energy scale and resolution	7	17
Recoil energy scale and resolution	6	5
Lepton tower removal	2	na
Backgrounds	3	2
PDFs	10	11
$p_T(W)$ model	5	2
Photon radiation	4	7
Statistical	12	13
Total	19	26

CDF and DØ have analysed a quarter and a half, respectively, of their full Run 2 dataset, and the full datasets are being analyzed. The PDF uncertainty is the dominant source of uncertainty and it is correlated between the two experiments. The PDFs can be constrained by the W boson charge asymmetry and the Z boson rapidity distributions. Including PDF constraints from the Tevatron and the LHC can help obtain a combined Tevatron measurement of M_W with a total uncertainty of 10 MeV.

6. Summary and Conclusions

The Tevatron experiments have published a variety of measurements using W and Z bosons which tested higher-order QCD calculations and the electroweak sector of the ST. The precision electroweak measurement program at the Tevatron has included M_W, Γ_W and the Z boson forward–backward asymmetry A_{FB}. In particular, the impact of the M_W measurement on the electroweak sector has been equivalent to the asymmetry measurements from LEP and SLD. Further improvement in the precision of M_W is anticipated with a goal of 10 MeV, and these will be legacy measurements from the Tevatron.

Beyond the Tevatron, the LHC promises nearly unlimited samples of W and Z bosons for the M_W measurement with vanishing statistical error. The challenge at the LHC will be to overcome the issues associated with various sources of systematic uncertainty. As we have learnt from the Tevatron experience, we can be optimistic that improvements in analysis techniques over time will continue to overcome the challenges.

Acknowledgments

The author would like to thank Bodhitha Jayatilaka, Paul Grannis, Willis Sakumoto, Heidi Schellman, Jadranka Sekaric, Marco Verzocchi and Hang Yin for useful comments and discussions.

We thank the Fermilab staff and technical staffs of the participating institutions for their vital contributions. We acknowledge support from the DOE and NSF (USA), ARC (Australia), CNPq, FAPERJ, FAPESP and FUNDUNESP (Brazil), NSERC (Canada), NSC, CAS and CNSF (China), Colciencias (Colombia), MSMT and GACR (Czech Republic), the Academy of Finland, CEA and CNRS/IN2P3 (France), BMBF and DFG (Germany), DAE and DST (India), SFI (Ireland), INFN (Italy), MEXT (Japan), the KoreanWorld Class University Program and NRF (Korea), CONACyT (Mexico), FOM (Netherlands), MON, NRC KI and RFBR (Russia), the Slovak R&D Agency, the Ministerio de Ciencia e Innovacion, and Programa Consolider–Ingenio 2010 (Spain), The Swedish Research Council (Sweden), SNSF (Switzerland), STFC and the Royal Society (United Kingdom), the A.P. Sloan Foundation (USA), and the EU community Marie Curie Fellowship contract 302103.

References

1. S. Glashow, *Nucl. Phys.* **22**, 579 (1961); A. Salam and J. C. Ward, *Phys. Lett.* **13**, 168 (1964); S. Weinberg, *Phys. Rev. Lett.* **19**, 1264 (1967).
2. G. Aad *et al.* (The ATLAS Collaboration), *Phys. Lett. B* **716**, 1 (2012); S. Chatrchyan *et al.* (The CMS Collaboration), *Phys. Lett. B* **716**, 30 (2012).
3. P. W. Anderson, *Phys. Rev.* **130**, 439 (1960); F. Englert and R. Brout, *Phys. Rev. Lett.* **13**, 321 (1964); P. W. Higgs, *Phys. Rev. Lett.* **13**, 508 (1964); G. S. Guralnik, C. R. Hagen, and T. W. B. Kibble, *Phys. Rev. Lett.* **13**, 585 (1964).
4. M. Ciuchini, E. Franco, S. Mishima, and L. Silvestrini, *J. High Energy Phys.* 08, 106 (2013).
5. G. Arnison *et al.* (UA1 Collaboration), *Phys. Lett. B* **122**, 103 (1983); UA2 Collaboration (M. Banner *et al.*), *Phys. Lett. B* **122**, 476 (1983); UA1 Collaboration (G. Arnison *et al.*), *Phys. Lett. B* **126**, 398 (1983); UA2 Collaboration (P. Bagnaia *et al.*), *Phys. Lett. B* **129**, 130 (1983).
6. CDF Collaboration (T. Affolder *et al.*), *Phys. Rev. D* **64**, 052001 (2001).
7. D0 Collaboration (V. M. Abazov *et al.*), *Phys. Rev. D* **66**, 012001 (2002); D0 Collaboration (B. Abbott *et al.*), *Phys. Rev. D* **62**, 092006 (2000); D0 Collaboration (B. Abbott *et al.*), *Phys. Rev. D* **58**, 092003 (1998).
8. D0 Collaboration (V. M. Abazov *et al.*) (CDF Collaboration, The D0 Collaboration, Tevatron Electroweak Working Group), *Phys. Rev. D* **70**, 092008 (2004).
9. ALEPH, CDF, D0, DELPHI, L3, OPAL, and SLD Collaborations, the LEP Electroweak Working Group, the Tevatron Electroweak Working Group, and the SLD Electroweak and Heavy Flavour Groups, CERN Report No. CERN-PH-EP-2010-095, and Fermilab Report No. FERMILAB-TM-2480-PPD, 2010.
10. ALEPH Collaboration (S. Schael *et al.*), *Eur. Phys. Jour. C* **47**, 309 (2006).
11. DELPHI Collaboration (J. Abdallah *et al.*), *Eur. Phys. Jour. C* **55**, 1 (2008).
12. L3 Collaboration (P. Achard *et al.*), *Eur. Phys. Jour. C* **45**, 569 (2006).
13. OPAL Collaboration (G. Abbiendi *et al.*), *Eur. Phys. Jour. C* **45**, 307 (2006).
14. D. Bandurin *et al.*, *Int. J. Mod. Phys. A* **30**, 1541001 (2015).
15. A. Sirlin, *Phys. Rev. D* **22**, 971 (1980).
16. M. E. Peskin and T. Takeuchi, *Phys. Rev. Lett.* **65**, 964 (1990); M. Golden and L. Randall, *Nucl. Phys. B* **361**, 3 (1990); B. Holdom and J. Terning, *Phys. Lett. B* **247**, 88 (1990); M. E. Peskin and T. Takeuchi, *Phys. Rev. D* **46**, 381 (1992); G. Altarelli and R. Barbieri, *Phys. Lett. B* **253**, 161 (1991); G. Altarelli, R. Barbieri and S. Jadach, *Nucl. Phys. B* **369**, 3 (1992) [Erratum-ibid. B **376**, 444 (1992)].
17. Gfitter Group Collaboration (M. Baak *et al.*), *Eur. Phys. J. C* **74**, 3046 (2014).
18. M Baak, M. Goebel, J. Haller, A. Hoecker, D. Kennedy, K. Mönig, M. Schott, and J. Stelzer (The Gfitter Group), *Eur. Phys. J. C* **72**, 2003 (2012).
19. CDF Collaboration (T. Aaltonen *et al.*), *Phys. Rev. Lett.* **99**, 151801 (2007); CDF Collaboration (T. Aaltonen *et al.*), *Phys. Rev. D* **77**, 112001 (2008).
20. CDF Collaboration (T. Aaltonen *et al.*), *Phys. Rev. Lett.* **108**, 151803 (2012).
21. D0 Collaboration (V. M. Abazov *et al.*), *Phys. Rev. Lett.* **103**, 141801 (2009).
22. D0 Collaboration (V. M. Abazov *et al.*), *Phys. Rev. D* **89** 012005 (2014).
23. CDF and D0 Collaborations (T. Aaltonen *et al.*), *Phys. Rev. D* **88**, 052018 (2013).
24. A. V. Kotwal and J. Stark, *Ann. Rev. Nucl. Part. Sci.* **58**, 147 (2008).
25. Particle Data Group (K. Nakamura *et al.*), *J. Phys. G* **37**, 075021 (2010).
26. R. Brun and F. Carminati, CERN Program Library Long Writeup, W5013, 1993 (unpublished), version 3.15.
27. CDF Collaboration (T. Aaltonen *et al.*), *Phys. Rev. D* **89** 072003 (2014).

28. T. Affolder *et al.*, *Nucl. Instrum. Methods Phys. Res. Sect. A* **526**, 249 (2004).
29. H. Minemura *et al.*, *Nucl. Instrum. Methods Phys. Res. Sect. A* **238**, 18 (1985).
30. L. Balka *et al.*, *Nucl. Instrum. Methods Phys. Res. Sect. A* **267**, 272 (1988).
31. K. Yasuoka, S. Mikamo, T. Kamon, and A. Yamashita, *Nucl. Instrum. Meth. Phys. Res. Sect. A* **267**, 315 (1987).
32. A. V. Kotwal, H. K. Gerberich and C. Hays, *Nucl. Instrum. Methods Phys. Res. Sect. A* **506**, 110 (2003).
33. A. V. Kotwal and C. Hays, *Nucl. Instrum. Methods Phys. Res. Sect. A* **762**, 85 (2014).
34. A. B. Migdal, *Phys. Rev* **103**, 1811 (1956); L. D. Landau and I. J. Pomeranchuk, *Dokl. Akad. Nauk. SSSR* **92**, 535 (1953); **92**, 735 (1953).
35. H. Bichsel, *Rev. Mod. Phys.* **60**, 663 (1988).
36. G. R. Lynch and O. I. Dahl, *Nucl. Instrum. Methods Phys. Res. Sect. B* **58**, 6 (1991).
37. MuScat Collaboration (D. Attwood *et al.*), *Nucl. Instrum. Methods Phys. Res. Sect. B* **251**, 41 (2006).
38. A. V. Kotwal and C. Hays, *Nucl. Intrum. Methods Phys. Res. Sect. A* **729**, 25 (2013).
39. P. Nadolsky, H. Lai, Q. Cao, J. Huston, J. Pumplin, D. Stump, W. Tung, and C.-P. Yuan, *Phys. Rev.* D **78**, 013004 (2008); J. Pumplin, D. Stump, J. Huston, H. Lai, P. Nadolsky, and W. Tung, *J. High Energy Phys.* **0207**, 012 (2002).
40. A. D. Martin, W. J. Stirling, R. S. Thorne and G. Watt, *Eur. Phys. J. C* **63**, 189 (2009).
41. F. Landry, R. Brock, P. Nadolsky, and C.-P. Yuan, *Phys. Rev. D* **67**, 073016 (2003).
42. P. Golonka and Z. Was, *Eur. Phys. J. C* **45**, 97 (2006).
43. F. Landry, R. Brock, P. M. Nadolsky, and C.-P. Yuan, *Phys. Rev. D* **67**, 073016 (2003); C. Balazs and C.-P. Yuan, *Phys. Rev.* D **56**, 5558 (1997); G. A. Ladinsky and C.-P. Yuan, *Phys. Rev. D* **50**, 4239 (1994).
44. C. M. Carloni Calame, G. Montagna, O. Nicrosini, and A. Vicini, *J. High Energy Phys.* **10** 109 (2007).
45. CDF Collaboration (T. Aaltonen *et al.*), *Phys. Rev. Lett.* **100**, 071801 (2008).
46. D0 Collaboration (V. M. Abazov *et al.*), *Phys. Rev. Lett.* **103**, 231802 (2009).

Chapter 11

Top Quark Mass

Martijn Mulders
CERN, Physics Department, Geneva, Switzerland
martijn.mulders@cern.ch

Ever since the discovery of the top quark at the Tevatron collider in 1995 the measurement of its mass has been a high priority. As one of the fundamental parameters of the Standard Theory of particle physics, the precise value of the top quark mass together with other inputs provides a test for the self-consistency of the theory, and has consequences for the stability of the Higgs field that permeates the Universe. In this review I will briefly summarize the experimental techniques used at the Tevatron and the LHC experiments throughout the years to measure the top quark mass with ever improving accuracy, and highlight the recent progress in combining all measurements in a single world average combination. As experimental measurements became more precise, the question of their theoretical interpretation has become important. The difficulty of relating the measured quantity to the fundamental top mass parameter has inspired alternative measurement methods that extract the top mass in complementary ways. I will discuss the status of those techniques and their results, and present a brief outlook of further improvements in the experimental determination of the top quark mass to be expected at the LHC and beyond.

1. A Brief History of the Top Quark

When the existence of the top quark was first postulated[1] in 1973 to explain the observation of CP symmetry in the kaon system, few could have imagined that its mass would turn out to be comparable to a gold atom, more than a hundred times heavier than the charm quark, the heaviest quark known at the time. Even today, after the recent discovery of a Higgs boson[2–4] with a mass of 125 GeV, the top quark remains the heaviest known elementary particle, a striking empirical fact for which the Standard Theory offers no explanation.

Searches at e^+e^- colliders during the 1970s and 80s looking for a narrow $t\bar{t}$ resonance failed to find the top quark, and a mass below 23.3 GeV or 30.2 GeV was ruled out at the PETRA and TRISTAN colliders respectively.[5] Even at the SLC and LEP colliders operating at the Z resonance energy no evidence was found for any $Z \rightarrow t\bar{t}$ decays, raising the bar to 45.8 GeV by Spring 1990.

In the meantime hadron collider experiments UA1 and UA2 at the ISR searching for the process $p\bar{p} \rightarrow W \rightarrow t\bar{b}$ had not fared any better, and by 1990 had set the limit at 69 GeV. By that time the Tevatron collider at Fermilab had started operations, and first results by CDF including the now dominant $p\bar{p} \rightarrow t\bar{t}$ process extended the exclusion up to 91 GeV, until finally in 1995 the CDF and D0 collaborations announced the discovery of the top quark[6,7] with a mass of 175 ± 8 GeV.

While the Standard Theory does not prescribe what should be the value of the top quark mass m_t, its value in turn has big consequences for the phenomenology of the Standard Theory as we know it, and its consistency with experimental observations.

The special value of m_t has a direct impact on the physics of on-shell top quarks, as will be discussed in Section 2. However, the top quark indirectly plays an important role in many other areas of the Standard Theory as well. In particular, the top quark is present in radiative quantum-loop corrections that appear everywhere in theoretical calculations, and often has a dominant effect due to the large value of m_t. For example, corrections to the W boson mass are proportional to m_t^2, a fact which was used in the early 1990s to constrain the mass of the top quark through its effect in virtual one-loop corrections before it was discovered. A comparison of the indirect prediction of m_t from a global electroweak fit with the direct measurements of m_t as function of time is shown in Fig. 1. A step-wise improvement in the indirect prediction of m_t is visible at the time of the Higgs boson discovery in 2012.

Precision measurements of the Higgs boson mass, W mass m_W, and m_t together with other electroweak parameters allow a strong test of the self-consistency of the

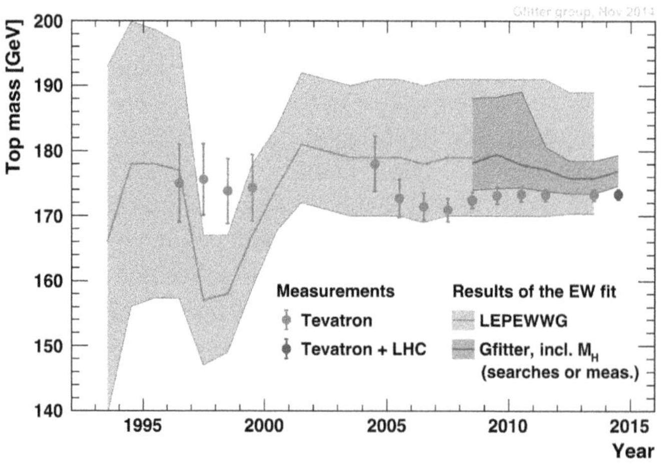

Fig. 1. Recent history of our knowledge of the top quark mass, as predicted by the global fit of electroweak variables[18,19] and from direct measurements.

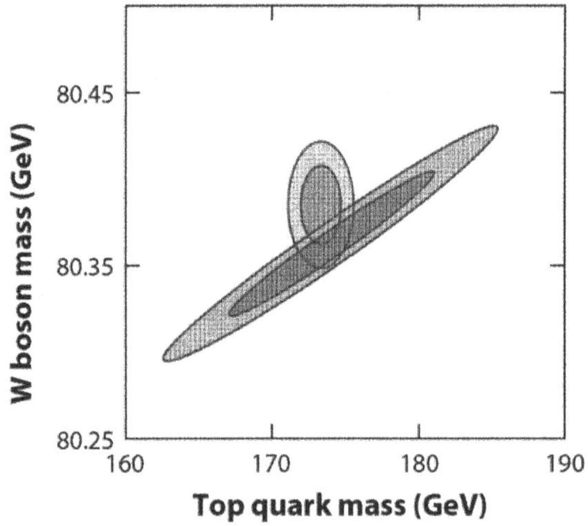

Fig. 2. Precision measurements of W boson and top quark mass (small ellipses), compared to the prediction (large ellipses) from calculations including the latest measurement of the Higgs boson mass.[4,19,20]

Standard Theory. The indirect predictions of m_W and m_t are compared to the direct measurements in Fig. 2. Radiative corrections and the global electroweak fit are disscussed in more detail elsewhere in this book.

While electroweak precision tests are currently limited more by the uncertainty in m_W than in m_t, an improvement in our knowledge of m_t would directly benefit calculations of the Higgs potential at high energy, allowing tighter constraints on the stability of the electroweak vacuum[8–10] and on some cosmological models.[11–14]

Through its strong Yukawa coupling $Y_t \approx 1$ the top quark also plays a special role in Higgs physics. While at the time of writing this review the associate production of $t\bar{t}$ with a Higgs boson has not yet been observed, the top quark has contributed already to the Higgs discovery and mass measurements by enabling the production of Higgs bosons in processes containing virtual top quark loops, such as the gluon fusion channel at the LHC and the subsequent decay to two-photon final states.

And finally, top quarks play an important role in flavour physics, as they contribute to many rare processes through virtual top quark loops. Thus the observation of $B^0 - \overline{B^0}$ oscillations in 1988 allowed the ARGUS collaboration to derive a lower limit[15] $m_t > 50$ GeV. Similarly, the predicted rate of a rare decay such as $B_s \to \mu\mu$, recently observed[16] at the LHC, depends strongly[17] on the value of m_t.

Thus many motivations exist to measure m_t with the best possible experimental precision. And as always, the quest for precise measurements doubles as a search for possible minute deviations from the predictions that could reveal signs of new physics processes beyond the Standard Theory.

2. The Short Life of a Top Quark

The large mass of the top quark allows it to decay to an on-shell W boson and a b quark, resulting in a very short lifetime ($\approx 5\cdot 10^{-25}$s) that prevents the formation of $t\bar{t}$ bound states like the characteristic J/ψ and Υ resonances for the charm and bottom quarks respectively, and it also means that top quarks decay before hadronizing into jets. However, the lifetime is still long enough that the corresponding decay width Γ_t is narrow compared to the top quark mass m_t. The prediction at NLO precision in quantum chromodynamics (QCD) is given by:[21]

$$\Gamma_t = \frac{G_F m_t^3}{8\pi\sqrt{2}}|V_{tb}|^2 \left(1 - \frac{m_W^2}{m_t^2}\right)^2 \left(1 + 2\frac{m_W^2}{m_t^2}\right) \left[1 - \frac{2\alpha_s}{3\pi}\left(\frac{2\pi^2}{3} - \frac{5}{2}\right)\right]. \quad (1)$$

The top-quark decay width depends on the top-quark mass (m_t), the W boson mass (m_W), the Fermi coupling constant (G_F), the strong coupling constant (α_s) and the magnitude of the top-to-bottom-quark coupling in the quark-mixing matrix (V_{tb}).[22–24] The most recent calculations at next-to-next-to leading order (NNLO) including QCD and electroweak corrections predict $\Gamma_t = 1.32$ GeV,[25] for m_t =172.5 GeV and other parameters fixed to their best-known values.[26]

Indeed, a relatively narrow mass peak can be observed experimentally by reconstructing the invariant mass of the top quark decay products, as shown in Fig. 3 both for doubly-resonant $t\bar{t}$ events and for singly produced top quarks. The width of the observed mass peak is dominated by the experimental resolution, and has been shown to be consistent with an underlying mass resonance with the narrow width predicted by the Standard Theory,[27] also confirmed by indirect estimates based on top quark production and decay rates.[28–30]

The position of the mass peak is strongly related to the value of m_t, which makes this a suitable observable for experimental determinations of the top quark

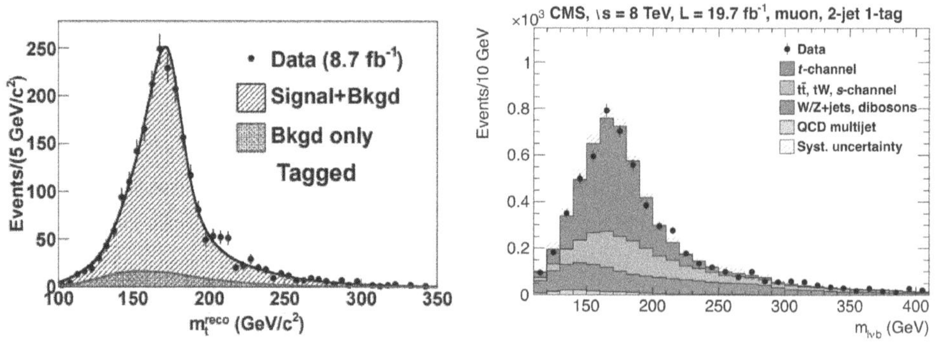

(a) (b)

Fig. 3. (a) Invariant mass of top-quark decay products in $t\bar{t}$ events reconstructed in CDF Template analysis using the full Tevatron Run II data set.[31] (b) Reconstructed top quark invariant mass in CMS single top t-channel cross-section analysis using the 2012 data set.[32]

mass. Generally Monte Carlo (MC) programs are used to model the full complexity of the candidate top quark events, and the top mass value that is measured is the MC top mass parameter m_t^{MC} for which the observable of interest (in this case the shape of the invariant mass peak) agrees best between the MC simulation and the experimental data.

However, to make a theoretically well-defined connection between a bare parameter like m_t and a physical observable, a mass renormalization scheme must be used in which higher-order quantum corrections are accounted for in the definition. Many possible choices for mass renormalization schemes exist. Two commonly used mass definitions are the pole mass scheme (m_t^{pole}) and the modified minimal subtraction scheme ($\overline{\mathrm{MS}}$) also known as a "running mass" scheme.

The $\overline{\mathrm{MS}}$ scheme is a so-called "short distance" mass, suitable for calculations involving high scales, such as the global electroweak fit or the calculation of the electroweak vacuum stability. The pole mass remains attractive, however, because m_t^{pole} has a numerical value close to the position of the top invariant mass peak, while $\overline{\mathrm{MS}}$ has a value that is about 10 GeV lower.

The relation between m_t^{pole} and other schemes can generally be expressed as

$$m_t^{\mathrm{pole}} = m_t(R,\mu) + \delta m_t(R,\mu);$$
$$\delta m_t(R,\mu) = R \sum_{n=1}^{\infty} \sum_{k=0}^{n} a_{nk} \left[\alpha_s(\mu)\right]^n \ln^k \left(\frac{\mu^2}{R^2}\right) \tag{2}$$

where R is the subtraction scale associated with the scheme; for the $\overline{\mathrm{MS}}$ scheme $R \sim m_t$. The translation between the pole and $\overline{\mathrm{MS}}$ masses was recently calculated at four-loop precision in QCD[33] with a convergence to about 200 MeV, the size of the four-loop term. Relations between proper short-distance mass schemes are known with even better precision. However, in the case of m_t^{pole} an additional non-perturbative term remains, called the renormalon ambiguity.

While transformations between field-theoretical mass schemes are well described, no formal connection exists between m_t^{MC} and any quantum-field theoretical mass definition. Based on analogies between the models in MC programs and field-theoretical QCD calculations it is often assumed that m_t^{pole} and m_t^{MC} are close to within about 1 GeV. It has been suggested[34] that m_t^{MC} may be similar to a mass scheme with a renormalization scale R close to the cut-off scale of the parton shower (typically 1–3 GeV) used in the MC simulation. Eventually a more quantitative numerical correspondence between m_t^{MC} and a suitably chosen theoretical short-distance low-scale mass may possibly be established by comparing the predictions of physical observables between the two approaches. Now that the most precise invariant-mass based measurements of m_t^{MC} have reached sub-GeV precision this is an important question, and a topic of active discussion.[34–36]

Measurements of m_t based on observables that can be calculated directly in a QCD calculation with a well-defined theoretical mass definition avoid this issue. An example of such an observable is the inclusive $t\bar{t}$ cross-section, or the $t\bar{t}$ production threshold at a potential future e^+e^- collider. A brief overview of the various approaches and their potential performance is given in the following sections.

3. Conventional Top Quark Mass Measurements at Hadron Colliders

Throughout the twenty years since the discovery of the top quark, the measurement of its mass has been an active field of development for novel analysis methods and techniques. In the quest for experimental precision, the initial focus was above all the improvement of the statistical precision of the top mass determination, by far the dominant source of uncertainty during the early years.

The experimentally most precise determinations of the top quark mass have traditionally been obtained with full kinematic reconstruction of $t\bar{t}$ events, and reconstruction of the invariant mass of the decay products of the top (and anti-top) quarks: $t \to Wb$. Depending on the decay modes of the two W bosons in the event, three final state topologies are possible, listed with their relative abundance assuming lepton universality:[26]

(a) dilepton: $t\bar{t} \to W^+bW^-\bar{b} \to \bar{\ell}\nu_\ell b\ell\bar{\nu}_\ell\bar{b}$ (10.5%)
(b) lepton+jets: $t\bar{t} \to W^+bW^-\bar{b} \to q\bar{q}b\ell\bar{\nu}_\ell\bar{b}$ or charge conjugate (43.8%)
(c) all-jets: $t\bar{t} \to W^+bW^-\bar{b} \to q\bar{q}b\bar{q}q\bar{b}$ (45.7%)

The b quarks and light-flavour quarks (q) hadronize and are observed as jets of particles. Additional jets may be produced through the radiation of energetic gluons from the hard scattering production process and in the top quark decay. The lepton ℓ can be an electron, muon or tau lepton but in top quark mass measurements tau leptons have hardly ever been used explicitely (with one exception[37]), due to more challenging detection and reduced kinematic information due to the presence of neutrinos in their decay. To reconstruct the invariant mass of the top decay products with the best possible resolution advanced statistical methods can be applied, often including one or more of the following commonly used techniques:

Kinematic constraint fits In the lepton+jets and all-jets channel a kinematic constrained fit can be used to improve the reconstruction of the event final state kinematics beyond the detector resolution. The five constraints that are typically used are equality of the top and anti-top mass, transverse momentum balance of the full event, and knowledge of the mass of the two W bosons in the event. In the case of the dilepton channel the final state is underconstrained, with 6 missing momentum components of the two energetic neutrinos, and dedicated techniques were developed to estimate as well as possible the top quark invariant mass in these events.

Monte Carlo Template method In this approach the distribution of a variable that is sensitive to the top quark mass is compared between data and the Monte Carlo (MC) prediction. The MC prediction (also called "Template") is produced for different values of the MC top mass parameter, and the top mass parameter is varied until the best agreement is obtained between data and the MC Template, using a maximum-likelihood fit. This method is conceptually simple and elegant, but the information that can be extracted event-by-event is limited. In its simplest form only a single variable per event is considered, typically the reconstructed top mass after a kinematic fit, but also 2-dimensional and 3-dimensional Template fits have been used. The Template method was the method of choice in the first generation of top mass measurements in the lepton+jets channel by CDF[38] and D0,[39] and has been used in many analyses at the Tevatron and LHC since.

Matrix Element and Ideogram techniques In these methods an event-by-event likelihood is calculated as function of the top quark mass, allowing to extract more statistical information from the events by including a more complete picture of the complexity of the $t\bar{t}$ events. The event likelihoods can take into account the probability that a given event is a background event, all relevant ambiguities such as multiple possible jet assignments, and an estimated mass resolution for every mass solution depending on the event topology and compatibility with the $t\bar{t}$ hypothesis. In the case of the Matrix Element (ME) method the $t\bar{t}$ hypothesis is defined by a (typically) LO theoretical matrix element, and the signal likelihood is calculated by integration over a multi-dimensional phase space taking into account detector resolutions. In the Ideogram approach the $t\bar{t}$ event hypothesis is defined by the constraints of a kinematic fit, and the signal likelihood is based on the post-fit variables including the goodness-of-fit. Both methods effectively allow events that are most signal-like and have less ambiguity or better resolution to have the a bigger impact on the top mass determination.

By employing a more detailed signal model contained in the matrix element, the ME method is statistically the more powerful technique, but the calculations involved make this method very CPU intensive. The Ideogram method is faster to execute and less model dependent. The ME technique is used in the most precise measurement at the Tevatron to date,[40,41] while the most precise determinations by CMS are based on the Ideogram approach.[42,43]

Regardless of the exact method used to reconstruct the top invariant mass, all measurements in this class of analyses have to rely on the same (or very similar) MC simulation programs to apply corrections for experimental and theoretical effects that play a role in the mass determination, including detector resolution, trigger effects, ambiguities and approximations in the event reconstruction and interpretation, and effects of perturbative and non-perturbative QCD as modeled by the MC simulation. For all mass determinations based on direct reconstruction of the invariant mass of the top decay products and calibrated this way one can therefore assume that within the uncertainties assigned, effectively the same "MC mass definition" (m_t^{MC}) applies.

3.1. *World average anno 2014*

In March 2014 the four collaborations ATLAS, CDF, CMS and D0 joined forces and prepared a combination of the most precise top quark mass measurements available at the time. Possible correlations between each source of systematic uncertainty of the different input analyses were evaluated and taken into account, yielding the following world-average value of the top quark mass[44] (using the m_t^{MC} definition):

$$m_t = 173.34 \pm 0.76 \text{ GeV}. \tag{3}$$

As shown in Fig. 4 the combination was also provided per experiment, per collider and per decay channel. It is interesting to see that at this point in time the combined measurements of the LHC had reached a precision equal to the Tevatron measurements, and that the central values are in good agreement in spite of the big change in collision energy from 1.96 TeV to 7 TeV. Measured top mass values also agree well between the various decay channels with very different event topologies and background conditions. While the lepton+jets channel still yields slightly better results than the dilepton and all-jets final states, they all have reached a relative experimental precision of well below 1%. The consistency of the results across different event topologies, collider energies, luminosity conditions and detectors is impressive, and provides a confirmation of the ability to understand the relevant physics effects and experimental conditions with great accuracy.

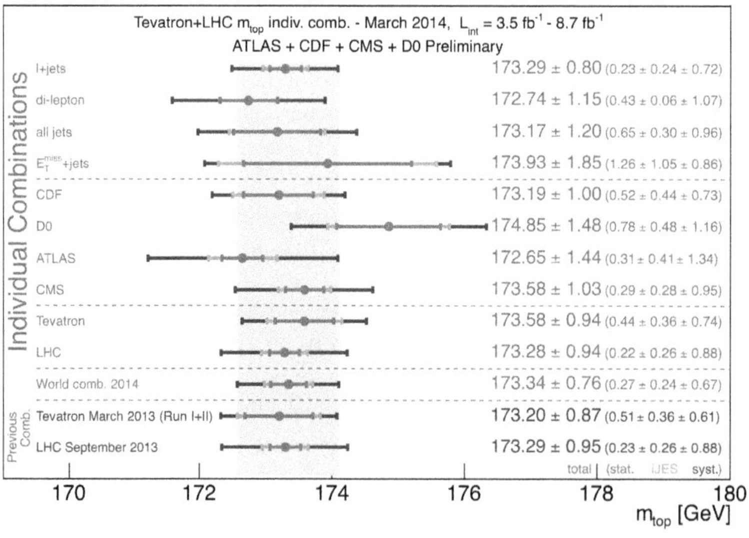

Fig. 4. World average combination[44] of top mass results from the LHC and Tevatron collaborations available in March 2014, and for subsets in different channels, experiments and colliders.

3.2. *New results in m_t^{MC} measurements since 2014*

The latest ATLAS measurements[45,46] using the full 7 TeV dataset were not yet included in the world average, but are in good agreement. The combination of the results in the dilepton and single lepton channel yields[46] $m_t = 172.99 \pm 0.91$ GeV.

The CDF collaboration published their final measurements with the full Tevatron Run II data set in the all-jets and di-lepton channel, and D0 in the lepton+jets channel. The D0 measurement uses the Matrix Element technique and reaches a precision equal to the 2014 world average and central value confirming earlier D0 measurements: $m_t = 174.98 \pm 0.76$ GeV.

The most precise measurements all use the hadronic decay of the $W(\to jj)$ and the known value of m_W to determine an overall jet-energy scale factor (JSF) *in situ* in the $t\bar{t}$ events. By correcting all measured jet energies with this factor, the main experimental systematic uncertainty due to jet energy calibration is reduced. However, the ATLAS analysis in the lepton+jets channel[46] goes one step further by performing a 3-dimensional fit of top quark mass, the overall JSF, and an additional scale factor bJSF describing any possible deviation between the jet response for jets from light-flavour quarks and b quarks. To constrain bJSF, a new variable is introduced probing the transverse momentum balance of b-tagged jets versus non-b-tagged jets. A simulated distribution of this variable, $R_{\mathrm{bq}}^{\mathrm{reco}}$, and its dependence on bJSF is shown in Fig. 5(a), and the reconstructed top mass in Fig. 5(b). This novel analysis approach reduces flavour-dependent uncertainties in jet energy calibration, at the cost of an increased statistical uncertainty. With larger data sets at 8 TeV and in the upcoming LHC Run 2 prospects are excellent for further reduction of the overall measurement uncertainties.

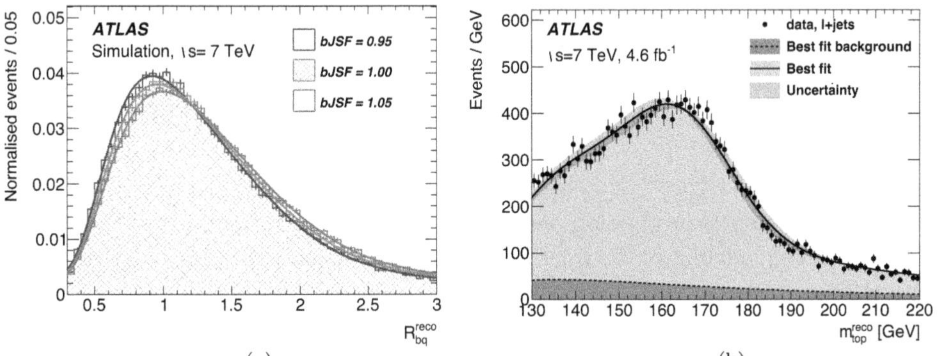

(a) (b)

Fig. 5. Mass analysis in the lepton+jets channel using the full 7 TeVdataset in ATLAS using transverse momentum balance of b-tagged jets versus other jets (a) to improve the jet energy scale calibration and reduce systematic uncertainty on the reconstructed top mass (b).

3.3. *Prospects for* m_t^{MC}

The prospects for further improvements in measurements of m_t in the m_t^{MC} definition are good. Experimental uncertainties are well understood and expected to become better with increasing statistics of data samples for calibration. While the modelling of QCD effects at perturbative and non-perturbative level is far from trivial, the MC simulation programs currently in use by experimental collaboration generally describe the data very well, and theoretical uncertainties on the determination of m_t^{MC} are of the order of 0.2 GeV or less per effect studied. The latest tools with full NLO matching between matrix elements and parton shower and calculations taking into account off-shell NLO effects, or calculations achieving NNLO+NNLL accuracy promise further improvements.

Increased top quark data samples will also allow more detailed studies of the stability of the top mass observable as a function of kinematic variables, which may help to pinpoint issues with QCD modeling and could possibly even help in the future to shed more light on the question of a correspondence between m_t^{MC} and well-defined theoretical mass definitions.[36]

3.4. *Extraction of* m_t^{MC} *with different observables*

The top mass determinations discussed so far were all based on full reconstruction of the invariant mass of top decay products. Various other observables have been proposed to minimize the effects of certain modeling uncertainties or experimental effects, or more generally to have different and therefore less correlated uncertainties.

One example is the use of only leptonic variables,[47] which has been proposed to minimize potentially poorly understood and modeled effects of non-perturbative QCD. Another approach advocates the use of the position of the peak in the b jet energy spectrum,[48] a method which is claimed to reduce sensitivity to the production mechanism, but does not mitigate uncertainties related to b jet reconstruction.

To avoid reliance on jet reconstruction and energy calibration, methods have been proposed that are purely based on charged particles: the L_{xy} method determines the boosts of b jets by measuring the b-hadron decay length, a method proposed initially in Ref. 49 and applied by CDF.[27,50] Another method is based on the invariant mass of a J/psi from the b-jet and the isolated lepton from the W decay of the same top,[51] a Lorentz-invariant quantity and potentially less sensitive to top quark production modeling. However, statistical uncertainties are large and the modeling of the b jet fragmentation becomes an important source of uncertainty.

Finally the use of the single top channel is being considered. As suggested by the invariant mass peak clearly visible in Fig. 3 it should be possible to perform a mass measurement using single top events at the LHC. The production mechanism is different so some of the theory modeling systematics would be different from measurements based on $t\bar{t}$ events.

4. Top Mass Extraction Using Other Top Mass Definitions

The idea to extract a well-defined (pole or $\overline{\text{MS}}$) top quark mass by comparing the $t\bar{t}$ cross section to a theoretical prediction as function of m_t was pioneered by D0.[52] Recently more precise measurements by CMS[53] and ATLAS[54] were obtained using the $t\bar{t}$ cross section calculated at NNLO+NNLL precision.[55] The combination of the ATLAS 7 and 8 TeV results yields $m_t = 172.9^{+2.5}_{-2.6}$ GeV, and the method is illustrated in Fig. 6(a). While this is a theoretically very clean method to extract the top quark mass, it is hard to imagine significant further improvements as this analysis already has excellent experimental precision on the cross-section with minimal dependence on the assumed m_t, and theoretical predictions beyond the current NNLO+NNLL accuracy are not expected in the near future.

Rather than using the inclusive production cross-section, differential distributions of observables that can be calculated using first-principles QCD may allow to obtain a theoretically well-defined top quark mass with improved precision. One example is the normalized inverse of the $t\bar{t}$ +one jet invariant mass which has been calculated to NLO precision in perturbative QCD[56] and was recently employed by the ATLAS collaboration in the lepton+jets channel to extract a top quark pole mass[57] $m_t^{\text{pole}} = 173.7 \pm 1.5(\text{stat}) \pm 1.4(\text{syst})^{+1.0}_{-0.5}$ GeV.

Another promising observable that may provide for theoretically well defined measurements is the $m_{\ell b}$ distribution which can be calculated with NLO precision in QCD,[58] also including off-shell non-factorizable corrections.[59–61]

The CMS collaboration has performed a radically different measurement using the kinematic endpoints of various distributions in the dilepton channel.[62] The endpoints of various lepton-related distributions are used, as well as the $m_{\ell b}$ distribution, shown in Fig. 6(b). No MC simulation is used to calibrate the measurement, nor does the analysis include any QCD corrections or calculations. The endpoints

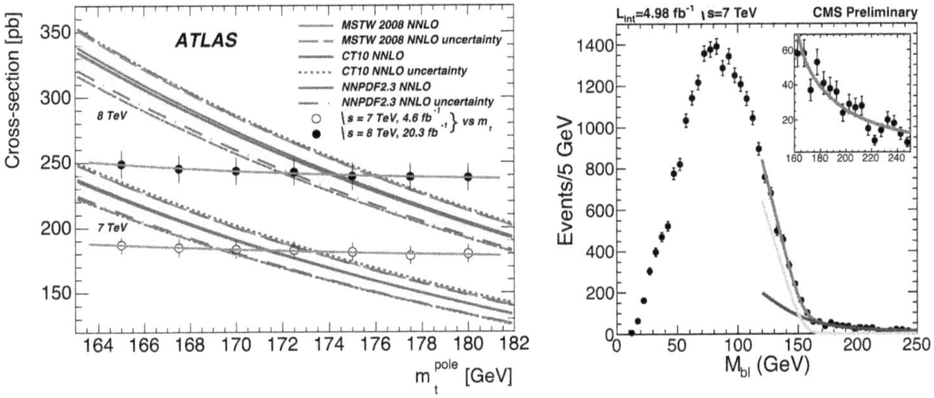

(a) (b)

Fig. 6. Extraction of the top pole mass from the $t\bar{t}$ cross-section by ATLAS (a) and the $m_{\ell b}$ mass distribution used in the CMS endpoint analysis (b).

are predicted based on the picture of a narrow mass resonance, extracting a value close to the position of the underlying mass peak. By definition the mass extracted is not m_t^{MC} nor is it m_t^{pole}. Yet the result is numerically in good agreement with the top mass measurements obtained with those definitions: $m_t = 172.9^{+1.9}_{-2.3}$ GeV.

5. Top Mass Prospects at Lepton Colliders

Prospects for top mass measurements at a possible future e^+e^- collider at or above the $t\bar{t}$ production threshold have been investigated in detail and several useful overviews and reports exist.[63–67]

Theoretical calculations of the $e^+e^- \to t\bar{t}$ threshold shape are advanced and studies show that a precise experimental scan of the cross-section shape around the threshold would allow a determination of m_t with a precision below 100 MeV in a well-defined short-distance scheme optimized for the threshold scan.

At energies above the threshold, it would also be possible to perform invariant mass measurements of the top decay very similar to methods used at hadron colliders before. This would have the advantage of a much cleaner e^+e^- environment without the additional underlying hadronic event activity from proton remains. The question of interpretation in a well-defined mass scheme, while not fundamentally different from hadron collider measurements, will be easier to address theoretically and perhaps better controlled for the e^+e^- case.[67–70]

6. Summary and Outlook

In the 20 years since the discovery of the top quark, measurements of its mass at hadron colliders have made huge strides, evolving towards a precision well below 1% and even approaching 0.5 GeV. A further step forwards is critically dependent on a detailed and accurate treatment of the effects of perturbative and non-perturbative QCD that link the m_t parameter to the physical particles observed in experiment. The more than 100-fold increase in top quark data sets anticipated at the LHC, innovative analysis approaches, and state-of-the-art theoretical calculations and MC tools will offer further opportunities to make a fundamental step fowards in measuring this fundamental constant of Nature. Reaching a precision of the order of Λ_{QCD} or better is not excluded at the LHC, and a future lepton collider would open a window to even greater levels of scrutiny and precision.

The question why the top mass is so much heavier than other known fermions remains an unexplained mystery of particle physics. Knowing its precise value allows to better calculate the experimental implications of the Standard Theory and test its consistency, but also to constrain possible effects of new physics beyond it.

References

1. M. Kobayashi and T. Maskawa, *Prog. Theor. Phys.* **49**, 652–657 (1973).
2. G. Aad *et al.* [ATLAS Collaboration], *Phys. Lett. B* **716**, 1–29 (2012).
3. S. Chatrchyan *et al.* [CMS Collaboration], *Phys. Lett. B* **716**, 30–61 (2012).
4. G. Aad *et al.* [ATLAS and CMS Collaborations], *Phys. Rev. Lett.* **114**, 191803 (2015).
5. C. Campagnari and M. Franklin, *Rev. Mod. Phys.* **69**, 137–212 (1997).
6. F. Abe *et al.* [CDF Collaboration], *Phys. Rev. Lett.* **74**, 2626–2631 (1995).
7. S. Abachi *et al.* [D0 Collaboration], *Phys. Rev. Lett.* **74**, 2632–2637 (1995).
8. G. Degrassi, S. Di Vita, J. Elias-Miro, J. R. Espinosa, G. F. Giudice, G. Isidori and A. Strumia, *JHEP* **08**, 098 (2012).
9. S. Alekhin, A. Djouadi and S. Moch, *Phys. Lett. B* **716**, 214–219 (2012).
10. D. Buttazzo, G. Degrassi, P. P. Giardino, G. F. Giudice, F. Sala, A. Salvio and A. Strumia, *JHEP* **12**, 089 (2013).
11. F. L. Bezrukov and M. Shaposhnikov, *Phys. Lett. B* **659**, 703–706 (2008).
12. A. De Simone, M. P. Hertzberg and F. Wilczek, *Phys. Lett. B* **678**, 1–8 (2009).
13. F. Bezrukov and M. Shaposhnikov, *J. Exp. Theor. Phys.* **120**, 335–343 (2015) [*Zh. Eksp. Teor. Fiz.* 147, 389 (2015)].
14. M. Fairbairn and R. Hogan, *Phys. Rev. Lett.* **112**, 201801 (2014).
15. H. Albrecht *et al.* [ARGUS Collaboration], *Phys. Lett. B* **192**, 245–252 (1987).
16. V. Khachatryan *et al.* [CMS and LHCb Collaborations], *Nature.* **522**, 68–72 (2015).
17. C. Bobeth, M. Gorbahn, T. Hermann, M. Misiak, E. Stamou and M. Steinhauser, *Phys. Rev. Lett.* **112**, 101801 (2014).
18. S. Schael *et al.* [ALEPH and DELPHI and L3 and OPAL and LEP Electroweak Collaborations], *Phys. Rept.* **532**, 119–244 (2013).
19. M. Baak, J. Cúth, J. Haller, A. Hoecker, R. Kogler, K. M onig, M. Schott and J. Stelzer, *Eur. Phys. J. C* **74**, 3046 (2014).
20. C. Quigg, *Physics.* **8**, 45 (2015).
21. M. Jezabek and J. H. Kuhn, *Nucl. Phys. B* **314**, 1 (1989).
22. A. Denner and T. Sack, *Nucl. Phys. B* **358**, 46–58 (1991).
23. A. Czarnecki and K. Melnikov, *Nucl. Phys. B* **544**, 520–531 (1999).
24. K. G. Chetyrkin, R. Harlander, T. Seidensticker and M. Steinhauser, *Phys. Rev. D* **60**, 114015 (1999).
25. J. Gao, C. S. Li and H. X. Zhu, *Phys. Rev. Lett.* **110**(4), 042001 (2013).
26. K. Olive *et al.*, Review of Particle Physics, *Chin. Phys. C* **38**, 090001 (2014).
27. A. Abulencia *et al.* [CDF Collaboration], *Phys. Rev. D* **75**, 071102 (2007).
28. V. M. Abazov *et al.* [D0 Collaboration], *Phys. Rev. Lett.* **106**, 022001 (2011).
29. V. M. Abazov *et al.* [D0 Collaboration], *Phys. Rev. D* **85**, 091104 (2012).
30. V. Khachatryan *et al.* [CMS Collaboration], *Phys. Lett. B* **736**, 33 (2014).
31. T. Aaltonen *et al.* [CDF Collaboration], *Phys. Rev. Lett.* **109**, 152003 (2012).
32. V. Khachatryan *et al.* [CMS Collaboration], *JHEP* **1406**, 090, (2014).
33. P. Marquard, A. V. Smirnov, V. A. Smirnov and M. Steinhauser, *Phys. Rev. Lett.* **114** (14), 142002 (2015).
34. A. H. Hoang, The top mass: Interpretation and theoretical uncertainties. arXiv:1412.3649 [hep-ph].
35. A. H. Hoang, A. Jain, I. Scimemi and I. W. Stewart, *Phys. Rev. Lett.* **101**, 151602 (2008).

36. S. Moch *et al.*, High precision fundamental constants at the TeV scale. arXiv:1405.4781 [hep-ph].
37. T. Aaltonen *et al.* [CDF Collaboration], *Phys. Rev. Lett.* **109**, 192001 (2012).
38. F. Abe *et al.* [CDF Collaboration], *Phys. Rev. Lett.* **80**, 2767–2772 (1998).
39. B. Abbott *et al.* [D0 Collaboration], *Phys. Rev. D* **58**, 052001 (1998).
40. V. M. Abazov *et al.* [D0 Collaboration], *Phys. Rev. Lett.* **113**, 032002 (2014).
41. V. M. Abazov *et al.* [D0 Collaboration], *Phys. Rev. D* **91**(11), 112003 (2015).
42. S Chatrchyan *et al.* [CMS Collaboration], *JHEP.* **1212**, 105 (2012).
43. S Chatrchyan *et al.* [CMS Collaboration], *Eur. Phys. J. C* **74**, 2758 (2014).
44. ATLAS, CDF, CMS and D0 Collaborations. First combination of Tevatron and LHC measurements of the top-quark mass, arXiv:1403.4427 [hep-ex].
45. G. Aad *et al.* [ATLAS Collaboration], *Eur. Phys. J. C* **75**(4), 158 (2015).
46. G. Aad *et al.* [ATLAS Collaboration], *Eur. Phys. J. C* **75**, no. 7, 330 (2015).
47. S. Frixione and A. Mitov, *JHEP* **1409**, 012 (2014).
48. K. Agashe, R. Franceschini and D. Kim, *Phys. Rev. D* **88**(5), 057701 (2013).
49. C. Hill, J. Incandela and J. Lamb, *Phys. Rev. D* **71**, 054029 (2005).
50. T. Aaltonen *et al.* [CDF Collaboration], *Phys. Rev. D* **81**, 032002 (2010).
51. A. Kharchilava, *Phys. Lett. B* **476**, 73–78 (2000).
52. V. M. Abazov *et al.* [D0 Collaboration], *Phys. Lett. B* **703**, 422–427 (2011).
53. S Chatrchyan *et al.* [CMS Collaboration], *Phys. Lett. B* **728**, 496 (2014) [Corrigendum: *ibid.* **738**, 526].
54. G. Aad *et al.* [ATLAS Collaboration], *Eur. Phys. J. C* **74**, 3109 (2014).
55. M. Czakon, P. Fiedler and A. Mitov, *Phys. Rev. Lett.* **110**, 252004 (2013).
56. S. Alioli *et al.*, *Eur. Phys. J. C* **73**, 2438 (2013).
57. G. Aad *et al.* [ATLAS Collaboration] *JHEP* **1510**, 121 (2015).
58. S. Biswas, K. Melnikov and M. Schulze, *JHEP* **1008**, 048 (2010).
59. G. Heinrich, A. Maier, R. Nisius, J. Schlenk and J. Winter, *JHEP* **06**, 158 (2014).
60. F. Cascioli, S. Kallweit, P. Maierhfer and S. Pozzorini, *Eur. Phys. J. C* **74**(3), 2783 (2014).
61. J. M. Campbell, R. K. Ellis, P. Nason and E. Re, *JHEP* **04**, 114 (2015).
62. S Chatrchyan *et al.* [CMS Collaboration], *Eur. Phys. J. C* **73**, 2494 (2013).
63. M. Bicer *et al.*, *JHEP* **01**, 164 (2014).
64. A. H. Hoang, A. V. Manohar, I. W. Stewart and T. Teubner, *Phys. Rev. D* **65**, 014014 (2002).
65. M. Martinez and R. Miquel, *Eur. Phys. J. C* **27**, 49–55 (2003).
66. S. Heinemeyer and G. Weiglein, Top, GigaZ, MegaW, arXiv:1007.5232 [hep-ph].
67. K. Seidel, F. Simon, M. Tesar and S. Poss, *Eur. Phys. J. C* **73**(8), 2530 (2013).
68. S. V. Chekanov, *Eur. Phys. J. C* **26**, 173 (2002).
69. H. Aihara, P. Burrows, M. Oreglia, E. L. Berger, V. Guarino, J. Repond, H. Weerts, L. Xia, J. Zhang, Q. Zhang, *et al.*, SiD Letter of Intent. (2009).
70. T. Abe *et al.*, The International Large Detector: Letter of Intent. (2010).

Chapter 12

Global Fits of the Electroweak Standard Theory: Past, Present and Future

M. Baak,* J. Haller† and K. Mönig‡

*CERN, Geneva, Switzerland

†Institut für Experimentalphysik, Universität Hamburg, Germany

‡DESY, Hamburg and Zeuthen, Germany

The last decades have seen tremendous progress in the experimental techniques for measuring key observables of the Standard Theory (ST) as well as in theoretical calculations that has led to highly precise predictions of these observables. Global electroweak fits of the ST compare the precision measurements of electroweak observables from lepton and hadron colliders at CERN and elsewhere with accurate theoretical predictions of the ST calculated at multi-loop level.

For a long time, global fits have been used to assess the validity of the ST and to constrain indirectly (by exploiting contributions from quantum loops) the remaining free ST parameters, like the masses of the top quark and Higgs boson before their direct discovery. With the discovery of the Higgs boson at the Large Hadron Collider (LHC), the electroweak sector of the ST is now complete and all fundamental ST parameters are known. Hence the global fits are a powerful tool to probe the internal consistency of the ST, to predict ST parameters with high precision, and to constrain theories describing physics beyond the ST.

In this chapter we review the global fits of the electroweak sector of the ST from an experimentalist's perspective. We briefly recall the most important achievements from the past (mainly driven by the precise measurements of Z pole observables), discuss the present situation after the accurate measurements of the top quark and Higgs boson masses, and present prospects of the fits as expected from new measurements at the LHC and future lepton colliders.

1. Introduction

Formulated during the 60s and 70s of the last century, the Standard Theory (ST) of particle physics describes the elementary particles and their electromagnetic, weak and strong interactions as a relativistic quantum field theory using a $U(1)_Y \times SU(2)_L \times SU(3)_C$ gauge theory.[1–5] The past few decades have seen tremendous progress in the experimental techniques for measuring key observables of the ST and at the same time theoretical progress which led to highly precise theoretical calculations of these observables in the ST. A dedicated world-wide effort to study the electroweak sector of the ST with precise measurements, at lepton colliders (LEP, SLC) around the Z pole and beyond during the 90s,[6,7] then the measurements of the top quark mass with increasing precision at hadron colliders (Tevatron, LHC),[8] and finally the discovery of the Higgs boson at the LHC[9,10] with the determination of its mass,[11] has turned this research area into a science of high precision.

The experimental and theoretical progress of the last decades make global fits of the electroweak sector an ideal concept for stringent tests of the ST validity and its internal consistency. Including the influence of loop effects in the calculations, physics can even be studied at much higher energy scales than available in the centre-of-mass energies of the existing particle colliders. Important achievements of the fits following this approach include the correct prediction of the masses of the top quark and the Higgs boson before their direct discovery.

Beginning with the description of the high-precision measurements of Z pole observables, during the last decades global fits have demonstrated impressively the consistency of the ST with the experimental data. After the direct measurements of the top quark and Higgs boson masses all electroweak ST parameters are measured experimentally and the influence of the loop corrections is fixed in the ST. As a result, global fits can indirectly predict key observables, like the mass of the W boson or the effective weak mixing angle, with a precision exceeding the direct experimental measurements. Hence such fits, in combination with precise measurements, form a powerful tool for global assessments of the accuracy of ST calculations.

At the same time multiple theoretical extensions of the ST have been proposed in the literature to solve known shortcomings of the ST, like e.g. the hierarchy problem or the existence of dark matter. These theories describing physics beyond the ST (BST) often lead to changes in the theoretical prediction of precision observables compared to the ST, e.g. from the effect of new hypothetical heavy particles to be considered in the calculation of the radiative corrections of the observables. In global fits these small changes can be exploited using the concept of oblique parameters which parametrise the deviation of the experimental data from the radiative corrections as calculated in the ST.[12,13] A comparison of the determined values of these parameters with their prediction in various BST models — normally dependent on new parameters — allows one to constrain the parameters of the new theory, and sometimes the BST theory can be excluded entirely.

In this document we review global fits of the electroweak sector of the ST from an experimentalist's point of view. The document is organised as follows. The next section presents the ingredients to the fits, including a short recap of the theoretical ST calculations of observables and their experimental measurements available today. In Section 3 we discuss important milestones in the history of electroweak fits, while Section 4 gives an overview of the current status of the fit after the Higgs discovery. Section 5 summarises the constraints on BST physics derived from the fits. In Section 6 the prospectives of the fits at future colliders, like an additional run at the LHC or lepton colliders, are discussed.

2. Ingredients of Electroweak Fits

There are two ingredients needed for global fits: accurate measurements and precise calculations. In global fits of the electroweak sector of the ST, both types have evolved together and influenced each other over the last decades: the calculations have adapted to what the experiments can measure and the measurements have adapted to what can be calculated best. The measurements as well as the calculations entering the fits have been reviewed in detail in dedicated chapters of this book. Here only a brief summary is presented.

Because of their clean experimental and theoretical environment, observables from experiments at the e^+e^- colliders LEP and SLC could be measured during the 90s with very high precision. For these electroweak precision measurements around the Z resonance the concept of pseudo-observables has been adapted.[14] In this approach a pseudo-observable is an observable in a world without initial and final state radiation and with only a Z boson exchanged in the process $e^+e^- \rightarrow f\bar{f}$. Typical pseudo-observables are the mass and width of the Z boson, the forward–backward asymmetry for a produced fermion or the polarisation of final state τ leptons. Since the experiments measure the observables including radiation and photon exchange after unavoidable experimental cuts, corrections must be applied which are, however, model independent in the sense that they only depend on QED effects and do not influence the radiative corrections that may contain new physics. The latter appear in the predictions of the pseudo-observables. Due to precise second-order calculations including exponentiation the uncertainties of the QED corrections are not relevant for the final precision.[15] Close to the Z pole only radiative corrections affecting the Z propagator are of interest while the contributions from weak box diagrams are numerically irrelevant.[14]

For these reasons electroweak corrections can be parametrised with only three parameters: $\Delta\rho_f$ normalising the absolute Z fermion couplings, $\Delta\kappa_f$ correcting the weak mixing angle obtained from the ratio of the vector to the axial-vector coupling, normally written as $\sin^2\theta^f_{\text{eff}} = (1 + \Delta\kappa_f)\sin^2\theta$, and Δr correcting the relation between the fine structure constant α, the Fermi constant in muon decays G_F, the mass of the Z boson M_Z and the mass of the W boson M_W. A dependence of these parameters on the fermion type f is introduced by vertex corrections. These are independent of most BST effects apart from the case of the b quark, where corrections containing a top quark can be important.

2.1. *Experimental measurements*

The coupling sector of the electroweak model is given by three parameters, so the three most precisely measured ones are used for the predictions.

These are[16]: α known to a relative precision of $\Delta\alpha/\alpha = 3 \cdot 10^{-10}$, G_F known to $\Delta G_F/G_F = 5 \cdot 10^{-7}$ and M_Z measured precisely in the scan of the Z resonance in e^+e^- collisions.

From measurements at CERN's LEP collider and SLC two classes of observables are used: the total and partial decay widths of the Z boson and asymmetries measured at the Z pole.[6] The total and partial widths are sensitive to the absolute size of the vector and axial vector couplings of the Z ($\propto (g_V^2 + g_A^2)$), while the asymmetries are sensitive to the ratio of these couplings (g_V/g_A) and thus to the effective weak mixing angle. The ratio of the Z partial decay widths to hadrons and leptons is also sensitive to the strong coupling constant $\alpha_s(M_Z^2)$. In addition, the fits use the mass and width of the W boson, both measured at LEP2 and the Tevatron.[17] For the determination of the loop corrections in principle the masses of all fermions and of the Higgs boson are needed. However, in practice only the mass of the top quark m_t, measured at the Tevatron and the LHC,[8] and the mass of the Higgs boson M_H, measured at the LHC,[11,18,19] are relevant. All relevant input observables that enter the electroweak fits are summarised in Table 1. The exact definitions for the Z observables and $\Delta\alpha_{\rm had}^{(5)}(M_Z^2)$ can be found in Ref. 6.

Table 1. Observables used in electroweak fits.

Observable	Description	Collider
Z resonance scan:		
M_Z	Mass of the Z boson	LEP
Γ_Z	Width of the Z boson	LEP
$\sigma_0^{\rm had}$	Hadronic pole cross section	LEP
R_ℓ^0	Ratio of hadronic and leptonic partial width	LEP
Z asymmetries:		
$A_{\rm FB}^{0,l}$	Forward–backward asymmetry for leptons	LEP
$\mathcal{A}_\ell$(LEP)	Asymmetry parameter from the τ-polarisation	LEP
$\mathcal{A}_\ell$(SLD)	Asymmetry parameter from the left–right asymmetry	SLC
$\sin^2\theta_{\rm eff}^\ell(Q_{FB})$	Weak mixing angle from inclusive quark asymmetries	LEP
$A_{\rm FB}^{0,b}$	Forward–backward asymmetry for b quarks	LEP
$A_{\rm FB}^{0,c}$	Forward–backward asymmetry for c quarks	LEP
$\mathcal{A}_b$	Asymmetry parameter for b quarks	SLC
$\mathcal{A}_c$	Asymmetry parameter for c quarks	SLC
Other inputs:		
R_b^0/R_c^0	Ratio of the Z partial width to b/c quarks to its hadronic width	LEP, SLC
M_W	Mass of the W boson	LEP, TEV
Γ_W	Width of the W boson	LEP, TEV
m_t	Mass of the top quark	TEV, LHC
M_H	Mass of the Higgs boson	LHC
$\Delta\alpha_{\rm had}^{(5)}(M_Z^2)$	Hadronic contribution to $\alpha(M_Z^2)$	

2.2. *Theoretical predictions*

In general, calculations at two-loop order precision are available for all pseudo-observables today. The effective mixing angle $\sin^2\theta^f_{\rm eff}$ is known up to two-loop order with leading three- and four-loop terms.[20–22] The W mass is calculated to the same order with the addition of four-loop QCD corrections.[23–26] For the Z partial decay widths also the full two-loop corrections are available apart from closed fermion loops.[27] Final state QED and QCD radiation is included via radiator functions known to $\mathcal{O}(\alpha_s^4)$ for massless final-state quarks, $\mathcal{O}(\alpha_s^3)$ for massive quarks,[28–30] and $\mathcal{O}(\alpha^2)$ for contributions with closed fermion loops.[31] The width of the W boson has only been calculated at one-loop order precision only,[32] but the experimental precision is much worse than for the other observables.

3. Important Milestones of the Electroweak Fit

Global fits of the electroweak sector of the ST are possible if the input parameters are over-constrained with a precision better than effects from new particles at the 1-loop level. As mentioned earlier the electroweak sector of the ST is determined by three parameters and the three most precise ones, α, G_F and M_Z were chosen. α and G_F were known with good precision already for a long time. With the start of data taking at LEP the value of M_Z was quickly measured with a much better precision than all electroweak observables (apart from α and G_F). All other observables could then be predicted using these three parameters plus the loop corrections from the new parameters of interest.

Already at the Singapore conference in 1990 the mass of the top quark could be predicted from a global fit to be $m_t = 139 \pm 32 \pm 20\,\text{GeV}$, where the first error is experimental and the second quantifies the impact of the unknown Higgs boson mass.[33] This fit used the very first data obtained at LEP as well as the ratio of the masses M_W/M_Z measured at CERN's UA2 experiment and the Tevatron and a value of the weak mixing angle $\sin^2\theta$ measured in lepton nucleon scattering experiments. At that time the LEP data already allowed to set a direct lower limit of 44 GeV on the Higgs boson mass. The above-mentioned uncertainty on m_t from the unknown Higgs boson mass was evaluated by varying M_H in the fits between the lower LEP limit and 1 TeV.

From 1993 onwards the LEP electroweak working group (LEPEW) published official fits every year.[34] For these fits the ST predictions for the central values of the (pseudo-)observables were obtained from calculations with the ZFITTER program.[35] The results were always cross-checked with the TOPAZ program,[36] giving identical results.

A major confirmation of the concept of global electroweak fits was obtained in 1995 when the top quark was discovered by CDF[37] and D0[38] with a mass of $m_t = 176 \pm 8(\text{stat.}) \pm 10(\text{sys.})\,\text{GeV}$ (CDF) and $m_t = 199^{+19}_{-21}(\text{stat.}) \pm 22(\text{syst.})\,\text{GeV}$ (D0). The prediction from electroweak fits at that time was $m_t = 178 \pm 8(\text{exp.}) \pm$

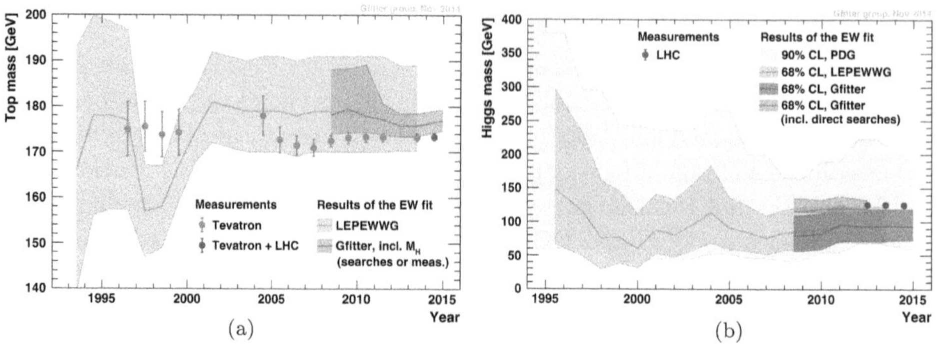

(a) (b)

Fig. 1. (a) Prediction of the top quark mass obtained from electroweak fits[33,34,39–41] compared with the direct measurements[42] as a function of time. (b) Prediction of the Higgs boson mass obtained from electroweak fits[34,41,42] compared with the direct measurements[9–11] as a function of time.

19(Higgs) GeV, beautifully confirming the ST at the loop level. From 1995 onwards the data were also precise enough to allow simultaneous fits of m_t and M_H. Figure 1(a) compares the prediction of the mass of the top quark obtained in global fits with the results of the direct measurements as a function of time. Good agreement is always seen. Starting from 2009, more direct information on the value of the mass of the Higgs boson — initially from searches at the Tevatron and CERN's LHC, later from the direct mass measurements at LHC — have allowed the global fits to determine the mass of the top quark with much better precision, as indicated by the dark band in Fig. 1(a) (cf. Section 4).

Once a reasonably precise top-mass measurement was available from CDF and D0, indirect limits on the mass of the Higgs boson could also be obtained from electroweak fits. The Higgs-mass prediction obtained in these fits as a function of time is shown in Fig. 1(b). From the beginning the fit results have shown nice agreement with the value of $M_H \approx 125\,\text{GeV}$ later found by CERN's ATLAS and CMS,[9,10] again confirming the ST at the loop-level. Figure 2(a) shows the $\Delta\chi^2$ profile of the last fit obtained by the LEP electroweak working group[34] containing all LEP data and an M_W value from the Tevatron close to the present precision.[7] The minimum is located around 100 GeV indicating the preference of a light Higgs. The absolute value of χ^2 of the fits has always been good, with a fit probability better than 10%. A detailed description of the fit and its results together with the data measured at LEP and the SLD can be found in Ref. 6.

Since a long time a slight tension is determined between $\sin^2\theta^{\ell}_{\text{eff}}$ obtained from the left–right asymmetry using polarised beams at SLD, A_{LR}, and the forward–backward asymmetry for b quarks measured at LEP, A^b_{FB}, with A_{LR} showing a preference for small M_H values, while the measured value of A^b_{FB} indicates larger Higgs masses. Many studies have been performed trying to clarify the situation. However the ST, including the precise measurements of the mass of the top quark and the Higgs boson, predicts a value almost exactly in the middle of the two measurements,

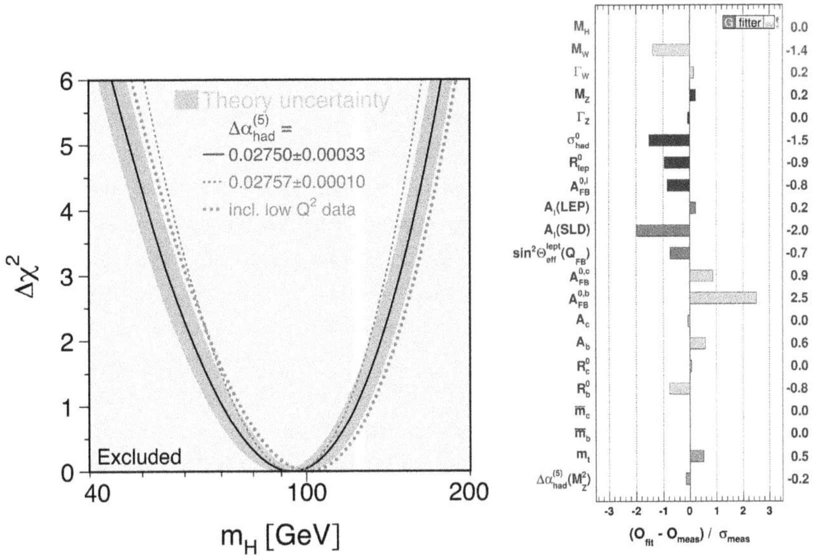

(a) (b)

Fig. 2. (a) $\Delta\chi^2$ profile of the global fit as a function of M_H as obtained by the LEP Electroweak Working Group.[7,34] (b) Comparison of the fit results with the direct measurements in units of the experimental uncertainty.[41,46]

supporting the interpretation that the tension is most likely a statistical fluctuation.

Over the years other groups have performed electroweak fits as well. The Particle Data Group (PDG) regularly performs a fit working strictly in the $\overline{\text{MS}}$ scheme.[16] The Gfitter group has written a fit package in C++ trying to include always the latest experimental data and theoretical predictions.[41] The groups obtained very similar results, demonstrating that the treatment of data and ST predictions is in general very well understood.

The LEPEW also tried to include electroweak measurements from lower energies like atomic parity violation,[43] Møller scattering[44] and neutrino–nucleon scattering.[45] These measurements contribute little to the precision, however in the case of neutrino–nucleon scattering large ununderstood systematic uncertainties contribute significantly; the low energy data are no longer used for the central results.

4. Current Status After the Higgs Discovery

The discovery of a new boson in 2012 by the ATLAS[9] and CMS[10] collaborations at the LHC and the subsequent confirmation that this boson has a spin of zero and that its properties are consistent with the properties of the ST Higgs boson[19,47–49] constitutes one of the greatest triumphs of the ST. A mass value around 125 GeV is in striking agreement with the indirect ST prediction obtained using the tool of global electroweak fits, which have always predicted a light Higgs by exploiting the contributions from quantum loops to electroweak precision observables (cf. Fig. 1(b)).

Assuming the new boson discovered by ATLAS and CMS is indeed the long sought ST Higgs boson, and interpreting its precise mass measurement as a measurement of M_H in the ST, has a dramatic impact on the scope and the interpretation of global fits of the electroweak sector of the ST. For the first time all fundamental parameters of the ST are known and all electroweak observables can be predicted at the loop level allowing stringent consistency tests. For observables like the top quark mass, the W boson mass and the weak mixing angle the knowledge of the mass of the Higgs boson leads to an enormous reduction of the uncertainty of the prediction of the observable[a] based on two-loop calculations in the ST. These more precise predictions can be confronted with the direct measurements and often the fit results are more precise. Furthermore models beyond the ST can be tested using the approach of oblique parameters without the additional uncertainty from the Higgs boson mass.

Due to the weak logarithmic dependence of the radiative corrections on M_H the results of the electroweak fit are insensitive to the exact combination method of the ATLAS and CMS mass measurements. In Ref. 46 a weighted average of $M_H = 125.14 \pm 0.24\,\text{GeV}$ was used which agrees within 1.3σ with the prediction of the fit. Consequently, including the M_H measurement in the list of input observables leads only to a small change of the goodness-of-fit compared to earlier results with global p-values around 0.20.[46] In Fig. 2(b) a comparison of the fit results with the direct measurements in units of the experimental uncertainty is shown for each observable. These pull values demonstrate impressively that the ST prediction is able to describe all input measurements consistently: no observable shows a pull value exceeding 3σ. The long-standing tension between leptonic and hadronic asymmetries is clearly visible (cf. Section 3).

The dark-grey region of Fig. 3(a) indicates the currently allowed regions at 68% and 95% CL in the (M_W vs. m_t) plane of the fit using the Higgs boson mass as measured by ATLAS and CMS.[11,18,19,b] For comparison the ST prediction for various values of M_H is also indicated as thin solid and dotted lines. The uncertainties on other input parameters, like e.g. $\Delta\alpha_{\text{had}}^{(5)}(M_Z^2)$, and uncertainties in the theoretical calculation, e.g. on M_W from missing higher orders, widen the allowed region around the thin line of $M_H = 125.14\,\text{GeV}$ such that the dark-grey area is obtained. The allowed region of a fit without using the M_H measurement is indicated in the figure by the light-grey regions. From a comparison of the dark-grey and light-grey regions it can be seen that the additional M_H information drastically reduces the

[a]The top quark mass cannot be predicted directly but only indirectly using the consistency of loop effects.

[b]Using the private Higgs-mass average instead of the official value, which was not yet available at the time of the fits, results in no visible difference.

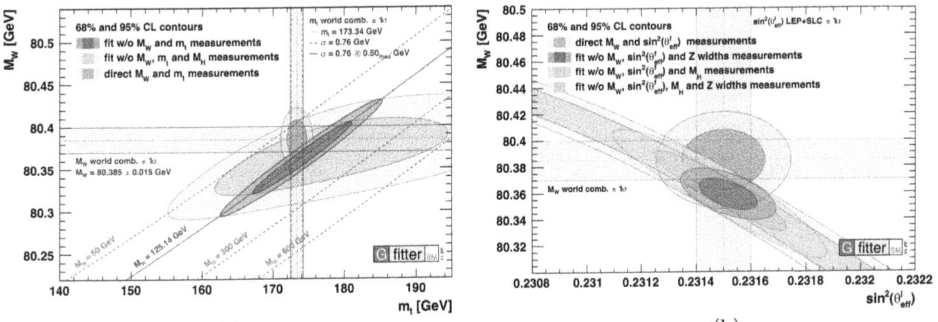

(a) (b)

Fig. 3. (a) Allowed regions in the (M_W vs. m_t) plane for fits including (dark) and excluding (light) the M_H measurement.[46] For comparison the direct M_W and m_t measurements are shown as vertical and horizontal bands and ellipses. The corresponding direct measurements are excluded from the fits. (b) Allowed regions in the (M_W vs. $\sin^2\theta^\ell_{\rm eff}$) plane with the same notation as used in the left figure.[46]

allowed values of the ST parameters m_t and M_W. The direct and indirect results on M_W and m_t are in agreement within their uncertainties which constitutes an important consistency test of the ST after the Higgs discovery. Similarly, in Fig. 3(b) the allowed regions of the fit are shown in the (M_W vs. $\sin^2\theta^\ell_{\rm eff}$) plane. Again the impact of the new M_H measurement is clearly visible.

Including the new M_H measurements in the prediction of the W mass in the electroweak fit (i.e. the direct M_W measurement is excluded) leads to:[46] $M_W = 80.358 \pm 0.008$ GeV, which exceeds the precision of the world average of the direct measurements with an uncertainty of 15 MeV.[17] For the indirect determination of the effective leptonic weak mixing angle $\sin^2\theta^\ell_{\rm eff}$, the fit obtains its minimum at $\sin^2\theta^\ell_{\rm eff} = 0.23149 \pm 0.00007$, which is again more precise than the average of the LEP/SLD measurements[6] with an uncertainty of 0.00016. Since the top quark mass enters the electroweak fit only in loop corrections the precision of the fit prediction is worse (± 2.4 GeV) than the direct measurement (currently ± 0.76 GeV[8]). These values demonstrate impressively the precision in the indirect determination of ST parameters that can be obtained using the tools of global fits after the Higgs discovery.

5. Constraints on Physics Beyond the ST

As explained the precision observables on the Z pole and M_W can be parametrised on loop level by three parameters: $\Delta\rho$, $\Delta\kappa$ and Δr, where the three parameters are largely correlated in terms of the input parameters.[50] In 1990, two similar parametrisations have been developed to take out these correlations: the STU parametrisation[12] and the ε parameters.[13] In both parameter sets one parameter (T, ε_1) absorbs the large isospin violating corrections proportional to ${m_t}^2$. Another parameter (S, ε_3) takes the rest of the corrections to $\sin^2\theta^\ell_{\rm eff}$ and the third parameter

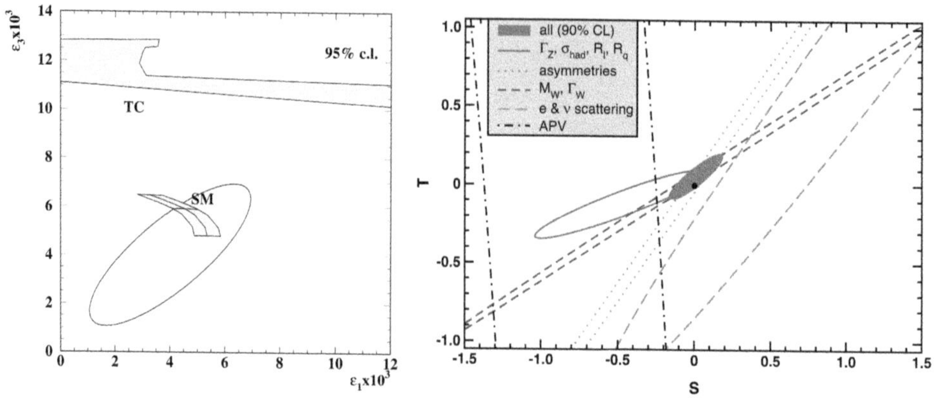

(a) (b)

Fig. 4. (a) Allowed regions in the (ε_3 vs. ε_1) plane as obtained from the LEP/SLC data in 1998 compared to prediction from the ST (grey sickle) and from technicolour models.[52,53] (b) Current experimental constraints on the oblique parameters S and T (with $U = 0$) using all observables (dark-grey) compared to the ST prediction (black dot).[16] Individual experimental constraints are illustrated by dashed and dotted ellipses.

the remainder in M_W. The ε parameters are defined such that they are zero on Born level and typical deviations are of the order of α, while the STU parameters are zero in the ST with assumed values of m_t and M_H, and typical deviations are of the order of one. An additional parameter can be defined to absorb the vertex corrections to R_b.

These parametrisations allow the experiments to publish their data in a simple but reasonably model independent way by publishing the experimental results on the STU or ε parameters as obtained from a global fit. Theorists can then easily interpret them by comparing the experimental values with the theoretical predictions obtained e.g. in models of physics beyond the ST. In the past decades this concept was heavily used to constrain various BST models. In Ref. 51 a collection of example results are given.

A first application of this idea using the ε parameters was an analysis which compared the experimental results with technicolour (TC) predictions.[52] Figure 4(a) shows the (ε_3 vs. ε_1) plane as measured (meaning obtained in a global fit) in 1998, compared to the prediction in the ST and the technicolour predictions using a simple copy of QCD at higher scales. While the ST prediction was in agreement with the experimental results, this family of TC models could clearly be excluded at this early stage.

In Fig. 4(b) the current experimental situation in the (T vs. S) plane is compared to the ST prediction whose uncertainty is much reduced by the availability of the M_H measurement. The influence of the various experimental observables is illustrated by the dotted and dashed ellipses. The values obtained currently for

$U = 0$[c] show a strong correlation between S and T and are compatible with zero within uncertainties.[16,46] Hence, the data are in impressive agreement with the ST prediction ($S = T = 0$). If only one of the three parameters is allowed to vary, this parameter would deviate at the 1.5σ level, reflecting the slight deviation in M_W.[16]

Most models tested in Ref. 51 are excluded by now by the discovery of the Higgs boson and the measurement of its properties. In recent analyses the STU parameters have been analysed together with the Higgs coupling measurements and they improve significantly limits on universal extra dimensions[54] or on the littlest Higgs model.[55]

6. Perspectives of the Electroweak Fit

Also in the coming years the interplay between precision experiments and precision theory can be used to probe the ST with unprecedented accuracy and constrain physics models beyond the ST. From experiments at the LHC and future colliders it is expected that some of the key observables entering the global ST fits and featuring a strong sensitivity to electroweak loop effects will be measured with increased experimental precision. At the LHC the datasets to be collected in the future can be used to reduce the systematic uncertainties of crucial measurements (e.g. the energy calibration of the detectors). The clean experimental environment of e^+e^- collider will allow for highly precise measurements with often small systematic and theoretical uncertainties. On the theoretical front multi-loop corrections to these observables should become available and the accuracy of the determination of the hadronic contribution to the fine-structure constant $\Delta\alpha_{\text{had}}^{(5)}(M_Z^2)$ will be increased. In Ref. 46 a detailed study of the prospects of the global ST fit for the LHC with $300\,\text{fb}^{-1}$ and ILC/GigaZ is presented. The interested reader is referred to the original literature. Here only the most important aspects relevant for the next two decades of particle physics are presented.

Among the key observables where experimental progress can be expected at the above-mentioned facilities are the W boson mass, the top quark mass and the effective weak mixing angle. The precision on the Higgs mass itself is of minor importance for the fit due to its weak logarithmic dependence in the loop calculations. For the W boson mass a combination of the LEP, Tevatron and LHC (with $300\,\text{fb}^{-1}$) could[56] optimistically lead to a total precision of 8 MeV, while at the ILC/GigaZ a precision of 5 MeV is assumed from cross-section measurements around the WW production threshold.[57] The top mass measurements at the LHC with $300\,\text{fb}^{-1}$ could reach a final experimental precision of 0.6 GeV. It is essential that the current theoretical uncertainties due to colour reconnection effects and the

[c]This is a reasonable choice since the contribution to U is negligible in most BST models.

mass definition are understood by then. At the moment an additional theory uncertainty of 0.5 GeV is assumed on the interpretation of the measured top quark mass. Scans of the $t\bar{t}$ production threshold at ILC should yield a precision of only 30 MeV where an additional theoretical uncertainty of about 100 MeV is conservatively taken into account.[57,58] While the LHC is unlikely be able to improve the precision of any electroweak observable related to the Z boson, the measurements of the left–right asymmetry A_{LR} of leptonic and hadronic Z decays at the ILC/GigaZ are expected to yield a precision of $1.3 \cdot 10^{-5}$ for $\sin^2\theta^\ell_{\rm eff}$ which represents an improvement of the present world average by more than a factor of 10.[57] The uncertainty on $\Delta\alpha^{(5)}_{\rm had}(M_Z^2)$ is expected to reduce by roughly a factor of two.[59] Since theoretical progress is difficult to quantify exactly it has been assumed in Ref. 46 that all theoretical uncertainties will be reduced by a common factor of four, which in many cases will require ambitious three-loop electroweak calculations.

With these expected improvements the prospects of global fits can be estimated and compared to the current precision. In these studies the central values of the observables are adjusted to values compatible with a Higgs boson mass of $M_H \simeq 125\,\text{GeV}$ to allow a fair comparison. With the above-mentioned improvements on M_W and m_t the LHC can improve the indirect constraint on M_H from presently (with $M_H \simeq 125\,\text{GeV}$) $^{+33}_{-27}\,\text{GeV}$ to $^{+21}_{-18}\,\text{GeV}$. An even more substantial improvement is expected for the ILC/GigaZ with an expected M_H uncertainty of about $\pm 7\,\text{GeV}$.[40] The expected improvements from LHC and ILC/GigaZ on both the fit prediction and the direct measurements in the (M_W vs. $\sin^2\theta^\ell_{\rm eff}$) plane are illustrated in Fig. 5(a). The improved indirect fit precision expected in the LHC scenario from theoretical improvements and improvements in $\Delta\alpha^{(5)}_{\rm had}(M_Z^2)$ and m_t would by far exceed the current precision of the direct measurements, which cannot be improved at the LHC significantly. A precise $\sin^2\theta^\ell_{\rm eff}$ measurement from the ILC would then be needed to confront the indirect fit determination with a direct measurement matching (or even exceeding) in precision. These studies at ILC/GigaZ

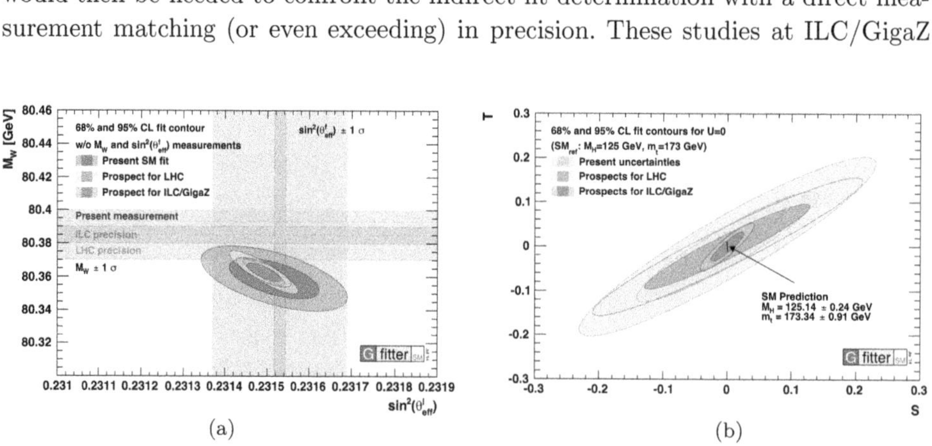

(a) (b)

Fig. 5. (a) Indirect fit constraints in the M_W vs. $\sin^2\theta^\ell_{\rm eff}$ plane for the present and extrapolated future scenarios compared to the direct measurements.[46] (b) Present and expected results on the oblique parameters S and T with U fixed to zero.[46]

will represent tests of the ST using global fits with unprecedented precision. At a future circular collider with about 100 km circumference, which is currently studied at CERN, also the uncertainties on the Z mass and width can be improved substantially. If beam polarisation can be established and the systematics can be brought under control there is also a potential to improve the measurement of the weak mixing angle by an order of magnitude.[60]

The expected future constraints on S and T for a fixed value of $U = 0$ are shown in Fig. 5(b) for the LHC only and for LHC + ILC. Again the fit assuming current uncertainties of the observables is shown for comparison. The shift in the position of the ellipses between the present data and future scenarios is caused by different central values used for the electroweak observables in these scenarios to obtain $M_H \simeq 125\,\mathrm{GeV}$. The future scenarios are by construction centred at $S = T = 0$. The ST prediction with uncertainties is also indicated. Only a minor improvement is expected for the LHC, while an improvement by a factor of three to four is expected with future lepton colliders. With these studies the e^+e^- colliders provide excellent indirect sensitivity to physics effects beyond the ST. The error on the top quark mass will be the limiting factor until a more precise value will become available from the threshold scan at a lepton collider. The error on the Higgs boson mass is irrelevant in any scenario using the present precision.

If the constraint $U = 0$ in Fig. 5(b) is dropped, the parameters S and T are only determined by $\sin^2\theta^{\ell}_{\mathrm{eff}}$ and the leptonic partial decay width of the Z boson Γ_l which is determined from a scan of the Z resonance. However, improvements in Γ_l are difficult to predict due to the unclear situation in the determination of the beam energy at a linear accelerator and due to beamstrahlung effects. Without further improvements in Γ_l the (T vs. S) ellipse in Fig. 5(b) will get rather narrow in one direction due to the $\sin^2\theta^{\ell}_{\mathrm{eff}}$ constraint but will remain very long in the other. At a future circular collider the beam energy will be measured precisely with resonant depolarisation similar to LEP[61] and beamstrahlung will be much weaker so that here significant progress on Γ_l can be expected[62] allowing a precise S and T measurement without the $U = 0$ constraint.

7. Conclusion

During the last decades global fits of the electroweak sector of the Standard Theory have been a crucial tool for highly accurate tests of the model's consistency and for indirect predictions of unmeasured model parameters. Among the greatest achievements in the past are certainly the correct prediction of the mass of the top quark and more recently the mass of the Higgs boson before their actual direct discovery.

This impressive success is based on the intense interplay of precision measurements in collider experiments at CERN and elsewhere and highly accurate theoretical calculations taking into account quantum loop effects which give access to physics at energy scales which are much higher than the energies directly available in the experimental facilities. The high precision of measurements of electroweak observables at the Z pole obtained at CERN's LEP collider, together with key measurements at the SLC and at hadron colliders are of crucial importance for these studies and allow tests of the theory with unprecedented accuracy.

With even more accurate measurements at future colliders and theoretical progress matching this precision, and e.g. including high-precision Higgs coupling measurements and predictions, the results of the electroweak fits will remain interesting for many years to come. Deeper insights into the fundamental building blocks of nature and their interactions will most certainly be obtained, and the effects of potential new physics phenomena may possibly be discovered.

References

1. S. Glashow, *Nucl. Phys.* **22**, 579–588 (1961).
2. S. Weinberg, *Phys. Rev. Lett.* **19**, 1264–1266 (1967).
3. A. Salam, *Conf. Proc. C* **680519**, 367–377 (1968).
4. H. Fritzsch, M. Gell-Mann, and H. Leutwyler, *Phys. Lett. B* **47**, 365–368 (1973).
5. D. J. Gross and F. Wilczek, *Phys. Rev. Lett.* **30**, 1343–1346 (1973).
6. The ALEPH, DELPHI, L3, OPAL, SLD Collaborations, the LEP Electroweak Working Group, the SLD Electroweak and Heavy Flavour Working Groups, 257 (2006).
7. The ALEPH, DELPHI, L3, OPAL Collaborations, and the LEP Electroweak Working Group, *Phys. Rept.* **532**, 119–244 (2013).
8. The ATLAS, CDF, CMS, D0 Collaborations, First combination of Tevatron and LHC measurements of the top-quark mass. (2014).
9. G. Aad *et al.* [ATLAS Collaboration], *Phys. Lett. B* **716**, 1–29 (2012).
10. S. Chatrchyan *et al.* [CMS Collaboration], *Phys. Lett. B* **716**, 30–61 (2012).
11. G. Aad *et al.* [ATLAS and CMS Collaborations], *Phys. Rev. Lett.* **114**, 191803 (2015).
12. M. E. Peskin and T. Takeuchi, *Phys. Rev. Lett.* **65**, 964–967 (1990).
13. G. Altarelli and R. Barbieri, *Phys. Lett. B* **253**, 161–167, (1991).
14. D. Y. Bardin and G. Passarino, *The Standard Model in the Making: Precision Study of the Electroweak Interactions* (Clarendon Press, 1999).
15. M. Kobel *et al.*, Two-Fermion Production in Electron-Positron Collisions. (2000).
16. K. Olive *et al.*, *Chin. Phys. C* **38**, 090001 (2014).
17. Tevatron Electroweak Working Group, 2012 Update of the Combination of CDF and D0 Results for the Mass of the W Boson. (2012).
18. G. Aad *et al.* [ATLAS Collaboration], *Phys. Rev. D* **90**(5), 052004 (2014).
19. V. Khachatryan *et al.* [CMS Collaboration], *Eur. Phys. J. C* **75**(5), 212 (2015).
20. M. Awramik, M. Czakon, A. Freitas, and G. Weiglein, *Phys. Rev. Lett.* **93**, 201805 (2004).
21. M. Awramik, M. Czakon, and A. Freitas, *JHEP* **11**, 048 (2006).
22. M. Awramik, M. Czakon, A. Freitas, and B. Kniehl, *Nucl. Phys. B* **813**, 174–187 (2009).
23. M. Awramik, M. Czakon, A. Freitas, and G. Weiglein, *Phys. Rev. D* **69**, 053006 (2004).
24. Y. Schröder and M. Steinhauser, *Phys. Lett. B* **622**, 124–130 (2005).

25. K. G. Chetyrkin, M. Faisst, J. H. Kühn, P. Maierhofer, and C. Sturm, *Phys. Rev. Lett.* **97**, 102003 (2006).
26. R. Boughezal and M. Czakon, *Nucl. Phys. B* **755**, 221–238 (2006).
27. A. Freitas, *JHEP* **1404**, 070 (2014).
28. K. Chetyrkin, J. H. Kühn, and A. Kwiatkowski, *Phys. Rept.* **277**, 189–281 (1996).
29. P. Baikov, K. Chetyrkin, and J. H. Kühn, *Phys. Rev. Lett.* **101**, 012002 (2008).
30. P. Baikov, K. Chetyrkin, J. Kühn, and J. Rittinger, *Phys. Rev. Lett.* **108**, 222003 (2012).
31. A. Kataev, *Phys. Lett. B* **287**, 209–212 (1992).
32. G.-C. Cho, K. Hagiwara, Y. Matsumoto, and D. Nomura, *JHEP* **1111**, 068 (2011).
33. F. Dydak, *Conf. Proc. C* **900802**, 3–32 (1990).
34. The yearly reports of the LEP electroweak working group can be found at http://lepewwg.web.cern.ch/LEPEWWG/stanmod/.
35. D. Y. Bardin *et al.*, *Comput. Phys. Commun.* **133**, 229–395 (2001).
36. G. Montagna, O. Nicrosini, F. Piccinini, and G. Passarino, *Comput. Phys. Commun.* **117**, 278–289 (1999).
37. F. Abe *et al.* [CDF Collaboration], *Phys. Rev. Lett.* **74**, 2626–2631 (1995).
38. S. Abachi *et al.* [D0 Collaboration], *Phys. Rev. Lett.* **74**, 2632–2637 (1995).
39. J. Carter, *Conf. Proc. C* **910725V2**, 3–26 (1991).
40. L. Rolandi, *AIP Conf. Proc.* **272**, 56–80 (1993).
41. http://cern.ch/gfitter.
42. All versions of the PDG particle listings and reviews can be found at http://pdg.lbl.gov/2014/html/rpp_archives.html.
43. J. Ginges and V. Flambaum, *Phys. Rept.* **397**, 63–154 (2004).
44. P. Anthony *et al.*, *Phys. Rev. Lett.* **95**, 081601 (2005).
45. G. Zeller *et al.*, *Phys. Rev. Lett.* **88**, 091802 (2002) [Erratum: *ibid.* **90** (2003) 239902].
46. M. Baak *et al. Eur. Phys. J. C* **74** (9), 3046 (2014).
47. G. Aad *et al.* [ATLAS Collaboration], *Phys. Lett. B* **726**, 120–144 (2013).
48. G. Aad *et al.* [ATLAS Collaboration], *Phys. Lett. B* **726**, 88–119 (2013).
49. V. Khachatryan *et al.* [CMS Collaboration], *Phys. Rev. D* **92**, no. 1, 012004 (2015).
50. W. F. L. Hollik, *Fortschr. Phys.* **38**, 165–260 (1990).
51. M. Baak *et al.*, *Eur. Phys. J. C* **72**, 2003 (2012).
52. J. R. Ellis, G. L. Fogli, and E. Lisi, *Phys. Lett. B* **343**, 282–290 (1995).
53. K. Mönig, *Rept. Prog. Phys.* **61**, 999–1043 (1998).
54. T. Kakuda, K. Nishiwaki, K.-y. Oda, N. Okuda, and R. Watanabe, Phenomenological constraints on universal extra dimensions at LHC and electroweak precision test. (2013).
55. J. Reuter, M. Tonini, and M. de Vries, *JHEP* **1402**, 053 (2014).
56. G. Bozzi, J. Rojo, and A. Vicini, *Phys. Rev. D* **83**, 113008 (2011).
57. H. Baer *et al.*, The International Linear Collider Technical Design Report — Volume 2: Physics. (2013).
58. A. Hoang *et al.*, *Eur. Phys. J. direct. C* **2**, 1 (2000).
59. M. Davier, private communication (2012).
60. M. Bicer *et al.*, *JHEP* **1401**, 164 (2014).
61. R. Assmann *et al.*, *Eur. Phys. J. C* **6**, 187–223 (1999).
62. A. Blondel, Chapter 8 of this Book.

Chapter 13

Production of Electroweak Bosons at Hadron Colliders: Theoretical Aspects

Michelangelo L. Mangano
CERN, TH Department, 1211 Geneva 23, Switzerland
michelangelo.mangano@cern.ch

Since the $W^{\pm}$ and Z^0 discovery, hadron colliders have provided a fertile ground, in which continuously improving measurements and theoretical predictions allow to precisely determine the gauge boson properties, and to probe the dynamics of electroweak and strong interactions. This article will review, from a theoretical perspective, the role played by the study, at hadron colliders, of electroweak boson production properties, from the better understanding of the proton structure, to the discovery and studies of the top quark and of the Higgs, to the searches for new phenomena beyond the Standard Model.

1. Introduction

All bosons of the electroweak (EW) sector of the Standard Model (SM), namely the gauge vector bosons $W^{\pm}$ and Z^0, and the scalar Higgs boson H^0, have been discovered at hadron colliders.[1–6] This well known fact is sufficient to underscore in the strongest terms the key role played by hadron colliders in the exploration of the EW sector of the SM.

In hadron colliders, the physics of EW gauge bosons has many facets. The abundant production rates via the Drell–Yan (DY) process[7] enables significant measurements of their properties, the best example being the so far unparalleled precision of the determination of the W boson mass, M_W, obtained at the Tevatron.[8] The production of EW gauge bosons in the decays of the top quark and of the Higgs boson, furthermore, makes them indispensable tools in the study of the properties of these particles. The presence of W and Z bosons in the final state of a hadronic production process acts as a tag of the underlying dynamics, singling out a limited number of production channels, which can then be studied with great precision, due to the experimental cleanness of the leptonic decay modes, and thanks to the high accuracy achieved by the theoretical calculations. Last but not least, W and Z bosons appear as final or intermediate states in the decay of most particles predicted in theories beyond the SM (BSM). Examples include the heavy bosons of new gauge interactions, supersymmetric particles, or heavy resonances featured in models of EW symmetry breaking alternative to the SM. The theoretical study and the measurements of EW gauge bosons are therefore a primary ingredient in the physics programme of hadron colliders.

It is impossible to provide, in this contribution, a complete historical overview of the development of this field, and to properly acknowledge the main contributions to both theoretical and experimental developments: on one side there are too many to fit in these few pages; on the other, the field is undergoing a continuous development, thanks to the multitude of new data that are arising from the LHC and to the rapid theoretical progress. Each of the topics briefly touched upon in this review is examined in the theoretical and experimental literature with a great degree of sophistication, and only an extended discussion would fully address the complexity and ramifications of their theoretical implications. In this review, I shall therefore limit myself to expose the great diversity of gauge boson physics in hadron collisions, through the qualitative discussion of the main ideas and results. Furthermore, I shall only cover the physics of vector gauge bosons, since the Higgs boson is covered in other chapters of this book.

I shall start from the general properties of inclusive W and Z production, focusing on the transverse and longitudinal dynamics and on the implications for the knowledge of the partonic content of the proton. I shall then discuss the phenomenological interest in the production of multiple gauge bosons. Finally, I shall overview the various mechanisms of associated production of gauge bosons and other SM particles, namely jets and heavy quarks.

2. QCD Aspects of Inclusive Vector Boson Production

The main feature of inclusive gauge boson production in hadronic collisions is that the leading-order (LO) amplitude, describing the elementary process $q\bar{q}^{(\prime)} \to V$ ($V = W, Z$) is purely EW. The dynamics of strong interactions, at this order, only enters indirectly through the parton distribution functions (PDFs), which parameterise in a phenomenological way the quark and gluon content of the proton.[a] At the large momentum scales typical of gauge boson production ($Q \sim M_V$), higher-order perturbative QCD corrections to the inclusive production are proportional to $\alpha_s(Q)$ and are typically small, in the range of 10–20%. They are known[9,10] today to next-to-next-to-leading order (NNLO), including the description of the differential distributions of the boson and of its decay leptons,[11–14] leaving theoretical uncertainties from higher-order QCD effects at the percent level. These results have been

[a]For the overview of the principles and tools of perturbative QCD and of the parton models, which are relevant to the physics of hadronic collisions, I refer to the Chapter in this Book by R. K. Ellis.

incorporated in full Monte Carlo calculations including the shower evolution, to give a complete description of the physical final states.[15–17] Next-to-leading-order (NLO) EW corrections are also known,[18–21] and play an important role both for precision measurements, and in the production rate of dilepton pairs at large p_T or with large mass, above the TeV, where they can be larger than 10%. Finally, progress towards a complete calculation of the mixed $\mathcal{O}(\alpha_s\alpha)$ corrections has been recently reported in Ref. 22.

When considering the first and second generation quarks that dominate the production of W and Z bosons, their weak couplings, including the CKM mixing parameters, are known experimentally with a precision better than a percent. This exceeds the accuracy of possible measurements in hadronic collisions, indicating that such measurements could not be possibly affected, at this level of precision, by the presence of new physics phenomena. They therefore provide an excellent ground to probe to percent precision the effects of higher-order QCD corrections and of PDFs.[23] To be more explicit, consider the leading-order (LO) cross section given by:

$$d\sigma(h_1h_2 \to V+X) = \int dx_1\, dx_2 \sum_{i,j} f_i(x_1,Q)\, f_j(x_2.Q)\, d\hat{\sigma}(ij \to V), \tag{1}$$

where $x_{1,2}$ are the fractions of the hadrons momenta and $f_{i,j}$ are the corresponding distributions of (anti)quark flavours (i,j). In the case of W production (a similar result holds for the Z), the LO partonic cross section is given by:

$$\hat{\sigma}(q_i\bar{q}_j \to W) = \pi\, \frac{\sqrt{2}}{3}\, |V_{ij}|^2\, G_F\, M_W^2\, \delta(\hat{s}-M_W^2) = A_{ij}M_W^2\, \delta(\hat{s}-M_W^2). \tag{2}$$

Here $\hat{s} = x_1x_2S$ is the partonic centre-of-mass energy squared, and V_{ij} is the element of the Cabibbo–Kobayashi–Maskawa (CKM) matrix.

Written in terms of $\tau = x_1x_2$ and of the rapidity $y = \log[(E_W + p_W^z)/(E_W - p_W^z)]^{1/2} \equiv \log(x_1/x_2)^{1/2}$, the differential and total cross sections are given by:

$$\frac{d\sigma_W}{dy} = \sum_{i,j} \frac{\pi A_{ij}}{M_W^2}\, \tau\, f_i(x_1)\, f_j(x_2), \quad x_{1,2} = \sqrt{\tau}e^{\pm y}, \tag{3}$$

$$\sigma_W = \sum_{i,j} \frac{\pi A_{ij}}{M_W^2}\tau \int_\tau^1 \frac{dx}{x}\, f_i(x)\, f_j\left(\frac{\tau}{x}\right) \equiv \sum_{i,j} \frac{\pi\, A_{ij}}{M_W^2}\tau L_{ij}(\tau) \tag{4}$$

where the function $L_{ij}(\tau)$ is usually called *partonic luminosity*. In the case of $u\bar{d}$ collisions, $\frac{\pi\, A_{ij}}{M_W^2} \sim 6.5\,\text{nb}$. It is interesting to study the partonic luminosity as a function of the hadronic centre-of-mass energy. This can be done by taking a simple approximation for the parton densities. Using the approximate behaviour $f_i(x) \sim 1/x^{1+\delta}$, with $\delta < 1$, one easily obtains:

$$L(\tau) = \frac{1}{\tau^{1+\delta}} \log\left(\frac{1}{\tau}\right) \quad \text{and} \quad \sigma_W \propto \left(\frac{S}{M_W^2}\right)^{\delta} \log\left(\frac{S}{M_W^2}\right). \tag{5}$$

The gauge boson production cross section grows therefore at least logarithmically with the hadronic centre-of-mass energy.

2.1. *Rapidity spectrum of W and Z bosons*

The features of the momentum distribution of vector bosons along the beam direction (z) are mostly controlled by properties of the parton PDFs. In particular, in the case of W bosons, the differences between the PDFs of up- and down-type quarks and antiquarks lead to interesting production asymmetries. Since the measurement of asymmetries is typically very accurate, due to the cancellation of many experimental and theoretical uncertainties, these play a fundamental role in the precision determination of quark and antiquark PDFs. Furthermore, the production asymmetries are modulated by the parity violation of the vector boson couplings, leading to further handles for the discrimination of quark and antiquark densities, and inducing a sensitivity to the weak mixing angle $\sin^2\theta_{\text{eff}}^{\text{lept}}$, which controls the vector and axial components of Z boson interactions.

2.1.1. *W charge asymmetries*

For $p\bar{p}$ collisions, and assuming for simplicity the dominance of u and d quarks, we have:

$$\frac{d\sigma_{W^+}}{dy} \propto f_u^p(x_1)\, f_{\bar{d}}^{\bar{p}}(x_2) + f_{\bar{d}}^p(x_1) f_u^{\bar{p}}(x_2), \tag{6}$$

$$\frac{d\sigma_{W^-}}{dy} \propto f_{\bar{u}}^p(x_1)\, f_d^{\bar{p}}(x_2) + f_d^p(x_1) f_{\bar{u}}^{\bar{p}}(x_2). \tag{7}$$

We can then construct the following charge asymmetry (using $f_q^{\bar{p}} = f_{\bar{q}}^p$ and assuming the dominance of the quark densities over the antiquark ones, which is valid in the kinematical region of interest for W production at the Tevatron):

$$A(y) = -A(-y) = \frac{\frac{d\sigma_{W^+}}{dy} - \frac{d\sigma_{W^-}}{dy}}{\frac{d\sigma_{W^+}}{dy} + \frac{d\sigma_{W^-}}{dy}} \sim \frac{f_u^p(x_1)\, f_d^p(x_2) - f_d^p(x_1)\, f_u^p(x_2)}{f_u^p(x_1)\, f_d^p(x_2) + f_d^p(x_1)\, f_u^p(x_2)}. \tag{8}$$

Setting $f_u(x) = f_d(x)R(x)$ we then get:

$$A(y) \sim \frac{R(x_1) - R(x_2)}{R(x_1) + R(x_2)}, \tag{9}$$

which gives an explicit relation between asymmetry and the functional dependence of the $u(x)/d(x)$ ratio. This ratio is close to 1 at small x, where the quark distributions arise mostly from sea quarks, and it increases at larger x, where the valence contribution dominates. At positive y, where $x_1 > x_2$, we therefore expect a positive asymmetry. This is confirmed in the left plot of Fig. 2, showing the asymmetry measured at the Tevatron by the CDF experiment,[27] and compared to the NNLO QCD prediction[13,21,28,29] and an estimate of the PDF uncertainty. When measuring the charged lepton from W decay, the W production asymmetry is however modulated by the W decay asymmetry caused by parity violation. The squared amplitude for the process $f_1\bar{f}_2 \to W \to f_3\bar{f}_4$ is proportional to $(p_1 \cdot p_4)(p_2 \cdot p_3)$, where $f_{1,3}$ are fermions and $f_{2,4}$ antifermions, of momenta $p_{1,\ldots,4}$. In the rest frame of this process, this is proportional to $(1 + \cos\theta)^2$, where θ is the scattering angle between final- and initial-state fermions. The momentum of the final-state fermion, therefore, points preferentially in the direction of the initial-state fermion's momentum, $\cos\theta \to 1$. For $d\bar{u} \to W^- \to \ell^-\bar{\nu}$ the charged lepton (a fermion) is more likely to move in the direction of the d quark, while for $u\bar{d} \to W^+ \to \ell^+\nu$ the charged lepton (an antifermion) is more likely to move backward. The rapidity distribution of charged leptons is therefore subject to a tension between the W production asymmetry, which at positive rapidity favours W^+ over W^-, and the decay asymmetry, which at positive rapidity favours ℓ^- over ℓ^+. The net result is a distribution that changes sign, becoming negative at large lepton rapidity. This is seen explicitly in the right plot of Fig. 2, from the D0 experiment,[30] which also shows the great sensitivity of this quantity to different PDF parameterisations, and the potential to improve their determination.

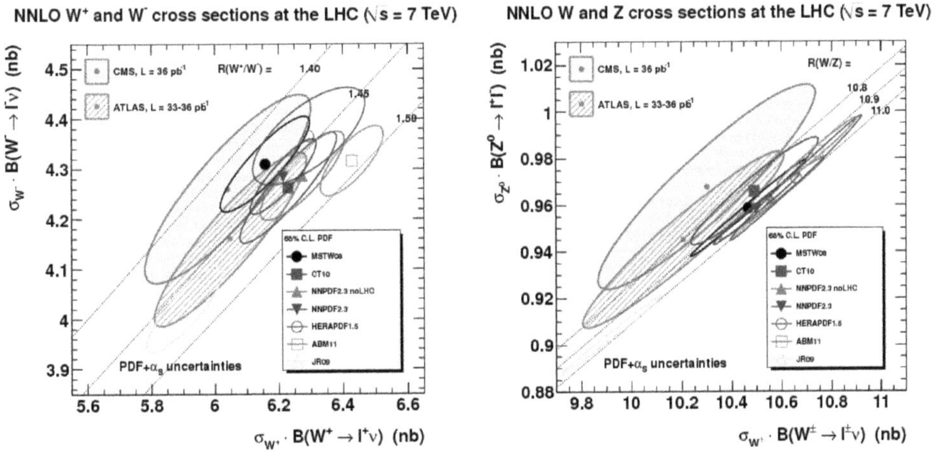

Fig. 1. W and Z boson cross sections in pp collisions at $\sqrt{S} = 7\,\text{TeV}$: ATLAS[24] and CMS[25] data, compared to NNLO predictions for various PDF sets.[23]

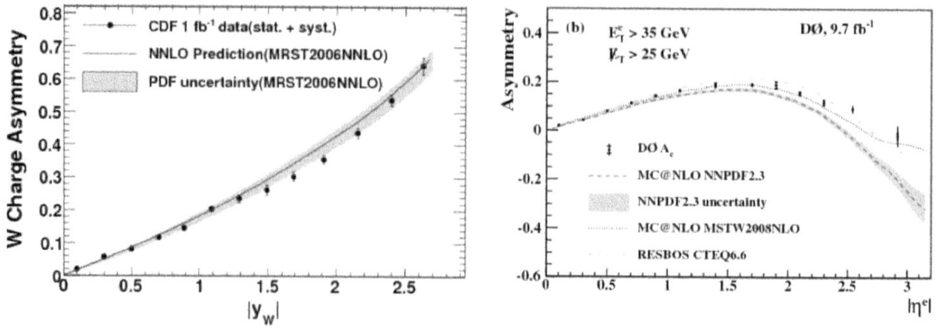

Fig. 2. Production[27] (left) and leptonic[30] (right) charge asymmetries of W bosons in $p\bar{p}$ collisions at the Tevatron, $\sqrt{S} = 1.96\,\text{TeV}$.

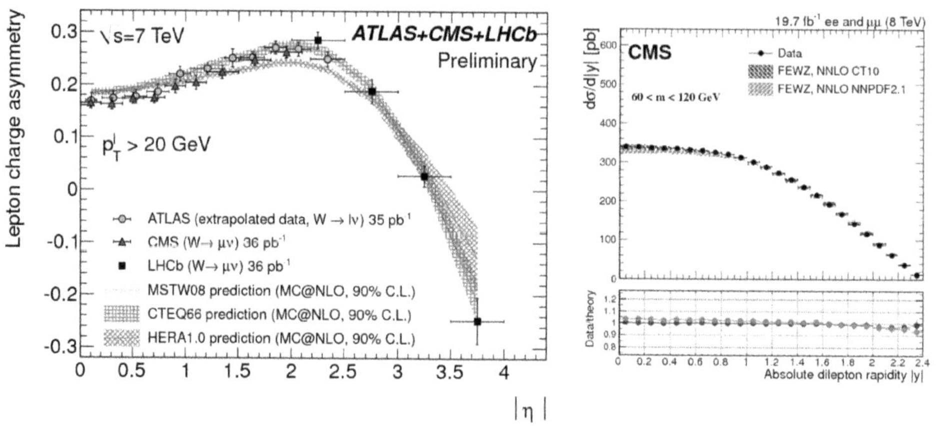

Fig. 3. Left: leptonic charge asymmetries in W production at the LHC ($\sqrt{S} = 7$ TeV), extracted from the measurements of the ATLAS,[31] CMS[32] and LHCb[26] experiments. Right: Z boson rapidity spectrum from CMS,[33] compared with NNLO predictions.[21]

In pp collisions, assuming again the dominance of the first generation of quarks and $f_q^p(x) \gg f_{\bar{q}}^p(x)$ $(q = u, d)$ at large x, the W charge asymmetry takes the form:[b]

$$A(y) = A(-y) \sim \frac{R(x_{\rm max}) - r(x_{\rm min})}{R(x_{\rm max}) + r(x_{\rm min})}, \tag{10}$$

where $x_{\rm max(min)} = \max(\min)(x_1, x_2)$ and $f_{\bar{u}}^p(x) = r(x) f_{\bar{d}}^p(x)$. The extended rapidity coverage offered by the combination of the ATLAS, CMS and LHCb detectors at the LHC, allows to fully exploit the potential of asymmetry measurements as a probe of the proton structure. This is highlighted in the left plot of Fig. 3, which summarises the LHC experimental results for the lepton charge asymmetry,

[b] It goes without saying that in actual analyses the contributions of all quark and antiquark flavours are taken into account. At the LHC, in particular, the contribution of strange and charm quarks is significant for the $W^{\pm}$ production rate, at the level of $\sim 30\%$.

obtained at $\sqrt{S} = 7$ TeV, compared to the theoretical predictions based on several sets of PDFs. In particular, notice the large spread of predictions in the largest rapidity regions, spread to be reduced once these data are included as new constraints in global PDF fits (see for example Refs. 34–37). The PDF sensitivity can be further enhanced by considering the W asymmetry at large rapidity in events produced in association with a high-p_T jet, as discussed in Ref. 38.

2.1.2. *Z rapidity spectrum and lepton charge asymmetries*

The measurement of the Z rapidity spectrum is very accurate, due to the precise reconstruction of its decay leptons. A comparison between CMS data[33] and the NNLO theoretical calculation is shown in the right plot of Fig. 3. The agreement is excellent, at the level of $\pm 5\%$, and on this scale one can detect differences between the two choices of PDFs, CT10[40] and NNPDF2.1,[39] confirming the power of these measurements to improve our knowledge of the quark distributions. Further inputs will arise from the Z production measurements performed by LHCb,[41–43] in the range $2 < y_Z < 4.25$.

As discussed above, parity violation effects lead to particular correlations in the decay directions of the final- and initial-state fermions. For Z^0 production, these correlations depend on the value of the weak mixing angle $\sin^2\theta_{\rm eff}^{\rm lept}$, which parameterises the relative strength of vector and axial couplings. In e^+e^- collisions at the Z pole, the measurement of these correlations is in principle straightforward, since we know which of the initial state particles is a fermion. The combination of such measurements, done at LEP and SLC using both leptonic and b-quark Z decays, and in particular using at SLC polarised electron beams, led[44] to the very precise determination of $\sin^2\theta_{\rm eff}^{\rm lept} = 0.23153 \pm 0.00016$. These measurements remain nevertheless puzzling, in view of a discrepancy between two of the most precise inputs into the global average, namely LEP's measurement of the forward–backward asymmetry of b quarks ($A_{\rm FB}^{0,b}$), $\sin^2\theta_{\rm eff}^{\rm lept} = 0.23221 \pm 0.00029$, and SLD's measurement of the polarised leptonic left–right asymmetry ($A_{\rm LR}$), $\sin^2\theta_{\rm eff}^{\rm lept} = 0.23098 \pm 0.00026$.

Due to the large statistics of Z bosons, experiments at hadron colliders have the potential to contribute to these measurements, and to address this puzzle. In practice, things are complicated by the lack of information, on a event-by-event basis, on which of the two initial-state partons is the quark, which is the antiquark, and what is their flavour (note that the initial state could also be qg, giving a different leptonic angular distribution). The problem is less severe in $p\bar{p}$ collisions than in pp collisions: in the former case, the most likely initial-state configuration has the quark coming from the p direction, and the antiquark coming from the $\bar{p}$ direction. A residual ambiguity remains, on whether the quark is of *up*-type or *down*-type, leading to a further slight dilution of the sensitivity. The CDF and D0 experiments at the Tevatron have presented the measurement of the weak mixing angle, from the analysis of their full dataset of Z decays (in the muonic channel for CDF, and electronic channel for D0). CDF published[45] $\sin^2\theta_{\rm eff}^{\rm lept} = 0.2315 \pm 0.0010$. D0 reports[46] $\sin^2\theta_{\rm eff}^{\rm lept} = 0.21106 \pm 0.00053$, with individual contributions of ± 0.0004 from statistics and ± 0.0003 from the PDFs. This is the most precise measurement from hadronic colliders to date, with an overall uncertainty less than a factor of 2 larger than the individual $A_{\rm FB}^{0,b}$ and $A_{\rm LR}$ determinations from LEP and SLD. These Tevatron measurements are consistent with the overall LEP+SLD average.

At the LHC, the extraction of the asymmetry is complicated by the reduced discrimination between the $q\bar{q}$ and the $\bar{q}q$ initial states, since both beams are protons. This reduced sensitivity is partly alleviated when considering events in which the Z boson is strongly boosted in either the forward or backward directions, since in this case it is more likely that the quark moves in the direction of the boosted Z, and that it is a u rather than a d. The extended rapidity coverage of the ATLAS and CMS experiments, and in particular the very forward coverage of LHCb, allows these measurements. The first result was reported by CMS,[47] based on the analysis of 1.1fb^{-1} of data at 7 TeV: $\sin^2\theta_{\rm eff}^{\rm lept} = 0.2287 \pm 0.0020({\rm stat}) \pm 0.0025({\rm syst})$. ATLAS[48] published a result based on the full 4.8fb^{-1} 7 TeV dataset: $\sin^2\theta_{\rm eff}^{\rm lept} = 0.2308 \pm 0.0005({\rm stat}) \pm 0.0006({\rm syst}) \pm 0.0009({\rm PDF}) = 0.2308 \pm 0.0012$. LHCb combined their measurements at both 7 and 8 TeV,[49] to obtain $\sin^2\theta_{\rm eff}^{\rm lept} = 0.23142 \pm 0.00073({\rm stat}) \pm 0.00052({\rm syst}) \pm 0.00056({\rm theory}) = 0.2314 \pm 0.0011$, where the theoretical uncertainty arises mostly from the PDF uncertainty. The size of the PDF systematics in both ATLAS and LHCb results underscores the importance of future progress that should emerge from the ongoing PDF determination programme,[50] based on LHC data. Current estimates suggest that the LHC experiments should eventually reach systematics at the level of today's world average uncertainty.

2.2. *Transverse momentum spectrum*

When QCD corrections to inclusive gauge boson production are considered, the most notable effect is the appearance of a transverse momentum, $p_{T,V}$. This is the result of parton-level processes such as $q\bar{q} \to Vg$ and $qg \to qV$. The former are typically dominant in $p\bar{p}$ collisions, the latter in pp collisions, as shown in Fig. 4. Depending on the value of $p_{T,V}$, relative to M_V, different dynamical and theoretical issues are exposed, as summarised in this Section.

The spectrum at small p_T is dominated by the multiple emission of soft gluons (soft with respect to the hard scale of the process, namely M_V). This leads to corrections to $d\sigma/dp_{T,V}^2$ proportional to $1/p_{T,V}^2 \alpha_S^n \log^m(M_V/p_{T,V})$ (where n is the number of soft gluons emitted, and $m \leq 2n-1$), which need to be resummed.[54–57] The leading-logarithmic soft-gluon resummation has been implemented in the context of the exact fixed-order NLO calculation,[62–65] and by now it has been extended to next-to-next-to-leading logarithmic accuracy (NNLL). For the most recent results, and a review of the existing literature on resummation, see Ref. 66.

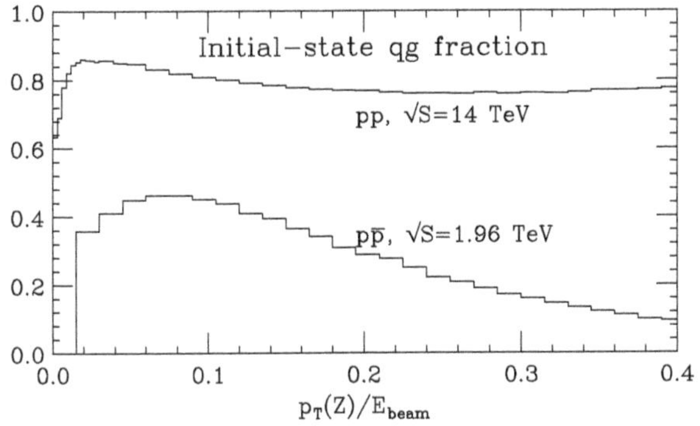

Fig. 4. Fraction of the Z bosons of transverse momentum $p_T(Z)$ produced by the quark–gluon initial state, at the Tevatron and LHC.

At the lowest end, where $p_T \sim \mathcal{O}(\text{GeV})$, the comparison of data with LO theoretical calculations has historically required the introduction of a modelling for the non-perturbative Fermi motion inside the hadron.[58] Most recently, the inclusion of exact higher-order perturbative effects up to the next-to-leading order and the resummation[59,60] of leading and sub-leading logarithms of p_T/M_V greatly reduced the need to introduce a phenomenological parameterisation of Fermi motion.[61]

The production dynamics for $p_{T,V}$ of $\mathcal{O}(M_V)$ and beyond is mostly controlled by purely perturbative physics, in addition of course to the required knowledge of the partonic densities of the proton. The comparison of data with theoretical predictions can therefore be used to improve the determination of the PDFs. The QCD corrections are known up to NLO,[62,64,65] and work is in progress towards a full NNLO result. In pp collisions the dominant process for high-p_T vector boson production is the Compton-like scattering $qg \to q'V$, as shown in Fig. 4 for the Z boson. This makes this process particularly sensitive to the gluon PDF, over a very large range of x values, as discussed in detail in Ref. 67. Such measurements should lead in the future to a more accurate determination of the gluon PDF, an essential step towards improving the precision of theoretical predictions for the total production rate of Higgs bosons.

Figure 5 shows the recent ATLAS results,[68] compared to theory, for the Z spectrum at $\sqrt{S} = 7$ TeV (similar results at 8 TeV have been reported by CMS).[69] Notice the reach of the measurement, extending up to p_T values of several hundred GeV, covering five orders of magnitude in rate. The upper right plot compares, on a linear scale, data and the results of NNLO QCD[13,21] (NNLO here and in Fig. 5 refers to $\mathcal{O}(\alpha_s^2)$, namely NNLO for the inclusive rate, but NLO for production at finite p_T). In the region of $p_T \gtrsim 20$ GeV, where the effect of the small-p_T logarithms discussed earlier is formally suppressed, data and theory agree to within 10%. At large p_T, 10% differences arise when changing the functional form of the renormalisation scale μ, from $\mu = M_W$ to $\mu = \sqrt{M_W^2 + {p_T}_W^2}$. Notice also the non-negligible effect of NLO EW corrections,[72] which grow at large p_T. For smaller p_T

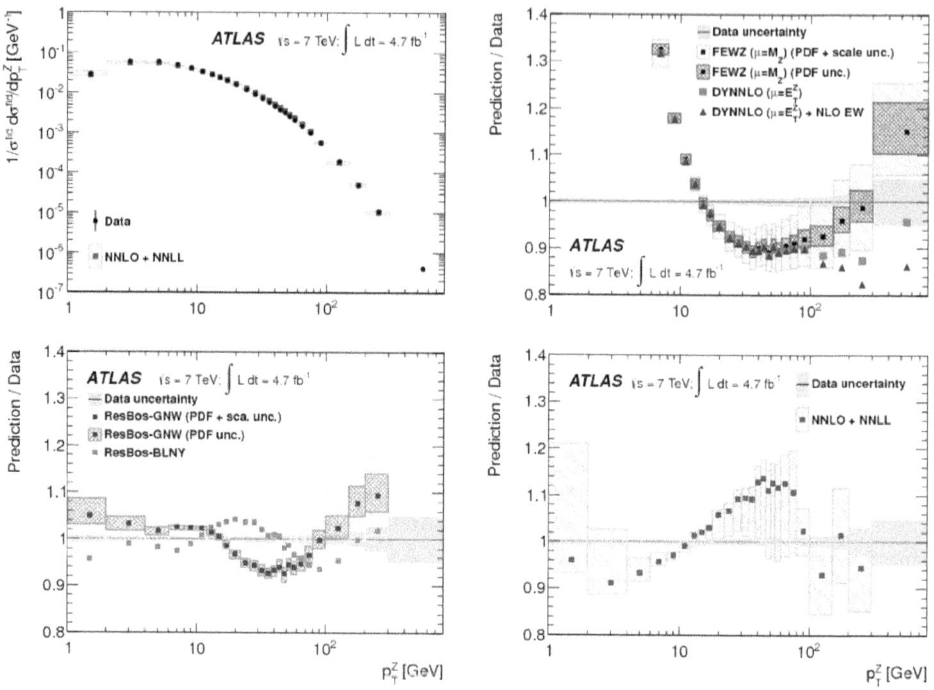

Fig. 5. Z boson p_T spectrum measured by ATLAS[68] (upper left), compared to various theoretical predictions: fixed-order NNLO (with different scales and with/without NLO EW corrections, upper right), resummed (N)NLO+(N)NLL (left[58,70] and right[71] lower plots).

values, where the fixed-order calculation is insufficient, a better agreement with the data shape is obtained by including the logarithmic resummation, an improvement included in the theoretical predictions shown in the bottom two plots. Overall, this comparison shows features in the pattern of the comparison between data and theory, and between different theoretical predictions, suggesting the need for further improvements before these very precise data can be used, for example, for improved determinations of the PDFs. Nevertheless, one should appreciate that the overall scale of deviations, which are compatible with the quoted uncertainties, is of order $\pm 10\%$, which remains quite impressive for a process in hadronic collisions.

2.3. *Off-shell gauge-boson production at large invariant mass*

The study of large-mass DY pairs is the primary probe in the search for new interactions, characterised by the existence of heavy Z' (W'), and detectable as resonances (or jacobian peaks) in the $\ell^+\ell^-$ mass (or $\ell\nu$ transverse mass) spectra. This process, furthermore, tests the pointlike nature of quarks and leptons or the possible existence of contact interactions, mediated by heavy states beyond the reach of on-shell production. In this case, BSM signals would appear as smooth deviations w.r.t. the SM behaviour in the tails of the mass spectrum. Interpreting such deviations

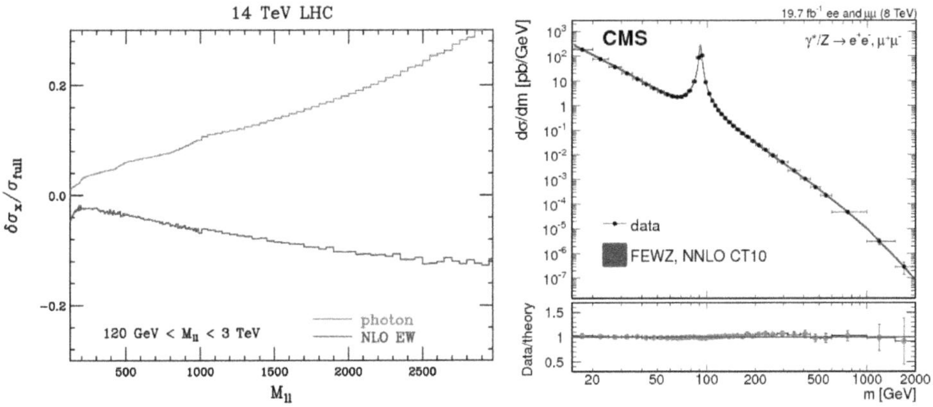

Fig. 6. Left: impact of EW and $\gamma\gamma \to \ell^+\ell^-$ corrections[75] on the DY mass spectrum at $\sqrt{S} =$ 14 TeV. Right: Drell–Yan mass spectra in pp collisions at $\sqrt{S} = 8$ TeV,[33] compared against the theoretical prediction[75] including NNLO QCD and NLO EW corrections, with PDFs from Ref. 40.

requires a reliable control of the SM prediction, including the precise knowledge of PDFs and of higher-order QCD and EW effects, including, as recently emphasised in the literature, of purely QED processes such as $\gamma\gamma \to \ell^+\ell^-$, which require as input the knowledge of the photon density inside the proton. An example of the impact of NLO EW and of the $\gamma\gamma$ corrections is given in Fig. 6, obtained in Ref. 75 using the photon PDF from the NNPDF analysis.[76] Notice the large compensation between the two opposite-sign contributions. Notice also that the $\gamma\gamma$ channel is particularly large for this observable, since the DY pair here is allowed to have small p_T ($\gamma\gamma$-induced final states are peaked at small p_T). The contribution to other large-Q^2 DY configurations, such as inclusive production at large p_T, is suppressed.

The excellent agreement between theory and data, at the few percent level, is shown in the right plot of Fig. 6, from a recent analysis of the CMS experiment.[33]

3. Multiple Production of Vector Bosons

Pair production of gauge bosons includes contributions from channels like $f\bar{f} \to \gamma/Z^* \to W^+W^-$ (with e.g. $f = e, q$), which probe directly gauge boson self-interactions, and are sensitive to deviations from the SM value of the relevant couplings (deviations known as "anomalous couplings"). Until recently the most accurate studies of these couplings came from LEP2 data above the WW threshold. In hadron colliders, one can explore a much broader range of momentum scales and of off-shell configurations (e.g. probing channels such as $q\bar{q}' \to W^* \to W\gamma$). The limited statistics and kinematical reach available at the Tevatron is nowadays largely surpassed at the LHC, whose sensitivity to anomalous couplings is quickly overtaking that of LEP2. The LHC will also have sensitivity to quartic gauge interactions, via the contributions to the $VV \to VV$ amplitude in the vector boson scattering processes $qq \to qqVV$.

Table 1. Production cross sections (pb) in pp collisions at 13 TeV for processes with multiple vector boson final states, from Ref. 77.

$W^\pm$	Z^0	W^+W^-	$W^\pm Z^0$	Z^0Z^0	$W^+W^-W^\pm$	$W^+W^-Z^0$
$1.7 \cdot 10^5$	$5.4 \cdot 10^4$	$1.0 \cdot 10^2$	$4.5 \cdot 10^1$	$1.4 \cdot 10^1$	$2.1 \cdot 10^{-1}$	$1.7 \cdot 10^{-1}$
$W^\pm Z^0Z^0$	$Z^0Z^0Z^0$	$W^+W^-W^+W^-$	$W^+W^-W^\pm Z^0$	$W^+W^-Z^0Z^0$	$W^\pm Z^0Z^0Z^0$	$Z^0Z^0Z^0Z^0$
$5.6 \cdot 10^{-2}$	$1.4 \cdot 10^{-2}$	$1.0 \cdot 10^{-3}$	$1.2 \cdot 10^{-3}$	$7.1 \cdot 10^{-4}$	$1.2 \cdot 10^{-4}$	$2.6 \cdot 10^{-5}$

Table 2. Cross section ratios in pp collisions at 13 TeV for processes with multiple vector boson final states.

$W^\pm/Z^0$	$W^+W^-/W^\pm Z^0$	$W^+W^-W^\pm/W^+W^-Z^0$	$W^+W^-W^+W^-/$ $W^+W^-W^\pm Z^0$
3.1	2.2	1.2	0.8
$W^+W^-/W^\pm$	$W^\pm W^+W^-/W^+W^-$	$W^+W^-W^+W^-/W^+W^-W^\pm$	
$0.6 \cdot 10^{-3}$	$2.1 \cdot 10^{-3}$	$4.8 \cdot 10^{-3}$	

Final states with gauge boson pairs are also fundamental signatures of Higgs decays, $H \to ZZ^*$ and $H \to WW^*$. The direct VV production is an important background to the isolation of these Higgs signals, as well as to the signals of associated Higgs production, $pp \to VH$, where Higgs decays such as $H \to b\bar{b}$ or $H \to \tau^+\tau^-$ are subject to the background coming from the tail of the Z^0 invariant mass distribution in $pp \to VZ^0$.

Furthermore, as a source of multilepton final states, multiple gauge boson production is a potential background to a large number of BSM searches, for example searches for the supersymmetric partners of gauge and Higgs bosons.

Table 1 collects the NLO production rates for all processes with up to four massive gauge bosons in the final state, taken from the comprehensive tabulation of NLO results for high-multiplicity final states in Ref. 77, which also lists the respective systematic uncertainties. Most of these processes should be eventually measurable at the LHC. The ratio $\sigma(W^\pm)/\sigma(Z^0) \sim 3$ (see Table 2) turns out to be rather independent of the energy, and of whether we consider pp or $p\bar{p}$ collisions. When considering leptonic final states, this leads to the well know factor of $\sigma(W^\pm)BR(W \to \ell\nu)/\sigma(Z^0)BR(Z \to \ell^+\ell^-) {\sim} 10$, which was observed at the $Sp\bar{p}S$, at the Tevatron and at the LHC. This ratio reflects primarily the nature and value of the couplings of W and Z bosons to the up and down quarks in the proton. When the number of final-state gauge bosons increases, the relative emission rate of further W or Z bosons gets closer to 1, as gauge boson selfcouplings become dominant with respect to the couplings to initial state quarks. This is seen, for example, in the first row of Table 2. Notice, finally, that the cost of emitting additional gauge bosons decreases with multiplicity: in part this is due to the larger number of sources to

Table 3. W^+W^- cross sections measured in pp collisions at 7 and 8 TeV. The first three measurements include the Higgs contribution, the fourth one subtracts it. NLO and NNLO theoretical predictions are from Ref. 79. The $gg \to WW$ process is included only in the NNLO contribution, and the Higgs contribution,[80] to be added to the NNLO result for the comparison with the data in the first three rows, is shown separately.

Experiment	Data (pb)	NLO	NNLO	$gg \to H \to WW^*$
ATLAS[81] (7 TeV, incl. H)	54.4 ± 6.0	$45.2^{+1.7}_{-1.3}$	$49.0^{+1.0}_{-0.9}$	$3.3^{+0.2}_{0.3}$
CMS[82] (7 TeV, incl. H)	52.4 ± 5.1			
ATLAS[83] (8 TeV, incl. H)	$71.4 \pm 1.2_{\rm stat}\,{}^{+5.6}_{-5.0}{}_{\rm tot}$	$54.8^{+2.0}_{-1.6}$	$59.8^{+1.3}_{-1.1}$	$4.1^{+0.3}_{-0.3}$
CMS[84] (8 TeV, no H)	$60.1 \pm 0.9_{\rm stat} \pm 3.1_{\rm th} \pm 3.5_{\rm exp+lum}$			

couple to, in part to the reduced relative increase in required energy (producing two gauge boson takes at least twice the energy than producing one, while producing four takes only an extra 30% more than producing 3).

At this time, the only final states measured with large statistics are those with two bosons (for a complete phenomenological study of boson pair production at the LHC, see e.g. Ref. 78). In the case of W^+W^- production, in particular, the statistical uncertainty of the LHC measurements at 8 TeV is already about half the size of the systematics one. Table 3 summarises the status of comparisons between data and theory for this channel. The consistency of theoretical predictions and data is greatly improved by the inclusion of the NNLO results.[79,c] While compatible with uncertainties, some small discrepancy is nevertheless present at 8 TeV, even between experimental results. New measurements at 13 TeV, and in particular the potentially more accurate measurement of cross section ratios[88] between 13 and 8 TeV, will certainly clarify the whole picture.

On the theoretical side, note that at $\mathcal{O}(\alpha_s^2)$ a new contribution appears, namely $gg \to WW$, mediated by a quark loop (since this gg channel enters for the first time at this order, its description is referred to as LO even though it enters through a loop diagram). Its size is significant, due to the large gg luminosity, and contributes toward improving the agreement between theory and data. The NLO correction to this new channel (therefore of $\mathcal{O}(\alpha_s^3)$) has recently been computed.[89] At the LHC, the correction relative to the LO $gg \to WW$ process can be large, up to 50%, leading to a further increase of the total cross section by about 2%. But the size of the cross section depends strongly on the kinematical cuts applied to the final state,[89] an element that should be taken into proper account in the comparison with the experimental measurements. This underscores the complexity of such high-precision tests of QCD dynamics, but it is encouraging that continuous progress is taking place in improving the theoretical calculations.

[c]A similar pattern is observed in the case of the $V\gamma$ production cross sections, where the very recent completion of the NNLO calculations[85] has improved the agreement with LHC data.[86,87]

4. Associated Production of Vector Bosons with Jets and Heavy Quarks

The associated production of gauge bosons and jets[90] is a natural manifestation of higher-order QCD corrections to inclusive production. The measurement of such final states has a long history, starting from the CERN $Sp\bar{p}S$ collider experiments,[91–93] which highlighted their role as backgrounds to new physics, and as a probe of α_S. Later studies at the Tevatron[94,95] have been crucial to test quantitatively the theoretical modelling, and to establish background rates for the discovery and study of the top quark and for the search of the Higgs boson. On the theory side, the last few years have seen remarkable progress, with the NLO calculations[96–101] of processes with up to 5 jets, the inclusion of NLO EW[102] corrections, and most recently of NNLO QCD corrections to $W + 1$ jet[103] and $Z + 1$ jet[104] production.

The latest LHC measurements[105–108] of $V+$ multijet production have reached multiplicities up to 7 jets, with a precision and an agreement with theoretical calculations that, at least for multiplicities up to 4 jets and for most kinematical distributions, reach the level of $\pm(10–20)\%$. This is shown for example in Fig. 7. Notice that a new feature of the production of gauge bosons with jets emerges at the LHC, given the large jet energies that can be reached: the probability of weak

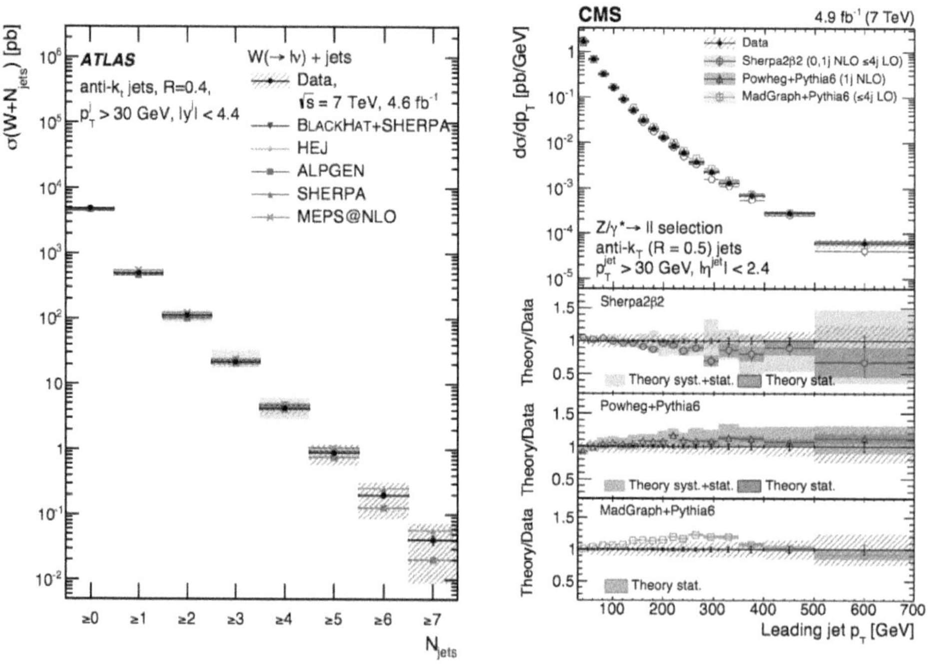

Fig. 7. Data vs. theory comparison for $V+$jets at the LHC. Left: $W + N$ jet rates at 7 TeV.[105] Right: leading-jet p_T spectrum in Z+jets at 7 TeV.[108]

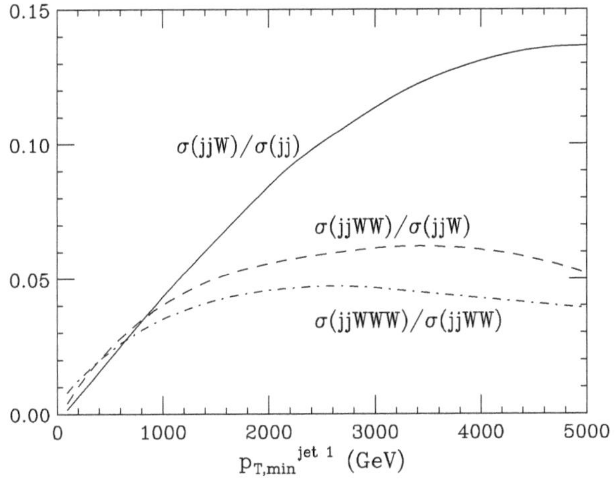

Fig. 8. Emission probability for W bosons at 14 TeV, in events with 2 or more jets where the leading jet has $p_T > p_{T,\mathrm{min}}$.

boson emission increases, from the 10^{-3} level of inclusive QCD processes, up to over 10% for jet transverse momenta of several TeV (see Fig. 8).

4.1. *W+charm quarks*

At the LO, the W+charm cross section is proportional to $\sin^2\theta_C d(x) + \cos^2\theta_C s(x)$, where $d(x)$ and $s(x)$ are the PDFs of the down and strange quarks and θ_C is the Cabibbo mixing angle. The Cabibbo-allowed process is dominant, and allows the measurement of the strange quark distribution.[110,111] The difference in the production rate of W^-c and $W^+\bar{c}$, after accounting for the small contribution of d and $\bar{d}$ quarks, is furthermore sensitive to the difference between $s(x)$ and $\bar{s}(x)$.[d] The first measurements at the LHC[112–114] have already led to useful constraints on PDF fits, but are still statistics limited, and there is still large room for improvements.

4.2. $V + Q\bar{Q}$, *with* $Q = c, b$

The associated production of vector bosons and heavy quark pairs is an interesting SM process, which has particular relevance as leading background to studies of the top quark, of the Higgs boson, and to many searches for physics beyond the SM. While the case of $c\bar{c}$ and $b\bar{b}$ production are similar from the theoretical point of view, we shall focus here on the case of b pairs, which has a larger phenomenological relevance, and for which more experimental data are available.

[d]The assumption $s(x) = \bar{s}(x)$, which has been used in the past in global PDF fits due to the lack of direct experimental discriminating observables, is not respected at the NNLO in the Q^2 evolution of PDFs.

$Wb\bar{b}$ + jets production gives rise to final states similar to those arising from $t\bar{t}$ production, and its presence, particularly at the Tevatron, was one of the main hurdles in the top quark discovery and in its precision studies. This is less so at the LHC, where its rate relative to the $t\bar{t}$ signal is much smaller than at the Tevatron. $Vb\bar{b}$ is an irreducible background to the associated production of gauge and Higgs bosons, in the leading Higgs decay channel $VH \to Vb\bar{b}$. As such, its understanding is nowadays the subject of many studies.

From the theoretical perspective, the Wbb and Zbb processes are rather different. In the former case, the only LO production channel is $q\bar{q}' \to Wb\bar{b}$. In the latter case, both $q\bar{q}$ and gg initial states contribute to the $Zb\bar{b}$ production. For $q\bar{q} \to Vb\bar{b}$, the $b\bar{b}$ pair is produced by the splitting of a final state gluon in the $q\bar{q} \to Vg$ process. The difference between $q\bar{q}$- and gg-initiated processes is particularly relevant when one considers final states where only one b jet is tagged: in the $gg \to Zb(\bar{b})$ case, in fact, there is a large contribution induced by processes in which one of the gluons in $gg \to Zbb$ undergoes a collinear splitting to $b\bar{b}$, and the b quark undergoes a hard scattering with the other gluon ($gb \to Zb$), leading to the high-p_T tagged b-jet. The $\bar{b}$ is preferentially emitted at small p_T, covering a wide rapidity range, and the integral over its full emission phase space leads to a large logarithm. One could describe this process by associating this large logarithm to the build up of a b quark PDF inside the proton, and the measurement of Zb final states provides therefore a powerful probe for the determination of the b PDF. In the case of $q\bar{q} \to Wbb$, on the other hand, the measurement of single b-jet production receives comparable contributions from the cases where the $\bar{b}$ is too soft to reconstruct a jet, and cases in which the pair produced by gluon splitting is collinear, and is merged within the same jet. In the latter case, one exposes potentially large logarithms $\log(p_T^{\rm jet}/m_b)$.

For what concerns the comparison of theory and data from the Tevatron[118–120] and from the LHC,[114,121–125] the agreement of data for $V + b$-jet with NLO fixed-order perturbative calculations[115–117] is often marginal (possibly due to the presence of large logarithms that call for improved resummed calculations), although consistent with the uncertainties. A better agreement is typically found in the comparisons with data where both b-jets are tagged. Future measurements at the LHC, with larger statistics and better control on the experimental systematics, will allow further improvements of the theoretical modelling.

4.3. $V + t\bar{t}$

The associated production of W and Z bosons with a pair of top quarks is a special case of the processes discussed in the previous subsection. In many respects, the theoretical description of $Vt\bar{t}$ production is however simpler: the mass of the top quark and of the gauge bosons are both large and of comparable size, so that we do not have the difficulties associated with the presence of largely different scales. For example, there are no large logarithms to be resummed, or assumptions to be made

Table 4. Production cross sections (pb) in pp collisions at 13 TeV for various top quark and vector boson final states, from Ref. 77.

$t\bar{t}$	$t\bar{t}W^{\pm}$	$t\bar{t}Z^0$	$t\bar{t}W^+W^-$	$t\bar{t}W^{\pm}Z^0$	$t\bar{t}Z^0Z^0$
674	0.57	0.76	$9.9 \cdot 10^{-3}$	$3.5 \cdot 10^{-3}$	$1.8 \cdot 10^{-3}$

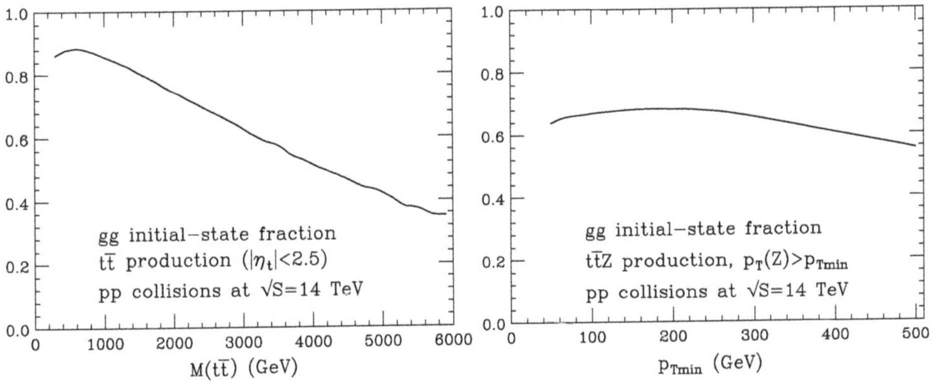

Fig. 9. Initial state gg fraction in inclusive production of $t\bar{t}$ (left) and $Zt\bar{t}$ (right), in pp collisions at $\sqrt{S} = 14$ TeV.

about the relevant heavy quark density of the proton; furthermore, the prediction for the basic process, namely the inclusive production of the heavy quark pair, is much more precise for top quarks than for bottom or charm quarks.

From the phenomenological perspective, the associated production with top quarks has interesting features. To start with, at LHC energies and above these processes are the main source of multiple gauge boson production. This is clear from Table 4, which reports the total rates for several final states involving top quarks and massive gauge bosons. Considering that each top and antitop quark produces a W boson in their decay, the comparison with the multi-V rates given in Table 1 shows that final states with W^+W^- pairs are more likely to arise from $t\bar{t}$ production and decay than from direct EW production. This is true as well of processes with production of additional gauge bosons, e.g. $t\bar{t}Z$ versus W^+W^-Z. This fact should be taken into account when extracting EW production rates from the data, and when estimating multiboson backgrounds to BSM signals.

Another interesting observation is that, in pp collisions at the LHC energies and above, the production of a Z boson is more frequent than the production of a W boson, contrary to the usual hierarchy of W vs. Z production rates. The reason is that, at LO, the $t\bar{t}W$ final state can only be produced starting from the $q\bar{q}'$ initial state, while $t\bar{t}Z$ can be produced from both $q\bar{q}$ and gg initial states. The gg fraction in $t\bar{t}Z$ production at 14 TeV is shown in Fig. 9, as a function of $p_{T,Z}$. Since inclusive $t\bar{t}$ production is dominated by the gg channel (85% of the rate at $\sqrt{S} = 14$ TeV, see also the left plot in Fig. 9), the emission of a W is suppressed with respect to the emission of a Z. This is shown explicitly in Table 4, where the rates of various processes with top quarks and gauge bosons are given. The usual hierarchy in rate between W and Z production is restored for associated production of W^+W^- versus ZZ, when the gg initial state is active for both processes.

The above considerations have several corollaries. The production of a W boson, singling out the $q\bar{q}'$ initial state, allows to scrutinise more closely the production mechanism $q\bar{q} \to t\bar{t}$, which is otherwise suppressed at the LHC. This may be useful[126] to enhance the sensitivity to possible new physics at the origin of the forward-backward production asymmetry reported at the Tevatron.[127–129] The study of $t\bar{t}W$ production at very large invariant mass of the $t\bar{t}$ system, furthermore, allows to probe directly the $t\bar{t}g$ vertex in the domain of gluon virtuality $Q \sim m_{tt}$. This is because the leading production diagram has a W emitted from the initial state, followed by the s-channel annihilation $q\bar{q} \to t\bar{t}$.

The final state $t\bar{t}Z$, on the other hand, allows to measure directly the $t\bar{t}Z$ vertex, since this is the coupling that drives the dominant $gg \to t\bar{t}Z$ contribution.[130] Future LHC data will allow to set stringent and model-independent limits on anomalous dipole contributions to the $t\bar{t}Z$ vertex, with sensitivity comparable to that obtained from indirect EW precision measurements at LEP.[131]

Both $t\bar{t}W$ and $t\bar{t}Z$ processes are known theoretically with full NLO accuracy in QCD,[132–135] which leads to an intrinsic systematic uncertainty of about $\pm 10\%$. In spite of low production rates at the energies of the first run of the LHC ($\sigma(t\bar{t}W^\pm) \sim \sigma(t\bar{t}Z) \sim 200 \pm 20$ fb at $\sqrt{S} = 8$ TeV), ATLAS and CMS have nevertheless obtained a signal evidence, for both processes, at the level of 5σ, or better. The first CMS results[136] have been updated recently,[137] leading to the measurements of $\sigma(t\bar{t}W^\pm) = 382^{+117}_{-102}$ fb (4.8σ) and $\sigma(t\bar{t}Z) = 242^{+65}_{-55}$ fb (6.4σ). ATLAS[138] measured $\sigma(t\bar{t}W^\pm) = 369^{+100}_{-91}$ fb (5.0σ) and $\sigma(t\bar{t}Z) = 176^{+58}_{-52}$ fb (4.2σ). All these results are well compatible with the SM predictions.

5. Conclusions

The production of vector gauge bosons in hadron collisions is like a Swiss knife: it is a versatile, reliable and robust tool for physics at the high-energy frontier. It exposes a vast variety of phenomena, covering most aspects of the dynamics of both EW and strong interactions. While contributing to our deeper understanding and consolidation of the SM, the knowledge acquired about the production mechanisms of gauge bosons is also essential to study the properties of the top quark and of the Higgs boson, and to refine the sensitivity of searches for BSM phenomena.

The precision of measurements and theoretical calculations has greatly improved in the past few years, and allows now comparisons of gauge boson production properties at the few percent level of precision. This precision is superior to what can be achieved in most other hard processes in hadronic collisions, and is liable to improve even further, through continued theoretical and experimental efforts and additional ingenuity. Already today, this precision can be used to improve the determination of the EW parameters and of the proton PDFs, competing with the pre-LHC state-of-the-art provided, respectively, by the results of the LEP and SLC e^+e^- colliders, and of the HERA *ep* collider. It is easy to predict that the physics of W and Z bosons at the LHC will continue surprising us for its richness for a long time to come.

References

1. G. Arnison *et al.* [UA1 Collaboration], *Phys. Lett. B* **122**, 103 (1983).
2. M. Banner *et al.* [UA2 Collaboration], *Phys. Lett. B* **122**, 476 (1983).
3. G. Arnison *et al.* [UA1 Collaboration], *Phys. Lett. B* **126**, 398 (1983).
4. P. Bagnaia *et al.* [UA2 Collaboration], *Phys. Lett. B* **129**, 130 (1983).
5. G. Aad *et al.* [ATLAS Collaboration], *Phys. Lett. B* **716**, 1 (2012) [arXiv:1207.7214 [hep-ex]].
6. S. Chatrchyan *et al.* [CMS Collaboration], *Phys. Lett. B* **716**, 30 (2012) [arXiv: 1207.7235 [hep-ex]].
7. S. D. Drell and T. M. Yan, *Phys. Rev. Lett.* **25**, 316 (1970) [*Phys. Rev. Lett.* **25**, 902 (1970)].
8. T. A. Aaltonen *et al.* [CDF and D0 Collaborations], *Phys. Rev. D* **88**, 5, 052018 (2013) [arXiv:1307.7627 [hep-ex]].
9. R. Hamberg, W. L. van Neerven and T. Matsuura, *Nucl. Phys. B* **359**, 343 (1991) [*Nucl. Phys. B* **644**, 403 (2002)].
10. R. V. Harlander and W. B. Kilgore, *Phys. Rev. Lett.* **88**, 201801 (2002) [hep-ph/0201206].
11. C. Anastasiou, L. J. Dixon, K. Melnikov and F. Petriello, *Phys. Rev. D* **69**, 094008 (2004) [hep-ph/0312266].
12. K. Melnikov and F. Petriello, *Phys. Rev. D* **74**, 114017 (2006) [hep-ph/0609070].
13. S. Catani, L. Cieri, G. Ferrera, D. de Florian and M. Grazzini, *Phys. Rev. Lett.* **103**, 082001 (2009) [arXiv:0903.2120 [hep-ph]].
14. R. Gavin, Y. Li, F. Petriello and S. Quackenbush, *Comput. Phys. Commun.* **182**, 2388 (2011) [arXiv:1011.3540 [hep-ph]].
15. A. Karlberg, E. Re and G. Zanderighi, *JHEP* **1409**, 134 (2014) [arXiv:1407.2940 [hep-ph]].
16. S. Höche, Y. Li and S. Prestel, *Phys. Rev. D* **91**, 7, 074015 (2015) [arXiv:1405.3607 [hep-ph]].
17. S. Alioli, C. W. Bauer, C. Berggren, F. J. Tackmann and J. R. Walsh, *Phys. Rev. D* **92**, 9, 094020 (2015) [arXiv:1508.01475 [hep-ph]].
18. S. Dittmaier and M. Kramer, *Phys. Rev. D* **65**, 073007 (2002) [hep-ph/0109062].
19. U. Baur and D. Wackeroth, *Phys. Rev. D* **70**, 073015 (2004) [hep-ph/0405191].
20. U. Baur, O. Brein, W. Hollik, C. Schappacher and D. Wackeroth, *Phys. Rev. D* **65**, 033007 (2002) [hep-ph/0108274].
21. Y. Li and F. Petriello, *Phys. Rev. D* **86**, 094034 (2012) [arXiv:1208.5967 [hep-ph]].
22. S. Dittmaier, A. Huss and C. Schwinn, arXiv:1511.08016 [hep-ph].
23. S. Forte and G. Watt, *Ann. Rev. Nucl. Part. Sci.* **63**, 291 (2013) [arXiv:1301.6754 [hep-ph]].
24. G. Aad *et al.* [ATLAS Collaboration], *Phys. Rev. D* **85**, 072004 (2012) [arXiv:1109.5141 [hep-ex]].

25. S. Chatrchyan *et al.* [CMS Collaboration], *JHEP* **1110**, 132 (2011) [arXiv:1107.4789 [hep-ex]].
26. R. Aaij *et al.* [LHCb Collaboration], *JHEP* **1412**, 079 (2014) [arXiv:1408.4354 [hep-ex]].
27. T. Aaltonen *et al.* [CDF Collaboration], *Phys. Rev. Lett.* **102**, 181801 (2009) [arXiv:0901.2169 [hep-ex]].
28. S. Catani, G. Ferrera and M. Grazzini, *JHEP* **1005**, 006 (2010) [arXiv:1002.3115 [hep-ph]].
29. R. Gavin, Y. Li, F. Petriello and S. Quackenbush, *Comput. Phys. Commun.* **184**, 208 (2013) [arXiv:1201.5896 [hep-ph]].
30. V. M. Abazov *et al.* [D0 Collaboration], *Phys. Rev. D* **91**, 3, 032007 (2015) [*Phys. Rev. D* **91**, 7, 079901 (2015)] [arXiv:1412.2862 [hep-ex]].
31. G. Aad *et al.* [ATLAS Collaboration], *Phys. Lett. B* **701**, 31 (2011) [arXiv:1103.2929 [hep-ex]].
32. S. Chatrchyan *et al.* [CMS Collaboration], *Phys. Rev. D* **90**, 3, 032004 (2014) [arXiv:1312.6283 [hep-ex]].
33. V. Khachatryan *et al.* [CMS Collaboration], *Eur. Phys. J. C* **75**, 4, 147 (2015) [arXiv:1412.1115 [hep-ex]].
34. S. Alekhin, J. Bluemlein and S. Moch, *Phys. Rev. D* **89**, 5, 054028 (2014) [arXiv:1310.3059 [hep-ph]].
35. R. D. Ball *et al.* [NNPDF Collaboration], *JHEP* **1504**, 040 (2015) [arXiv:1410.8849 [hep-ph]].
36. L. A. Harland-Lang, A. D. Martin, P. Motylinski and R. S. Thorne, *Eur. Phys. J. C* **75**, 5, 204 (2015) [arXiv:1412.3989 [hep-ph]].
37. S. Dulat *et al.*, arXiv:1506.07443 [hep-ph].
38. S. Farry and R. Gauld, arXiv:1505.01399 [hep-ph].
39. R. D. Ball, V. Bertone, F. Cerutti, L. Del Debbio, S. Forte, A. Guffanti, J. I. Latorre and J. Rojo *et al.*, *Nucl. Phys. B* **849**, 296 (2011) [arXiv:1101.1300 [hep-ph]].
40. J. Gao, M. Guzzi, J. Huston, H. L. Lai, Z. Li, P. Nadolsky, J. Pumplin and D. Stump *et al.*, *Phys. Rev. D* **89**, 3, 033009 (2014) [arXiv:1302.6246 [hep-ph]].
41. R. Aaij *et al.* [LHCb Collaboration], *JHEP* **1508**, 039 (2015) [arXiv:1505.07024 [hep-ex]].
42. R. Aaij *et al.* [LHCb Collaboration], arXiv:1503.00963 [hep-ex].
43. R. Aaij *et al.* [LHCb Collaboration], arXiv:1511.08039 [hep-ex].
44. S. Schael *et al.* [ALEPH and DELPHI and L3 and OPAL and SLD and LEP Electroweak Working Group and SLD Electroweak Group and SLD Heavy Flavour Group Collaborations], *Phys. Rept.* **427**, 257 (2006) [hep-ex/0509008].
45. T. A. Aaltonen *et al.* [CDF Collaboration], *Phys. Rev. D* **89**, 7, 072005 (2014) [arXiv:1402.2239 [hep-ex]].
46. S. Yang *et al.* (D0 Collaboration), D0 Note 6426-CONF 2014.
47. S. Chatrchyan *et al.* [CMS Collaboration], *Phys. Rev. D* **84**, 112002 (2011) [arXiv:1110.2682 [hep-ex]].
48. G. Aad *et al.* [ATLAS Collaboration], arXiv:1503.03709 [hep-ex].
49. R. Aaij *et al.* [LHCb Collaboration], arXiv:1509.07645 [hep-ex].
50. J. Rojo *et al.*, *J. Phys. G* **42**, 103103 (2015) [arXiv:1507.00556 [hep-ph]].
51. J. Kubar-Andre and F. E. Paige, *Phys. Rev. D* **19**, 221 (1979).
52. G. Altarelli, R. K. Ellis and G. Martinelli, *Nucl. Phys. B* **143**, 521 (1978) [*Nucl. Phys. B* **146**, 544 (1978)].
53. G. Altarelli, R. K. Ellis and G. Martinelli, *Nucl. Phys. B* **157**, 461 (1979).
54. Y. L. Dokshitzer, D. Diakonov and S. I. Troian, *Phys. Rept.* **58**, 269 (1980).

55. G. Parisi and R. Petronzio, *Nucl. Phys. B* **154**, 427 (1979).
56. G. Curci, M. Greco and Y. Srivastava, *Nucl. Phys. B* **159**, 451 (1979).
57. J. C. Collins and D. E. Soper, *Nucl. Phys. B* **193**, 381 (1981) [*Nucl. Phys. B* **213**, 545 (1983)].
58. C. Balazs and C. P. Yuan, *Phys. Rev. D* **56**, 5558 (1997) [hep-ph/9704258].
59. G. Bozzi, S. Catani, G. Ferrera, D. de Florian and M. Grazzini, *Phys. Lett. B* **696**, 207 (2011) [arXiv:1007.2351 [hep-ph]].
60. T. Becher, M. Neubert and D. Wilhelm, *JHEP* **1202**, 124 (2012) [arXiv:1109.6027 [hep-ph]].
61. V. M. Abazov *et al.* [D0 Collaboration], *Phys. Rev. Lett.* **106**, 122001 (2011) [arXiv:1010.0262 [hep-ex]].
62. R. K. Ellis, G. Martinelli and R. Petronzio, *Nucl. Phys. B* **211**, 106 (1983).
63. G. Altarelli, R. K. Ellis, M. Greco and G. Martinelli, *Nucl. Phys. B* **246**, 12 (1984).
64. R. J. Gonsalves, J. Pawlowski and C. F. Wai, *Phys. Rev. D* **40**, 2245 (1989).
65. P. B. Arnold and M. H. Reno, *Nucl. Phys. B* **319**, 37 (1989) [*Nucl. Phys. B* **330**, 284 (1990)].
66. S. Catani, D. de Florian, G. Ferrera and M. Grazzini, arXiv:1507.06937 [hep-ph].
67. S. A. Malik and G. Watt, *JHEP* **1402**, 025 (2014) [arXiv:1304.2424 [hep-ph]].
68. G. Aad *et al.* [ATLAS Collaboration], *JHEP* **1409**, 145 (2014) [arXiv:1406.3660 [hep-ex]].
69. V. Khachatryan *et al.* [CMS Collaboration], arXiv:1504.03511 [hep-ex].
70. M. Guzzi, P. M. Nadolsky and B. Wang, *Phys. Rev. D* **90**, 1, 014030 (2014) [arXiv:1309.1393 [hep-ph]].
71. A. Banfi, M. Dasgupta, S. Marzani and L. Tomlinson, *Phys. Lett. B* **715**, 152 (2012) [arXiv:1205.4760 [hep-ph]].
72. A. Denner, S. Dittmaier, T. Kasprzik and A. Muck, *JHEP* **1106**, 069 (2011) [arXiv:1103.0914 [hep-ph]].
73. K. Hagiwara, R. D. Peccei, D. Zeppenfeld and K. Hikasa, *Nucl. Phys. B* **282**, 253 (1987).
74. U. Baur and E. L. Berger, *Phys. Rev. D* **47**, 4889 (1993).
75. R. Boughezal, Y. Li and F. Petriello, *Phys. Rev. D* **89**, 3, 034030 (2014) [arXiv:1312.3972 [hep-ph]].
76. R. D. Ball *et al.* [NNPDF Collaboration], *Nucl. Phys. B* **877**, 290 (2013) [arXiv:1308.0598 [hep-ph]].
77. J. Alwall, R. Frederix, S. Frixione, V. Hirschi, F. Maltoni, O. Mattelaer, H.-S. Shao and T. Stelzer *et al.*, *JHEP* **1407**, 079 (2014) [arXiv:1405.0301 [hep-ph]].
78. J. M. Campbell, R. K. Ellis and C. Williams, *JHEP* **1107**, 018 (2011) [arXiv:1105.0020 [hep-ph]].
79. T. Gehrmann, M. Grazzini, S. Kallweit, P. Maierhfer, A. von Manteuffel, S. Pozzorini, D. Rathlev and L. Tancredi, *Phys. Rev. Lett.* **113**, 21, 212001 (2014) [arXiv:1408.5243 [hep-ph]].
80. S. Heinemeyer *et al.* [LHC Higgs Cross Section Working Group Collaboration], arXiv:1307.1347 [hep-ph].
81. G. Aad *et al.* [ATLAS Collaboration], *Phys. Lett. B* **712**, 289 (2012) [arXiv:1203.6232 [hep-ex]].
82. S. Chatrchyan *et al.* [CMS Collaboration], *Eur. Phys. J. C* **73**, 10, 2610 (2013) [arXiv:1306.1126 [hep-ex]].
83. The ATLAS Collaboration, ATLAS-CONF-2014-033, ATLAS-COM-CONF-2014-045.
84. CMS Collaboration [CMS Collaboration], CMS-PAS-SMP-14-016.

85. M. Grazzini, S. Kallweit and D. Rathlev, arXiv:1504.01330 [hep-ph].
86. G. Aad *et al.* [ATLAS Collaboration], *Phys. Rev. D* **87**, 11, 112003 (2013) [*Phys. Rev. D* **91**, 11, 119901 (2015)] [arXiv:1302.1283 [hep-ex]].
87. V. Khachatryan *et al.* [CMS Collaboration], *JHEP* **1504**, 164 (2015) [arXiv:1502.05664 [hep-ex]].
88. M. L. Mangano and J. Rojo, *JHEP* **1208**, 010 (2012) [arXiv:1206.3557 [hep-ph]].
89. F. Caola, K. Melnikov, R. Rntsch and L. Tancredi, arXiv:1511.08617 [hep-ph].
90. J. Campbell and M. Mangano, *Ann. Rev. Nucl. Part. Sci.* **61**, 311 (2011).
91. G. Arnison *et al.* [UA1 Collaboration], *Phys. Lett. B* **147**, 493 (1984).
92. P. Bagnaia *et al.* [Bern-CERN-Copenhagen-Orsay-Pavia-Saclay and UA2 Collaborations], *Phys. Lett. B* **139**, 105 (1984).
93. R. Ansari *et al.* [UA2 Collaboration], *Phys. Lett. B* **215**, 175 (1988).
94. T. Aaltonen *et al.* [CDF Collaboration], *Phys. Rev. D* **77**, 011108 (2008) [arXiv:0711.4044 [hep-ex]].
95. V. M. Abazov *et al.* [D0 Collaboration], *Phys. Rev. D* **88**, 9, 092001 (2013) [arXiv:1302.6508 [hep-ex]].
96. R. K. Ellis, K. Melnikov and G. Zanderighi, *Phys. Rev. D* **80**, 094002 (2009) [arXiv:0906.1445 [hep-ph]].
97. C. F. Berger, Z. Bern, L. J. Dixon, F. Febres Cordero, D. Forde, T. Gleisberg, H. Ita and D. A. Kosower *et al.*, *Phys. Rev. D* **80**, 074036 (2009) [arXiv:0907.1984 [hep-ph]].
98. C. F. Berger, Z. Bern, L. J. Dixon, F. Febres Cordero, D. Forde, T. Gleisberg, H. Ita and D. A. Kosower *et al.*, *Phys. Rev. D* **82**, 074002 (2010) [arXiv:1004.1659 [hep-ph]].
99. C. F. Berger, Z. Bern, L. J. Dixon, F. Febres Cordero, D. Forde, T. Gleisberg, H. Ita and D. A. Kosower *et al.*, *Phys. Rev. Lett.* **106**, 092001 (2011) [arXiv:1009.2338 [hep-ph]].
100. H. Ita, Z. Bern, L. J. Dixon, F. Febres Cordero, D. A. Kosower and D. Maitre, *Phys. Rev. D* **85**, 031501 (2012) [arXiv:1108.2229 [hep-ph]].
101. Z. Bern, L. J. Dixon, F. Febres Cordero, S. Hche, H. Ita, D. A. Kosower, D. Matre and K. J. Ozeren, *Phys. Rev. D* **88**, 1, 014025 (2013) [arXiv:1304.1253 [hep-ph]].
102. A. Denner, S. Dittmaier, T. Kasprzik and A. Muck, *JHEP* **0908**, 075 (2009) [arXiv:0906.1656 [hep-ph]].
103. R. Boughezal, C. Focke, X. Liu and F. Petriello, arXiv:1504.02131 [hep-ph].
104. A. Gehrmann-De Ridder, T. Gehrmann, E. W. N. Glover, A. Huss and T. A. Morgan, arXiv:1507.02850 [hep-ph].
105. G. Aad *et al.* [ATLAS Collaboration], *Eur. Phys. J. C* **75**, 2, 82 (2015) [arXiv:1409.8639 [hep-ex]].
106. V. Khachatryan *et al.* [CMS Collaboration], *Phys. Lett. B* **741**, 12 (2015) [arXiv:1406.7533 [hep-ex]].
107. G. Aad *et al.* [ATLAS Collaboration], *JHEP* **1307**, 032 (2013) [arXiv:1304.7098 [hep-ex]].
108. V. Khachatryan *et al.* [CMS Collaboration], *Phys. Rev. D* **91**, 5, 052008 (2015) [arXiv:1408.3104 [hep-ex]].
109. G. Aad *et al.* [ATLAS Collaboration], *Eur. Phys. J. C* **74**, 12, 3168 (2014) [arXiv:1408.6510 [hep-ex]].
110. U. Baur, F. Halzen, S. Keller, M. L. Mangano and K. Riesselmann, *Phys. Lett. B* **318**, 544 (1993) [hep-ph/9308370].
111. W. T. Giele, S. Keller and E. Laenen, *Phys. Lett. B* **372**, 141 (1996) [hep-ph/9511449].
112. G. Aad *et al.* [ATLAS Collaboration], *JHEP* **1405**, 068 (2014) [arXiv:1402.6263 [hep-ex]].

113. S. Chatrchyan *et al.* [CMS Collaboration], *JHEP* **1402**, 013 (2014) [arXiv:1310.1138 [hep-ex]].
114. R. Aaij *et al.* [LHCb Collaboration], arXiv:1505.04051 [hep-ex].
115. J. M. Campbell, R. K. Ellis, F. Maltoni and S. Willenbrock, *Phys. Rev. D* **73**, 054007 (2006) [*Phys. Rev. D* **77**, 019903 (2008)] [hep-ph/0510362].
116. J. M. Campbell, R. K. Ellis, F. Maltoni and S. Willenbrock, *Phys. Rev. D* **75**, 054015 (2007) [hep-ph/0611348].
117. F. Febres Cordero, L. Reina and D. Wackeroth, *Phys. Rev. D* **80**, 034015 (2009) [arXiv:0906.1923 [hep-ph]].
118. T. Aaltonen *et al.* [CDF Collaboration], *Phys. Rev. D* **79**, 052008 (2009) [arXiv:0812.4458 [hep-ex]].
119. V. M. Abazov *et al.* [D0 Collaboration], *Phys. Rev. D* **87**, 9, 092010 (2013) [arXiv:1301.2233 [hep-ex]].
120. V. M. Abazov *et al.* [D0 Collaboration], *Phys. Rev. D* **91**, 5, 052010 (2015) [arXiv:1501.05325 [hep-ex]].
121. G. Aad *et al.* [ATLAS Collaboration], *JHEP* **1306**, 084 (2013) [arXiv:1302.2929 [hep-ex]].
122. S. Chatrchyan *et al.* [CMS Collaboration], *Phys. Lett. B* **735**, 204 (2014) [arXiv:1312.6608 [hep-ex]].
123. R. Aaij *et al.* [LHCb Collaboration], *JHEP* **1501**, 064 (2015) [arXiv:1411.1264 [hep-ex]].
124. G. Aad *et al.* [ATLAS Collaboration], *JHEP* **1410**, 141 (2014) [arXiv:1407.3643 [hep-ex]].
125. S. Chatrchyan *et al.* [CMS Collaboration], *JHEP* **1406**, 120 (2014) [arXiv:1402.1521 [hep-ex]].
126. F. Maltoni, M. L. Mangano, I. Tsinikos and M. Zaro, *Phys. Lett. B* **736**, 252 (2014) [arXiv:1406.3262 [hep-ph]].
127. T. Aaltonen *et al.* [CDF Collaboration], *Phys. Rev. D* **87**, 9, 092002 (2013) [arXiv:1211.1003 [hep-ex]].
128. V. M. Abazov *et al.* [D0 Collaboration], *Phys. Rev. D* **90**, 7, 072001 (2014) [arXiv:1403.1294 [hep-ex]].
129. J. A. Aguilar-Saavedra, D. Amidei, A. Juste and M. Perez-Victoria, arXiv:1406.1798 [hep-ph].
130. U. Baur, A. Juste, L. H. Orr and D. Rainwater, *Phys. Rev. D* **71**, 054013 (2005) [hep-ph/0412021].
131. R. Röntsch and M. Schulze, arXiv:1501.05939 [hep-ph].
132. A. Lazopoulos, T. McElmurry, K. Melnikov and F. Petriello, *Phys. Lett. B* **666**, 62 (2008) [arXiv:0804.2220 [hep-ph]].
133. A. Kardos, Z. Trocsanyi and C. Papadopoulos, *Phys. Rev. D* **85**, 054015 (2012) [arXiv:1111.0610 [hep-ph]].
134. J. M. Campbell and R. K. Ellis, *JHEP* **1207**, 052 (2012) [arXiv:1204.5678 [hep-ph]].
135. M. V. Garzelli, A. Kardos, C. G. Papadopoulos and Z. Trocsanyi, *JHEP* **1211**, 056 (2012) [arXiv:1208.2665 [hep-ph]].
136. V. Khachatryan *et al.* [CMS Collaboration], *Eur. Phys. J. C* **74**, 9, 3060 (2014) [arXiv:1406.7830 [hep-ex]].
137. V. Khachatryan *et al.* [CMS Collaboration], arXiv:1510.01131 [hep-ex].
138. G. Aad *et al.* [ATLAS Collaboration], arXiv:1509.05276 [hep-ex].

Chapter 14

A Historical Profile of the Higgs Boson

John Ellis,* Mary K. Gaillard,† Dimitri V. Nanopoulos‡

**Theoretical Particle Physics and Cosmology Group, Department of Physics, King's College London, London WC2R 2LS, UK*
Theoretical Physics Department, CERN, CH-1211 Geneva 23, Switzerland

†Department of Physics, University of California and Theoretical Physics Group, Bldg. 50A5104, Lawrence Berkeley National Laboratory Berkeley, CA 94720, USA

‡George P. and Cynthia W. Mitchell Institute for Fundamental Physics and Astronomy, Texas A&M University, College Station, TX 77843, USA
Astroparticle Physics Group, Houston Advanced Research Center (HARC), Mitchell Campus, Woodlands, TX 77381, USA
Academy of Athens, Division of Natural Sciences, 28 Panepistimiou Avenue, Athens 10679, Greece

The Higgs boson was postulated in 1964, and phenomenological studies of its possible production and decays started in the early 1970s, followed by studies of its possible production in e^+e^-, $\bar{p}p$ and pp collisions, in particular. Until recently, the most sensitive searches for the Higgs boson were at LEP between 1989 and 2000, which were complemented by searches at the Fermilab Tevatron. Then the LHC experiments ATLAS and CMS entered the hunt, announcing on July 4, 2012 the discovery of a "Higgs-like" particle with a mass of about 125 GeV. This identification has been supported by subsequent measurements of its spin, parity and coupling properties. It was widely anticipated that the Higgs boson would be accompanied by supersymmetry, although other options, like compositeness, were not completely excluded. So far there are no signs of any new physics, and the measured properties of the Higgs boson are consistent with the predictions of the minimal Standard Model. This article reviews some of the key historical developments in Higgs physics over the past half-century.

1. Introduction

The Standard Model of particle physics codifies the properties and interactions of the fundamental constituents of all the visible matter in the Universe. It describes successfully the results of myriads of accelerator experiments, some of them to a very high degree of precision. However, for quite some time the Standard Model resembled a jigsaw puzzle with one piece missing: the Higgs boson. It, or something capable of replacing it, was essential for the calculability of the Standard Model and its consistency with experimental data. The last piece of the puzzle, at times (somewhat dubiously) termed the "Holy Grail" of particle physics, or even the "God Particle", was finally put into place with the July 4, 2012, announcement of the discovery[1] at the CERN Large Hadron Collider (LHC) of a "Higgs-like" particle at a mass of approximately 125 GeV. Subsequently, measurements of its properties by the ATLAS and CMS collaborations have shown more detailed consistency with predictions for the Higgs boson of the Standard Model, but searches for possible discrepancies indicative of new physics beyond the Standard Model are continuing.

The existence of the Higgs boson was first postulated in 1964,[2] following earlier theoretical work that introduced spontaneous symmetry breaking into condensed-matter[3] and particle physics.[4–6] It was incorporated into the Standard Model in 1967,[7,8] and shown in 1971[9] to lead to a calculable and predictive unified theory of the weak and electromagnetic interactions. With the discovery of neutral currents in 1973,[10] the discovery of charmonium in 1974,[11] the discoveries of the $W^{\pm}$ and Z^0 particles in 1983[12] and subsequent detailed measurements, the predictions of the Standard Model have been crowned with a series of successes.

Already in 1975, before the experimental discovery of charm was confirmed, the authors considered that the discovery of the Higgs boson would be the culmination of the experimental verification of the Standard Model, and we published a paper outlining its phenomenological profile.[13] At the time, the ideas of spontaneously-broken gauge theories were still generally regarded as quite hypothetical, and the Higgs boson was not on the experimental agenda. However, its star rose over the subsequent years, first in e^+e^- collisions[14] and subsequently in $\bar{p}p$ and pp collisions,[15,16] until it became widely (though incompletely) perceived as the primary objective of experiments at the LHC. The 2012 ATLAS and CMS discovery has finally provided closure on half a century of theoretical conjecture, and set the stage for a new phase of searches for physics beyond the Standard Model.

In this paper we trace the trajectory of the Higgs boson from its humble theoretical origins, through its rise to phenomenological prominence, to its experimental apotheosis. However, its discovery raises as many questions as it answers.

2. Prehistory

The physicist's concept of the vacuum does not correspond to the naive idea of 'empty' space. Instead, a physicist recognises that even in the absence of physical particles there are quantum effects due to 'virtual' particles fluctuating in the vacuum. For a physicist, the vacuum is the lowest-energy state, after taking these quantum effects into account. This lowest-energy state may not possess all the symmetries of the underlying equations of the physical system, a phenomenon known as 'spontaneous' symmetry breaking, or 'hidden' symmetry.

This mechanism of spontaneous symmetry breaking first came to prominence in the phenomenon of superconductivity, as described in the theory of Bardeen, Cooper and Schrieffer.[17] According to this theory, the photon acquires an effective mass when it propagates through certain materials at sufficiently low temperatures, as discussed earlier by Ginzburg and Landau.[18] In free space, the masslessness of the photon is guaranteed by Lorentz invariance and U(1) gauge symmetry. A superconductor has a well-defined rest frame, so Lorentz invariance is broken explicitly. However, the gauge symmetry is still present, though 'hidden' by the condensation of Cooper pairs of electrons[19] in the lowest-energy state (vacuum). It was explicitly shown by Anderson[3] how the interactions with the photon of the Cooper pairs inside a superconductor caused the former to acquire an effective mass.

The idea of spontaneous symmetry breaking was introduced into particle physics by Nambu[4] in 1960. He suggested that the small mass and low-energy interactions of pions could be understood as a reflection of a spontaneously-broken global chiral symmetry, which would have been exact if the up and down quarks were massless. His suggestion was that light quarks condense in the vacuum, much like the Cooper pairs of superconductivity. When this happens, the 'hidden' chiral symmetry causes the pions' masses to vanish, and fixes their low-energy couplings to protons, neutrons and each other.

A simple model of spontaneous global U(1) symmetry breaking was introduced by Goldstone[20] in 1961, with a single complex field ϕ as illustrated in Fig. 1. The effective potential

$$V(|\phi|) = \frac{\lambda}{4}(|\phi|^2 - v^2)^2 \tag{1}$$

is unstable at the origin where $\langle|\phi|\rangle = 0$. Instead, the lowest-energy state, the vacuum, is at the bottom of the brim of the 'Mexican hat', with

$$\langle|\phi|\rangle = v \neq 0\,. \tag{2}$$

The phase of ϕ is, however, not determined, and all choices are equivalent with the same energy. The system must choose some particular value of the phase, but changing the phase would cost no energy. Hence the system has a massless degree of freedom corresponding to rotational fluctuations of the field around the brim of the Mexican hat. It is a general theorem, proven later in 1961 by Goldstone, Salam and Weinberg[21] that spontaneous breaking of a global symmetry such as chiral symmetry must be accompanied by the appearance of one or more such Nambu–Goldstone bosons.

However, this is not necessarily the case if it is a gauge symmetry that is broken, as in the non-relativistic case of superconductivity.[3] Anderson conjectured that it should be possible to extend this mechanism to the relativistic case, as did Klein

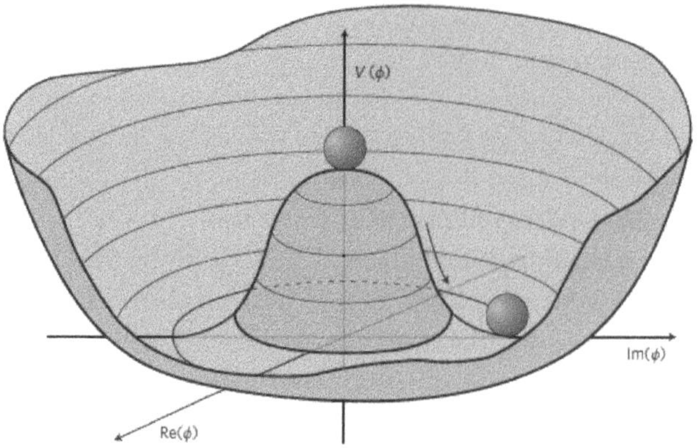

Fig. 1. A prototypical effective 'Mexican hat' potential that leads to 'spontaneous' symmetry breaking. The vacuum, i.e., the lowest-energy state, is described by a randomly-chosen point around the bottom of the brim of the hat. In a 'global' symmetry, movements around the bottom of the hat corresponds to a massless spin-zero 'Nambu–Goldstone' boson.[4,20] In the case of a local (gauge) symmetry, as was pointed out by Englert and Brout,[5] by Higgs[2] and by Guralnik, Hagen and Kibble,[24] this boson combines with a massless spin-one boson to yield a massive spin-one particle. The Higgs boson[2] is a massive spin-zero particle corresponding to quantum fluctuations in the radial direction, oscillating between the centre and the side of the hat.

and Lee,[22] but it was argued by Gilbert[23] that Lorentz invariance would forbid this.

3. And Then There Was Higgs

Spontaneous breaking of gauge symmetry was introduced into particle physics in 1964 by Englert and Brout,[5] followed independently by Higgs,[2,6] and subsequently by Guralnik, Hagen and Kibble.[24] They demonstrated how one could dispose simultaneously of two unwanted massless bosons, a spinless Nambu–Goldstone boson and a gauge boson of an exact local symmetry, by combining them into a single massive vector boson in a fully relativistic theory. The two polarisation states of a massless vector boson are combined with the single degree of freedom of a spin-zero particle to yield the three degrees of freedom of a massive spin-one particle V with mass:

$$m_V = g_V \frac{v}{\sqrt{2}}, \tag{3}$$

where g_V is the corresponding gauge coupling constant.

Englert and Brout[5] considered explicitly a non-Abelian Yang–Mills theory, assumed the formation of a vacuum expectation value (vev) of a non-singlet scalar field, and used a diagrammatic approach to demonstrate mass generation for the gauge field. The first paper by Higgs[6] demonstrated that gauge symmetry provides a loophole in the 'no-go' theorem of Gilbert mentioned above, and his second paper[2] exploited this loophole to demonstrate mass generation in the Abelian case. The subsequent paper by Guralnik, Hagen and Kibble[24] referred in its text to the

Englert/Brout and Higgs papers, and also demonstrated mass generation in the Abelian case.

The second paper by Higgs[2] is the only one of the 1964 papers to mention explicitly [his equation (2b)] the existence of a massive scalar particle associated with the curvature of the effective potential (1) that determines the vev v of the charged field:

$$m_H = \sqrt{2\lambda} v. \tag{4}$$

Englert and Brout[5] do not discuss the spectrum of physical scalars, whilst Guralnik, Hagen and Kibble[24] mention a massless scalar that decoupled from the massive excitations in their model.

Also worthy of note is a remarkable paper written in ignorance of these papers by Migdal and Polyakov in 1965,[25] while they were still students,[a] in which they discuss partial spontaneous symmetry breaking in the non-Abelian case. The year 1966 also saw a further important paper by Higgs,[26] in which he discussed in detail the formulation of the spontaneously-broken Abelian theory. In particular, he derived explicitly the Feynman rules for processes involving what has come to be known as the massive Higgs boson, discussing its decay into 2 massive vector bosons, as well as vector-scalar and scalar–scalar scattering. Another important paper by Kibble[27] discussed in detail the non-Abelian case, including partial spontaneous symmetry breaking, and also mentioned the appearance of massive scalar bosons à la Higgs.

The next important step was the incorporation by Weinberg[7] and by Salam[8] of non-Abelian spontaneous symmetry breaking into Glashow's[28] unified SU(2) × U(1) model of the weak and electromagnetic interactions. The paper by Weinberg was the first to observe that the scalar field vev could also give masses to fundamental fermions f that are proportional to their coupling to the Higgs boson:

$$m_f = g_{f\bar{f}H} v. \tag{5}$$

However, the seminal papers on spontaneous breaking of gauge symmetries and electroweak unification were largely ignored by the particle physics community until the renormalisability of spontaneously-broken gauge theories was demonstrated by 't Hooft and Veltman.[9] These ideas then joined the mainstream very rapidly, thanks in particular to a series of influential papers by B. W. Lee and collaborators.[29,30]

4. A Phenomenological Profile of the Higgs Boson

B. W. Lee also carries much of the responsibility for calling the Higgs boson the Higgs boson, mentioning repeatedly 'Higgs scalar fields' in a review talk at the

[a]It was finally published in 1966 after a substantial delay caused by the scepticism of Soviet academicians.

International Conference on High-Energy Physics in 1972.[31] However, in the early 1970s there were only a few suggestions how to constrain or exclude the possible existence of a physical Higgs boson. One paper considered the possible effect of Higgs exchange on neutron- and deuteron-electron scattering and derived a lower bound $m_H > 0.6$ MeV,[32] and another constrained Higgs emission from neutron stars, yielding the lower bound $m_H > 0.7$ MeV.[33] There was also a theoretical discussion of possible Higgs production in $0^+ \to 0^+$ nuclear transitions,[34] and its non-observation in excited ^{16}O and ^{4}He decays led to the Higgs being excluded in the mass range 1.03 MeV $< m_H <$ 18.3 MeV.[35] In parallel, data on neutron-nucleus scattering were used to constrain $m_H > 15$ MeV.[36] The two latter were the strongest limits obtained in this period.

This was the context in which we embarked in 1975 on the first systematic study of possible Higgs phenomenology.[13] Neutral currents had been discovered,[10] and the J/ψ particle[11] was thought to be charmonium, though doubts remained and the discovery of open charm still lay in the future. The search for the intermediate vector bosons $W^\pm$ and Z^0 was appearing on the experimental agenda, but the CERN $\bar{p}p$ collider that was to discover them had not yet been proposed. However, it seemed to us that the clinching test of the spontaneous symmetry-breaking paradigm underlying the Standard Model would be discovering the Higgs boson.

To this end, we considered the decay modes of the Higgs boson if it weighed up to 100 GeV, calculating for the first time the loop-induced Higgs decays to photon pairs.[13] The dominant mechanism for this decay is an anomalous triangle diagram with a $W^\pm$ loop, and there are also subdominant diagrams with massive quarks:[b,37]

$$\Gamma(H \to \gamma\gamma) = \frac{G_\mu \alpha^2 m_H^3}{128\sqrt{2}\pi^3} \left| A_1(r_W) + \sum_f N_c Q_f^2 A_{1/2}(r_f) \right|^2, \tag{6}$$

where

$$A_1(r) \equiv -[2r^2 + 3r + 3(2r-1)f(r)]/r^2, \quad A_{1/2}(r) \equiv 2[r + (r-1)f(r)]/r^2 \tag{7}$$

and $f(r) \equiv \arcsin^2 \sqrt{r}$ for $r_{W,f} \equiv m_H^2/4m_{W,f}^2$. We also estimated the cross sections for many different mechanisms for producing the Higgs boson, intending to cover the full allowed mass range from $\mathcal{O}(15)$ MeV upwards. In addition to considering the production of a relatively light Higgs boson in hadron decays and interactions, we also considered production in e^+e^- collisions, including Z^0 decays and Higgs-strahlung processes such as $e^+e^- \to Z^0 + H$.[13,c]

Back in 1975, the likelihood of a definitive search for the Higgs boson seemed somewhat remote. That was why, rather tongue-in-cheek, we closed our paper[13]

---

[b]We did not calculate these: the t and b had not been discovered at that time.

[c]The latter processes were also considered independently in Refs. 38 and 39, respectively.

with the following modest words: "*We should perhaps finish our paper with an apology and a caution. We apologise to experimentalists for having no idea what is the mass of the Higgs boson, ..., and for not being sure of its couplings to other particles, except that they are probably all very small. For these reasons, we do not want to encourage big experimental searches for the Higgs boson, but we do feel that people doing experiments vulnerable to the Higgs boson should know how it may turn up.*"

In those early days, there was very little theoretical guidance as to the possible mass of the Higgs boson, which is one reason why these early studies also included very low Higgs masses. One possibility that attracted attention was that the Higgs mass was entirely due to quantum corrections, which would have yielded $m_H \sim 10\,\text{GeV}$ in the absence of heavy fermions.[40,d] At the other end of the mass scale, it was emphasised that the Higgs self-interactions would become strong for $m_H \sim 1\,\text{TeV}$.[41]

5. Searches for the Higgs Boson at LEP

In addition to Ref. 13, there was an early discussion of searches for the Higgs boson in e^+e^- collisions in Ref. 14. There are three important processes for producing the Higgs boson at an e^+e^- collider: in Z^0 decay $-Z^0 \to H + \bar{f}f$,[14,38] in association with the $Z^0 - e^+e^- \to Z^0 + H$[13] with the cross-section[39,42]

$$\sigma(e^+e^- \to Z + H) = \frac{\pi\alpha^2}{24}\left(\frac{2p}{\sqrt{s}}\right)\left(\frac{p^2 + 3m_Z^2}{(s - m_Z^2)^2}\right)\left(\frac{1 - 4\sin^2\theta_W + 8\sin^4\theta_W}{\sin^2\theta_W(1 - \sin^2\theta_W)^2}\right), \tag{8}$$

where p is the momentum of the final-state particles, and via W^+W^- or Z^0Z^0 fusion $-e^+e^- \to \bar{\nu}H\nu, e^+He^-$.[43] The direct process $e^+e^- \to H$ is negligible because — see (5) — of the small H coupling to e^+e^-,[13] though the corresponding reaction at a muon collider, $\mu^+\mu^- \to H$, may be interesting.[44] We also note that high-intensity lasers may be able to convert an e^+e^- collider into a high-luminosity $\gamma\gamma$ collider, which might also be an interesting Higgs factory.[45]

It should be noted that in the early CERN reports on LEP physics the only discussions of Higgs production were theoretical.[14,46] The concerns of our experimental colleagues lay elsewhere, and these early CERN reports contain no experimental discussions of possible searches for the Higgs boson. The first written experimental discussion of which we are aware was in an unpublished 1979 report for ECFA,[47] compiled by a joint working group of theorists and experimentalists. This was followed in 1985 by a more detailed study by a joint theoretical and experimental working group in a CERN report[48] published in 1986. Thereafter, Higgs searches

[d]This was long before it was recognised that the top quark weighed $> m_W$.

were firmly in the experimental sights of the LEP collaborations, as seen in later CERN reports on LEP physics.[49]

In parallel with the searches for the Higgs boson, notably that for the process $Z^0 \to H + \bar{f}f$[14,38] at LEP 1 and that for $e^+e^- \to Z^0 + H$[13,39,42] at LEP 2, the high-precision electroweak data obtained at LEP, the SLC and elsewhere made it possible for the first time to estimate the possible mass of the Higgs boson within the framework of the Standard Model.[50] The dominant m_H-dependent corrections had been calculated earlier,[51] but the possibility of using them in conjunction with LEP data to constrain m_H was not discussed before LEP start-up, perhaps because the precision of LEP data exceeded all previous expectations. The constraint on m_H was relatively weak before the top quark was discovered[52] (with a mass that agreed with predictions based on electroweak data), but the measurement of m_t allowed more accurate estimates of m_H to be made. Values < 300 GeV were favoured already in the early days of such studies,[50] which eventually matured to indicate that $m_H \sim 100 \pm 30$ GeV.[53] This constraint is combined with the (negative) results of the direct Higgs searches in Fig. 2.

The non-appearance of the Higgs boson in searches in Z^0 decays required $m_H >$ 58 GeV.[54] Thereafter, successive increases in the LEP energy during the LEP 2 era prompted recurrent hopes that the Higgs discovery might lie just around the corner, but instead the lower limit on m_H kept rising inexorably. Finally, in 2000 the LEP

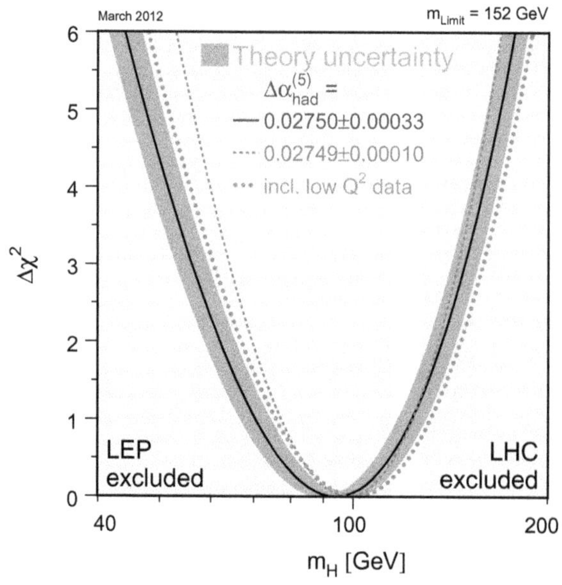

Fig. 2. A compilation of information about the possible mass of the Higgs boson just prior to the LHC discovery.[53] The yellow-shaded regions had been excluded by searches at LEP,[56] the Tevatron collider[57] and the LHC.[58,59] The black line (and the blue band) is the χ^2 function for the precision electroweak data as a function of m_H (and its theoretical uncertainty). The dotted lines are obtained using alternative treatments of the precision electroweak data.[53]

centre-of-mass energy was pushed to 206 GeV, and a few Higgs-like events were observed, corresponding to a mass $\sim$ 115 GeV.[55] To the disappointment of many physicists, it was not possible to push the LEP energy higher, and the difficult decision was taken to shut LEP down at the end of the year 2000, leaving the lower limit $m_H > 114.4$ GeV at the 95% CL.[56] There was much speculation at the time that this decision forced LEP to miss out on the Higgs discovery. However, with $m_H \approx 125$ GeV as has now been established[1] at the LHC, substantial extra investment in accelerating cavities would have been necessary back in the 1990s in order to be able to push LEP to sufficiently high energies to produce it.

6. Searches for the Higgs Boson at Hadron Colliders

The production of the Higgs boson at hadron colliders is more problematic than in e^+e^- collisions. On the one hand, the backgrounds from other physical processes are large, and on the other hand direct production via the dominant quark constituents in the proton is small, because they have very small masses.[13] There is in addition, however, production by gluon–gluon fusion via anomalous triangle diagrams:[60] as first discussed in Ref. 15:

$$\frac{d\sigma}{dy} = \left(\frac{\alpha_s}{\pi}\right)^2 \left(\frac{\pi G_\mu}{288\sqrt{2}}\right) rG(\sqrt{r}e^y)G(\sqrt{r}e^{-y}), \tag{9}$$

where y is the rapidity, $r \equiv m_H^2/s$ and G is the gluon distribution function within the proton. This is the dominant Higgs production mechanism at the LHC, and it is ironic that this and one of the most distinctive Higgs decays, that into $\gamma\gamma$,[13] are both due to similar quantum effects. Another important production mechanism is Higgs-strahlung in association with a $W^\pm$ or Z^0, as was first discussed in Ref. 16, which was the dominant Higgs production mechanism at the Tevatron. A third important mechanism is W^+W^- (and Z^0Z^0) fusion, as first discussed in Ref. 61.

A comprehensive theoretical survey of the new physics possibilities at the Super-conducting SuperCollider (SSC) was provided in Ref. 62, and the search for the Higgs boson naturally took pride of place. The same was true in the survey of the new physics possibilities at the LHC provided in Ref. 63, which also included Higgs production in association with a $\bar{t}t$ pair.[64] In anticipation of Higgs searches at the SSC, in particular, a comprehensive survey of the theory and phenomenology of the Higgs boson was published.[65] It served as the Bible for many subsequent Higgs hunters, also at LEP and the LHC following the much-lamented demise of the SSC.[37]

After the shutdown of LEP, the lead in Higgs searches was taken by the CDF and D0 experiments, at the Tevatron collider, where the dominant production mechanism was Higgs-strahlung in association with a $W^\pm$ or Z^0.[16] As the analysed Tevatron luminosity accummulated, CDF and D0 became able to exclude a range of Higgs masses between 156 and 177 GeV,[57] as well as a range of lower masses in the range excluded by LEP. There was a small excess of Higgs candidate events in a range around 130 to 140 GeV, though not strong enough to be considered a hint, let alone significant evidence. Since the most important Higgs decay channels for the Tevatron experiments are $H \to \bar{b}b$ and W^+W^-, which have relatively poor mass resolution, the excess did not provide much information what value m_H might have. Unfortunately, the Tevatron was shut down in September 2011, before it could realise its full potential for Higgs searches.

The LHC started producing collisions in late 2009, initially at low energies, and starting at 7 TeV in the centre of mass in March 2010. By the end of 2011, the ATLAS and CMS experiments had each accumulated $\sim 5/\text{fb}$ of data, and in 2012 they accumulated $\sim 20/\text{fb}$ of data at 8 TeV in the centre of mass. Already at the end of 2011 optimists could see a hint in their data of a new particle with a mass ~ 120 to 125 GeV, and this was followed by the announcement on July 4th, 2012 of the discovery of a new particle weighing ~ 125 GeV that resembled, *prima facie*, a Higgs boson. Later analyses of the Tevatron data subsequently supported the LHC discovery, albeit with much lower level of significance.[66]

7. Is It Really a/the Higgs Boson?

Although this new particle was widely expected to be a/the Higgs boson, and its measured mass[67]

$$m_H = 125.09 \pm 0.24\,\text{GeV} \tag{10}$$

is certainly consistent with the previous indications from precision electroweak and other data, it was important to check its properties and exclude possible alternatives. The following is a check-list of some properties that needed to be verified:

- What is its spin? A Higgs boson must have spin 0, and this is consistent with its observation in the $\gamma\gamma$ final state, which excludes spin 1. However, integer spins ≥ 2 remained a possibility, albeit unexpected.
- Is it scalar or pseudoscalar? In the Standard Model the Higgs would necessarily be a scalar, but many models have a family of Higgs-like particles, with at least one being a pseudoscalar, e.g., supersymmetry, and these could mix in the presence of CP violation.
- Is it elementary to the same extent as the other Standard Model particles, or does it show signs of being a composite particle, like a Cooper pair or a pion?
- Does it couple to other particles in proportion to their masses? This would be a 'smoking gun' for the new particle's connection with the origin of particle masses.
- Are quantum effects in the new particle couplings consistent with calculations within the Standard Model?

As we now discuss, so far the new particle discovered by ATLAS and CMS has passed all these tests with flying colours.

Many probes of the putative Higgs spin have been proposed and used, including kinematic correlations in $H \to WW^*$ and ZZ^* decays, the kinematics of associated $H + W/Z$ production and the energy dependence of the H production cross-section.[68–70] Here we mention just one example, the angular distribution of the final-state photons in $gg \to H \to \gamma\gamma$.[71] In the case of a spin-0 particle, this angular distribution would be isotropic, but the same would not be true for a spin-2 particle with graviton-like couplings. It would be produced by gluon pairs with parallel spins in an equal admixture of states with spin ± 2 along the collision axis. This non-trivial initial spin state would be imprinted on the angular distribution of the decay photons:

$$\frac{d\sigma}{d\Omega} \propto \frac{1}{4} + \frac{3}{2}\cos^2\theta + \frac{1}{4}\cos^4\theta, \tag{11}$$

and analysis of the LHC data strongly disfavours (11) compared with the isotropic hypothesis. Many other tests also support spin 0 and disfavour a wide range of alternative hypotheses.

Many of the same strategies also distinguish between the scalar and pseudoscalar hypotheses. One example is provided by the distribution of the invariant mass M_{HV} in associated $H + W/Z(\equiv V)$ production.[72] In the Standard Model the final-state particles are produced in a relative S-wave, and the cross-section grows like β just above the threshold $M_{HV} \to m_H + m_V$. However, if the new particle were pseudoscalar, the cross-section would grow as β^3 just above threshold. As a consequence, the distributions in both M_{HV} and the transverse momenta of the H and V would be much broader in the pseudoscalar case. Data from the Tevatron are consistent with the scalar hypothesis, and many LHC measurements of other Higgs channels also disfavour strongly the pure pseudoscalar hypothesis. However, in the presence of CP violation there could be channel-dependent admixtures of pseudoscalar couplings, so these tests should be continued.

One way to explore any possible composite nature of the 'Higgs' boson is to parametrise its couplings to bosons and fermions as follows:

$$\mathcal{L} = \frac{v^2}{4}\mathrm{Tr}D_\mu\Sigma^\dagger D^\mu\Sigma\left(1 + 2a\frac{H}{v} + \dots\right) - \bar{\psi}^i_L\Sigma\psi_R\left(1 + c\frac{H}{v} + \dots\right), \tag{12}$$

where Σ is a 2×2 matrix containing the 3 Goldstone fields that are 'eaten' by the massive gauge bosons appearing in the gauge-covariant derivatives D_μ. As seen in Fig. 3(a), the data are completely consistent with the Standard Model prediction $a = c = 1$: no sign of any significant deviation that would require new strongly-interacting physics at a low energy scale, as would be expected in a composite Higgs model.

One way to probe the mass dependence of the 'Higgs' couplings is to parametrise those to fermions λ_f and massive bosons g_V in the forms[73]

$$\lambda_f = \sqrt{2}\left(\frac{m_f}{M}\right)^{1+\epsilon}, \quad g_V = 2\left(\frac{m_V^{2(1+\epsilon)}}{M^{1+2\epsilon}}\right), \tag{13}$$

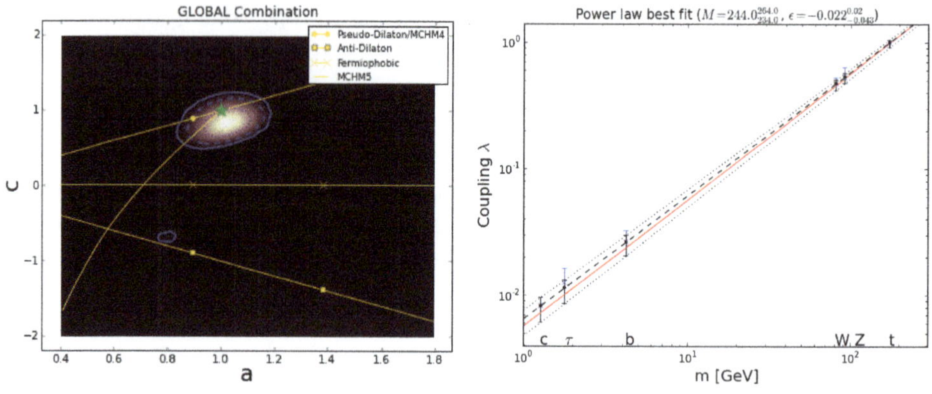

(a) (b)

Fig. 3. (a) A global fit to bosonic and fermionic H couplings (12) rescaled by factors a and c, respectively. The SM prediction $a = c = 1$ is shown as the green star,[73] and the yellow lines show the possible predictions of some composite models. (b) A global fit to the H couplings of the form (13) (central values as dashed and $\pm 1\sigma$ values as dotted lines), which is very consistent with the linear mass dependence for fermions and quadratic mass dependence for bosons (solid red line) expected in the Standard Model.[73]

where one would expect the power $\epsilon = 0$ and the scaling coefficient $M = v = 246\,\text{GeV}$ in the Standard Model. Figure 3(b) shows the result of a fit to the two parameters (M, ϵ) as a dashed line, with the one-σ range indicated by dotted lines. The points with error bars are the predictions of the two-parameter fit, and the solid red line is the prediction of the Standard Model. The data are completely compatible with the Standard Model.[e]

As already commented, it is ironic that the dominant Higgs production mechanism $gg \to H$ and one of its most prominent decay modes $H \to \gamma\gamma$ are quantum (loop) effects due to anomalous triangle diagrams. (How else could it couple to massless particles?) The good news is that this implies that one has good sensitivity to any possible additional massive particles circulating in the loops. Figure 3(b) shows that these loop couplings do not deviate significantly from the Standard Model, confirming that it works at the quantum level and constraining possible extensions of the Standard Model.

In awarding the Nobel Prize to Francois Englert and Peter Higgs, the Royal Swedish Academy of Sciences agreed[75] with[73] that *"beyond any reasonable doubt, it is a Higgs boson."*[f] There are good prospects that future runs of the LHC experiments will pin down further many of the Higgs couplings, including those to $\bar{t}t$ and $\mu^+\mu^-$, but more detailed studies would require an e^+e^- collider — running either at low energy in 'Higgs factory' mode,[76,77] or at higher energies where Higgs

[e] We note also that direct evidence has been presented for 'Higgs' decays to fermions.[74]

[f] Even though this phrase was removed from the published version of Ref. 73 at the request of the referee, who considered it "unscientific".

production would be more copious and more Higgs couplings could be measured, including its self-couplings,[78] or a higher-energy proton–proton collider.[79]

8. More Higgs, Less Higgs? More than Higgs?

So far, we have focused on the least adventurous hypothesis of a single Standard Model-like Higgs boson, but alternatives abound, with most theorists expecting supplements to the minimal Higgs sector of the Standard Model.

One of the simplest possibilities is that there are two complex doublets of Higgs bosons, in which case there would be five physical Higgs bosons: three neutral and two charged $H^{\pm}$. The most natural framework for such a possibility is supersymmetry.[80] In simple supersymmetric models the lightest of the three neutral Higgs boson often has couplings similar to those of the Higgs boson in the Standard Model, whereas one of the heavier neutral Higgs bosons would be a pseudoscalar.

The mass of the Higgs boson is linked to the magnitude of its self-coupling, which would be fixed by supersymmetry in terms of the electroweak gauge couplings. For this reason, supersymmetry predicts a restricted range for the mass of the lightest Higgs boson:

$$m_H = \frac{1}{2}m_Z^2 + m_A^2 - \sqrt{(m_Z^2 + m_A^2)^2 - 4m_Z^2 m_A^2 \cos^2 2\beta}, \tag{14}$$

at the classical (tree) level, where m_A is the mass of the pseudoscalar Higgs boson A and $\tan\beta$ is the ratio of the v.e.v.s of the two Higgs doublets. Equation (14) gives $m_H < m_Z$. However, nearly 25 years ago it was realised that this prediction would be subject to important radiative corrections due to the heavy top quark:

$$\Delta m_H^2 = \frac{3g^2}{4\pi^2}\frac{m_t^4}{m_W^2}\ln\frac{m_{\tilde{t}}}{m_t}, \tag{15}$$

where g is the SU(2) gauge coupling and $m_{\tilde{t}}$ is the mass of the stop squark, which could push m_H up to $\sim$ 130 GeV in simple supersymmetric models.[81] In such a supersymmetric scenario there are no significant restrictions on the masses of the heavier Higgs bosons such as the pseudoscalar A. The measured mass of the Higgs boson lies comfortably within the range of Higgs masses favoured in simple supersymmetric models,[82] and simple supersymmetric models also predicted successfully that its couplings would be very similar to those in the Standard Model. However, so far the LHC has found no evidence of any supersymmetric particles and, as long as this is the case, theorists will consider other possibilities.

An alternative to an elementary Higgs boson of the type found in supersymmetric models would be that the spontaneous breaking of the electroweak gauge symmetry is due to a condensate in the vacuum of pairs of new, strongly-interacting fermions,[83] analogous to the Cooper pairs of superconductivity.[19] In this case, there would in general be a composite scalar particle that might be accessible to experiment. This would not necessarily correspond to a strongly-interacting Higgs boson, which would have to confront issues with the precision electroweak data. For example, if the composite scalar is a (relatively) light pseudo-Goldstone boson of some higher-level broken symmetry, such as a larger chiral symmetry[84] or approximate scale invariance, it would have weak interactions and could mimic a Standard-Model-like Higgs boson to some extent. It would need to because, as we have seen, the data give no indication of any deviation from the Standard Model.

One such example would be a pseudo-dilaton of approximate scale invariance,[85] which would have tree-level couplings similar to those of the Higgs boson, but rescaled (and probably suppressed) by a universal factor. The loop-induced couplings of the pseudo-dilaton to gluon and photon pairs might not share this universal rescaling. This model provides a straw person to compare with the Standard Model Higgs scenario. However, as we have seen, the data provide no encouragement for such a scenario, and a Higgs-like particle with suppressed couplings would not fulfill all the functions of the Higgs boson of the Standard Model, e.g., in unitarising W^+W^- scattering, and would have to be supplemented by some other detectable degrees of freedom in the TeV energy range[86] none of which has been seen so far.

Another scenario for a light Higgs-like particle is the radion,[87] the quantum of the degree of freedom corresponding to rescaling an extra dimension. Models with extra dimensions offered many other possibilities, including the possibility that there was no Higgs boson at all,[88] but those particular models are now out of fashion!

Since the new particle discovered in 2012 looks so much like the Higgs boson of the Standard Model, a popular way to parametrise our ignorance what may lie beyond it is to assume that the Higgs and all the other Standard Model particles have exactly the Standard Model interactions, supplemented by higher-dimensional operators O_i composed of these Standard Model fields:

$$\Delta\mathcal{L} = \sum_i C_i O_i, \tag{16}$$

where the unknown coefficients C_i have mass dimensions 4 — [Dim O_i] and hence are likely to be suppressed by the corresponding powers of some higher mass scale. Expressing these coefficients as $\bar{c}_i/m_W^2$, one can use present data (precision electroweak data, Higgs data, triple-gauge couplings (TGCs), etc.) to constrain the dimensionless reduced coefficients c_i. The results of one such analysis of dimension-6 operators are shown in Fig. 4: there are no indications that any deviate significantly from zero. The Standard Model still rules OK!

9. Après Higgs

The discovery of the Higgs boson marks a watershed in particle physics. In the future, the calendar of particle physics will surely be divided into BH (before Higgs) and AH (after Higgs), with 2012 being year 0. The Higgs boson will signpost the

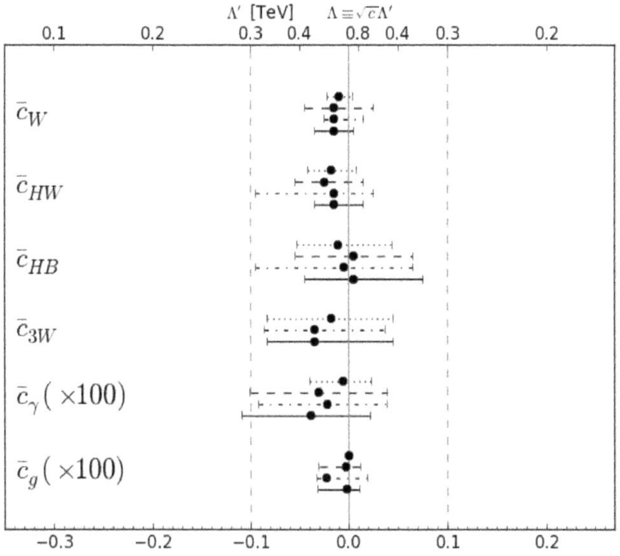

Fig. 4. The 95% CL constraints for single-coefficient fits (dotted lines), and the marginalised 95% ranges for the LHC Higgs signal-strength data combined with the kinematic distributions for associated $H + V$ production measured by ATLAS and D0 (dash lines), with the LHC TGC data (dot-dashed lines), and the global combination with both the associated production and TGC data (solid lines).[90]

direction that both theoretical and experimental physics will take in the decades to come.

We can be optimistic, because there are already many arguments that there must be new physics. One of them is provided by measured value of the Higgs mass (10). Coupled with the world average value of the mass of the top quark: $m_t = 173.34 \pm 0.76\,\mathrm{GeV}$. Extrapolating to high renormalisation scales indicates[91] that the effective Higgs quartic coupling λ would become negative at a Higgs scale Λ, as seen in Fig. 5(a):

$$\log_{10}\left(\frac{\Lambda}{\mathrm{GeV}}\right) = 11.3 + 1.0\left(\frac{m_H}{\mathrm{GeV}} - 125.66\right) - 1.2\left(\frac{m_t}{\mathrm{GeV}} - 173.10\right) + 0.4\left(\frac{\alpha_s(M_Z) - 0.1184}{0.0007}\right). \tag{17}$$

Using the world average values of m_t, m_H and $\alpha_s(M_Z) = 0.1185 \pm 0.0006$, this formula yields

$$\Lambda = 10^{10.6\pm1.0}\ \mathrm{GeV}. \tag{18}$$

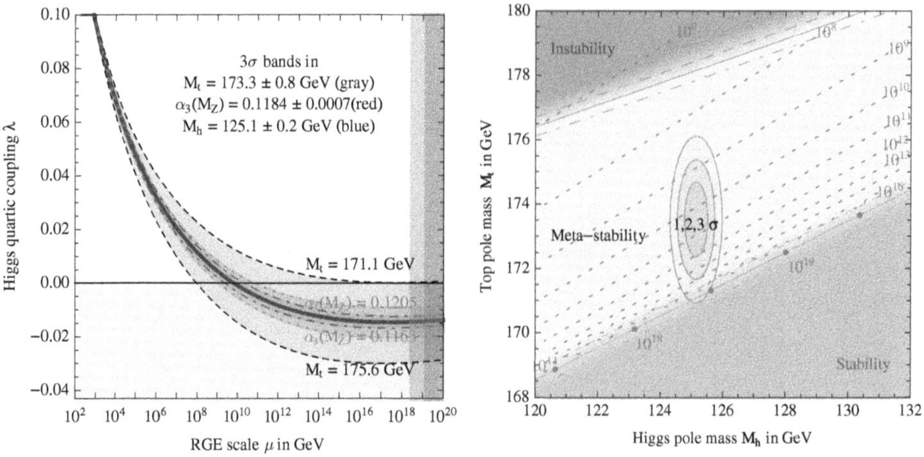

(a) (b)

Fig. 5. (a) Within the SM, renormalisation by the top quark appears to drive the Higgs self-coupling $\lambda < 0$ at large scales, destabilising the electroweak vacuum. (b) Regions of vacuum stability, metastability and instability in the (m_H, m_t) plane. Taken from Ref. 91.

As seen in Fig. 5(b), the most important uncertainty in this calculation is that associated with m_t, but all the indications are that there is an instability in the electroweak vacuum. The lifetime for tunnelling from our present electroweak vacuum to a state with a Higgs v.e.v. larger than (18) is probably longer than the age of the Universe, so one might be tempted to ignore this problem. However, if the Universe was once very hot and dense as suggested by conventional Big Bang theory and cosmological inflation, most of the early Universe would have got stuck in this unphysical state, and would never have reached the present electroweak vacuum. This indicates that there should be some new physics below the scale (18) that is capable of stabilising the electroweak vacuum, such as supersymmetry.[92] This is a third reason that Run 1 of the LHC has given us to favour supersymmetry, in addition to its successful prediction of the Higgs mass and the similarity of its couplings to those in the Standard Model.

There are many other reasons why there must be physics beyond the Standard Model, and the Higgs boson may play an important role in most of them. What is the origin of the matter in the Universe — is it due to a first-order electroweak phase transition? What is the dark matter — could the Higgs boson serve as a portal towards it? How to stabilise the Higgs mass and hence the electroweak scale so far below the Planck scale? Are neutrino masses due to the conventional Higgs, or some different mechanism? What is the dark energy, and why is it much smaller than the electroweak scale? Is the great size and age of the Universe, and its near-flatness, due to a primordial epoch of cosmological inflation, driven by the energy in some scalar field like the Higgs, perhaps even the Higgs itself?

With the discovery of the Higgs boson in 2012, many new directions for physics have opened up. On the one hand, there is a need for detailed investigation of the

Higgs, to see whether it conforms to the Standard Model paradigm or whether it exhibits deviations due to new physics. On the other hand, the hunt will be on for whatever new physics complements the Higgs boson, be it supersymmetry or ...? We look forward to the years AH *(Anno Higgsi)* > 0.

Acknowledgements

The work of J.E. was supported in part by the London Centre for Terauniverse Studies (LCTS), using funding from the European Research Council via the Advanced Investigator Grant 267352 and from the STFC (UK) via the research grants ST/J002798/1 and ST/L000326/1. The work of M.K.G. was supported in part by the Director, Office of Science, Office of High Energy and Nuclear Physics, Division of High Energy Physics, of the U.S. Department of Energy under Contract DE-AC02-05CH11231, in part by the National Science Foundation under grant PHY-1002399. The work of D.V.N. was supported in part by the DOE grant DE-FG02-13ER42020.

References

1. G. Aad *et al.* [ATLAS Collaboration], *Phys. Lett. B* **716**, 1 (2012) [arXiv:1207.7214 [hep-ex]]; S. Chatrchyan *et al.* [CMS Collaboration], *Phys. Lett. B* **716**, 30 (2012) [arXiv:1207.7235 [hep-ex]].
2. P. W. Higgs, *Phys. Rev. Lett.* **13**, 508 (1964).
3. P. W. Anderson, *Phys. Rev.* **130**, 439 (1963); see also Y. Nambu, *Phys. Rev.* **117**, 648 (1960).
4. Y. Nambu, *Phys. Rev. Lett.* **4**, 380 (1960).
5. F. Englert and R. Brout, *Phys. Rev. Lett.* **13**, 321 (1964).
6. P. W. Higgs, *Phys. Lett.* **12**, 132 (1964).
7. S. Weinberg, *Phys. Rev. Lett.* **19**, 1264 (1967). This seems to be the first paper in which it is shown that matter particles can also acquire their masses from 'spontaneous' symmetry breaking.
8. A. Salam, in *Proceedings of 8th Nobel Symposium*, Lerum, Sweden, 19–25 May 1968, pp. 367–377.
9. G. 't Hooft, *Nucl. Phys. B* **35**, 167 (1971); G. 't Hooft and M. J. G. Veltman, *Nucl. Phys. B* **44**, 189 (1972).
10. F. J. Hasert *et al.* [Gargamelle Neutrino Collaboration], *Phys. Lett. B* **46**, 121 (1973) and *Phys. Lett. B* **46**, 138 (1973).
11. J. J. Aubert *et al.* [E598 Collaboration], *Phys. Rev. Lett.* **33**, 1404 (1974); J. E. Augustin *et al.* [SLAC-SP-017 Collaboration], *Phys. Rev. Lett.* **33**, 1406 (1974).
12. G. Arnison *et al.* [UA1 Collaboration], *Phys. Lett. B* **122**, 103 (1983) and *Phys. Lett. B* **126**, 398 (1983).
13. J. R. Ellis, M. K. Gaillard and D. V. Nanopoulos, *Nucl. Phys. B* **106**, 292 (1976).
14. J. Ellis and M. K. Gaillard, *Theoretical remarks*, in L. Camilleri *et al.*, *Physics with very high-energy* e^+e^- *colliding beams*, CERN report 76–18 (Nov. 1976), pp. 21–94.
15. H. M. Georgi, S. L. Glashow, M. E. Machacek and D. V. Nanopoulos, *Phys. Rev. Lett.* **40**, 692 (1978).
16. S. L. Glashow, D. V. Nanopoulos and A. Yildiz, *Phys. Rev. D* **18**, 1724 (1978).
17. J. Bardeen, L. N. Cooper and J. R. Schrieffer, *Phys. Rev.* **106**, 162 (1957).

18. V. L. Ginzburg and L. D. Landau, *Zh. Eksp. Teor. Fiz.* **20**, 1064 (1950) and *Theory of superconductivity, Phys. Rev.* **108**, 1175 (1957).
19. L. N. Cooper, *Phys. Rev.* **104**, 1189 (1956).
20. J. Goldstone, *Nuovo Cim.* **19**, 154 (1961).
21. J. Goldstone, A. Salam and S. Weinberg, *Phys. Rev.* **127**, 965 (1962).
22. A. Klein and B. W. Lee, *Phys. Rev. Lett.* **12**, 266 (1964).
23. W. Gilbert, *Phys. Rev. Lett.* **12**, 713 (1964).
24. G. S. Guralnik, C. R. Hagen and T. W. B. Kibble, *Phys. Rev. Lett.* **13**, 585 (1964).
25. A. A. Migdal and A. M. Polyakov, *J. Expt. Theor. Phys. (U.S.S.R.)* **51**, 135 (1966).
26. P. W. Higgs, *Phys. Rev.* **145**, 1156 (1966).
27. T. W. B. Kibble, *Phys. Rev.* **155**, 1554 (1967).
28. S. L. Glashow, *Nucl. Phys.* **22**, 579 (1961).
29. B. W. Lee and J. Zinn-Justin, *Phys. Rev. D* **5**, 3121 (1972); *Phys. Rev. D* **5**, 3137 (1972) [Erratum *ibid.* **8**, 4654 (1973)] [Erratum *ibid.* **8**, 4654 (1973)]; *Phys. Rev. D* **5**, 3155 (1972); *Phys. Rev. D* **7**, 1049 (1973).
30. K. Fujikawa, B. W. Lee and A. I. Sanda, *Phys. Rev. D* **6**, 2923 (1972).
31. B. W. Lee, *Perspectives on theory of weak interactions,* Talk at the 16th International Conference on High-Energy Physics, Batavia, Illinois, 6–13 Sep 1972, eConf **C720906V4** (1972) 249.
32. S. L. Adler, R. F. Dashen and S. B. Treiman, *Phys. Rev. D* **10**, 3728 (1974).
33. K. Sato and H. Sato, *Prog. Theor. Phys.* **54**, 1564 (1975).
34. L. Resnick, M. K. Sundaresan and P. J. S. Watson, *Phys. Rev. D* **8**, 172 (1973).
35. D. Kohler, B. A. Watson and J. A. Becker, *Phys. Rev. Lett.* **33**, 1628 (1974).
36. R. Barbieri and T. E. O. Ericson, *Phys. Lett. B* **57**, 270 (1975).
37. A. Djouadi, *Phys. Rept.* **457**, 1 (2008) [arXiv:hep-ph/0503172].
38. J. D. Bjorken, *Weak interaction theory and neutral currents,* in the Proceedings of 4th SLAC Summer Institute on Particle Physics: *Weak Interactions at High Energies & the Production of New Particles*, Stanford, California, 2–10 Aug 1976, SLAC Report 198 (Nov. 1976), pp. 1–84, also available as SLAC-PUB-1866, Jan. 1977.
39. B. L. Ioffe and V. A. Khoze, *Sov. J. Part. Nucl.* **9**, 50 (1978) [*Fiz. Elem. Chast. Atom. Yadra* **9**, 118 (1978)], also available as LENINGRAD-76-274, Nov 1976.
40. S. R. Coleman and E. J. Weinberg, *Phys. Rev. D* **7**, 1888 (1973).
41. C. E. Vayonakis, Print-77-0432 (ATHENS); M. J. G. Veltman, *Phys. Lett. B* **70**, 253 (1977); B. W. Lee, C. Quigg and H. B. Thacker, *Phys. Rev. Lett.* **38**, 883 (1977); *Phys. Rev. D* **16**, 1519 (1977).
42. B. W. Lee, C. Quigg and H. B. Thacker, in Ref. 41.
43. D. R. T. Jones and S. T. Petcov, *Phys. Lett. B* **84**, 440 (1979).
44. C. Blöchinger *et al.*, *Physics opportunities at $\mu^+\mu^-$ Higgs Factories*, in CERN Report 2004–002, http://cdsweb.cern.ch/record/735196/files/ECFA-04-230.pdf, p. 336.
45. See, for example, D. Asner *et al.*, *Eur. Phys. J. C* **28**, 27 (2003) [arXiv:hep-ex/0111056].
46. J. Ellis, *Zedology* and M. K. Gaillard, *Weak interactions beyond the Z pole(s)*, in *Proc. LEP Summer Study*, Les Houches, France, Sept. 1978, CERN Report 79–02 (1979), Vol. 2, http://cdsweb.cern.ch/record/133077/files/cern-79-01-v2.pdf, pp. 637–661.
47. G. Barbiellini, G. Bonneaud, G. Coignet, J. Ellis, J. F. Grivaz, M. K. Gaillard, C. Matteuzzi and B. H. Wiik, *The production and detection of Higgs particles at LEP,* DESY 79/27, ECFA/LEP SSG/9/4, May 1979.

48. H. Baer *et al.*, *New particles*, in *Physics at LEP*, *eds.* J. Ellis and R. D. Peccei, CERN Report 86–02 (1986), Vol. 1, http://cdsweb.cern.ch/record/166310/files/CERN-86-02-V-1.pdf, pp. 297–348.
49. J. Boucrot *et al.*, *Search for neutral Higgs at LEP 200,* in *Proceedings of the ECFA workshop on LEP 200,* Aachen 1986, eds. A. Böhm and W. Hoogland, CERN Report 87–08 (1987), Vol. 2, http://cdsweb.cern.ch/record/180188/files/CERN-87-08-V-2.pdf, p 312; P. J. Franzini and P. Taxil, *Higgs search,* in CERN Report 89–08 (1989), Vol. 2, http://cdsweb.cern.ch/record/116932/files/CERN-89-08-V-1.pdf, p. 59; E. Accommando *et al.*, *Higgs physics*, in CERN Report 96-01 (1996), Vol. 1, http://cdsweb.cern.ch/record/300671/files/CERN-96-01-V1.pdf, p. 351.
50. J. R. Ellis and G. L. Fogli, *Phys. Lett. B* **249**, 543 (1990), J. R. Ellis, G. L. Fogli and E. Lisi, *Phys. Lett. B* **274**, 456 (1992) and *Phys. Lett. B* **318**, 148 (1993).
51. M. J. G. Veltman, *Acta Phys. Polon. B* **8**, 475 (1977).
52. F. Abe *et al.* [CDF Collaboration], *Phys. Rev. Lett.* **74**, 2626 (1995) [hep-ex/9503002].
53. LEP Electroweak Working Group, http://lepewwg.web.cern.ch/LEPEWWG/.
54. D. Buskulic *et al.* [ALEPH Collaboration], *Phys. Lett. B* **313**, 299 (1993).
55. R. Barate *et al.* [ALEPH Collaboration], at ALEPH, *Phys. Lett. B* **495**, 1 (2000) [arXiv:hep-ex/0011045].
56. R. Barate *et al.* [LEP Working Group for Higgs boson searches and ALEPH and DELPHI and L3 and OPAL Collaborations], *Phys. Lett. B* **565**, 61 (2003) [hep-ex/0306033]; for a recent review of Higgs searches at LEP and the LHC, see W. D. Schlatter and P. M. Zerwas, arXiv:1112.5127 [physics.hist-ph].
57. Tevatron New Phenomena and Higgs Working Group and CDF and D0 Collaborations, arXiv:1107.5518 [hep-ex].
58. CMS Collaboration, *Combination of SM Higgs Searches*, http://cdsweb.cern.ch/record/1406347/files/HIG-11-032-pas.pdf.
59. ATLAS Collaboration, *Combination of Higgs boson searches with up to 4.9 fb-1 of pp collision data taken at sqrt(s)=7 TeV with the ATLAS experiment at the LHC*, https://atlas.web.cern.ch/Atlas/GROUPS/PHYSICS/CONFNOTES/ATLAS-CONF-2011-163/.
60. F. Wilczek, *Phys. Rev. Lett.* **39**, 1304 (1977).
61. R. N. Cahn and S. Dawson, *Phys. Lett. B* **136**, 196 (1984) [Erratum *ibid.* **138**, 464 (1984)].
62. E. Eichten, I. Hinchliffe, K. D. Lane and C. Quigg, *Rev. Mod. Phys.* **56**, 579 (1984) [Addendum *ibid.* **58**, 1065 (1986)].
63. J. R. Ellis, G. Gelmini and H. Kowalski, *New particles and their experimental signatures,* in *Proceedings of the ECFA/CERN Workshop on the Possibility of a Large Hadron Collider*, Lausanne and Geneva, Mar. 21–27, 1984, CERN Report 84-10, ECFA 84/85, Vol. 2, pp. 393–454, also available as DESY 84/071, CERN-TH-3493/84 (1984).
64. Z. Kunszt, *Nucl. Phys. B* **247**, 339 (1984).
65. J. F. Gunion, H. E. Haber, G. L. Kane and S. Dawson, *Front. Phys.* **80**, 1 (2000).
66. CDF and D0 Collaborations, http://tevnphwg.fnal.gov.
67. G. Aad *et al.* [ATLAS and CMS Collaborations], arXiv:1503.07589 [hep-ex].
68. G. Aad *et al.* [ATLAS Collaboration], *Phys. Lett. B* **726**, 120 (2013) [arXiv:1307.1432 [hep-ex]].
69. V. Khachatryan *et al.* [CMS Collaboration], arXiv:1411.3441 [hep-ex].
70. T. Aaltonen *et al.* [CDF and D0 Collaborations], arXiv:1502.00967 [hep-ex].
71. J. Ellis and D. S. Hwang, *JHEP* **1209**, 071 (2012) [arXiv:1202.6660 [hep-ph]]; J. Ellis, R. Fok, D. S. Hwang, V. Sanz and T. You, *Eur. Phys. J. C* **73**, 2488 (2013) [arXiv:1210.5229 [hep-ph]].

72. J. Ellis, D. S. Hwang, V. Sanz and T. You, *JHEP* **1211**, 134 (2012) [arXiv:1208.6002 [hep-ph]].
73. J. Ellis and T. You, *JHEP* **1306**, 103 (2013) [arXiv:1303.3879 [hep-ph]].
74. S. Chatrchyan *et al.* [CMS Collaboration], *Nature Phys.* **10**, 557 (2014) [arXiv:1401.6527 [hep-ex]].
75. Class for Physics of the Royal Swedish Academy of Sciences, *The BEH-Mechanism, Interactions with Short-Range Forces and Scalar Particles*, http://www.nobelprize.org/nobel_prizes/physics/laureates/2013/advanced-physicsprize2013.pdf.
76. G. Aarons *et al.* [ILC Collaboration], arXiv:0709.1893 [hep-ph].
77. A. Blondel and F. Zimmermann, *A High luminosity* e^+e^- *collider in the LHC tunnel to study the Higgs boson,* arXiv:1112.2518 [hep-ex]; M. Bicer *et al.* [TLEP Design Study Working Group Collaboration], *JHEP* **1401**, 164 (2014) [arXiv:1308.6176 [hep-ex]].
78. E. Accomando *et al.* [CLIC Physics Working Group], *Physics at the CLIC multi-TeV linear collider,* arXiv:hep-ph/0412251.
79. https://espace2013.cern.ch/fcc/Pages/default.aspx.
80. J. Wess and B. Zumino, *Phys. Lett. B* **49**, 52 (1974) and *Nucl. Phys. B* **70**, 39 (1974).
81. J. R. Ellis, G. Ridolfi and F. Zwirner, *Phys. Lett. B* **257**, 83 (1991); H. E. Haber and R. Hempfling, *Phys. Rev. Lett.* **66**, 1815 (1991); Y. Okada, M. Yamaguchi and T. Yanagida, *Prog. Theor. Phys.* **85**, 1 (1991).
82. See, for example J. R. Ellis, D. V. Nanopoulos, K. A. Olive and Y. Santoso, *Phys. Lett. B* **633**, 583 (2006) [arXiv:hep-ph/0509331]; T. Li, J. A. Maxin, D. V. Nanopoulos and J. W. Walker, arXiv:1109.2110 [hep-ph]; O. Buchmueller *et al.*, arXiv:1110.3568 [hep-ph].
83. S. Weinberg, *Phys. Rev. D* **13**, 974 (1976).
84. N. Arkani-Hamed, A. G. Cohen, E. Katz and A. E. Nelson, *JHEP* **0207**, 034 (2002) [arXiv:hep-ph/0206021].
85. W. D. Goldberger, B. Grinstein and W. Skiba, *Phys. Rev. Lett.* **100**, 111802 (2008) [arXiv:0708.1463 [hep-ph]].
86. See, for example R. Contino, C. Grojean, M. Moretti, F. Piccinini and R. Rattazzi, *JHEP* **1005**, 089 (2010) [arXiv:1002.1011 [hep-ph]].
87. C. Csaki, J. Hubisz and S. J. Lee, *Phys. Rev. D* **76**, 125015 (2007) [arXiv:0705.3844 [hep-ph]].
88. C. Csaki, C. Grojean, L. Pilo and J. Terning, *Phys. Rev. Lett.* **92**, 101802 (2004) [arXiv:hep-ph/0308038].
89. W. Buchmuller and D. Wyler, *Nucl. Phys. B* **268**, 621 (1986).
90. J. Ellis, V. Sanz and T. You, arXiv:1410.7703 [hep-ph]. References to the original literature can be found here.
91. D. Buttazzo, G. Degrassi, P. P. Giardino, G. F. Giudice, F. Sala, A. Salvio and A. Strumia, *JHEP* **1312**, 089 (2013) [arXiv:1307.3536].
92. J. R. Ellis and D. Ross, *Phys. Lett. B* **506**, 331 (2001) [arXiv:hep-ph/0012067].

Chapter 15

The Higgs Boson Search and Discovery

Gregorio Bernardi* and Jacobo Konigsberg†

*LPNHE, Universités Paris VI & VII, CNRS/IN2P3, Paris, France

† University of Florida, Gainesville, Florida, USA

We present a brief account of the search for the Higgs boson at the three major colliders that have operated over the last three decades: LEP, the Tevatron, and the LHC. The experimental challenges encountered stemmed from the distinct event phenomenology as determined by the colliders energy and the possible values for the Higgs boson mass, and from the capability of these colliders to deliver as much collision data as possible to fully explore the mass spectrum within their reach. Focusing more on the hadron collider searches during the last decade, we discuss how the search for the Higgs boson was advanced through mastering the experimental signatures of standard theory backgrounds, through the comprehensive utilization of the features of the detectors involved in the searches, and by means of advanced data analysis techniques. The search culminated in 2012 with the discovery, by the ATLAS and CMS collaborations, of a Higgs-like particle with mass close to 125 GeV, confirmed more recently to have properties consistent with those expected from the standard theory Higgs boson.

1. Overview

Ever since the seminal papers by Englert, Brout, Higgs, Guralnik, Hagen, and Kibble appeared in 1964,[1] the search for the Higgs boson, a neutral scalar particle of unknown mass, m_H, has preoccupied several generations of experimentalists. A publication, in 1976, outlining the phenomenological profile of the Higgs boson[2] described the lower bounds on m_H that existed at the time (18 MeV), coming mainly from analysis of star emissions and from neutron scattering experiments. It also showed that masses of up to 4 GeV could be accessible in "high-energy" hadron colliders (Fig. 1(a)). In that publication, the decay branching fractions for Higgs boson masses up to 100 GeV were presented (see Fig. 1(b)). It is interesting to notice the many particles (which we now know exist, and into which the Higgs boson can actually decay) that are missing from the picture, and to compare it with the current picture shown in Fig. 2. In that publication a couple of

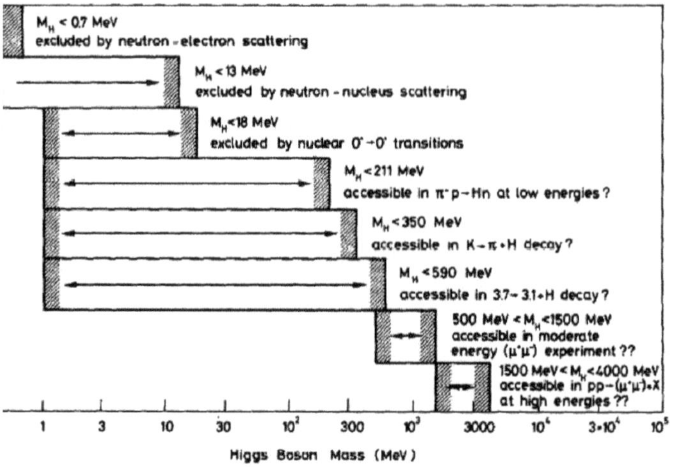

(a)

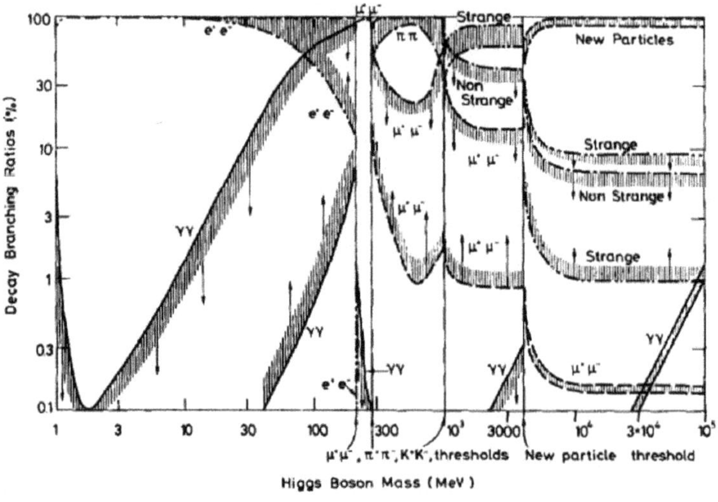

(b)

Fig. 1. (a) Higgs boson limits as a function of m_H in 1976. (b) Higgs boson branching ratios as expected in 1975.

whimsical statements were made: (i) "*The situation with regard to the Higgs boson is unsatisfactory. First it should be stressed that they may well not exist...*", and, at its conclusion, (ii) "*We apologize to experimentalists for having no idea what is the mass of the Higgs boson, unlike the case with charm and for not being sure of its couplings to other particles, except that they are probably all very small. For these reasons we do not want to encourage big experimental searches for the Higgs boson, but we do feel that people performing experiments vulnerable to the Higgs boson should know how it may turn up.*"

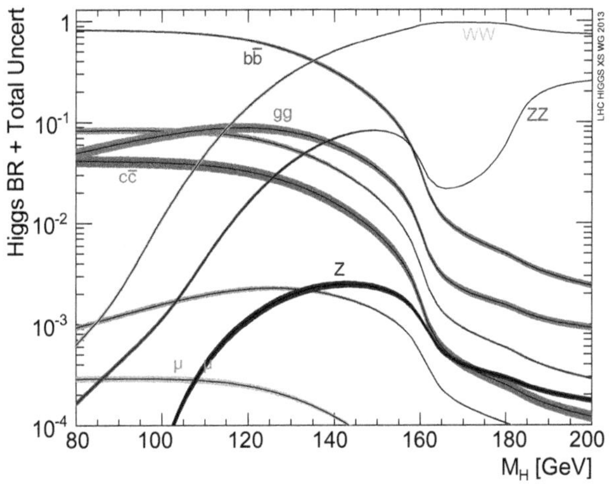

Fig. 2. Higgs boson branching ratios as expected in 2012.

As it happened, things played out somewhat differently. And, notwithstanding that message, after the discovery of the W and Z bosons at CERN in 1983, the attention to the search for the Higgs boson gained significant momentum. This quest was also spurred by several writings on the topic, amongst them was the "*The Higgs Hunter Guide*" book,[3] published in 1990, which contained an exhaustive account of the phenomenology of the Higgs boson at lepton and hadron colliders. In fact, the Higgs boson search was very much on the menu of the experiments at the *Large Electron Positron* (LEP) collider at CERN,[4,5] and front and center in the program of the ill-fated *Superconducting Super Collider* (SSC) in the United States.

During the LEP1 era (1989–1995), the e^+e^- center-of-mass collision energy, $\sqrt{s}$, was in the 89–93 GeV range, around the Z boson mass. During the LEP2 era (1996–2000) the collision energy increased above the W-pair production threshold of 160 GeV, and up to a maximum of 209 GeV. Amongst their many achievements, the four LEP experiments: Aleph, Delphi, L3, and Opal, all collaborations made up of hundreds of physicists, determined that there are exactly three types of (light) neutrinos.

Since the beginning the LEP experiments diligently searched for direct evidence for the Higgs boson, setting lower limits on its possible mass along the way. As shown in Fig. 3(a), at LEP the SM Higgs boson was mainly produced through Higgs-strahlung in the s-channel, $e^+e^- \to HZ$,[6,7] where the Z boson in the final state is either virtual (LEP1), or on-shell (LEP2). The SM Higgs boson could also be produced by WW and ZZ fusion in the t-channel,[8] but these processes have small cross sections and would have contributed minimally to the sensitivity, therefore no dedicated searches for these processes were performed. The sensitivity of the LEP searches to the Higgs boson strongly depended on the center-of-mass energy, $E_{\rm CM}$. For $m_H < E_{\rm CM} - m_Z$, the cross section is of order 1 pb or more, while for

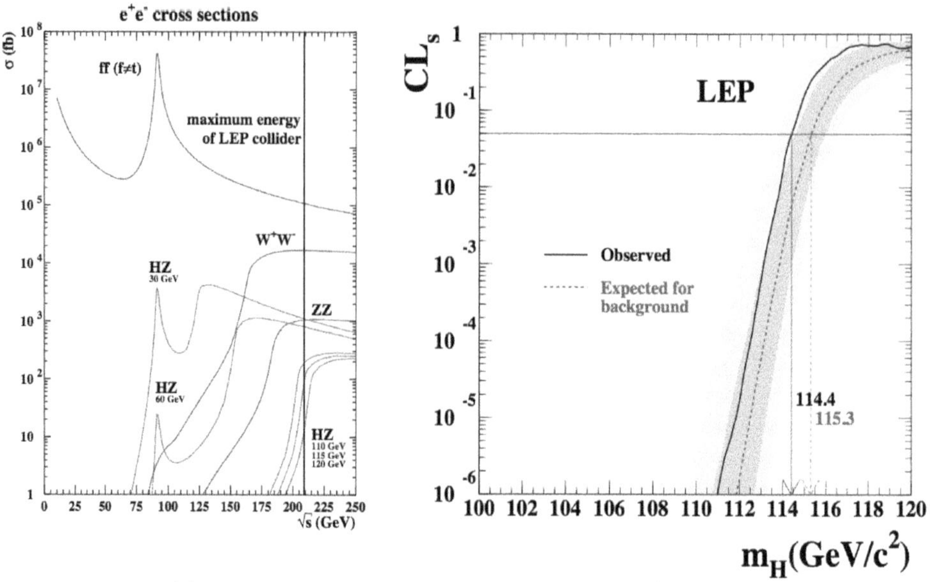

(a) (b)

Fig. 3. (a) Higgs boson production cross sections at LEP compared to WW, ZZ and fermion–antifermion cross sections. (b) CLs values as function of m_H for data and expectations from the signal + background hypothesis and background-only hypothesis with 68% and 95% CL error bands.

$m_H > E_{\rm CM} - m_Z$, the cross section is smaller by at least an order of magnitude, However, in practice, LEP did not have sensitivity for $m_H > E_{\rm CM} - m_Z$.

At LEP2 searches were performed in four different channels: (i) the four-jet channel $ZH \to q\bar{q}, q\bar{q}$ (which included specific analysis for $ZH \to q\bar{q}, b\bar{b}$), (ii) the missing-energy channel $ZH \to \nu\bar{\nu}, q\bar{q}$ (including $ZH \to \nu\bar{\nu}, b\bar{b}$), (iii) the tau channels $ZH \to q\bar{q}, \tau^+\tau^-$ and $ZH \to \tau^+\tau^-, q\bar{q}$ (including $ZH \to \tau^+\tau^-, b\bar{b}$), and (iv) the lepton channels $ZH \to e^+e^- q\bar{q}$ and $ZH \to \mu^+\mu^-, q\bar{q}$ (with $ZH \to e^+e^-, b\bar{b}$ and $ZH \to \mu^+\mu^-, b\bar{b}$). At each of the LEP experiments, the different production and decay modes were analyzed separately and the data recorded at each center-of-mass energy were studied independently. The results from the four LEP experiments were then combined.

With all data in hand, strong upper bounds on the $e^+e^- \to ZH$ cross section were obtained for Higgs boson masses between 1 keV and $\simeq$ 115 GeV. The final combination of the LEP data yielded a 95% C.L. lower bound of 114.4 GeV for the mass of the SM Higgs boson.[9] The median limit expected in a large ensemble of identical experiments when no signal is present was 115.3 GeV, and was limited by the collision energy achieved by the accelerator. Figure 3(b) shows these results. It was not possible to push the collider energy further and, to the disappointment of many, in particular since there was a small excess around a mass of $\sim$ 115 GeV,

LEP was shut down at the end of the year 2000. A short account of the progression and excitement of the enterprise can be found in Ref. 10.

The SSC, on the other hand, never got to take any data. The accelerator would have collided protons in a ring of 87 kilometers in circumference at a center-of-mass energy of 40 TeV. However, the project was plagued by cost overruns and managerial issues, all during a rough political period in the U.S.[11,12] After initial construction, cancelation ensued in 1993. It was a turning point for High Energy Physics in the U.S. and an opportunity for Europe to purse at full steam the construction of the Large Hadron Collider (LHC) in the LEP tunnel.

Between the shutdown of LEP and the advent of the LHC the Tevatron experiments, CDF and D0, had the playing field to themselves in the search for the Higgs boson. The Tevatron was a proton–antiproton collider configured to collide beams of 36 bunches at a center-of-mass energy of 1.96 TeV.

The value of the Higgs boson mass was not specified by theory. However, electroweak radiative corrections are sensitive to the mass of the particles that may play a role in them, and therefore the consistency between different electroweak measurements can constrain the values of those masses. Before the discovery of the top quark at the Tevatron[13] in 1995, which was another epic search saga in itself, the value of the Higgs boson mass was not very constrained experimentally either. The precision measurements of the top-quark and W-boson masses that were made throughout the Tevatron program considerably narrowed the range for the possible value of the Higgs boson mass. The Higgs boson contributes to the observed W and Z gauge boson masses through loop diagrams, leading to a logarithmic dependence of the ratio of the W and Z masses on the Higgs boson mass. The top quark contributes to the observed W boson mass through loop effects that depend quadratically on the top mass. A global fit to precision electroweak data at the time of the Higgs boson discovery (accumulated over the last two decades mainly at LEP, SLC and

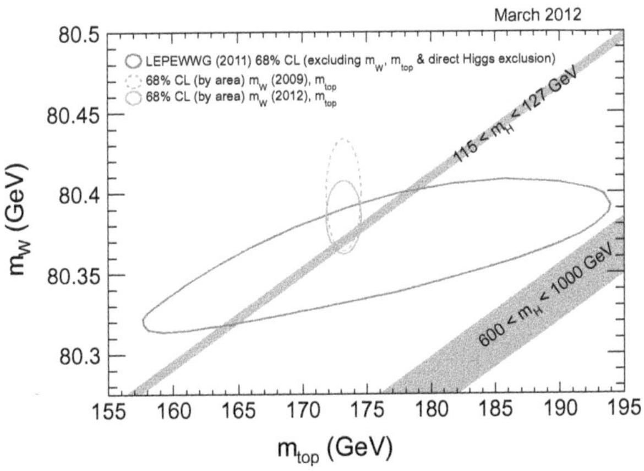

Fig. 4. Constraints from the top quark mass, the W boson mass and other SM measurements compared to the allowed Higgs search range in March 2012.

the Tevatron), obtained that $m_H = 94^{+29}_{-24}$ GeV, or $m_H < 152$ GeV at 95% C.L.[14] The measurements of the top-quark mass and of the W boson mass used in these fits were 173.2 ± 0.9 GeV,[15] and 80.385 ± 0.015 GeV,[16] respectively. Figure 4 shows the constraints to the Higgs boson mass value from these measurements as they stood in March 2012, while they were weaker when the Run II of the Tevatron started in 2002.

In the year 2000 a working group explored the possibilities for discovery of the Higgs boson at the Tevatron,[17] concluding the following:

"*Based on a simple detector simulation, we have determined the integrated luminosity necessary to discover the SM Higgs in the mass range 100–190 GeV. The first phase of the Run 2 Higgs search, with a total integrated luminosity of 2 fb^{-1} per detector, will provide a 95% CL exclusion sensitivity comparable to that expected at the end of the LEP2 run. With 10 fb^{-1} per detector, this exclusion will extend up to Higgs masses of 180 GeV, and a tantalizing 3-sigma effect will be visible if the Higgs mass lies below 125 GeV. With 25 fb^{-1} of integrated luminosity per detector, evidence for SM Higgs production at the 3-sigma level is possible for Higgs masses up to 180 GeV. However, the discovery reach is much less impressive for achieving a 5-sigma Higgs boson signal. Even with 30 fb^{-1} per detector, only Higgs bosons with masses up to about 130 GeV can be detected with 5-sigma significance.*" Figure 5(a) shows graphically these results.

Key to the success of the Tevatron's Higgs search program was the capacity of the accelerator to deliver ever more intense proton and antiproton beams. Through tremendous effort and inventiveness, the instantaneous luminosity was increased about one order of magnitude with respect to its maximum during Run 1, when the top quark discovery was made. Figure 5(b) shows the instantaneous luminosity delivered by the Tevatron over its 25 years of lifetime, reaching a maximum of $4 \times 10^{32}\,\mathrm{cm}^2\,\mathrm{s}^{-1}$ by the time it shut down. Tevatron operations focused not only on records for maximum instantaneous luminosity but also on maximizing the integrated luminosity delivered. As the luminosity improved, the mission to search for the Higgs boson ultimately became the main driver of the Tevatron's Run 2 program, which lasted from 2001 until 2011.

An important aspect in the search of the Higgs boson at the Tevatron was the continuous interplay of cooperation and healthy competition between the CDF and D0 collaborations. It was clear from the start that combining their datasets would ultimately give the Tevatron the best chance for signal sensitivity in the widest range of Higgs boson masses. At the same time, each experiment strived in besting each other in sensitivity by implementing innovative analysis techniques and incorporating as many channels as possible into the search. This led to a cross-pollination with the result that each experiment truly maximized its potential. A Tevatron Higgs Working Group was established in 2006 with the goal of combining the results from the two experiments, sharing discussions on the proper treatment of uncertainties, both experimental and theoretical, and on statistical methods for

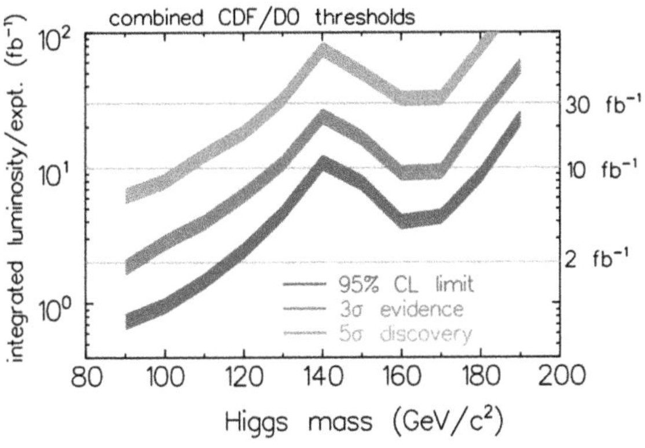

(a)

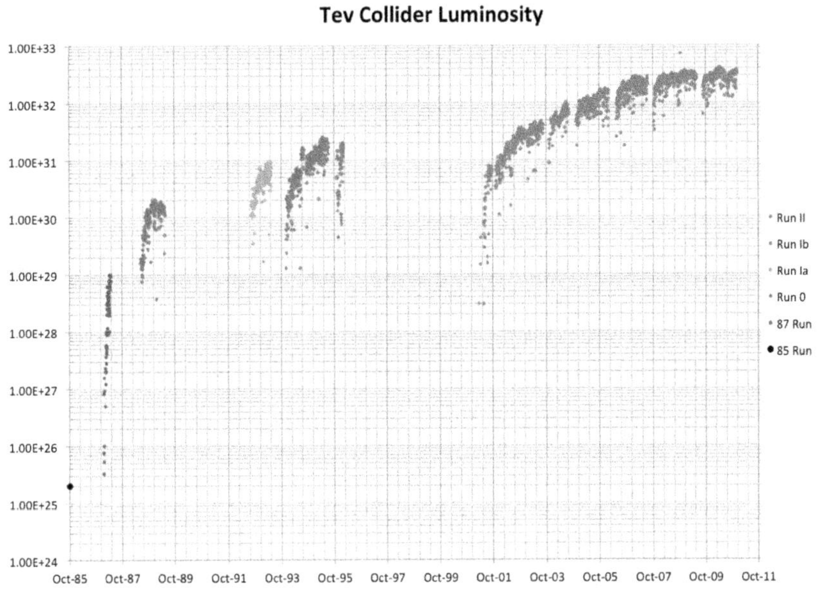

(b)

Fig. 5. (a) The integrated luminosity required per experiment as estimated in 2000, to either exclude a SM Higgs boson at 95% CL (curve at the bottom) or discover it at the 3-sigma (curve in the middle) or 5-sigma level (curve at the top), as a function of the Higgs mass. These projections were based on the Higgs production cross sections available at that time, sometimes based on Leading Order calculations only. (b) Instantaneous luminosity at the Tevatron collider over its 25 years of operation.

the combination. In fact, as time went on, the sensitivity of the searches improved far more than what was expected from increased luminosity alone.

At the end of Run 2 each of the Tevatron experiments collected datasets corresponding to a total integrated luminosity of $10\,\text{fb}^{-1}$. And for the first time since the final LEP limits were reported, new direct constraints were made on the mass of the Higgs boson, excluding the 156–177 GeV mass range.[18] With time the exclusion window in this "high-mass" region, dominated by the $H \to WW$ channel, was extended.[19] Had the Higgs boson mass been in this range, the discovery story would have been very different! These results, in combination with the precision electroweak measurements were able to establish, before the LHC produced significant results, that a SM Higgs boson with mass above $\simeq$145 GeV was excluded at 95% C.L.[20] In the remaining window at the lower mass range, dominated by the $H \to b\bar{b}$ channel, and where the sensitivity was actually worse, and not better as predicted by the 2000 Higgs Working Group, the Tevatron experiments could ultimately not exclude the presence of a Higgs boson. In July 2012, an excess of events, at a mass consistent with 125 GeV, corresponding to a significance near 3 standard deviations, was reported.[21] As it was, to reach much higher sensitivity in the lower mass range the experiments would have had to integrate several times more luminosity through an extended run. Much consideration was given to this idea, however, with the advent of the LHC this possibility did not materialize. Still, due to the difficulty to unequivocally observe the $H \to b\bar{b}$ channel at the LHC, until considerably more data is accumulated in the 13 TeV run, the contributions of the Tevatron to the Higgs search and discovery were timely and significant.

The Lausanne Workshop in 1984[22] can be considered as the event that gave birth to the LHC, although its actual construction started only in 1998. An interesting account of the construction and commissioning of the LHC can be found in Ref. 23. The goal was to discover a Higgs boson with a mass anywhere in the 0.1 to 1 TeV range, which is where it was theoretically expected. With that in mind, the LHC was built to collide protons at a center-of-mass energy of $\sqrt{s} = 14$ TeV with a nominal instantaneous luminosity of $L = 1 \times 10^{34}$ cm^{-2}s^{-1} and a bunch spacing of 25 ns. It can also collide heavy ions. Four main experiments were installed around the ring: two general-purpose detectors, ATLAS[24] and CMS[25] a detector dedicated to study heavy flavor physics, LHCb,[26] and a detector dedicated to the study of heavy ion collisions, ALICE.[27]

At hadron colliders the Higgs boson can be produced through several processes that are initiated either by quark–antiquark annihilation or by gluon–gluon fusion. Figure 6 shows these processes. Due to its higher energy relative to the Tevatron, the LHC, at 7 TeV for example, would produce particles with masses of 100 GeV at a rate of about 4 times higher for quark–antiquark induced processes, and at a rate of about 20 times higher for gluon–gluon processes. This is shown in Fig. 7. However, given that the cross sections for all background processes also increase with the increasing energy, this did not make the search for the Higgs boson at the LHC much easier. Figure 8 shows the Higgs production cross section being several orders of magnitude smaller than those of important standard theory background processes.

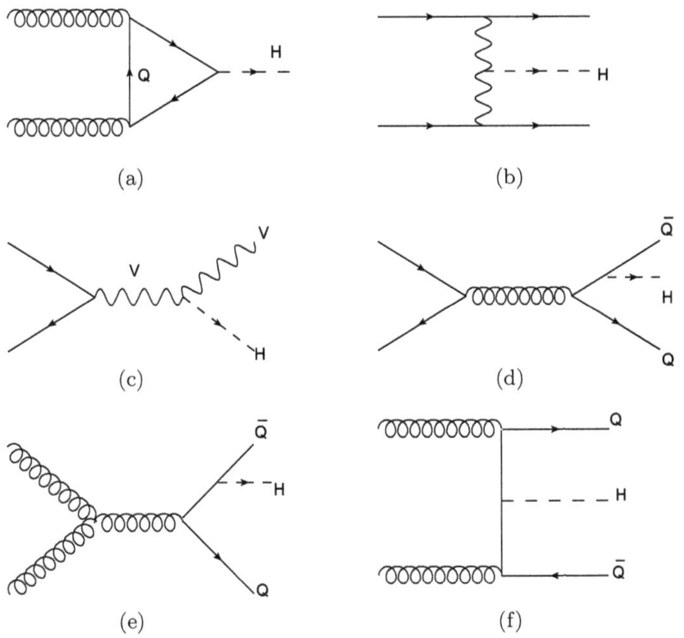

Fig. 6. Higgs production processes at hadron colliders.

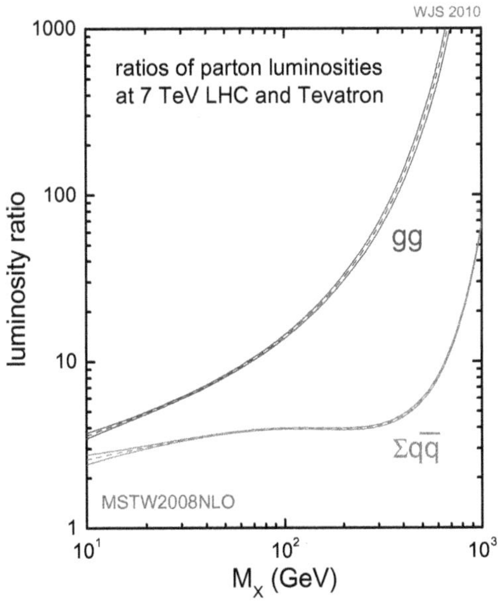

Fig. 7. Ratios of parton luminosities at 7 TeV LHC and at the Tevatron.

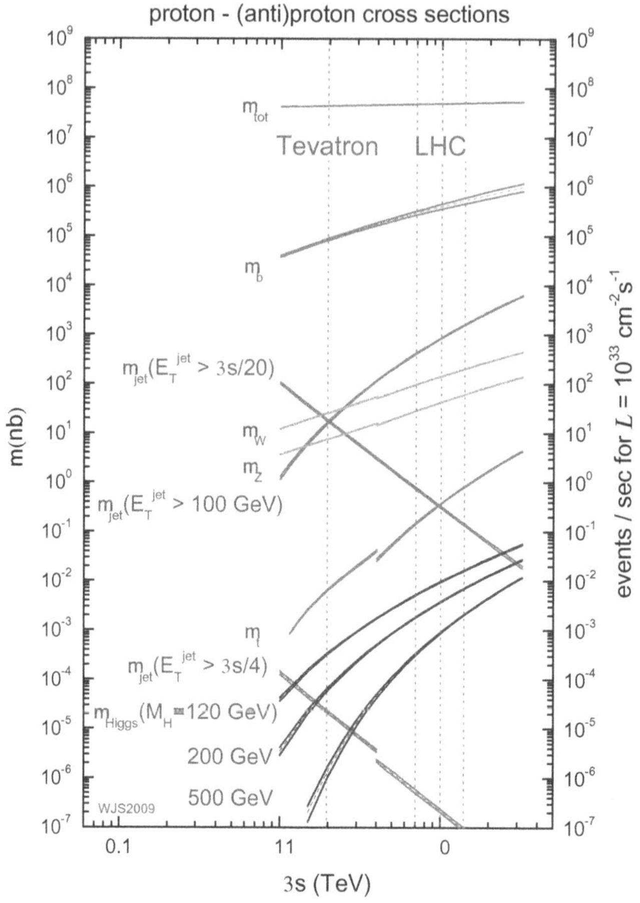

Fig. 8. Cross sections vs energy for different processes.

On September 10th, 2008, the LHC began the final preparations towards operation but unfortunately soon after an electrical fault with the interconnection between two dipole magnets caused a Helium explosion that damaged a large section of the accelerator.[28] It took more than a year to repair the damage, understand the problem, and establish proper measures to avoid recurrence. It was decided to limit the current to the magnets, which in turn limited the beams energy to 7 TeV, until the connections between all magnets could be fixed properly in a future long shutdown period. Meanwhile during 2010 and 2011 (Run 1) protons collided at that energy, and a total of $5.6\,\text{fb}^{-1}$ integrated luminosity was delivered. In 2012 the energy was safely increased to 8 TeV, and a total of $23.3\,\text{fb}^{-1}$ integrated luminosity was delivered. Figures 9 and 10 show the peak instantaneous luminosity and the cumulative integrated luminosity for both the 7 and 8 TeV running periods. The peak instantaneous luminosity that was ultimately reached was about $7 \times 10^{33}\ \text{cm}^{-2}\text{s}^{-1}$.

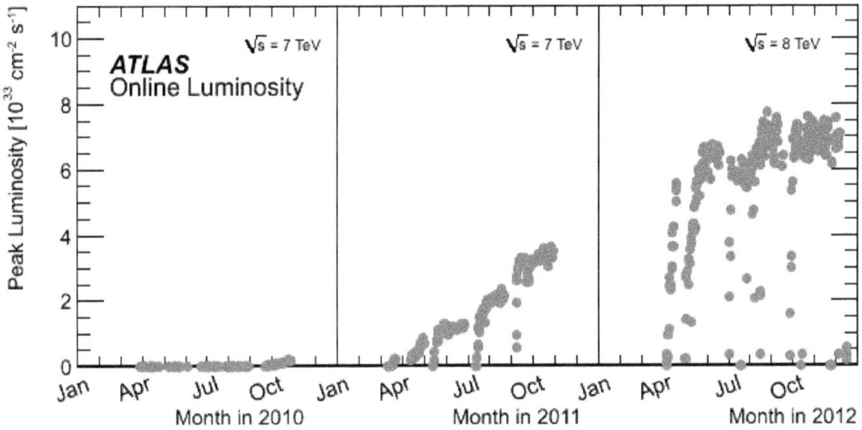

Fig. 9. LHC instantaneous luminosities as recorded by the ATLAS experiment in 2010, 2011 and 2012.

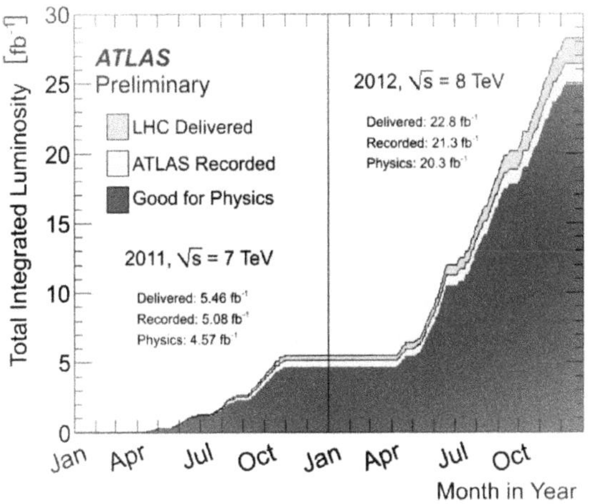

Fig. 10. LHC integrated luminosities as recorded by the ATLAS experiment in 2012.

At the LHC the search for the Higgs boson proceeded in parallel with the ATLAS and CMS experiments competing with each other to reach better sensitivity. Notwithstanding the Tevatron results, the LHC experiments approached the search without any *a priori* bias on the value of the Higgs boson mass and all possible channels were addressed. As detailed in Sec. 3, at high masses the sensitivity was dominated by the $H \to ZZ$ and $H \to W^+W^-$ channels while at the lower masses, near 125 GeV, the $H \to \gamma\gamma$ and the $H \to ZZ^{(*)} \to llll$ dominated the sensitivity.

By August 2011, with an integrated luminosity of about 2 fb^{-1} (ATLAS) and 1.5 fb^{-1} (CMS), the mass range of 145–450 GeV was excluded,[29] surpassing for the first time the Tevatron results. First small excesses were seen in July, but diminished in August. In December 2011, with about 5 fb^{-1} of data collected at $\sqrt{s} = 7$ TeV the following mass ranges were excluded[30,31] at the 95% CL by either ATLAS or CMS: $110.0 < m_H < 117.5$, $118.5 < m_H < 122.5$, and $127.5 < m_H < 600$. A small excess, with a local significance of $\sim 2.5\sigma$, was observed by both experiments at a mass of about 125 GeV, fueling much speculation that a discovery was in the cards. In the following Winter conferences (March 2012) the Tevatron reported an excess of similar strength in a wider but compatible mass region. The excess was coming mainly from the associated production channel, with the Higgs boson decaying to a pair of b quarks, which was not observable at that time by the LHC experiments.

On July 4th, 2012, the search for the Higgs boson culminated with the announcement that ATLAS and CMS had independently discovered a new particle with mass close to 125 GeV.[32–34] The discovery was clear, the signal being seen in several channels with high statistical significance, for a combined significance of 5σ or more by both experiments independently. CDF and D0 then released their combination on the full Tevatron dataset showing evidence for the decay in $b\bar{b}$ of a particle compatible with the one discovered at CERN.[21] The data gathered at the LHC in the months that followed helped strengthen the signals found and allowed for the first measurements of various properties of the new particle to be made. An account of these measurements can be found in a dedicated Chapter in this book.

In March 2013 the LHC entered a shutdown period planned for machine maintenance and to perform the needed interventions to bring the beams to collide a a center-of-mass energy of 13 TeV, close to the original design. At the time of this writing the first collisions at this new record energy were recorded, bringing great expectations for new discoveries.

In the following sections we discuss in more detail, both for the Tevatron and the LHC, the factors that contributed to the search and discovery of the Higgs boson, and the results obtained at the time of the discovery.

2. Higgs Searches at the Tevatron

While the Run 1 of the Tevatron (1992–95) had not enough luminosity to search for the SM Higgs boson, the luminosity gain expected in Run 2 changed the situation completely, allowing for the Higgs boson search to become a major physics goal of the CDF and D0 experiments. However, as mentioned earlier, it was clear from the beginning that new advances in reconstruction, analysis and combination would be needed to have a chance to detect the Higgs boson. Before describing the paths followed, let us briefly remind the properties of these two experiments, while referring to Refs. 35 and 36 for the detailed characteristics of the CDF and D0 detectors.

The D0 tracking system was contained within a solenoidal magnet with a 2T field and consisted of an inner silicon micro-strip tracker surrounded by an outer central scintillating fiber tracker. Track momentum measurements were made up to $|\eta| \sim 2.5$. Liquid argon calorimeters reach up to $|\eta| = 4.2$ and were used for the identification and energy measurement of electrons, photons, and jets. Muon detectors covered up to $|\eta| < 2$, and a 1.8 T toroidal magnet made it possible to determine muon momenta within the muon system alone.

The CDF tracking system consisted of a silicon micro-strip tracker surrounded by an open-cell drift chamber, all immersed in a 1.4 T solenoidal magnetic field. Tracking coverage extended up to $|\eta| \sim 2.0$. The sampling-scintillator calorimeters that surround the tracking system covered the pseudorapidity region of $|\eta| < 3.64$ and were based on a projective tower geometry. They used lead and iron as absorbers in the electromagnetic and hadronic modules, respectively.

At low masses ($m_H < 135$ GeV), the dominant decay of the Higgs boson is $H \to b\bar{b}$, but the search for direct ($gg \to H$) Higgs boson production in this decay mode, was not performed due to overwhelming multijet background. However, $q\bar{q}$ annihilation results in associated vector boson-Higgs boson production (VH) in $p\bar{p}$ collisions with a better signal to background ratio than available at a pp collider. The "primary" channels for searching for a low mass Higgs boson at the Tevatron, WH and ZH production, are best studied in the $\ell\nu b\bar{b}$, $\ell\ell b\bar{b}$ or $\nu\nu b\bar{b}$ final states. The lower branching ratios or poor signal to background of the other decay modes rendered their sensitivity smaller than these primary channels. They were also searched for to provide additional sensitivity in the combination of all channels, but are not described here, see for instance Ref. 37 for more details. In all the analyses, the data were separated into multiple orthogonal search samples of varying sensitivities.

Vector boson fusion production, associated production with vector bosons and direct Higgs boson production can also be exploited when the Higgs boson decays to a $\tau\tau$ pair, by making use of the kinematics of the potential additional jets in the final state. These processes suffer from significant background and so were considered as secondary channels at the Tevatron. Another secondary channel which was exploited at the Tevatron is the inclusive production of a Higgs boson with Higgs boson decay to two photons. The sensitivity of this channel is low due to the small branching ratio of this decay, typically smaller than 0.2%. The $t\bar{t}H$ production was also searched for, but has lower sensitivity.

Since the low mass analyses used advanced multi-variate techniques for separating signal and backgrounds and were obtained from low signal-to-background search samples, a crucial test was performed considering WZ and ZZ diboson production with Z decays to heavy flavor as signal, mimicking the final state of WH and ZH including the essential feature of resonant dijet production. These analyses, and their combination, were performed in the same way as their Higgs counterparts. The CDF + D0 combination displayed strong evidence (4.6σ) for such production, with a measured cross section $\sigma(VZ) = 4.47 \pm 0.97$ pb consistent with the SM prediction.[38]

2.1. *Low mass Higgs boson searches*

The search for the processes $q\bar{q} \to VH + X$ in which a quark–antiquark pair leads to the production of the Higgs boson in association with a Vector boson (W or Z) was based on a total integrated luminosity $\mathcal{L} \simeq 10$ fb^{-1} of collision data collected by both the CDF and D0 detectors at the Fermilab Tevatron $p\bar{p}$ collider between 2002 and 2011. In the WH channel, candidate W boson events were preselected via their decays to an electron or a muon plus a neutrino ($W \to e$ or $\mu\nu$) while the Higgs boson was identified though its decay mode into a pair of b-quarks ($H \to b\bar{b}$).[39,40] The experimental signature was a single isolated lepton, missing transverse energy, and either two or three (to accommodate additional gluon radiation in the hard collision) jets, at least one of which was required to be consistent with having been initiated by a b-quark.

To increase signal acceptance, the lepton identification criteria were as loose as possible. This resulted in backgrounds originating from multijet (MJ) events, in which one of the jets is misidentified as an isolated lepton. In CDF, the MJ background was strongly reduced by kinematic cuts and by using a dedicated multivariate technique to reject this background.[39] The remaining MJ background contribution was modeled from the data using side-band techniques. In D0, the MJ background contributions passing the preselection criteria in each sample was determined from the data using an unbinned matrix method approach.[40] The "physics" backgrounds with similar event topologies were modeled using Monte Carlo event generators. The SM predictions were used to set the relative normalizations of all the generated samples, with additional normalization factors applied to samples of $W + n$ partons generated using the ALPGEN Monte Carlo event generator. These factors were determined from data at the preselection stage where the SM Higgs boson contribution was negligible. The predicted backgrounds modeled the data well in the high statistics sample before a b-tagging algorithm was applied.

The analyses proceeded by subdividing the selected sample into orthogonal subsamples based on how many of the jets in the event, one or two, are consistent with having been initiated by a heavy b-quark, and at what level ("loose" or "tight") of confidence, resulting in 6 (D0) or 7 (CDF) independent samples. In two b-tagged jet events, the dominant remaining backgrounds were from $Wb\bar{b}$, $t\bar{t}$, and single top quark production. In single b-tagged jet events the dominant backgrounds were $W+$ light or c-quark jet production as well as MJ background events. To further discriminate the remaining backgrounds from the signal, MVA techniques were applied to each subsample.

The systematic uncertainties taken into account affected not only the normalization of the signal and backgrounds, but also the shape of the MVA output distributions. Uncertainties in the efficiencies of selection, on jet calibration and on the b-tagging criteria affecting the precision at which the background modeling is known were considered. The uncertainties on the parton density functions and the effect of renormalization and factorization scales on signal and background simulation were also taken into account.

In the leptonic ZH channel, candidate Z boson events were preselected via their decays into e^+e^- or $\mu^+\mu^-$ pairs, and the associated Higgs boson was identified through its decay into a pair of heavy b-quarks ($H \to b\bar{b}$).[41,42] Candidate events were required to have two or three jets, at least one of which was identified as a b-jet. In this final state, which required two leptons, the MJ background was negligible. The "physics" backgrounds were modeled using the same Monte Carlo event generators used in the WH analysis.

To maximize the lepton acceptance and benefit from higher quality lepton categories, the events were classified according to the lepton types. Those having both leptons identified with high confidence were treated separately from the others which contain loosely identified, forward, or track-based leptons. These samples were analyzed independently, allowing for an optimal sensitivity of the search. In addition, multivariate lepton selections were used. In CDF, to enhance the discriminating power of the dijet invariant mass, a neural network (NN) derived energy correction was applied to the jets. This correction depends on the missing transverse energy and its orientation with respect to the jets. In D0, jet energy resolution improvements were obtained through a kinematic fit of the complete event, since all particles can be detected in this process. These analyses also proceeded by subdividing the selected sample into 6 (D0) or 8 (CDF) orthogonal subsamples based on the number and the quality of the b-tagged jets in the event. CDF employed two NNs to simultaneously separate signal events from the dominant $Z+$jets and kinematically different $t\bar{t}$ backgrounds. In single and double b-tagged jet events, the dominant remaining background was $Zb\bar{b}$. To suppress the remaining background MVA techniques were applied to each subsample. Systematic uncertainties were overall less important than for WH since no missing transverse energy was involved, but most of the other systematic uncertainties were of similar magnitude to those of the WH analyses.

The remaining $b\bar{b}$ analysis was built to detect the $ZH \to \nu\nu b\bar{b}$ process but was also sensitive to WH events in which the charged lepton was not identified, hence its alternate label as $VH \to \not{E}_T\, b\bar{b}$.[43,44] Since this final state contains no leptons, triggering on these events and modeling the effects of the trigger requirements on the event selection were significant challenges. The triggers were based on $\not{E}_T$, with or without accompanying jets. The analyses were performed while studying in parallel several control samples to monitor the understanding of the background. Events were required to have significant $\not{E}_T$ and two or three jets, well separated from the $\not{E}_T$ direction. For the preselection, multivariate approaches were also applied to the events to remove a large part of the MJ background. For the final selection b-tagging was employed. For this analysis, the preselection plays a crucial role, given the size of the MJ background. As for the other $b\bar{b}$ analyses, the "physics" backgrounds were taken into account using Monte Carlo event generators. The preselected samples were subdivided into 6 (CDF) or 2 (D0) orthogonal subsamples based on the

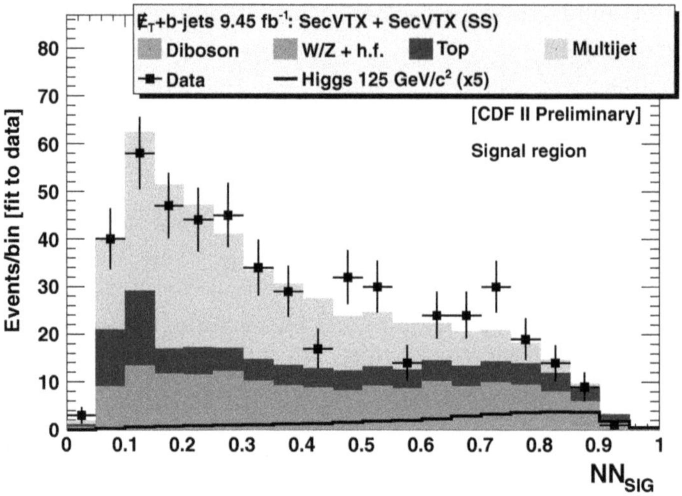

Fig. 11. Distribution of the NN discriminant in the CDF $\not{E}_T\, bb$ analysis for the sample with two or three jets where two of the jets are tagged as b-jets by the secondary vertex algorithm. The data are well described by the sum of all the SM backgrounds. The simulated signal is also represented.

number and the quality of the b-tagged jets in the event. To suppress the remaining backgrounds multivariate discriminant techniques were applied to each subsample as shown in Fig. 11.

The results of CDF and D0 are presented here separately while the CDF + D0 combination is given in Section 4. The statistical techniques used are also described in Section 4. They allow for the extraction of the limit on the signal cross section normalized to the SM expectation, or, in case of excess, determine the p-value of the background fluctuation. At the lowest searches masses, when combining the three $H \to b\bar{b}$ topologies, CDF and D0 exclude at 95% C.L. Higgs bosons with masses smaller than 96 GeV and 102 GeV, respectively. However, in the results from both collaborations an observed limit above the background-only expectation was obtained for the $\simeq$120–140 GeV range. The local p-values were calculated and found to be minimal for a Higgs boson mass of 135 GeV at CDF[45] and D0[46] searches, where the local significance of these deviations with respect to the background-only hypothesis corresponded to 2.5 (1.5) σ global significances.

Higgs boson searches using tau leptons decaying hadronically (τ_h) complemented those using electrons and muons. CDF performed a generic analysis searching for Higgs bosons decaying to τ lepton pairs originating from direct $gg \to H$ production, associated WH or ZH production, and vector boson fusion production. The D0 $\ell\tau_h jj$ analyses also included direct $gg \to H$ production, associated WH or ZH production, and vector boson fusion production.[47] Decays of the Higgs boson to tau, W, and Z boson pairs were considered. A final state consisting of one leptonic τ decay, one hadronic tau decay, and two jets were required. Both CDF[48] and D0[49] also searched for Higgs bosons decaying into diphoton pairs with the full statistics

(10 fb^{-1}) but their sensitivities remained limited. The CDF analysis searches for a signal peak in the diphoton invariant mass spectrum above the smooth background originating from QCD production in several detector based categories with different signal to background ratios. In the D0 analysis the contribution of jets misidentified as photons was reduced by combining information sensitive to differences in the energy deposition from real or false photons in the tracker and in the calorimeter in a neural network output (NN_o). The $t\bar{t}H$ production is interesting for the direct tH coupling it involves, however its cross section was too small at the Tevatron to contribute to the overall search sensitivity.

2.2. *High mass Higgs boson searches*

The SM Higgs boson has a strong coupling to both massive electroweak bosons. For $m_H > 135$ GeV the decay to a pair of W bosons is dominant since even below the threshold to produce on-shell W bosons the decay rate to one real and one virtual W boson is substantial.

The $H \to W^+W^{-(*)}$ decay was searched in final states with at least one charged lepton and from all production processes with substantial cross section. In addition, searches for the sub-dominant decay $H \to ZZ$ were performed. The Tevatron experiments observed all of the direct diboson production processes with pairs of heavy gauge boson in final states that were topologically similar to those used in the Higgs boson search.

The diboson production analyses in final states with charged leptons were performed with the same techniques used in the high mass Higgs boson search, also providing measurements of the WW, WZ, and ZZ cross sections.[50–54]

The Higgs analysis typically required a triggered lepton with $p_T > 20$ GeV, possibly additional leptons with lower thresholds, and significant $\not{E}_T$ that was not aligned along the direction of other physics objects in the events. The events were then categorized into a large number of topologies that were consistent with various Higgs boson production and decay modes. These topologies were characterized by the number of charged leptons, whether the leptons were same or opposite charge, and the number of jets (n_j). Each topology involved a limited set of dominant signals and backgrounds allowing for optimal discrimination, and was therefore analyzed separately. The most sensitive analysis topology involving zero jets and leptonic $H \to W^+W^-$ is described below, while the subdominant modes are briefly discussed afterwards.

The signature for $ggH \to W^+W^- \to \ell^+\nu\ell^-\bar{\nu} + n_j$ jet when $n_j = 0$ is two opposite sign leptons, $\not{E}_T$, and no observed jets.[55,56] The signal in this final state was almost 100% produced by the ggH process. The dominant background was from SM direct WW production with minor contributions from Drell–Yan production, the WZ and ZZ diboson processes where one or more charged leptons were not detected, and $W+$jets or $W+\gamma$ where a jet was misidentified as a lepton or the γ converts to an electron-positron pair, only one of which is detected. The strongest discriminant was the opening angle between the leptons in two dimensions, $\Delta\phi$, or three dimensions, ΔR, due to the spin correlation between the two spin one W bosons when decaying from the scalar Higgs boson. The D0 experiment further subdivided this mode by lepton flavor. CDF subdivided this analysis into modes with two well identified leptons, or one well identified lepton plus and an isolated track, and analyzes events with low dilepton invariant masses as a separate category. When $n_j = 1, 2$, the signature is two opposite sign leptons, $\not\!\!E_T$, and observed jets. These topologies had substantial contributions from VH or VBF where the jets were observed from either one of the vector boson decays or final state quarks respectively and additional background from top pair production.

Associated production $VH \to VW^+W^-$ was also studied in events with either same sign leptons or trileptons when the associated vector boson decays to charged leptons. The background included $W+$jets with a misidentified lepton in the same sign mode and SM direct WZ production in the trilepton case. Higgs boson production were also searched for inclusively in events where one of the W boson decays leptonically and the other W boson decays to two quark jets,[57] i.e. $(ggH, VH, VBF) \to H \to WW \to \ell\nu + \geq 2$ jet. The dominant backgrounds were from $W+$jets and multijet background where a jet is misidentified as a lepton. The experiments also considered modes where one W boson decays to a τ lepton which decays hadronically. A search for the Higgs boson was also performed in the $H \to ZZ$ mode where both Z bosons decay to charged leptons.[58] The only significant background in this mode was SM direct ZZ production but the sensitivity remained low given the cross section times BF.

All these search channels would have selected a total of approximately 75 Higgs boson events per detector for m_H =165 GeV, with a good signal to background ratio using MVA discriminants as illustrated in Fig. 12. The largest discriminating power was coming from the spin correlation variable, due to the unique scalar nature of the Higgs boson. No significant excess was seen in these high mass searches and limits were thus extracted, taking into account systematic uncertainties. The experiments each achieved expected sensitivity within a factor of approximately 1.5 of the SM cross section, in a mass range from m_H = 140–185 GeV. The CDF (D0) analysis excluded the SM Higgs boson in the mass range m_H = 148–175 GeV (157–172 GeV).[59,60] The CDF + D0 combined results are presented in Section 4.

3. Higgs Searches at the LHC

The search for the Higgs boson by the ATLAS and CMS experiments targeted from the start a very broad mass range. It was important to confirm the excluded range from the Tevatron experiments and to extend with direct searches the mass range to the highest accessible masses. It was paramount to explore the low mass window that had not been excluded by LEP or by the Tevatron. This meant that a large variety of processes needed to be studied. Figure 13(a) shows the cross sections for

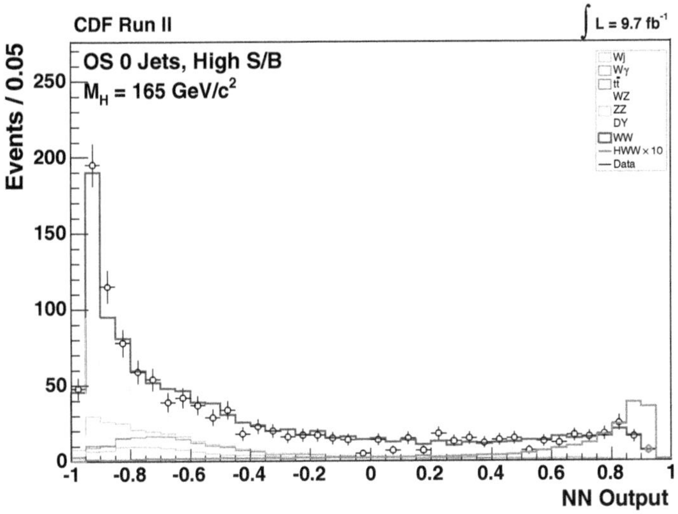

Fig. 12. NN discriminant for $ggH \to WW \to \ell\nu\ell\nu + 0$ jet at CDF. The distribution for background and a $H \to WW$ signal with $m_H = 165$ GeV (multiplied by 10) are compared.

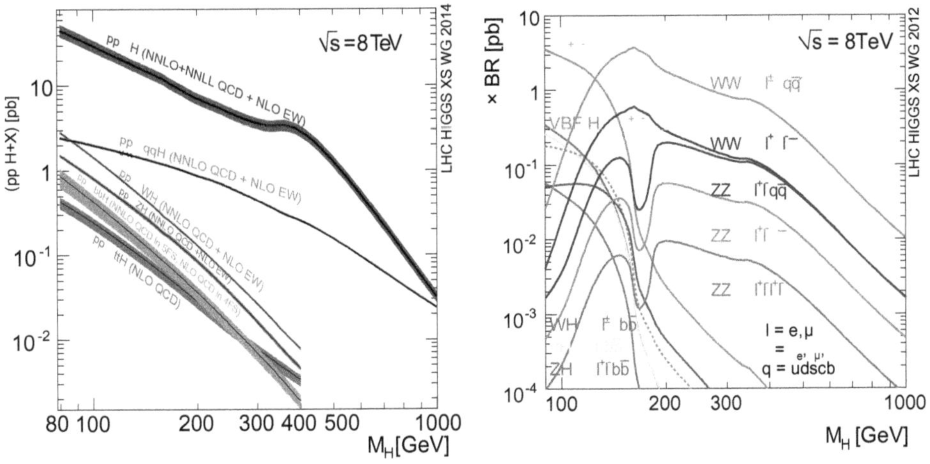

(a) (b)

Fig. 13. (a) Higgs production cross sections at the LHC at Ecm = 8 TeV. (b) Standard Theory Higgs boson production cross section times branching ratio at Ecm = 8 TeV.

the various productions mechanisms for a Higgs boson at the LHC as a function of its mass at 8 TeV. The convolution of these cross sections with the possible decay modes results in many final states with very different yields depending on m_H, as exemplified in Fig. 13(b).

At large masses ($\simeq 200$ GeV) the searches were performed in channels where the Higgs is produced through gluon fusion and decays to ZZ or W^+W^- and, as seen in the figure, included all channels in which at least one of the bosons decayed leptonically. These channels had the greatest sensitivity as, for the most part, they did not have to contend with the copious backgrounds that purely hadronic channels suffer from. At lower masses a wide variety of search channels were accessible from all production mechanisms: gluon-fusion (gg), vector-boson-fusion (VBF), and production in association with vector-bosons (VH) or top-quark pairs (ttH). In this mass range, and especially, at around 125 GeV the Higgs boson has many possible decays with significant branching fraction such that the various final states included essentially all possible detectable physics objects: leptons, photons, tau-leptons (taus), b quarks and generic jets.

Of course Fig. 13(b) by itself does not represent the actual sensitivity of each of the channels represented there as these are inclusive yields and triggering bandwidth and different event reconstruction efficiencies and background levels affect the sensitivities significantly. Before describing in more detail the channels studied and the data analysis challenges they represented, we describe in brief the characteristics of the multi-purpose ATLAS and CMS detectors that allowed the experimentalists to explore this large set of final states. The details of the design of these magnificent detectors is described in Ref. 24 and in Ref. 25.

The ATLAS detector includes an inner tracker that consists of a silicon pixel detector, a silicon micro-strip detector, and a transition radiation tracker. It is inside a solenoidal magnet with a 2 T field, and measures the trajectories of charged particles and in the $|\eta| < 2.5$ pseudorapidity range. The inner tracker is surrounded by a high-granularity liquid-argon sampling electromagnetic calorimeter which provides electron and photon identification in the $|\eta| \sim< 2.4$ range. An iron-scintillator tile calorimeter provides hadronic coverage in the central pseudorapidity range. The end-cap and forward regions extend the calorimetry coverage to $|\eta| < 4.9$, and both their electromagnetic and hadronic components are instrumented with liquid argon. The muon spectrometer covers the $|\eta| < 2.4$ range and surrounds the calorimeters. It consists of three large super conducting toroids, and a system of precision tracking chambers and specialized detectors for triggering.

The CMS detector comprises, in successive layers outwards from the collision region, a silicon pixel and strip tracker, a lead tungstate crystal electromagnetic calorimeter, a brass/scintillator hadron calorimeter, a super-conducting solenoid, and gas-ionization chambers embedded in the steel return yoke for the detection of muons. The silicon detector provides charged particle tracking in the pseudorapidity range of $|\eta| < 2.5$. The electromagnetic calorimeter and the muon chambers provide coverage to $|\eta| < 3$ and $|\eta| < 2.4$, for electron and muon identification, respectively. The return field of the magnet allows independent momentum measurement and triggering in the muon chambers. Jet finding can be performed up to an expanded pseudorapidity range of $|\eta| < 5.0$ using forward calorimeters.

These detectors need to overcome the challenges resulting from the high intensity proton beams that the LHC delivered. During Run 1 the LHC proton bunch spacing

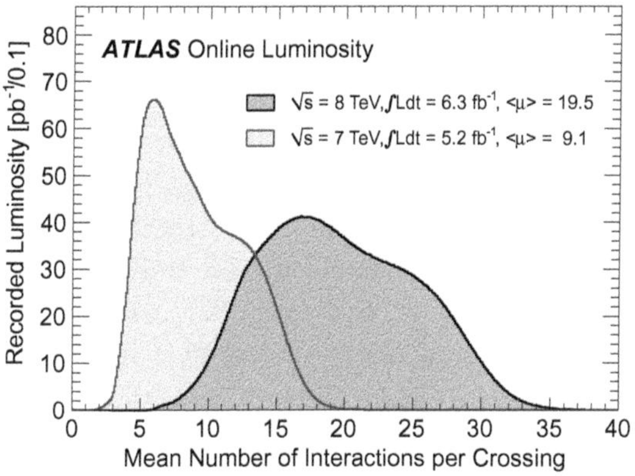

Fig. 14. Luminosity-weighted distribution of the mean number of interaction per crossing.

Table 1. The five main decay channels in the low-mass Higgs boson search and discovery at the LHC.

Decay channel	Reconstructed mass resolution
$H \to \gamma\gamma$	1–2%
$H \to ZZ \to \ell^+\ell^-\ell'^+\ell'^-$	1–2%
$H \to W^+W^- \to \ell^+\nu\ell'^-\nu$	~20%
$H \to \tau^+\tau^-$	~15%
$H \to b\bar{b}$	~10%

was 50 ns and therefore proton–proton collision events occurred at a rate of approximately 20 MHz. Figure 9 shows the instantaneous luminosity during Run 1, which reached a peak of 7.7×10^{37} cm^{2}s^{-1} by 2012. As a consequence multiple proton–proton inelastic collisions occurred within a single bunch crossing. The cross section for inelastic proton–proton interactions is $\sigma \approx 60$ mb, and the resulting multiplicity profile for these interactions (called "*pileup*"), as measured by the ATLAS detector, is shown in Fig. 14. At 8 TeV the average number of pileup collisions per event is about twenty, resulting in about 4×10^8 collisions per second seen by the detectors. The digitized information from each collision is about one Mbyte and recording 4×10^8 Mbytes/second over roughly 10^7 second of effective running during a year (roughly 30% of the time) would result in about 4 million Petabytes of data, and clearly impossible to record it all. To get to a manageable rate at which to record events, of a few hundred Hz, a multi-level triggering system is implemented. At each stage a decision is made on whether there might be a collision worth examining further until the event gets finally recorded. Still, even for interesting events, in order to maintain the trigger rates low enough, thresholds on all physics objects needed to be imposed at the trigger level resulting in acceptance losses. It was therefore imperative for ATLAS and CMS to maintained a so-called trigger menu that was optimized for the Higgs discovery, not an easy feat.

In what follows we describe the analyses as they were done at the time of the announcement of the Higgs boson discovery, including the figures with the available data at the time. We naturally focus on the searches at low mass where the discovery was made. In this region the five main decays modes studied are listed on Table 1. The first three channels are bosonic decays while the last two are fermionic decays. The Higgs boson mass resolution of the reconstructed final state is also shown on the Table. The excellent 1–2% resolution for fully reconstructed photon and lepton final states is a big factor on why these channels had the best sensitivity in the search. We present the highlights of the searches in these channels first. Only when germane we present distinctions between ATLAS and CMS.

3.1. *Searches for $H \to \gamma\gamma$*

The $H \to \gamma\gamma$ decay branching ratio is small as it can only proceed through loop diagrams. However the large inclusive production cross section and the distinct signature makes this a viable mode for a Higgs boson observation at low mass (< 160 GeV). At higher mass as can be seen in Fig. 2 there is basically no rate and therefore no sensitivity. The signal contains two isolated, high p_T (usually > 30 GeV), photons and possibly other objects from VBF and production mechanisms associated with vector bosons or top quarks. Both the ATLAS and CMS detectors were designed with excellent electromagnetic calorimeters that allow for precise measurements of photon (and electron) energies. This is paramount in the $H \to \gamma\gamma$ search as the continuum two-photon background, originating from multi-jet, multijet + photon and true di-photon production, is large relative to the number of signal events expected. Data is collected with di-photon triggers. Energy and isolation requirements are made on the photons to reduce backgrounds. Converted photons are also used and provide good energy and position resolution when electron pairs can be reconstructed by the trackers. The ATLAS experiment reduces background by using the longitudinal segmentation of the calorimeters to select photons that point to the hard interaction vertex. The CMS calorimeter is not segmented longitudinally and the experiment relied on information from the tracking system to resolve the most likely collision point from which the photons originated.

Several effects affect the precision with which the photon energy and direction are measured and the level of background relative to signal events. These include the various criteria used to identify photons, the photon p_T, photon conversions, isolation, the location of the photon in the detectors, and the presence of other objects, such as jets, in the event. To optimize the sensitivity of the search the experiments divide the two-photon event sample in multiple categories characterized by these properties. The search for the Higgs boson was performed as a fit to the di-photon invariant mass distribution using a smooth background functional form plus a narrow peak to describe the possible contribution for signal. This exercise is repeated at the different assumptions for Higgs boson mass. Each category is

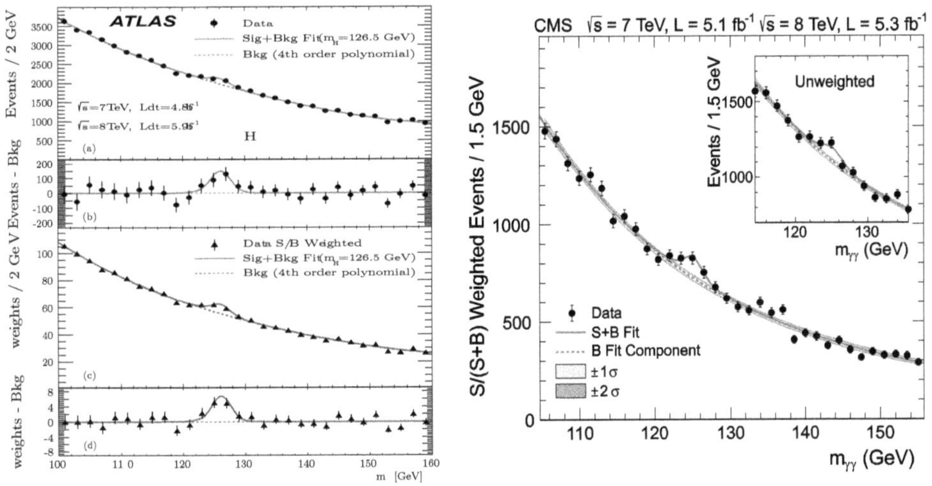

(a) (b)

Fig. 15. (a) Invariant mass of diphoton events in the ATLAS experiment. The results are presented with and without event weighting by expected signal to background ratio. (b) Invariant mass of diphoton events in the CMS experiment with events weighted by expected signal to background ratio and the unweighted distribution shown as an inset.

treated separately and the final result of the search is the combination of the results from all categories. The most challenging part of the analysis from the instrumental point of view was to maintain the precision of the calibration of the electromagnetic calorimeters in a high radiation environment. This required careful and continuous monitoring and recalibration. From the analysis side, in addition to the optimization by categories, the challenge was to describe the background shape in a robust manner. This was accomplished directly in data by testing several different choices of functional forms and by careful study of the systematic uncertainty associated with that process.

Initially the experiments were able to exclude the production of a Higgs boson at the rate predicted by the SM at several mass ranges in which there were no excess of events above the background expectation. However in the mass region around 125 GeV there was an excess of events in both CMS and ATLAS that became larger as the dataset sizes increased. The ATLAS di-photon invariant mass distributions, both weighted by signal to background ratio and unweighted are shown in Fig. 15(a). The corresponding CMS diphoton invariant mass distributions are shown in Fig. 15(b). The small signal-to-background ratio was compensated by the sharp resolution of the mass peak and a few hundred events consistent with the decay of a new particle can be clearly seen.

A statistical analysis to quantify these excesses was performed and in July 2012, the ATLAS experiment observed a 4.5σ excess of events compatible with a narrow resonance of mass 126.5 GeV with a signal strength of 1.9 ± 0.5 times the SM expectation. The CMS experiment observed a 4.1σ excess of events compatible with a narrow resonance of mass 125 GeV with a signal strength of 1.6 ± 0.4 times the SM expectation. The strong evidence seen by both experiments in this single $H \to \gamma\gamma$ decay mode indicated that a new boson had been observed and, very importantly, strongly disfavored its spin value to be 1. The next piece of evidence for the existence of such a particle came from another low decay rate process, $H \to ZZ$, as described here below.

3.2. *Searches for $H \to ZZ^{(*)} \to llll$*

The $H \to ZZ \to \ell^+\ell^-\ell'^+\ell'^-$ channel was used at the LHC to search for the Higgs boson in a broad mass range, roughly from one hundred and up to a few hundred GeV. The 4-lepton mode (where electrons and muons are considered) is one of the Higgs boson decays with smallest rate but it has the advantage that it can be fully reconstructed. The main background process are: electroweak ZZ production, $Z+$jets, and $t\bar{t}$ production. At low mass the background yield is relatively small, after adequate requirements are made on the leptons p_{T}, and comparable to what is expected from a 125 GeV Higgs boson. The Higgs boson signal is a narrow resonance in the 4-lepton invariant mass distribution while background processes have a smooth continuous shape. To maximize the number of signal events, because of the four leptons in the final state, the challenges are to use the largest possible fiducial regions for lepton detection and to use advanced techniques in lepton identification, such as multivariate discriminants.

Data is collected using single and dilepton (CMS) triggers. The CMS experiment improves mass resolution by using an algorithm designed to detect and recover the momentum of final state photons radiated by the leptons. ATLAS incorporates a similar technique as part of its electron momentum fit. The main SM ZZ background is estimated from simulation normalized to NLO calculations. The backgrounds from $Z+$jets and $t\bar{t}$, where some of the leptons originated from b-quark decays or from fakes, are estimated from control regions in data. The performance of the four-lepton search including selection efficiency estimates and the scale and resolution of the four lepton invariant mass is tested by both experiments by searching for the four-lepton final state produced by single Z-boson production where one of the decay leptons radiates a photon that converts to a lepton and anti-lepton pair, $Z \to 4\ell$. Both experiments observed this process at a rate consistent with SM expectations and therefore validate nicely all analysis steps towards the $H \to ZZ \to \ell^+\ell^-\ell'^+\ell'^-$ search.

In addition to the reconstructed 4-lepton invariant mass the CMS experiment implemented a matrix element likelihood discriminant (MELA) that uses the angular information of the 4-lepton final state, as expected from the spin zero and even parity nature of the Higgs boson.

Large regions at high mass were initially excluded: the 131–162 GeV and 170–460 GeV ranges for ATLAS, and the 131–162 GeV and 172–525 GeV ranges for CMS. The regions that were not excluded were just above the on-shell WW

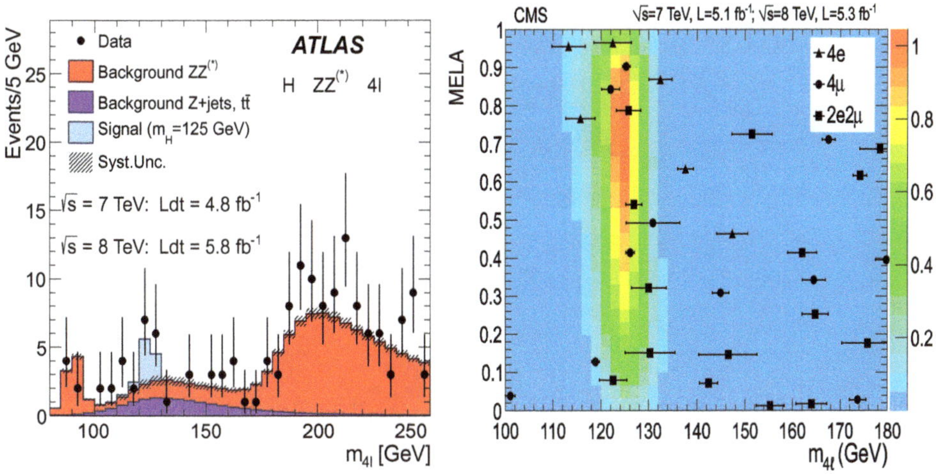

(a) (b)

Fig. 16. (a) The 4-lepton invariant mass distribution from the ATLAS experiment. The data is displayed as point and the background expectation as a histogram. Several SM Higgs boson signal contributions are included for different hypothetical Higgs boson masses. Background $Z+$jets and $t\bar{t}$ bottom, background ZZ middle and Higgs boson signal top. (b) A two dimensional plot of matrix element likelihood versus four lepton invariant mass from the CMS experiment. Data is shown with event by event mass uncertainties while the expectation of a 125 GeV SM Higgs boson is superimposed as a temperature plot. The central region around 125 GeV is highest in probability.

production threshold, where the WW branching ratio dominates, and the low mass region where a small excess, of about a dozen events per experiment, was observed.

The ATLAS experiment estimated a significance of 3.4 sigma for this excess, compatible with a narrow resonance of mass 125 GeV with a signal strength of 1.3 times the SM expectation. Figure 16(a) shows the ATLAS 4-lepton invariant mass distribution.

The evidence presented by both experiments of a narrow resonance with decays to ZZ indicated the observation of a new boson. The angular information used by CMS weakly favored the spin zero and even parity quantum numbers for the new boson though no definitive measurement of these properties was possible with the data available at the time. Figure 17 shows a candidate event for $H \to ZZ \to e^+e^-\mu^+\mu^-$ from the CMS experiment.

3.3. *Searches in* $H \to W^+W^- \to \ell^+\nu\ell^-\bar{\nu}$

The $H \to W^+W^- \to \ell^+\nu\ell^-\bar{\nu}$ channel has a large rate but the neutrinos in the final state preclude the full reconstruction of m_H. Still, some features of the signals can be exploited to result in relative good sensitivity for this search. The helicity analysis of a scalar Higgs boson decaying into two W-bosons implies that the two leptons tend to be collinear, with a small azimuthal separation. Kinematical distributions such

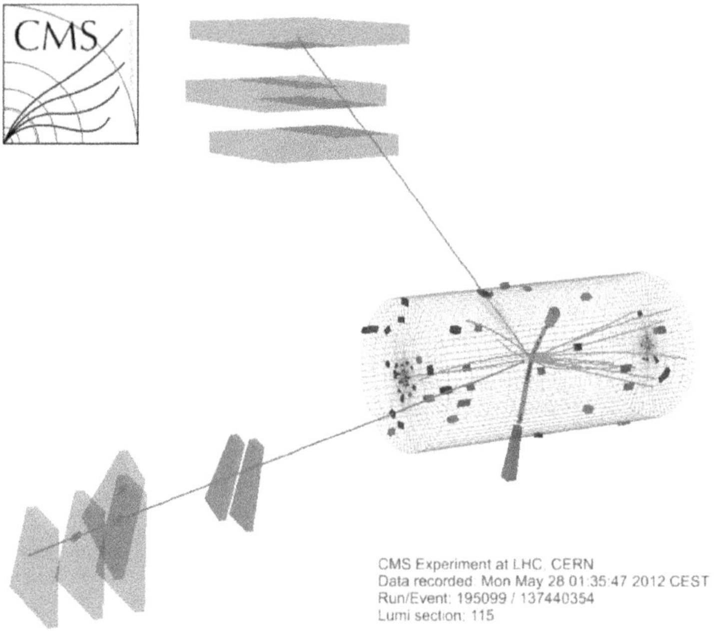

Fig. 17. A CMS candidate event for $H \to ZZ \to e^+e^-\mu^+\mu^-$. The two muons are identified in the outer muon chambers and the two electrons in the electromagnetic calorimeter.

as the invariant mass of the two leptons, $m_{\ell\ell}$, and the transverse mass of the leptons and $\not\!\!E_T$, m_T, also provide some separation power from background processes. The primary backgrounds are SM direct diboson production, $W +$jets, Drell–Yan, single top, and $t\bar{t}$. The reduction of the various backgrounds and the accurate estimate of their contribution are the major challenges in this channel.

Data is collected using triggers that require one or two leptons above a certain $p_{\rm T}$ threshold determined by the necessity to maintain the trigger rates at a reasonable level. The $\not\!\!E_T$ is required to be relatively large in order to reduced Drell–Yan and multijet backgrounds. The analysis is performed in categories that distinguish the different lepton flavor content and the number of accompanying jets in the event. Jets consistent with originating from b-quarks are vetoed to reduce contribution from top-quark processes. The jet content of the events helps separate productions processes, hence different level of signal to backgrounds subsamples can be selected for a more optimal sensitivity. For example, the two-jet category is used to optimize for VBF production by selecting jets with a large gap in pseudorapidity. In events with no additional jets the dominant background is non-resonant electroweak WW production. In events with same-flavor leptons the Drell–Yan background is largest, while $t\bar{t}$ and $W +$jets contribute to all categories. Other backgrounds are smaller in comparison. Control regions in data, depleted from signal, are used to ensure that the background composition is well-understood.

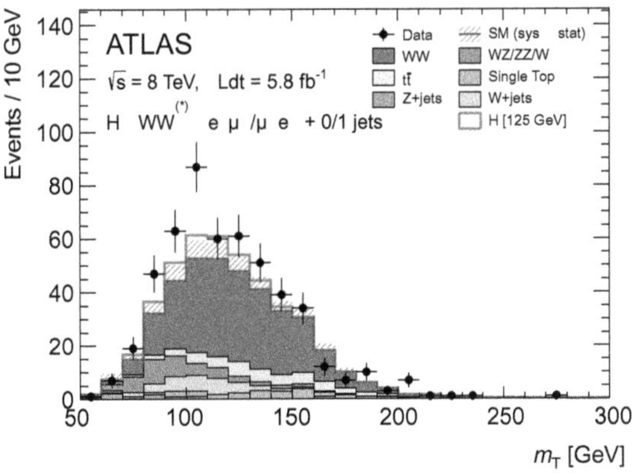

Fig. 18. Distribution of the transverse mass in the 0-jet and 1-jet WW analyses of ATLAS, with both $e\mu$ and μe channels combined. The expected signal for $m_H = 125\,\text{GeV}$ is shown stacked on top of the background prediction.

As a final discriminant, the ATLAS experiment used the m_T distribution, shown in Fig. 18, while CMS used a multivariate discriminant. With this channel the two experiments extended the sensitivity for a Higgs boson all the way down from the several hundred GeV and to the lower masses, near 125 GeV. ATLAS excluded the 137–261 GeV mass range and CMS the 129–520 GeV range. Both experiments observed a small excess of events at $\simeq$125 GeV. The ATLAS excess had a local significance of 2.8σ corresponding to a signal strength of 1.4 ± 0.5 times the expected SM rate, assuming a Higgs boson mass of 125 GeV. The CMS experiment local significance was 1.6σ. Even though these excesses were not as compelling as those found in the $H \to \gamma\gamma$ and $H \to ZZ$ channels, the fact that they occurred in the mass region around 125 GeV added confidence in the combined result.

3.4. *Searches in $H \to \tau^+\tau^-$ and in $H \to b\bar{b}$*

The attempt to observe directly the fermionic decays of the Higgs boson into particles of the 3rd generation (i.e. in pairs of τ leptons or b quarks) was an essential part of the search at the LHC, as it would help complete the understanding of the properties of the new boson by measuring its Yukawa couplings to these fermions.

The event yields in the $H \to \tau^+\tau^-$ channel are quite large at low m_H, between 100–150 GeV. The high-efficiency identification of τ leptons that decay hadronically (64% of the time) is one challenge that the LHC detectors have been able to meet, resulting in good sensitivity for the searches in these modes. Another challenge is to reconstruct as well as possible the mass of a di-tau resonance. The application of multivariate analysis techniques, that take into account the $\not{E}_T$, has enabled to reconstruct such states with mass resolutions of ~15%. The $Z \to \tau^+\tau^-$ provided an excellent calibration process in which to test the mass reconstruction methods and measure the tau identification efficiency. The experiments studied separately events with zero, one, or two associated jets and oppositely charged τ lepton pairs in the following $\tau\tau$ decay final states: ee (ATLAS), $\mu\mu$, $e\tau_h$, $\mu\tau_h$, and $\tau_h\tau_h$ (ATLAS), where e indicates $\tau \to e\nu\nu$, μ indicates $\tau \to \mu\nu\nu$, and τ_h indicates the τ lepton decaying to hadrons and a τ neutrino. In events with two jets the experiments targeted the VBF production process by requiring the jets to be in the forward regions of the detector and to have a rapidity gap between them. This is the mode with best sensitivity.

Data were collected on triggers that required one or two charged leptons and in the ATLAS experiment one trigger required two high p_T τ_h candidates ($\tau_h\tau_h$). The main backgrounds are Drell–Yan production and $W+$jets production where one jet is misidentified as a τ lepton. Multijet QCD processes and $t\bar{t}$ production are minor backgrounds. To enhance the presence of a possible signal relative to backgrounds, the leptons or sum of hadronic decay products are required to have substantial transverse energy and the $\not{E}_T$ from undetected neutrinos is required to be collinear with the direction of the dilepton system.

With the datasets analyzed at the time of the discovery the experiments reached good sensitivities but not enough to have detected the presence of a Higgs boson in these channels. The $\tau\tau$ invariant mass obtained in the CMS vector boson fusion analysis is shown in Fig. 19.

With about 5 fb^{-1} of 7 TeV data, and for Higgs boson masses in the 110–150 GeV range, the ATLAS experiment reached sensitivities to set limits on the $H \to \tau^+\tau^-$ production rate in the range of 3–11 times that of the expected production rate from the SM. The CMS experiment used in addition about 5 fb^{-1} of 8 TeV data to reach sensitivities in the 110–145 GeV mass range to set limits at a level 1.3–2.4 times the SM rate. At 125 GeV the actual limit set by CMS was 1.1 times the SM

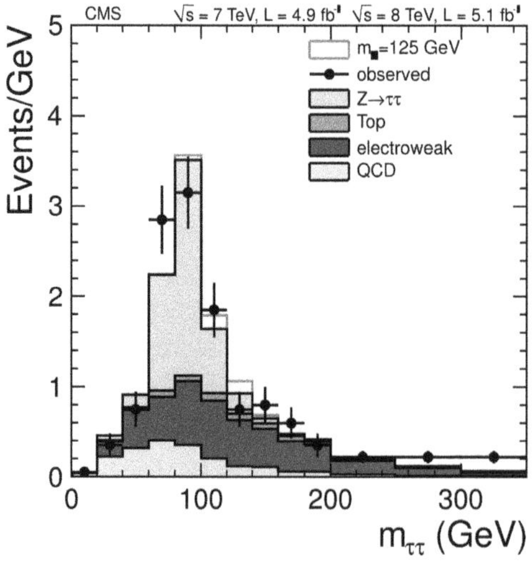

Fig. 19. Distribution of the $\tau\tau$ invariant mass in the combined 7 and 8 TeV data sets for the VBF category of the CMS $\tau^+\tau^-$ analysis. The expected signal for $m_H = 125$ GeV is shown stacked on top of the background prediction.

expected rate. With the full statistics of the Run 1, accumulated by the end of 2013, first evidence for $H \to \tau\tau$ was then obtained by both collaborations.

The $H \to b\bar{b}$ decay occurs with the largest branching fraction, about 58% at a Higgs boson mass of 125 GeV. However, the QCD $b\bar{b}$ production yields at the LHC are seven orders of magnitude larger than those from direct Higgs boson production. Therefore in their search for $H \to b\bar{b}$ the LHC experiments resorted mainly to the associated production mode, VH. Three channels, with leptons and $\not{E}_T$ in the final state, were considered: $WH \to \ell\nu b\bar{b}$, $ZH \to \ell^+\ell^- b\bar{b}$, and $ZH \to \not{E}_T\, b\bar{b}$ where the Z decays to neutrinos, "observed" as $\not{E}_T$. Even though these final states are rather distinct, several background processes with similar final states, but with significantly larger yields, present a formidable challenge. These processes include $V+$jets, $t\bar{t}$, single top, dibosons, and QCD multijet production with misidentified leptons.

Data were collected through several trigger paths, including single or double leptons, $\not{E}_T$, and requiring jets when needed. Two "b-jets", which are jets tagged by an algorithm as originating from b quarks, were required in all channels, reducing backgrounds very significantly. Another requirement made for all channels that helped reduce backgrounds further was to impose a significant p_{T} "boost" in the reconstructed $H \to b\bar{b}$ and vector bosons. In order to maximize the sensitivity, the analyses were performed separately in different boost regions. In channels with $\not{E}_T$ the multijet background is reduced to acceptable levels by requiring that the $\not{E}_T$ is not near any of the jets in the event. The most differentiating variable between signal and background is the invariant mass, $m_{b\bar{b}}$, of the two b-jet system. For $H \to b\bar{b}$ a peak in $m_{b\bar{b}}$ consistent with the jet energy resolution is expected, whereas for all backgrounds the resulting distribution is smooth without any structure. The ATLAS experiment used this distribution as the final discriminant in which to search for the $H \to b\bar{b}$ signal. The CMS experiment used a regression technique that improves on the energy resolution for generic jets in order to sharpen the mass peak, and used a multivariate discriminant, that included $m_{b\bar{b}}$ together with several other kinematical variables, as a final discriminant in the search. The ATLAS (CMS) experiment searched for a Higgs boson in the range 110–130 GeV (100–135 GeV) using 4.7–5.0 fb^{-1} of 7 TeV collision data per experiment, while the CMS experiment additionally included 5.1 fb^{-1} of 8 TeV collision data. The CMS analyses had sensitivity to set limits on order one to five times the expected SM Higgs boson cross section, depending on the mass, achieving a sensitivity of 1.6σ at a Higgs boson mass of 125 GeV. In this channel, observation will have to wait for more data, and should be achieved in the Run 2 of the LHC.

4. The Discovery of the Higgs Boson

At the end of 2011, based on ~5 fb^{-1} of 7 TeV data, ATLAS and CMS presented results at the CERN council with some small signal at the 2.5σ level around 125 GeV, so 2012 data became eagerly expected. At the Moriond Conferences in

March 2012, the LHC experiments did not update their Higgs boson results but the Tevatron experiments presented a combination based on $\sim$10 fb^{-1} of data which also showed a $\sim$2.5σ excess, but in a region of mass much wider (120–140 GeV), which was understandable since the signal was expected to be dominated by $H \to b\bar{b}$ decays which forms experimentally a wide resonance.

But the real breakthrough came on the 4th of July 2012. In a special seminar at CERN, the independent discovery of the Higgs boson was announced by the ATLAS and CMS experiments as detailed here below. To better characterize this discovery, we first present the last steps of the analyses.

Before the discovery, all Tevatron and LHC experiments were computing limits on the production cross section for a standard theory Higgs as a function of its mass, combining different searched processes. To facilitate comparisons with the SM and to accommodate analyses with different degrees of sensitivity and acceptance for more than one signal production mechanism, the limits were divided by the SM Higgs boson production cross section, as a function of Higgs boson mass, for test masses for which the experiments performed dedicated searches in different channels. A value of the combined limit ratio which is less than or equal to one indicates that particular Higgs boson mass is excluded at the 95% C.L. At the LHC the limits are calculated using the modified frequentist CL_s approach. At the Tevatron, both CL_s and a Bayesian technique were used. The limits were generally determined using the MVA output distributions or the invariant mass distributions, together with their associated uncertainties, as discriminating inputs to the limit setting procedure.

Systematic uncertainties were accounted for by nuisance parameters which were assigned an *a priori* probability distribution. Correlations between the uncertainties were taken into account. To diminish the impact of the nuisance parameters the profile likelihood distribution were maximized over the nuisance parameters. Each background was allowed to vary within its uncertainties by varying the nuisance parameters in the fitting procedure, while the fit was constrained to lie within the uncertainties.

This framework can also produce statistical results quantifying the properties and the expectation for a signal. Given the SM expectation for signal contributions, the expected p-value or probability for backgrounds to fluctuate to the statistical significance of the expected signal can be computed. Similarly, given an excess in the data, the observed p-value can be computed and evidence or discovery will be obtained if the p-value reaches the equivalent probability of 3 or 5 standard deviations.

4.1. *ATLAS and CMS discoveries*

For the discovery, each analysis channel was analyzed separately and within each experiment the Higgs boson search results were then combined.[32,33]

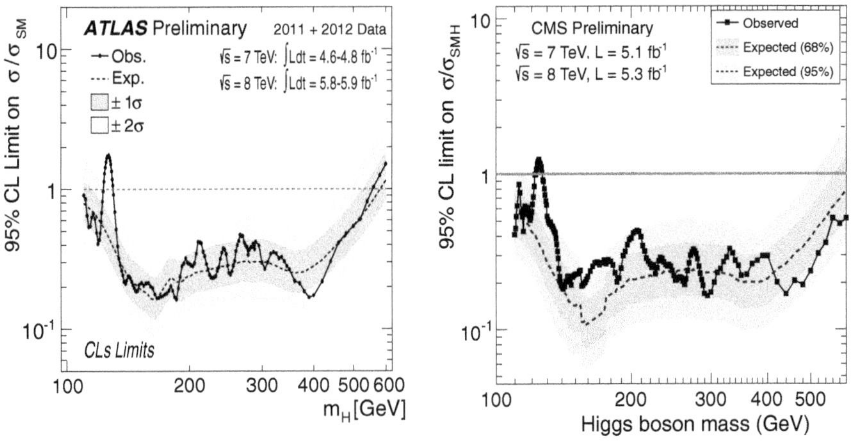

(a) (b)

Fig. 20. (a) The ATLAS experiments combined upper limit as a function of the Higgs boson mass between 100 and 600 GeV Solid black: observed limit/SM; Dashed black: median expected limit/SM in the background only hypothesis. Colored bands: $\pm 1, 2\sigma$ distributions around median expected limit. The excess around $m_H = 125\,\text{GeV}$ is clearly visible. (b) The CMS experiments combined upper limit as a function of the Higgs boson mass between 100 and 600 GeV Solid black: observed limit/SM; Dashed black: CMS expected limit/SM in the background only hypothesis. Colored bands: $\pm 1, 2\sigma$ distributions around median expected limit. The excess around $m_H = 125\,\text{GeV}$ is clearly visible.

Over a large region of masses the LHC experiments observed no evidence for a Higgs boson. The LHC data showed a consistent picture with the exclusion, typically by multiple channels, of a high mass SM Higgs boson. At high mass the ATLAS experiment excluded the production of a SM Higgs boson with masses from 131 to 559 GeV at 95% C.L. and the CMS experiment excluded a region from 128 to 600 GeV at 95% C.L., where 600 GeV was the limit of the search range. At low mass the ATLAS experiment excluded the production of a SM Higgs boson with masses from 111 to 122 GeV at 95% C.L., while the CMS experiment excluded the region from 110 to 122.5 GeV. The combined ATLAS and CMS limits are presented in Figs. 20(a) and 20(b) respectively.

At low masses the experiments had the sensitivity to exclude or observe the Higgs boson. The sensitivities for observation of a signal were quantified as an expected p-value for the background to fluctuate to a signal as large as the median expectation for a SM Higgs boson. The combined expected p-value at $m_H = 125\,\text{GeV}$ was 4.9σ for the ATLAS experiment and 5.8σ for the CMS experiment. In the region around 125 GeV both experiments observed an excess of events in the bosonic final state search channels. The experiments evaluated the p-values for each channel separately and for the entire combination and compared those values with the expected background only p-values given a SM Higgs boson as a function of mass (see Figs. 21(a), 21(b)). Information quantifying the most significant excesses in the individual search channels is summarized along with the most significant combined excess from each

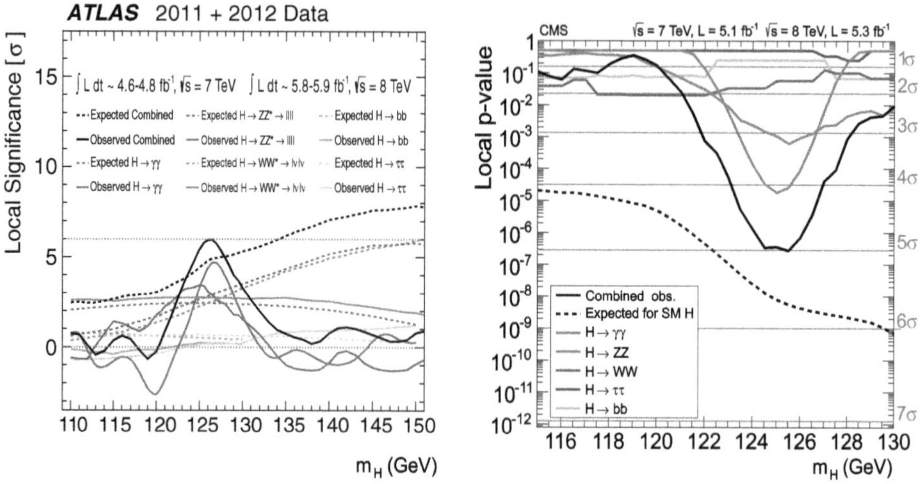

(a) (b)

Fig. 21. (a) ATLAS local significance[32] for each search channel and the combination. The observed significance are shown with solid curves, and the median expected significance assuming a signal is present at the SM strength are shown with dashed curves. A dotted line indicates the 6σ threshold. The highest local significances of the ZZ and $\gamma\gamma$ channels are 3.4σ and 4.5σ respectively while the combined significance of all channels 5.9σ. (b) CMS local p-values.[33] The observed p-values are shown with solid curves, and the median expected p-value for the combined search assuming a signal is present at the SM strength is shown with a dashed curve. Horizontal lines indicate the $1\sigma - 7\sigma$ thresholds. The highest local significances of the ZZ and $\gamma\gamma$ channels are 3.1σ and 4.1σ respectively while the combined significance of all channels 5.1σ.

Table 2. The most significant excesses seen in ATLAS and CMS results in the different channels and at the masses where they were seen, and their combined significances.

Topology	ATLAS Significance and Mass	CMS Significance and Mass
$H \to WW \to \ell\nu\ell\nu$	2.8σ 125.0 GeV	1.6σ 125.0 GeV
$H \to ZZ \to 4\ell$	3.4σ 125.0 GeV	3.1σ 125.6 GeV
$H \to \gamma\gamma$	4.5σ 126.5 GeV	4.1σ 125.0 GeV
Combined Significance	5.9σ 126.0 GeV	5.0σ 125.3 GeV

experiment in Table 2. Both experiments observed a Higgs boson signal with local significances above the evidence level of 3σ in the ZZ and $\gamma\gamma$ decay modes and combined significances of 5.9σ at $m_H = 126\,\text{GeV}$ for the ATLAS experiment and 5.0σ at $m_H = 125.3$ for the CMS experiment. These two observations characterized in an unambiguous way the discovery of a new particle with mass of approximately 125 GeV. The decay modes in which the particle was seen also indicated that the particle is a boson and plays a role in the mechanism of electroweak symmetry breaking.

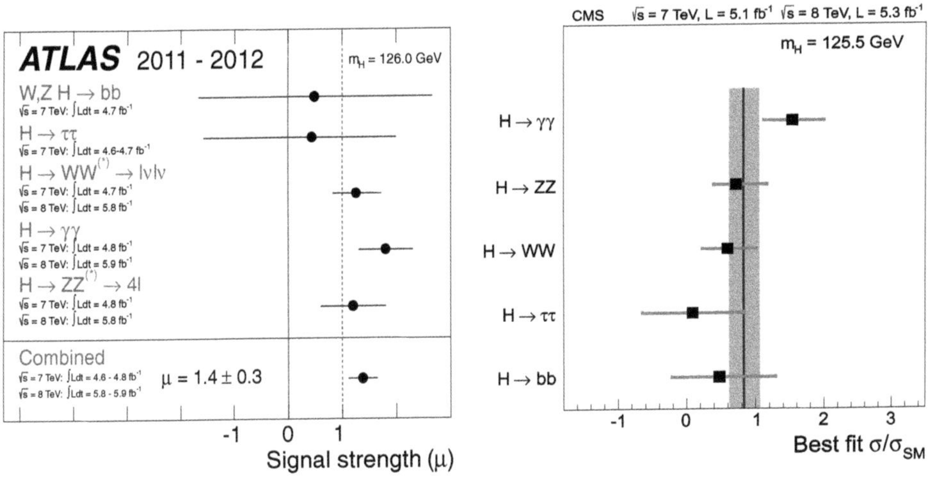

(a) (b)

Fig. 22. (a) ATLAS best-fit signal strength for all SM Higgs boson decays, for $m_H = 125$ GeV/c^2. (b) CMS best-fit signal strength for all SM Higgs boson decays, for $m_H = 125$ GeV/c^2. The shaded band corresponds to the $\pm 1\sigma$ uncertainty on the full combination.

The CMS and ATLAS collaborations measured right away several properties to understand the compatibility of the observed boson with the SM Higgs boson and presented the results in their papers reporting the observations.[32,33] The experiments measured the cross section for Higgs boson production in each decay channel and also globally combining all decay channels. The results were presented as a ratio to the expected SM values in Figs. 22(a) and 22(b) for the ATLAS and CMS experiment respectively. The combined signal strengths measured by the experiments were 1.4 ± 0.3 for ATLAS and 0.87 ± 0.23 for CMS compatible with the SM Higgs boson expectation. Individual signal strengths in the most sensitive modes where discussed in the sections on individual searches.

The fully reconstructed decays of the Higgs boson $H \to \gamma\gamma$ and $H \to ZZ \to \ell^+\ell^-\ell^+\ell^-$ have excellent mass resolution. The $H \to W^+W^- \to \ell^+\nu\ell^-\bar{\nu}$ decay mode has substantial rate but poor mass resolution due to the two neutrinos in the final state. The ATLAS experiment measured a mass for the observed boson of $m_H = 126.0 \pm 0.4(\text{stat}) \pm 0.4(\text{sys})$ GeV using all three decay modes. The individual fits in a two dimensional analysis of signal strength versus mass are shown in Fig. 23(a). The CMS experiment used the fully reconstructed $H \to \gamma\gamma$ and $H \to ZZ \to \ell^+\ell^-\ell^+\ell^-$ modes to measure a mass of $m = 125.3 \pm 0.4(\text{stat}) \pm 0.5(\text{sys})$ GeV. The individual and combined fits are shown in Fig. 23(b). The results were compatible with limits from previous searches, and the prediction of the SM Higgs boson mass from constraints derived from electroweak measurements.

If the observed boson is involved in the mechanism of electroweak symmetry breaking, the measurement of its coupling to the W and Z bosons is a crucial

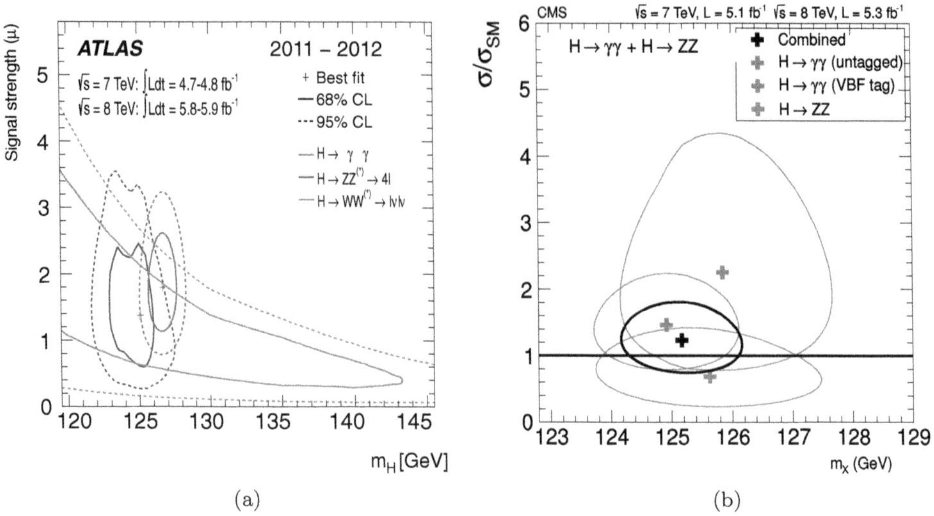

Fig. 23. (a) The ATLAS two dimensional fit for the cross section compared to the SM expectation and the Higgs boson mass for highest significance decay channels. (b) The CMS two dimensional fit for the cross section compared to the SM expectation and the Higgs boson mass for highest significance decay channels and the combined fit using those channels.

discriminant. The production and decay rates measured by the experiments were compatible with the SM. The ratio of the W and Z couplings can be computed by dividing the production times decay rates for $H \to WW$ and $H \to ZZ$ since the production of the Higgs boson takes place via the same mechanisms. The ATLAS experiment measured $R_{WZ} = 1.07^{+0.35}_{-0.27}$ and the CMS experiment measured $R_{WZ} = 0.9^{+1.1}_{-0.6}$ consistent with the SM expectation where both experiments normalized the measurement so that the expected value in the SM is one.

In summary, the LHC experiments extended the LEP exclusion to 122.5 GeV and further excluded a SM Higgs boson with mass between 128 and 600 GeV. The ATLAS and CMS experiments both observed a significant excess of events in the region around 125 GeV with evidence for the production of a new boson in the ZZ and $\gamma\gamma$ decay modes, with observed local significances of 4.5σ and 4.1σ in the $\gamma\gamma$ mode, 3.4σ and 3.1σ in the ZZ mode. Significant signals (2.8σ and 1.6σ) were also observed in the $H \to WW$ decay mode, while the observed significance in the fermionic modes ($H \to \tau\tau$ and $H \to b\bar{b}$) was weak, which was not unexpected given the low expected significance in these modes with that luminosity. When combining all their channels, both experiments independently reported the discovery of a new boson and provided first measurements of its fundamental properties, in agreement with those expected from a SM Higgs boson with a mass close to 125 GeV.

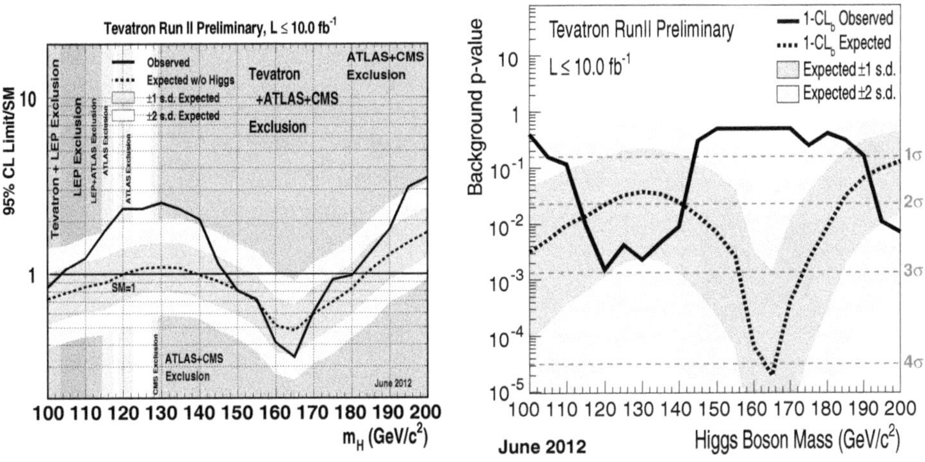

(a) (b)

Fig. 24. (a) Observed and expected 95% C.L. upper limits on the ratios to the SM cross section, as functions of the Higgs boson mass for the combined CDF and D0 analyses. The bands indicate the 68% and 95% probability regions where the limits can fluctuate, in the absence of signal. (b) The background p-values 1-CL_b as a function of the Higgs boson mass (in steps of 5 GeV), for the combination of the CDF and D0 analyses. The two bands correspond respectively to the regions enclosing 1σ and 2σ fluctuations around the median prediction in the signal-plus-background hypothesis at each value of m_H.

4.2. *Tevatron combined results*

When at the end of June 2012 CERN announced a special seminar for July 4th, the Tevatron released its latest CDF + D0 combination. These results were also pointing to an excess in the low mass region, but with much less significance as shown in Fig. 24(a) which shows the ratios of the 95% C.L. expected and observed limits to the SM cross section, and in Fig. 24(b) which shows the p-value 1-CL_b as a function of m_H. The smallest observed p-value corresponded to a Higgs boson mass of 120 GeV and had a local significance of 3.0σ. The width of the dip from 115 to 135 GeV was consistent with the combined resolution of the $H \to b\bar{b}$ and $H \to W^+W^-$ channels. The effective resolution of this search came from two independent sources: the reconstructed candidate masses, which directly constrain m_H, and the expected cross sections times the relevant branching ratios for the $H \to b\bar{b}$ and $H \to W^+W^-$ channels, which are functions of m_H in the SM. The observed excess in the $H \to b\bar{b}$ channels coupled with the outcome in the $H \to W^+W^-$ channels determined the shape of the observed p-value as a function of m_H. The strongest sensitivity at low mass comes from the $H \to b\bar{b}$ channels. The largest local significance in the combination of $H \to b\bar{b}$ channels was 3.3σ at a mass of 135 GeV, while it was 2.8σ at 125 GeV.[21]

The signal strength was allowed to vary as a function of m_H in the fit of the signal-plus-background hypothesis to the observed data over the full mass range.

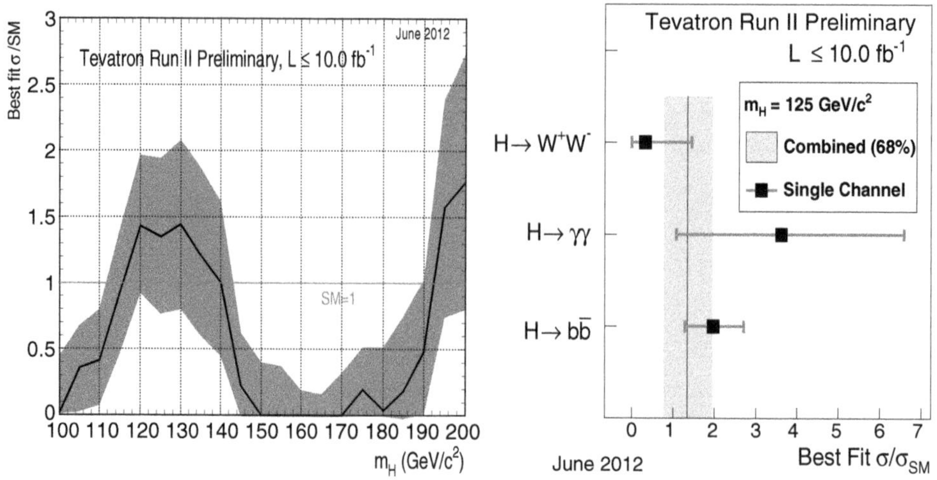

(a) (b)

Fig. 25. (a) The best fit signal cross section of all CDF and D0 search channels combined shown as a ratio to the SM cross section as a function of the tested Higgs boson mass. The horizontal line at 1 represents the signal strength expected for a SM Higgs boson hypothesis. The grey band shows the 1σ uncertainty on the signal fit. (b) Best-fit signal strength for the three most sensitive boson decay modes at the Tevatron, for $m_H = 125\,\text{GeV}/c^2$. The shaded band corresponds to the $\pm 1\sigma$ uncertainty on the full combination.

The resulting best-fit signal strength is shown in Fig. 25(a), normalized to the SM prediction and is within 1σ of the SM expectation for a Higgs boson signal in the range $115 < m_H < 140$ GeV. The largest signal fit in this range, normalized to the SM prediction, was obtained between 120 and 130 GeV. The excess in signal-strength around 200 GeV occurred in a region of low expected sensitivity ($\sim 1\sigma$) and with an unphysically narrow mass range, it thus could not be attributed to a SM Higgs boson signal at high mass.

The global significance of approximately 3.1σ obtained by this combination, provided evidence for the production of a resonance in the $b\bar{b}$ dijet mass distribution, produced in association with a massive vector boson, compatible with the Higgs boson discovered at CERN, given the mass resolution in this final state. This also provided the first evidence for fermionic decays of this boson. This was particularly interesting since the signals in the fermionic channels were then insignificant at the LHC, hence the Tevatron result supported the interpretation of a SM-like Higgs boson for this newly discovered particle.

The measured cross section times branching ratio $\sigma_{WH} + \sigma_{ZH} \times \mathcal{B}(H \to b\bar{b})$ had a value of $0.23^{+0.09}_{-0.08}$ pb for $m_H = 125$ GeV, consistent with the corresponding SM prediction of 0.12 ± 0.01 pb. The best fit signal cross section from the combined CDF and D0 analyses separated into the different Higgs boson decay channels is shown in Fig. 25(b), assuming $m_H = 125\,\text{GeV}$. The three decay channels shown are all in agreement with the standard theory prediction.

In summary, at the time of the Higgs discovery by the LHC experiments, the Tevatron had an excess of data events with respect to the background estimation in the mass range $115 < m_H < 135\,\text{GeV}$. The p-value for a background fluctuation to produce such an excess corresponded to a significance of 3σ at $125\,\text{GeV}$. The largest excess was observed in the $H \to b\bar{b}$ channels, with a global significance of $\approx 3.1\sigma$. The CDF and D0 collaborations could thus report evidence "right away" for the production of a resonance in the b flavored dijet mass distribution produced in association with a massive vector boson, consistent with the new boson observed by the LHC collaborations. The measured cross section for this process was consistent with the cross section expected for a SM Higgs boson of $125\,\text{GeV}$ produced in association with a W or a Z boson. Unfortunately, since the data taking was already stopped when the LHC discovery was made, this evidence could not be strengthened further. It showed however that the boson newly discovered trough its bosonic decay modes was behaving as expected in the fermionic decays, in spite of the limited statistics. Final results consistent with the results presented above, with a slightly lower measured production cross section, were published in 2013.[61]

5. Conclusion and Prospects

The search for the Higgs boson lasted for more than 30 years and was performed at the three last major colliders. In July 2012, the LHC experiments finally discovered a new boson with mass around 125 GeV, with evidence for this particle in several bosonic decay modes. The production and decay modes that were observed indicated right away that this boson plays a role in the mechanism of electroweak symmetry breaking and also in the mass generation for the quarks. The Tevatron experiments reported evidence for a particle, produced in association with W or Z bosons and which decays to $b\bar{b}$, with a mass compatible to that reported by the LHC experiments. The properties of this boson were compatible with those expected for a SM Higgs boson but more study were required to fully explore the nature of this discovery. This research program continued until the end of Run 1 (2014) and allowed to accumulate more precise results as those presented here, as is discussed in the other experimental contribution on Higgs properties in this Book. Furthermore it will continue for many more years at the upgraded LHC and later, with upgraded detectors. While the discovery of a new boson with properties indicating that it plays a role in electroweak symmetry breaking is a major breakthrough in fundamental physics, we all hope that physics beyond the standard model will be revealed by the precise study of this particle and its associated potential.

References

1. P. W. Higgs, *Phys. Rev. Lett.* **12**, 132 (1964); *Phys. Rev.* **145**, 1156 (1966); F. Englert and R. Brout, *Phys. Rev. Lett.* **13**, 321 (1964); G. S. Guralnik, C. R. Hagen, and T. W. Kibble, *Phys. Rev. Lett.* **13**, 585 (1964).

2. J. Ellis, M. K. Gaillard and D. V. Nanopoulos, *Nucl. Phys. B* **106**, 292 (1976).
3. J. F. Gunion, H. E. Haber, G. L. Kane, and S. Dawson, *The Higgs Hunter's Guide* (Addison-Wesley, 1990).
4. J. Ellis, R. Peccei, (eds.), *Physics at LEP 1*, CERN-86-02-V-1 (1986).
5. J. Ellis, R. Peccei, (eds.), *Physics at LEP 2*, CERN-86-02-V-2 (1986).
6. B. L. Ioffe and V. A. Khoze, *Nucl. Phys.* **9**, 50 (1978).
7. J. Ellis *et al.*, *Nucl. Phys. B* **106**, 292 (1976).
8. D. R. T. Jones and S. T. Petcov, *Phys. Lett. B* **84**, 440 (1979); R. N. Cahn and S. Dawson, *Phys. Lett. B* **136**, 96 (1984); *ibid.*, **138B**, 464 (1984); W. Kilian, M. Krmer and P. Zerwas, *Phys. Lett. B* **373**, 135 (1996) arXiv:hep-ph/9512355.
9. ALEPH, DELPHI, L3, and OPAL Collaborations, The LEP Working Group for Higgs Boson Searches, *Phys. Lett. B* **565**, 61 (2003) arXiv:hep-ex/0306033.
10. B. Clare, LEP pursues Higgs boson and greater W precision, http://cerncourier.com/cws/article/cern/28110 (1999).
11. D. Appell, The supercollider that never was, *Scientific American*, October (2013).
12. M. Riordan, L. Hoddeson, A. Kolb, *Tunnel Visions: The Rise and Fall of the Superconducting Super Collider*, (University of Chicago Press, 2015).
13. CDF collaboration, *Phys. Rev. Lett.* **74**, 2626 (1995); D0 collaboration, *Phys. Rev. Lett.* **74**, 2632 (1995).
14. LEP Electroweak Working Group, status of March 2012, http://lepewwg.web.cern.ch/LEPEWWG/, The ALEPH, CDF, DØ, DELPHI, L3, OPAL, SLD Collaborations, the LEP Electroweak Working Group, the Tevatron Electroweak Working Group, and the SLD Electroweak and Heavy Flavor groups, LEPEWWG/2009–01 (2009), arXiv:0911.2604 [hep-ex], J. Erler and P. Langacker, *Phys. Lett. B* **592**, 1 (2004), arXiv:hep-ph/0407097.
15. CDF and D0 Collaborations, and the Tevatron Electroweak Working Group, (2011), arXiv:1107.5255
16. Tevatron Electroweak Working Group, CDF and D0 Collaborations, (2012), arXiv:1204.0042 [hep-ex].
17. M. Carena *et al.*, Report of the Tevatron Higgs working group, arXiv:hep-ph/0010338v2 (2000).
18. CDF and D0 Collaborations, *Phys. Rev. Lett.* **104**, 061802 (2010), arXiv:1001.4162 [hep-ex].
19. CDF and D0 Collaborations, arXiv:1007.4587 [hep-ex], arXiv:1107.5518 [hep-ex].
20. GFitter collaboration, M. Baak *et al.*, *Eur. Phys. J. C* **72**, 2003 (2012), arXiv:1107.0975 [hep-ph].
21. CDF and D0 Collaborations, *Phys. Rev. Lett.* **109**, 071804 (2012), arXiv:1207.6436 [hep-ex].
22. A. Asner *et al.*, *ECFA-CERN Workshop on large hadron collider in the LEP tunnel*, Lausanne and CERN, Geneva, Switzerland, CERN-84-10-V-1, ECFA-84-85, (1984).
23. L. Evans, The Large Hadron Collider from conception to commissioning: A personal recollection, *Reviews of Accelerator Science and Technlogy*, **3**, 261 (2010).
24. ATLAS Collaboration, *JINST* **03**, S08003 (2008).
25. CMS Collaboration, *JINST* **03**, S08004 (2008).
26. LHCb Collaboration, *Int. J. Mod. Phys. A* **30**, 1530022 (2015), arXiv:1412.6352 [hep-ex]
27. ALICE Collaboration, *JINST* **03**, S08002 (2008).
28. *The LHC Incident*, http://press.web.cern.ch/press-releases/2008/10/cern-releases-analysis-lhc-incident.

29. ATLAS and CMS Collaborations, ATLAS-CONF-2011-135, CMS-PAS-HIG-11-022 (2011).
30. ATLAS Collaboration, ATLAS-CONF-2012-019, http://cds.cern.ch/record/1430033.
31. CMS Collaboration, CMS-PAS-HIG-12-008.
32. ATLAS Collaboration, *Phys. Lett. B* **716**, 1 (2012), arXiv:1207.7214 [hep-ex].
33. CMS Collaboration, *Phys. Lett. B* **716**, 30 (2012), arXiv:1207.7235 [hep-ex].
34. *The LHC Discovery*, http://press.web.cern.ch/press-releases/2012/07/cern-experiments-observe-particle-consisten-long-sought-higgs-boson.
35. CDF Collaboration, *J. Phys. G* **34**, 2457 (2007), arXiv:hep-ex/0508029.
36. D0 Collaboration, *Nucl. Instrum. Methods, Phys. Res. A* **565**, 463 (2006), arXiv:physics/0507191.
37. G. Bernardi and M. Herndon, *Rev. Mod. Phys.* **86**, 479, (2014), arXiv:1210.0021 [hep-ex].
38. CDF and D0 Collaborations, Combined CDF and D0 measurement of WZ and ZZ production in final states with b-tagged jets, arXiv:1203.3782 [hep-ex].
39. CDF Collaboration, *Phys. Rev. Lett.* **109**, 111804 (2012), arXiv:1207.1703 [hep-ex].
40. D0 Collaboration, *Phys. Rev. Lett.* **109**, 121804 (2012), arXiv:1208.0653 [hep-ex].
41. CDF Collaboration, *Phys. Rev. Lett.* **109**, 111803 (2012), arXiv:1207.1704 [hep-ex].
42. D0 Collaboration, *Phys. Rev. Lett.* **109**, 121803 (2012), arXiv:1207.5819 [hep-ex].
43. CDF Collaboration, *Phys. Rev. Lett.* **109**, 111805 (2012), arXiv:1207.1711 [hep-ex].
44. D0 Collaboration, *Phys. Lett. B* **716**, 285 (2012), arXiv:1207.5689 [hep-ex].
45. CDF Collaboration, *Phys. Rev. Lett.* **109**, 111802 (2012), arXiv:1207.1707 [hep-ex].
46. D0 Collaboration, *Phys. Rev. Lett.* **109**, 121802 (2012), arXiv:1207.6631 [hep-ex].
47. D0 Collaboration, *Phys. Lett. B* **714**, 237 (2012), arXiv:1203.4443 [hep-ex].
48. CDF Collaboration, *Phys. Lett. B* **717**, 173 (2012), arXiv:1207.6386 [hep-ex].
49. D0 Collaboration, *Phys. Rev. D* **88**, 052007 (2013), arXiv:1301.5358 [hep-ex].
50. D0 Collaboration, *Phys. Rev. Lett.* **103**, 191801 (2009), arXiv:0904.0673 [hep-ex].
51. CDF Collaboration, *Phys. Rev. Lett.* **104**, 201801 (2010), arXiv:0912.4500 [hep-ex].
52. CDF Collaboration, *Phys. Rev. D* **86**, 031104 (2012), arXiv:1202.6629 [hep-ex].
53. D0 Collaboration, *Phys. Rev. D* **85**, 112005 (2012), arXiv:1201.5652 [hep-ex].
54. CDF Collaboration, *Phys. Rev. Lett.* **108**, 101801 (2012), arXiv:1112.2978 [hep-ex].
55. CDF Collaboration, *Phys. Rev. D* **88**, 052012 (2013), arXiv:1306.0023 [hep-ex].
56. D0 Collaboration, *Phys. Rev. D* **88**, 052006 (2013), arXiv:1301.1243 [hep-ex].
57. D0 Collaboration, *Phys. Rev. Lett.* **106**, 171802 (2011), arXiv:1001.6079v2 [hep-ph].
58. CDF Collaboration, *Phys. Rev. D* **86**, 072012 (2012), arXiv:1207.5016 [hep-ex].
59. CDF Collaboration, *Phys. Rev. D* **88**, 052013 (2013), arXiv:1301.6668 [hep-ex].
60. D0 Collaboration, *Phys. Rev. D* **88**, 052011 (2013), arXiv:1303.0823 [hep-ex].
61. CDF and D0 Collaborations, *Phys. Rev. D* **88**, 052014 (2013), arXiv:1303.6346 [hep-ex].

Chapter 16

Higgs Boson Properties

André David and Michael Dührssen
Experimental Physics Department,
CERN, CH-1211 Geneva 23, Switzerland
andre.david@cern.ch, michael.duehrssen@cern.ch

This chapter presents an overview of the measured properties of the Higgs boson discovered in 2012 by the ATLAS and CMS collaborations at the CERN LHC. Searches for deviations from the properties predicted by the standard theory are also summarised. The present status corresponds to the combined analysis of the full Run 1 data sets of collisions collected at centre-of-mass energies of 7 and 8 TeV.

1. Introduction

The discovery of a Higgs boson at the LHC[1–3] and the subsequent measurements of its properties have yielded no significant deviations from the standard theory expectations.

This brings the status of the theoretical cornerstone of elementary particle physics from a model to a complete — renormalisable — theory, that seems to be self-consistent to energy scales that can reach the Planck scale.[4] Large efforts are being poured into direct searches that may point to how the standard theory may be supplemented in order to account for observed phenomena for which it makes no prediction, such as dark matter or neutrino masses.

On the other hand, precision measurements of the Higgs boson properties may reveal indirect evidence for physical phenomena beyond the standard theory. The measurement of the Higgs boson mass constrains a fundamental parameter in the standard theory and the 0.2% precision of the present measurement allows to test the internal consistency of the standard theory, against existing measurements of the W boson and top quark masses, for instance. Furthermore, the standard theory predicts all other properties of the Higgs boson once its mass is specified. Consequently, such a precise measurement of its mass sets the stage for more precise searches for deviations from the standard theory predictions in other Higgs boson properties.

2. Overview of Analyses Used

The ATLAS and CMS collaborations have designed and built the two largest and general-purpose detectors at the LHC.[5,6] Although their design requirements were similar because of the similar physics goals, the technologies used to fulfil those requirements are very different in all aspects, ranging from the geometry of the detectors to the offline software frameworks used.[7]

From 2010 to 2013, during the LHC Run 1, there were proton–proton collisions at centre-of-mass energies of 7 TeV and 8 TeV. The integrated luminosity available for analysis in each experiment was about 5/fb at 7 TeV and 20/fb at 8 TeV.

The combination of Higgs boson searches and measurements[8–14] is based on a large number of analyses. Depending on the combinations being performed and properties being probed, different channels and analyses are used. Table 1 provides a schematic overview of which Higgs boson production and decay modes enter in the various combinations.

The Higgs boson searches that lead to the discovery of the Higgs boson[1–3] were dominated by the gluon fusion production process and the $H \to \gamma\gamma$, $H \to ZZ^{(*)} \to llll$, and $H \to WW^{(*)} \to l\nu l\nu$ decay modes. Dedicated searches for other production and decay modes, as listed in Table 1, were also included and improved the overall expected sensitivity.

The combined measurement of the Higgs boson mass is based on the two channels where a narrow mass resonance can be observed: $H \to \gamma\gamma$ and $H \to ZZ^{(*)} \to llll$. The differential measurements of kinematic production properties in the $H \to \gamma\gamma$, $H \to ZZ^{(*)} \to llll$ and $H \to WW^{(*)} \to l\nu l\nu$ channels are mainly based on events produced via gluon fusion and vector-boson fusion (VBF). The measurements of kinematic decay properties for spin and parity hypothesis tests in the $H \to \gamma\gamma$, $H \to ZZ^{(*)} \to llll$, and $H \to WW^{(*)} \to l\nu l\nu$ channels aim to be independent of the production modes. As the selection criteria used are rather inclusive, the gluon fusion process dominates within the standard model (SM). Finally, the combined measurements of signal and coupling strengths make use of almost all channels analysed by the ATLAS and CMS collaboration for the 125 GeV Higgs boson, which cover almost all combinations of Higgs boson production and decay modes accessible at the LHC.

2.1. *Rare decays*

In addition to the Higgs boson decay modes summarised in Table 1, there are other, rarer, decays predicted by the SM for a 125 GeV Higgs boson. Those searched for include $H \to \mu\mu$,[40,41] $H \to ee$,[41] $H \to Z\gamma$,[42,43] $H \to \gamma^*\gamma$,[44] and $H \to V\gamma$[44,45] with $V = J/\psi, \Upsilon$. Even with small expected branching fractions in the SM, effects beyond the SM (BSM) could result in enhanced branching fractions and thus an observable signal. However, no signal from these decay modes was found with the Run 1 dataset, and 95% confidence level (CL) upper limits on the signal

Table 1. Overview of where different Higgs boson production and decay modes enter into the different Higgs combinations. The "5σ" dark grey triangles indicate channels for which dedicated searches were used in the discovery of the Higgs boson.[1–3] The "m_H" light grey triangles indicate channels contributing to the mass measurement.[8–10] The "J" and "J^{CP}" grey triangles indicate channels contributing to spin and CP hypothesis tests using kinematic decay properties.[11–13] The "$\frac{d\sigma}{dx}$" dark grey triangles indicate channels where differential measurements of kinematic production properties were performed.[15–21] The black check marks indicate channels entering the combined signal and coupling strength measurements.[14] These channels are documented in the references given in the table. References noted with † describe analyses not included in the combined analysis of both ATLAS and CMS data.

	$H \to \gamma\gamma$ Refs. 22, 23	$H \to ZZ^{(*)}$ Refs. 24, 25	$H \to WW^{(*)}$ Refs. 26, 27	$H \to \tau\tau$ Refs. 28, 29	$H \to b\bar{b}$
$gg \to H$	5σ, m_H, J, $\frac{d\sigma}{dx}$, $\checkmark$	5σ, m_H, J^{CP}, $\frac{d\sigma}{dx}$, $\checkmark$	5σ, J^{CP}, $\frac{d\sigma}{dx}$, $\checkmark$	5σ, $\checkmark$	
VBF	5σ, m_H, $\frac{d\sigma}{dx}$, $\checkmark$	m_H, $\frac{d\sigma}{dx}$, $\checkmark$	5σ, $\checkmark$	5σ, $\checkmark$	Refs. 30†, 31†
WH, ZH	$\checkmark$	$\checkmark$	$\checkmark$ Ref. 32	$\checkmark$ Ref. 33†	5σ, $\checkmark$ Refs. 34, 35
$t\bar{t}H$	$\checkmark$		$\checkmark$ Refs. 36, 37	$\checkmark$ Refs. 36, 37	$\checkmark$ Refs. 38†, 37, 39

strength, $\mu = (\sigma \times \text{BR})/(\sigma_{\text{SM}} \times \text{BR}_{\text{SM}})$, were reported by the ATLAS and CMS experiments.

The most notable limit is that on the $H \to \mu\mu$ decay with the observed (expected) limits $\mu < 7.4\,(6.5)$ and $\mu < 7.0\,(7.2)$ for CMS and ATLAS, respectively. For $H \to ee$, CMS also finds an upper limit for μ of about 3.7×10^5. Generation-universal couplings of the Higgs boson to leptons would imply a striking signal strength of $\mu \sim 280$ for $H \to \mu\mu$ and $\mu > 10^6$ for $H \to ee$. Universal couplings to leptons are therefore strongly excluded, something that is in contrast with the observed decays of the W and Z bosons.

Also of interest is the $H \to Z\gamma$ decay that is loop-induced in the SM and could be enhanced by the presence of (virtual) BSM particles. The observed (expected) limits from ATLAS and CMS are $\mu < 11\,(9)$ and $\mu < 9.5\,(10)$, respectively.

2.2. *BSM decays*

The ATLAS and CMS experiments have also performed direct searches for some BSM Higgs boson decay modes. Higgs boson decays to weakly interacting particles (*invisible* particles), that leave no signature in the detector, are not expected in the SM beyond a per-mille level contribution from $H \to ZZ^{(*)} \to 4\nu$ decays. Decays to invisible particles can be experimentally detected using events featuring large missing transverse momentum. Both experiments searched for such invisible decays in the VBF, WH, and ZH production modes.[46,47] The observed (expected) 95% CL limits on an invisible branching fraction $\mathrm{BR}_{\mathrm{inv.}}$ are <0.28 (0.31) and <0.58 (0.44), in ATLAS and CMS, respectively. In the future, analyses using boosted gluon-fusion topologies, where a single jet is detected, will increase the overall sensitivity to invisible decays.

Decays of the Higgs boson into lepton-flavour-violating final states are also not expected in the SM. ATLAS and CMS searched for the $H \to \tau\mu$ and $H \to \tau e$ decay, and both reported small excesses in $H \to \tau\mu$ with respect to the background expectations. For a 125 GeV Higgs boson, 95% CL observed (expected) limits on $\mathrm{BR}(H \to \tau\mu)$ was set to be <1.51% (0.75%) and <1.43% (1.01%) by CMS and ATLAS, respectively.[48,49] For a 125 GeV Higgs boson, the 95% CL observed (expected) limit on $\mathrm{BR}(H \to \tau e)$ was set to be <1.04% (1.21%) by ATLAS.[49]

3. Measurements

3.1. *Mass*

The mass of the Higgs boson, m_H, is a parameter of the standard theory that plays an important role in the stability of the electroweak vacuum, or lack thereof.[4]

Given the precision electroweak data[50] and the standard theory structure,[51] m_H can be determined indirectly to be 93^{+25}_{-21} GeV.[52] The large uncertainty reflects the fact that m_H enters (at one loop) only in logarithmic corrections, reducing the ability for the electroweak precision data to constrain it.

A precise measurement of m_H is fundamental in testing the compatibility of the observed properties of the Higgs boson with those predicted by the standard theory, since the standard theory can make precise predictions for all the Higgs boson properties for a given value of m_H.

For all these reasons, it is not surprising that the first combined Higgs result of the ATLAS and CMS collaborations was a measurement of m_H, resulting in a determination with a precision of 0.2%.[10]

The decay channels used in measuring the Higgs boson mass were $H \to \gamma\gamma$ and $H \to ZZ^{(*)} \to llll$. In both these final states it is possible to perform a precise measurement of all particles in the decay, leading to an expected relative mass resolution for the Higgs boson between 1 and 2%.[22–25] This is in contrast with other final states where the hadronisation into jets of particles ($H \to b\bar{b}$ and $H \to WW^{(*)}$)

or the presence of neutrinos in the final states ($H \to \tau\tau$ and $H \to WW^{(*)}$) smears the reconstructed mass, pushing their expected relative mass resolutions to values ranging between 10 and 20%.[26–29,34,35] An excellent experimental resolution can also be achieved from the measurement of the $H \to \mu\mu$ decay, once there is enough data to observe that decay.[40,41]

The precision of the m_H measurement in the $H \to \gamma\gamma$ decay channel essentially depends on the precision in the measurement of the energy of each photon and of the angle between the two photons. Here, the complementarity of the ATLAS and CMS detectors shines: given their different design, the electromagnetic (EM) calorimeters of ATLAS and CMS have different strengths. For instance, the ATLAS EM calorimeter is able to reconstruct the direction of the EM shower and therefore have a precise measurement of the diphoton opening angle,[22] while the CMS EM calorimeter is homogeneous, allowing for a better precision on the individual photon energy measurement.[53,54] Overall, the precision achieved by both experiments is comparable, the differences providing robustness to the combined measurement.

In the $H \to ZZ^{(*)} \to llll$ decay channel, the precision depends on the flavour of the leptons, with the $H \to ZZ^{(*)} \to \mu\mu\mu\mu$ final state providing the most precise measurements.

In both ATLAS and CMS, the energy scale is calibrated against the $Z \to ll$ "standard candle". In the $H \to ZZ^{(*)} \to llll$ decay channel, there are also cross-check measurements performed using the $Z \to llll$ final state, where one lepton pair is the result of QED radiation. In the $H \to \gamma\gamma$ decay channel, the energy scale is set from the study of $Z \to ee$ decays. In this case, some "electron-to-photon" extrapolation is involved, since the two particles are differently affected, for instance, by discrepancies between the simulated and actual material in front of the EM calorimeters.

The mass measurements in the $H \to ZZ^{(*)} \to llll$ and $H \to \gamma\gamma$ final states are performed using observables that are not necessarily the same used in the analyses targeting the measurement of yields or cross-sections. For instance, for the measurement of m_H the ATLAS $H \to \gamma\gamma$ analysis categorises events so as to minimise the experimental systematic uncertainties, and the CMS $H \to ZZ^{(*)} \to llll$ analysis uses an event-by-event estimator of the mass resolution. The results of the four individual analysis are shown in the top part of Fig. 1.

In order to reduce the dependence of the measurement on standard theory assumptions, which is used to model the expected yields in each event category, three separate signal strength modifiers are floated simultaneously. The factors separately scale gluon-fusion and $t\bar{t}H$ in $H \to \gamma\gamma$, VBF and VH in $H \to \gamma\gamma$, and the overall $H \to ZZ^{(*)} \to llll$ signal. Figure 1 also shows the result of the combined analysis of ATLAS and CMS data for the two decay channels. The fully-combined measurement yields $m_H = 125.09 \pm 0.21$ (stat) ± 0.11 (syst) GeV. The breakdown of the systematic uncertainty yields ± 0.11 (scale) $\pm$ 0.02 (other) $\pm$ 0.01 (theory) GeV. This shows that the largest contribution to the systematic uncertainty is due to

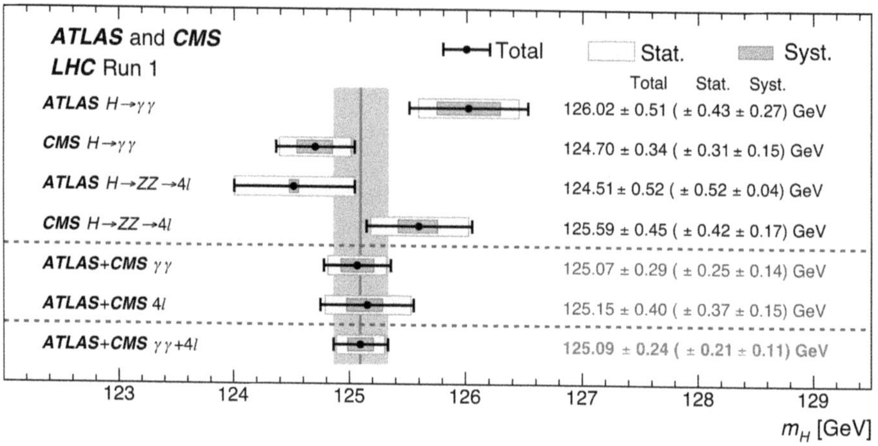

Fig. 1. Summary of the ATLAS and CMS m_H measurements, partial combinations per decay channel, and the result of the combined analysis. Systematic uncertainties are denoted by the thinner boxes with darker shade, the statistical uncertainties by the thicker boxes with lighter shade, and the total uncertainties by the black bars. (Figure reproduced from Ref. 10.)

the uncertainty on the energy or momentum scales, all other contributions being negligible in comparison.

The compatibility between the four individual measurements with the combined one was estimated from a fit with four separate masses. The asymptotic p-value is found to be 10%, changing to 7% if the production rates are also decorrelated between ATLAS and CMS.

It is worth noting that less precise measurements from other decay channels, such as $H \to WW^{(*)}$[26,27] or $H \to \tau\tau$,[29] are in agreement with the determination from $H \to \gamma\gamma$ and $H \to ZZ^{(*)} \to llll$, within their uncertainties, which are comparably larger.

Finally, given how the statistical uncertainty is about twice the systematic uncertainty, the existing results on the Higgs boson mass allow to confidently say that the LHC Run 2 has the potential to reduce the total uncertainty by a significant factor.

3.2. *Total width*

For $m_H \sim 125$ GeV the SM Higgs boson is predicted to be narrow, with a total width $\Gamma_{\mathrm{SM}} \sim 4$ MeV. From the study of off-shell Higgs boson production and decay into the $ZZ \to 4\ell$ and $VV \to 2\ell\,2\nu$ final states, ATLAS[55] and CMS[56] have set an indirect observed (expected) 95% CL limit on the total width in the range of $\Gamma_{\mathrm{tot}}/\Gamma_{\mathrm{SM}} < 4.5$–$7.5\,(6.5$–$11.2)$ for ATLAS, and $\Gamma_{\mathrm{tot}} < 13\,(26)\,\mathrm{MeV}$ for CMS. While that result is about two orders of magnitude better than the experimental mass resolution, it relies on assumptions about the underlying theory, such as the absence of contributions to Higgs boson off-shell production from BSM particles.

In contrast, a direct limit does not rely on such assumptions and is only limited by the experimental resolution.

The best experimental mass resolution, achieved in the $H \to \gamma\gamma$ and $H \to ZZ^{(*)} \to llll$ analyses, is typically between 1 and 3 GeV. The resolution depends on the energy, rapidity, and azimuthal angle of the decay products, and on the flavour of the leptons in the case of $H \to ZZ^{(*)} \to llll$ decays. If found inconsistent with the expected detector resolution, the total width measured in data could suggest the production of a resonance with a larger intrinsic width or the presence of two quasi-degenerate states.

To perform a measurement of the total width, the signal models in the $H \to \gamma\gamma$ and $H \to ZZ^{(*)} \to llll$ analyses are modified to allow for a natural width using the relativistic Breit–Wigner distribution, as described in Refs. 8, 25 and 23. For the $H \to \gamma\gamma$ channel the observed (expected) upper limit at the 95% CL is 2.4 (3.1) GeV for CMS[9] and 5.0 (6.2) GeV for ATLAS.[8] In the $H \to ZZ^{(*)} \to llll$ channel the observed (expected) upper limit at the 95% CL is 3.4 (2.8) GeV for CMS[9] and 2.6 (6.2) GeV for ATLAS.[8] For the combination of the two CMS analyses, the observed (expected) upper limit at the 95% CL is 1.7 (2.3) GeV.[9]

From the study of the flight distance in $H \to ZZ^{(*)} \to llll$ decays, CMS set a 95% CL observed (expected) lower limit on the total width of $\Gamma_{\text{tot}} > 3.5 \times 10^{-9}$ (3.6×10^{-9}) MeV.[57]

3.3. *Differential and fiducial cross-sections*

The very clean final state signatures of the $H \to \gamma\gamma$ and $H \to ZZ^{(*)} \to llll$ decays do not only allow to measure the mass and width of the Higgs boson, but also allow rather model-independent differential measurements of the Higgs boson production kinematic properties. These include the transverse momentum and rapidity of the Higgs boson, and the associated number of jets as well as the transverse momentum distribution of the jets. Comparisons between the measured production properties and predictions for different Higgs boson production modes can be used to draw conclusions about QCD aspects of Higgs boson production and possible contributions from BSM physics in Higgs boson production.

Both experiments measured the production properties in the two decay channels both as inclusive as well as differential fiducial cross sections.[15,16,18,19] ATLAS also combined the $H \to \gamma\gamma$ and $H \to ZZ^{(*)} \to llll$ channels in a common fiducial volume.[17] Measurements using the $H \to WW^{(*)}$ decay channel were also performed[20,21] but not considered in the following examples.

As examples of the observables considered, the measurements of the Higgs boson transverse momentum $p_T(H)$ and the number of jets N_{jets} associated with the Higgs boson are shown in Fig. 2. The measurements are shown together with predictions from the most accurate theory calculations at the time. For ATLAS, the combination of the cross-section-normalised distributions for $p_T(H)$ and N_{jets} are shown,

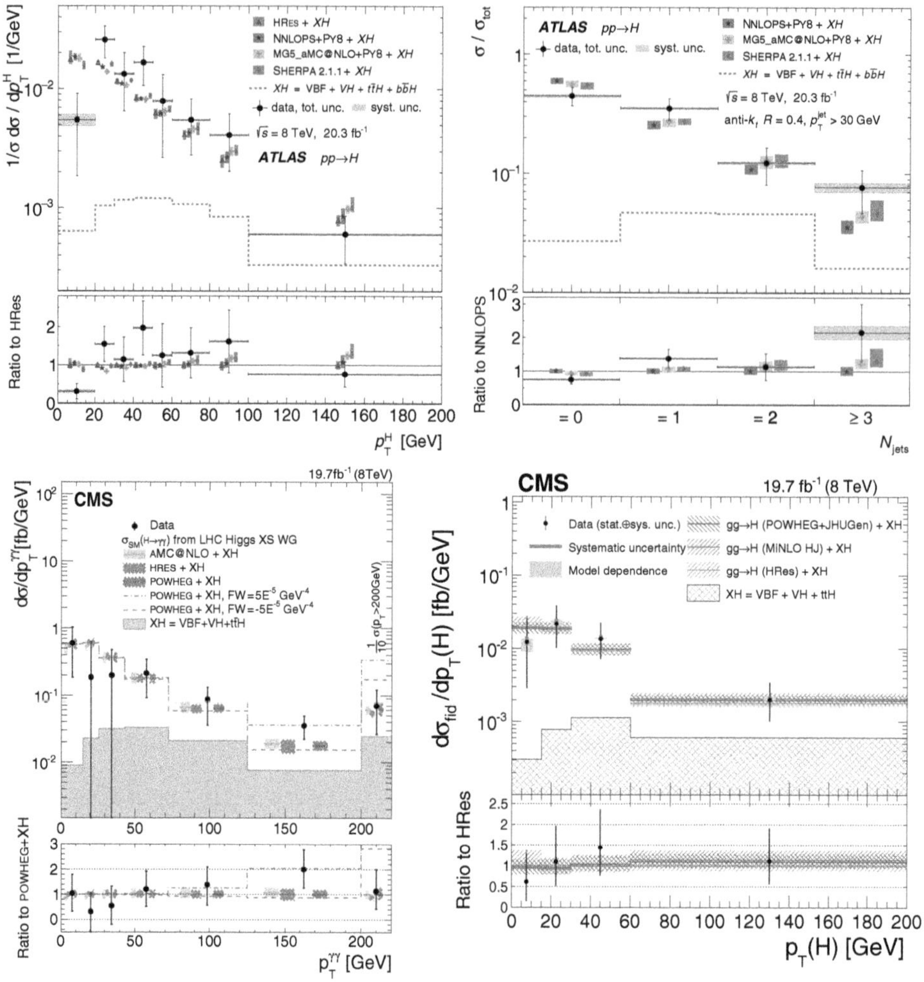

Fig. 2. Comparison of the predictions of several SM theory calculations with the measured distributions of the Higgs boson transverse momentum, p_T^H (top left and bottom row) and the number of jets produced in association with the Higgs boson, N_{jets} (top right). In the top row, ATLAS measurements combining the $H \to \gamma\gamma$ and $H \to ZZ^{(*)} \to llll$ channels, normalised to the cross section. In the bottom row, CMS measurements in the $H \to \gamma\gamma$ (left) and $H \to ZZ^{(*)} \to llll$ (right) channels. At the time, HRES[58] and NNLOPS[59] provided the most accurate SM predictions for p_T^H and N_{jets}, respectively, and are therefore used as theory references. (Figures reproduced from Refs. 17–19.)

as this combination is independent of assumptions on the Higgs boson branching fractions. For CMS, the p_T^H distribution is shown separately for the $H \to \gamma\gamma$ and $H \to ZZ^{(*)} \to llll$ channels.

Within the SM, the $gg \to H$ production mode dominates and these measurements are especially sensitive to the intricate QCD aspects of this process. However, some sensitivity to the VBF production is also reached in the high p_T and

$N_{\text{jets}} = 2$ part of the phase space, where dedicated measurements for VBF-related observables were done in the $H \to \gamma\gamma$ analyses.[15,18] With increasing luminosity, the sensitivity to other production modes can be expected to increase.

The overall agreement of the observed distributions and the SM predictions is good within the current uncertainties of the measurements and of the theory predictions. The most noticeable difference with respect to the SM predictions is in the lowest-p_T^H bins in ATLAS analyses and the CMS $H \to ZZ^{(*)} \to llll$ channel, where the theory seems to predict larger yields than observed. This is not visible in the p_T^H distribution of the CMS $H \to \gamma\gamma$ analysis. Besides statistical fluctuations, QCD effects unaccounted for in theory calculations could contribute a deviation. BSM modifications of Higgs boson production are mostly expected to appear in the high p_T and high jet multiplicity part of the phase space, where the agreement with the SM predictions is good within the uncertainties.

The analyses also include model-independent measurements of cross sections in fiducial volumes as well as comparisons with SM theory predictions within the corresponding fiducial regions. Examples are shown in Fig. 3 and more information can be found in Refs. 15–19.

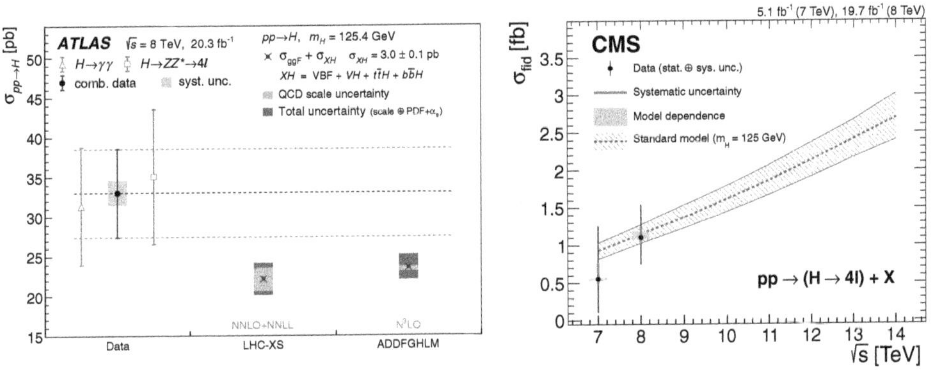

Fig. 3. Fiducial cross section measurements, for $H \to \gamma\gamma$, $H \to ZZ^{(*)} \to llll$, and combination at 8 TeV from ATLAS (left) and as a function of $\sqrt{s}$ for $H \to ZZ^{(*)} \to llll$ from CMS (right). In both cases a comparison to theory predictions is provided. (Figures reproduced from Refs. 17 and 19.)

4. Searches for Deviations

Measurements of the Higgs boson properties that are almost model independent are so far only possible in the cleanest Higgs boson channels, $H \to \gamma\gamma$ and $H \to ZZ^{(*)} \to llll$, as discussed in the previous section. The analyses in the $H \to WW^{(*)} \to l\nu l\nu$, $H \to \tau\tau$ and $H \to b\bar{b}$ channels retain some dependence on the SM, as this is needed for a good separation from background processes. Studies of Higgs boson properties involving these channels are therefore better understood as searches for deviations from the SM.

These searches for deviations in Higgs boson properties from the SM can be broadly grouped in two categories of comparisons to the SM predictions:

(1) Higgs boson signal yields in different Higgs boson production and decay modes or alternative comparisons of the Higgs boson coupling strength to the SM expectation. These comparisons assume for each Higgs boson production and decay mode the kinematic structure of the SM.
(2) Kinematic (tensor) structure of Higgs boson decays. While these analyses are designed to be independent of the observed signal yield, they assume production kinematics similar to the SM in some cases.

As the latter justifies — to some degree — the assumptions of the former, the following discussion will start with the kinematic structure of Higgs boson decays.

4.1. *Compatibility in decay kinematics*

The SM predicts a spin-0 Higgs boson with quantum numbers $J^{CP} = 0^{++}$. This results in characteristic kinematic distributions for Higgs boson decay products and kinematic correlations between them, which can be used to search for deviations from the SM. As the SM Higgs boson has spin 0, in the SM no kinematic correlations are expected between production and decay.

ATLAS[11,12] and CMS[13] have used the $H \to \gamma\gamma$, $H \to ZZ^{(*)} \to llll$, and $H \to WW^{(*)} \to l\nu l\nu$ decay channels to search for such deviations from the SM by testing several non-SM hypotheses of spin and parity J^P. This allows to determine J^{CP} assuming that either C or CP is conserved. In addition, non-SM effective Lagrangian operators or non-SM anomalous coupling contributions were explored in the case of a spin-0 Higgs boson.

4.1.1. *Hypothesis tests on the spin of the new boson*

Hypothesis tests for pure $J^P = 0^-$, 1^+, and 1^- states were performed in the $H \to ZZ^{(*)} \to llll$ and $H \to WW^{(*)} \to l\nu l\nu$ channels by analysing kinematic decay properties. The inclusive $H \to \gamma\gamma$ channel is not sensitive to the parity for a spin-0 particle since the two photons decay back-to-back in the rest frame of the Higgs boson and yield no additional information. The observation of the $H \to \gamma\gamma$ final state excludes a single spin-1 boson as a consequence of the Landau–Yang theorem.[60,61] Hypothesis tests for a new spin-2 boson were performed for a large number of models using the $H \to ZZ^{(*)} \to llll$, $H \to WW^{(*)} \to l\nu l\nu$ and $H \to \gamma\gamma$ channels. For spin-2 hypotheses, the angular correlations between the production and the decay of the new boson can also be exploited, as is done in the $H \to \gamma\gamma$ channel by measuring the θ^* angle between the production and decay system.

In order to be less dependent on the implementations of concrete BSM models in excluding J^P states, the overall observed signal yields were floated in the hypothesis tests. Consequently, the information on total predicted yield is not used. This is especially motivated in spin-2 models with universal couplings that predict branching fractions otherwise not compatible with the observed signal yields in different final states. This also ensures that conclusions drawn from these studies are valid when deviations from the SM in the Higgs boson signal and coupling strengths are allowed.

For the combined spin hypothesis tests[12,13] using the three channels, the best-motivated spin-2 model corresponding to a massive graviton-like particle (2^+_m) is taken here as a spin-2 example, even if many others were tested. Table 2 shows that all tested J^P states are excluded with CL_s values[62] $\ll$1% when compared with the SM hypothesis, $J^P = 0^+$.

Table 2. Observed CL_s values from ATLAS and CMS for different J^P hypotheses tested against the SM Higgs boson hypothesis of $J^P = 0^+$. The loosest limit is given whenever different production mode hypotheses were tested. Smaller values of CL_s disfavour more strongly the non-SM hypotheses.

J^P	ATLAS	CMS
0^-	<0.026%	<0.01%
1^+	0.03%	0.004%
1^-	0.27%	<0.001%
2^+_m	0.011%	0.13%

A model-independent test of all spin-2 structures may eventually become possible with more data. States with a spin larger than 2 were not considered by either experiment, and are even more challenging to motivate for an elementary particle.

4.1.2. *Kinematic decay structure of a* $J = 0$ *boson*

Deviations from the SM predictions in kinematic distributions for a scalar boson (with spin 0) can be analysed within different frameworks as extensively discussed in Ref. 63. Among the frameworks, the effective Lagrangian and anomalous coupling approaches are the most common. On the one hand, the most general effective Lagrangian compatible with Lorentz and gauge invariance can be used as a basis for calculations. On the other hand, in the anomalous coupling approach, the most general amplitude compatible with Lorentz and gauge invariance is introduced. As the latter does not assume a hierarchy in scales, the couplings become momentum-dependent form factors. However, no common and agreed-upon approach exists and also the notations differ substantially. ATLAS[12] used an effective Lagrangian approach, but provides results also in an anomalous coupling formulation. CMS[13] used an anomalous couplings approach, providing also results in terms of an effective Lagrangian.

For a better comparison, the results from both collaborations for the $H \to ZZ^{(*)} \to llll$ and $H \to WW^{(*)} \to l\nu l\nu$ channels are therefore translated in terms of the following effective Lagrangian, that describes a spin-0 on-shell scalar boson decaying to $H \to VV$ ($V = W^{\pm}, Z$):

$$\begin{aligned}\mathcal{L}_{HVV} \sim\ & \kappa_Z \frac{m_Z^2}{v} H Z^{\mu} Z_{\mu} + \kappa_W \frac{2m_W^2}{v} H W^{\mu} W_{\mu} \\ & + \frac{\alpha_Z}{v} H Z^{\mu} \Box Z_{\mu} + \frac{\alpha_W}{v} \left(H W^{+\mu} \Box W_{\mu}^{-} + H W^{-\mu} \Box W_{\mu}^{+} \right) \\ & + \frac{\beta_Z}{v} H Z^{\mu\nu} Z_{\mu\nu} + \frac{2\beta_W}{v} H W^{\mu\nu} W_{\mu\nu} \\ & + \frac{\gamma_Z}{v} H Z^{\mu\nu} \tilde{Z}_{\mu\nu} + \frac{2\gamma_W}{v} H W^{\mu\nu} \tilde{W}_{\mu\nu} \end{aligned} \quad (1)$$

where v is the SM Higgs field vacuum expectation value, V_{μ} represents the V boson field, $V_{\mu\nu} = \partial_{\mu} V_{\nu} - \partial_{\nu} V_{\mu}$ is the field strength tensor, and $\tilde{V}_{\mu\nu} = \frac{1}{2}\epsilon_{\mu\nu\rho\sigma} V^{\rho\sigma}$ is the dual field-strength tensor. The first two terms are the minimal dimension-3 operator and, with $\kappa_V = 1$, correspond to the SM Higgs boson ($J^P = 0^+$). The six terms in the last three rows of Eq. (1) are dimension-5 operators: the α- and β-terms are parity-even interactions, that can be kinematically distinguished from the SM Higgs boson. The β terms are commonly referred to as 0_h^+. The γ terms correspond to a pseudo-scalar interaction (0^-). In principle, also SM particles contribute to these last six terms due to higher-order corrections, but at a very small rate.

The Lagrangian in Eq. (1) allows for both CP-even and CP-odd tensor structures in addition to the SM gauge boson interactions. Within the effective Lagrangian calculations, a non-vanishing value of the parameters α_V, β_V, and γ_V leads to changes of the kinematic distributions as well as the overall signal yields, with respect to the SM. The SM tensor structure is scaled by κ_V, which modifies the SM signal yields, and correspond to the κ_W and κ_Z scalar coupling rescaling factors defined for Higgs boson couplings measurements, as discussed in Section 4.3.

As done in the J^P hypothesis tests, the overall observed signal yields are not used in the analysis for the discrimination from the SM. This not only simplifies the analysis, but also allows for a more general interpretation of the results. This implies that only ratios of Lagrangian parameters are probed. The ratios are chosen to be α_V/κ_V, β_V/κ_V, and γ_V/κ_V. Without further assumptions, these ratios of parameters may be different in the HZZ and HWW interaction vertexes. For the combination of measurements from the $H \to ZZ^{(*)} \to llll$ and $H \to WW^{(*)} \to l\nu l\nu$ channels, the same values of these ratios are assumed in the HZZ and HWW interactions: $\alpha^{ZZ}/\kappa^{ZZ} = \alpha^{WW}/\kappa^{WW}$, $\beta^{ZZ}/\kappa^{ZZ} = \beta^{WW}/\kappa^{WW}$, and $\gamma^{ZZ}/\kappa^{ZZ} = \gamma^{WW}/\kappa^{WW}$. The allowed 95% CL confidence intervals for these ratios are listed in Table 3 for

Table 3. The expected and observed 95% CL confidence intervals for non-SM spin-0 interactions with gauge bosons, as described in Eq. (1). The results are presented for the combination of the $H \to ZZ^{(*)} \to llll$ and $H \to WW^{(*)} \to l\nu l\nu$ channels. The results obtained by ATLAS and CMS in Refs. 12 and 13 are translated to the notation used in this Section.

		α/κ	β/κ	γ/κ
ATLAS	observed	-	$[-0.63, 0.73]$	$[-0.83, 2.18]$
	expected	-	$[-4.80, 0.55]$	$[-2.30, 2.33]$
CMS	observed	$[-1.66, 1.57]$	$[-0.76, 0.58]$	$[-1.57, 1.54]$
	expected	$[-\infty, 1.24] \cup [9.0, +\infty]$	$[-1.67, 0.45]$	$[-2.65, 2.65]$

the combined $H \to ZZ^{(*)} \to llll$ and $H \to WW^{(*)} \to l\nu l\nu$ analyses of the full LHC Run 1 data. In all cases, very small non-vanishing values for the α/κ, β/κ, and γ/κ ratios are also expected in the SM from higher-order loop contributions, but the sensitivity was not sufficient to be probe such contributions.

The structure of the HVV vertex can also be probed using the VBF and VH production modes. The analyses in Refs. 64 and 65 search directly for anomalous CP-odd tensor structures in addition to the SM gauge boson interactions. In Ref. 66 the differential cross section measurements described in Section 3.3 are used to probe for non-SM CP-even and CP-odd contributions in Higgs boson production modes.

4.2. *Compatibility in signal yields*

The signal yields for Higgs boson production and decay processes can be measured from experimental analyses optimised for each of the 25 combinations of five production processes and five decay modes briefly illustrated in Table 1. While the selection of the Higgs boson final state decay products allows for a rather clean experimental separation of decay modes, the separation of Higgs boson production modes poses a bigger challenge. In many cases, several Higgs boson production processes are expected to contribute to the different experimental analysis categories. Different Higgs boson production processes can still be separated, if the same acceptance as for the SM processes (within uncertainties) is assumed for each production process and decay mode. This is equivalent to assuming, for each process, the kinematic distributions predicted by the SM. A combined fit can then estimate the contributions from individual production and decay processes from the observed signal yields in different experimental analysis categories with different compositions.

The most general result in this context would be a measurement of all 25 cross section combinations $\sigma_i \times \mathrm{BR}^f$ for the production process i and decay mode f listed in Table 1. Such a measurement is not yet possible in full generality, but 20 combinations were determined, with varying precision, from a combined analysis

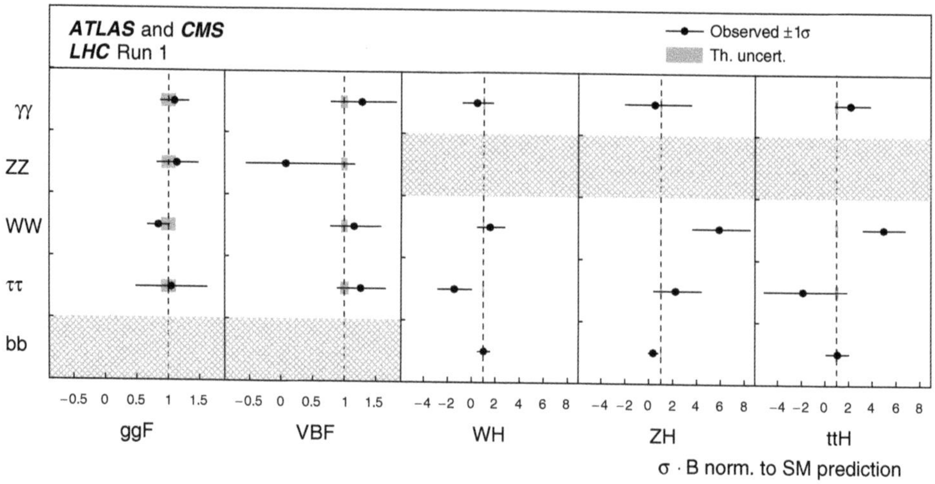

Fig. 4. Best-fit values of the combined analysis of ATLAS and CMS for twenty $\sigma_i \times \mathrm{BR}^f$ combinations. The results are expressed in terms of the SM predictions. The uncertainties in the points correspond to all sources of uncertainty, except for the SM prediction uncertainties on the overall signal yields, which are represented as the grey bars around the prediction. As could be expected, the Run 1 dataset did not allow to set precise bounds on some combinations of production processes and decay modes (cross-hatched). (Figure reproduced from Ref. 14.)

of ATLAS and CMS data,[14] as illustrated in Fig. 4. Among the missing combinations are some that could not be meaningfully constrained with the combined ATLAS and CMS Run 1 datasets for lack of statistics.

On the other hand, if it is assumed that all signals originate from a single narrow Higgs boson resonance, one can exploit the fact that different rows and columns in Table 1 and Fig. 4 share either common ratios of production cross sections σ_i or common ratios of branching fractions BR^f. This then allows to choose a reference cross section, such as $\sigma_{\mathrm{ggF}} \times \mathrm{BR}^{ZZ}$ for the $gg \to H \to ZZ$ process, and eight ratios, one for each of the other Higgs boson production and decay modes. Figure 5 shows the result for the ATLAS and CMS measurements,[14] and their combination. The results are compared to the SM theory predictions.[63] The most notable differences with respect to the SM are $\sigma_{ttH}/\sigma_{\mathrm{ggF}}$, which is about 3.0σ above the SM expectation, and $\mathrm{BR}^{bb}/\mathrm{BR}^{ZZ}$, which is about 2.5σ below the SM expectation. The p-value of the observed data under the SM hypothesis is 16%.

More sensitive, but less general, tests of the SM hypothesis are possible if additional assumptions are introduced. For instance, if the signal strength in a decay mode f, defined as $\mu_i^f = (\sigma_i \times \mathrm{BR}^f)/(\sigma_{i,\mathrm{SM}} \times \mathrm{BR}^f_{\mathrm{SM}})$, is assumed to be identical for all vector-boson-mediated Higgs production modes ($\mu^f_{\mathrm{VBF}+VH} = \mu^f_{\mathrm{VBF}} = \mu^f_{WH} = \mu^f_{ZH}$) and all fermion-mediated processes ($\mu^f_{\mathrm{ggF}+ttH} = \mu^f_{\mathrm{ggF}} = \mu^f_{ttH}$), deviations from the SM can be explored in the 2D plane of $\mu^f_{\mathrm{VBF}+VH}$ versus $\mu^f_{\mathrm{ggF}+ttH}$ for each decay mode f. Figure 6 shows the results of the combined analysis of ATLAS and CMS, which agree, within uncertainties, with the SM hypothesis for each decay mode studied.

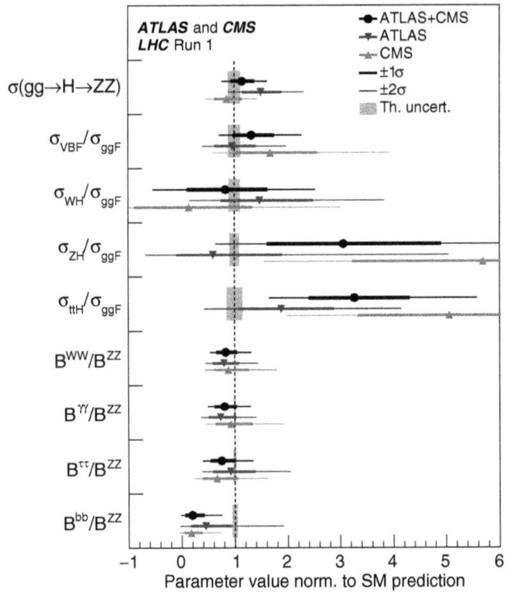

Fig. 5. Best-fit values for the ATLAS, CMS, and combined analyses of Higgs boson production and decay modes. The results are expressed in terms of ratios and compared to the SM predictions. The uncertainties in the points correspond to all sources of uncertainty, except for the SM prediction uncertainties on the overall signal yields, which are represented as grey bars around the prediction. (Figure from Ref. 14.)

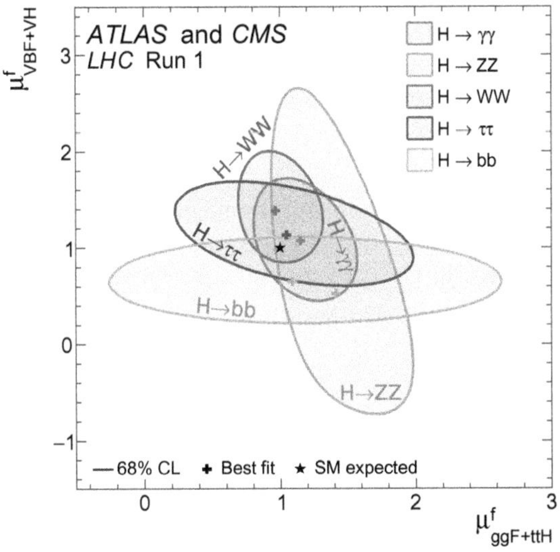

Fig. 6. Confidence regions for signal strength multipliers of the fermion-mediated and vector-boson-mediated productions processes. The regions are determined for each of the five main decay modes studied. Only the combined analysis of the ATLAS and CMS data is shown. (Figure adapted from Ref. 14.)

The most sensitive test, at the cost of the strongest assumptions, is the determination of a common signal strength $\mu = \mu_i^f$ for all Higgs boson production modes i and decay modes f. The combined ATLAS and CMS analysis yields $\mu = 1.09^{+0.11}_{-0.10}$ in agreement with the SM hypothesis. The detailed breakdown of the uncertainty yields ± 0.07 (stat) $\pm\, 0.04$ (expt) $\pm\, 0.03$ (thbgd) $^{+0.07}_{-0.06}$ (thsig). This shows that the statistical uncertainty for this test is already at the same level of the systematic uncertainty from theory calculations of the signal (thsig) and background (thbgd), as well as experimental sources of uncertainty (expt). Therefore, improvements in the theory systematic uncertainties will be needed for LHC Run 2, where substantially more data are expected than in Run 1, entailing even smaller statistical uncertainties.

The search for a second, mass-degenerate, state was also performed via a statistical test of the structure of the matrix of the different production and decay mode signal strengths.[67] The result from the combined analysis of the ATLAS and CMS datasets is compatible with SM prediction of a single particle.

4.3. *Compatibility in couplings*

In the previous section, the production and decay processes of this Higgs boson were treated as independent. Given that the couplings of the Higgs boson to SM particles are involved both in its production as well as in its decay, a framework was proposed for the simultaneous interpretation of all the information.[63,68] In this framework, the coupling of the Higgs boson with each of the SM particles is scaled by a multiplicative factor κ_i. These factors represent the ratios of production cross sections and partial decay width to the SM expectation: $\kappa_i^2 = \sigma_i/\sigma_i^{\mathrm{SM}}$ or $\kappa_i^2 = \Gamma_i/\Gamma_i^{\mathrm{SM}}$.

The fundamental assumptions underlying this framework are that there is one single state with $J^P = 0^+$ and a very narrow width, such that production and decay factorise, $\sigma_i \cdot \mathrm{BR}_f = \sigma_i \cdot \Gamma_f/\Gamma_{\mathrm{tot}}$.

In this framework, the loop-induced gluon and photon couplings to the Higgs boson can either be resolved into the SM amplitudes or treated as effective couplings. This allows to search for the effect of new particles "running" in the loops. As an example, the $H \to \gamma\gamma$ decay modifier can be expressed in a "resolved" way as $\kappa_\gamma^2 \sim 0.07\kappa_{\mathrm{t}} - 0.66\kappa_W \cdot \kappa_{\mathrm{t}} + 1.59\kappa_W^2$, where the (destructive) interference between the loops mediated by W bosons and top quarks can be seen. The effect of this structure can be clearly seen in Fig. 7 that shows the confidence regions allowed by each decay channel for two parameters, $\kappa_{\mathrm{V}} = \kappa_W = \kappa_Z$ and $\kappa_{\mathrm{F}} = \kappa_{\mathrm{b}} = \kappa_{\mathrm{t}} = \kappa_\tau$. The interference effects in the tHq[69–72] and $gg \to ZH$[71,73] subdominant production processes lead to the observations in some individual decays channels to prefer very slightly the $\kappa_{\mathrm{V}}\kappa_{\mathrm{F}} < 0$ BSM scenario over the $\kappa_{\mathrm{V}}\kappa_{\mathrm{F}} > 0$ SM-like region (largest deviation is 1.3σ). But because of the interference effect in the $H \to \gamma\gamma$ decay, the combination of all channels excludes the $\kappa_{\mathrm{V}}\kappa_{\mathrm{F}} < 0$ BSM scenario by almost 5σ.

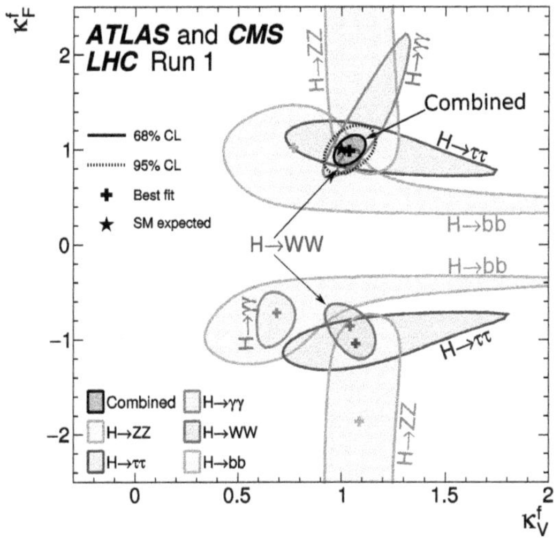

Fig. 7. Allowed confidence regions for each decay channel for two parameters that collect the scaling factors to massive bosons (κ_V) and fermions (κ_F). While most decay channels exclude regions that are almost symmetric, the interference of the top- and W-amplitudes in the $H \to \gamma\gamma$ decay make the contour for that channel very asymmetric. For this reason, the combination of channels strongly excludes the possibility that $\kappa_V = -\kappa_F$. (Figure adapted from Ref. 14.)

The results for a general fit with one scaling factor per SM particle as well as the gluon and photon interactions treated with an effective scaling parameter are shown in Fig. 8. In this figure, the two different sets of results correspond to different assumptions on the treatment of the total width.

The lighter points in Fig. 8 correspond to the case when the total width is a dependent function of all the other κ_i that scale the corresponding partial widths: $\Gamma_{\text{tot}}(\kappa_i) = \sum_i (\kappa_i^2 \Gamma_i^{\text{SM}})$. From the result of the fit to data, it can be seen that having separate gluon and top-quark modifiers reveals an excess, due to the results in ttH searches. The p-value for the data under the SM hypothesis in this model is 11%.

The assumption on the total width can be relaxed by allowing for the possibility of decays to invisible or undetected particles, whose branching fraction is denoted BR_{BSM}. In this case, the total width scales as $\Gamma_{\text{tot}}(\kappa_i) = \frac{\sum_i(\kappa_i^2\Gamma_i^{\text{SM}})}{1-\text{BR}_{\text{BSM}}}$. In this scenario, an additional constraint is assumed, namely $\kappa_V \leq 1$, something that is well-motivated in a large class of BSM models. The results of this fit are represented by the black points in Fig. 8, and yield a 95% CL observed (expected) upper limit $\text{BR}_{\text{BSM}} < 0.34\,(0.35)$.

Without assumptions on either the total width or the allowed range of any coupling only ratios $\lambda_{ij} = \kappa_i/\kappa_j$ of coupling modifiers can be determined. This is similar to what was done in the measurement of ratios of cross sections and ratios of branching fractions in the previous section. In addition to the λ_{ij} ratios, also one overall scale factor that modifies the rate of all processes by the same amount

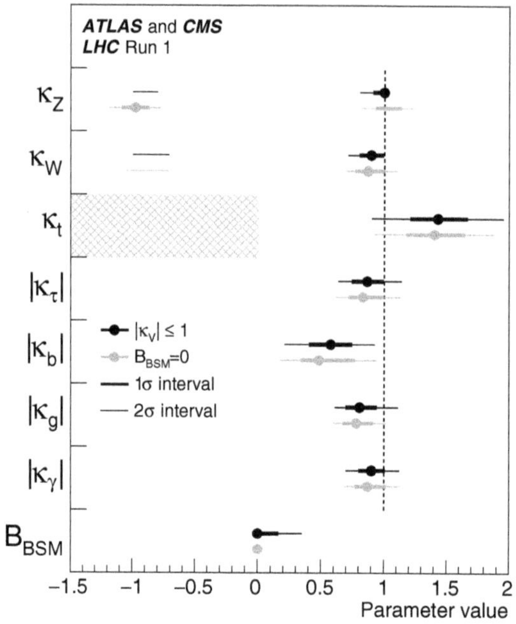

Fig. 8. General search for coupling strength deviations, where the gluon and photon couplings to the Higgs boson are scaled with effective coupling modifiers. The light points correspond to a scenario where the total width is scaled according to the κ_i, while for the black points the decay to BSM particles is probed under the assumption that $\kappa_V \leq 1$. (Figure from Ref. 14.)

is determined: $\kappa_{gZ}^2 = \sigma_{\text{ggF}}/\sigma_{\text{ggF}}^{\text{SM}} \times \text{BR}_{ZZ}/\text{BR}_{ZZ}^{\text{SM}}$, where again the $gg \to H \to ZZ$ process is chosen as reference.

The results of this fit are shown in Fig. 9. All other determinations of coupling modifiers can be derived from these results, provided the correct correlations are taken into account.

Assuming the SM structure for the loop amplitudes and no BSM Higgs decays, deviations from the expected tree-level couplings to the SM particles can be probed. In this case the gluon and photon coupling modifiers become functions of the other κ_i and the top coupling deviation is mostly probed by the gluon-fusion cross section measurement. The result is presented as a function of the mass of the SM particles in Fig. 10, where it can be seen that it is very likely that the Higgs boson has non-universal interactions, or else the muon coupling modifier would have been measured to be much larger than observed.

Other results of the tests include the relative strength of the couplings to W and Z bosons, $\lambda_{WZ} = 0.88^{+0.10}_{-0.09}$, as well as tests with fewer parameters that probe the relative up-down and lepton-quark coupling strengths, $\lambda_{du} = 0.92 \pm 0.12$ and $|\lambda_{lq}| = 1.06^{+0.15}_{-0.14}$, which are expected to deviate from unity in models with two Higgs doublets.

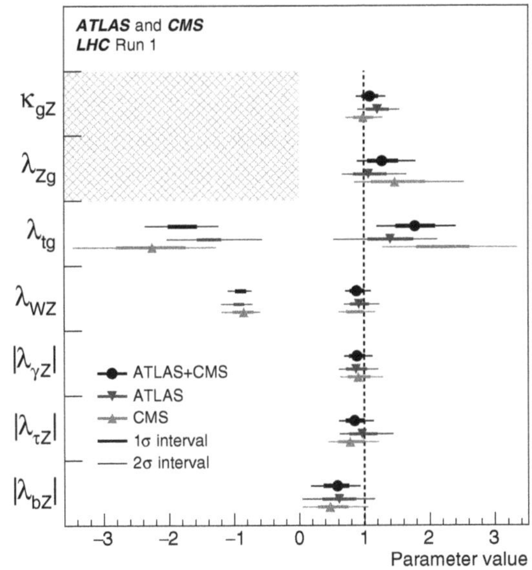

Fig. 9. General search for coupling strength deviations, where the gluon and photon couplings to the Higgs boson are scaled as effective coupling modifiers. No assumptions on either the total width or the allowed range of any coupling is made, which then allows to determine only ratios of coupling modifiers $\lambda_{ij} = \kappa_i/\kappa_j$ and one overall rate modifier $\kappa_{gZ}^2 = \sigma_{\text{ggF}}/\sigma_{\text{ggF}}^{\text{SM}} \times \text{BR}_{ZZ}/\text{BR}_{ZZ}^{\text{SM}}$. (Figure from Ref. 14.)

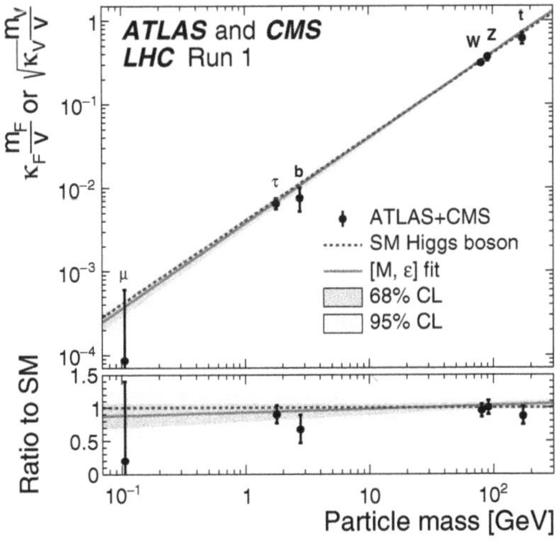

Fig. 10. Depiction of the result of the fit for deviations in the coupling of the Higgs boson to SM particles, presented as a function of the mass of the given particle. It can be seen that the Higgs boson has non-universal couplings with leptons, a highly distinctive feature of the standard theory Higgs boson. The coupling modifiers to the W and Z bosons were transformed with a square root in order to be plotted together with fermions and on the same (dotted) line, that represents the prediction of the standard theory. That transformation implies halving the uncertainties on κ_W and κ_Z. The solid line and bands indicate the result of the fit using the (M, ϵ) parameterization of Ref. 74. (Figure from Ref. 14.)

5. Summary

The data collected during the LHC Run 1 have allowed both ATLAS and CMS to characterise the observed Higgs boson with a mass of $m_H = 125.09 \pm 0.21$ (stat) ± 0.11 (syst) GeV with enough detail to rule out many non-standard alternatives. Measurements and searches for deviation from the SM in the Higgs boson signal yields, in the Higgs boson couplings as well as in Higgs boson production and decay differential distributions provided results consistent with the SM. These results have also brought intense discussion between the theory and experiment communities, demonstrating the importance of more accurate predictions.

With more data, the LHC experiments will improve on the detailed characterisation of the properties of this Higgs boson. While the ultimate goal remains to find a deviation from the standard theory predictions, every measurement along the way will be a tribute to the resilience of the standard theory.

References

1. G. Aad *et al.* [ATLAS], Observation of a new particle in the search for the Standard Model Higgs boson with the ATLAS detector at the LHC, *Phys. Lett. B.* **716**, 1 (2012), `arXiv:1207.7214 [hep-ex]`.
2. S. Chatrchyan *et al.* [CMS], Observation of a new boson at a mass of 125 GeV with the CMS experiment at the LHC, *Phys. Lett. B.* **716**, 30 (2012), `arXiv:1207.7235 [hep-ex]`.
3. S. Chatrchyan *et al.* [CMS], Observation of a new boson with mass near 125 GeV in pp collisions at $\sqrt{s} = 7$ and 8 TeV, *J. High Energy Phys.* **06**, 081 (2013), `arXiv:1303.4571 [hep-ex]`.
4. D. Buttazzo, G. Degrassi, P. P. Giardino, G. F. Giudice, F. Sala, A. Salvio, and A. Strumia, Investigating the near-criticality of the Higgs boson, *J. High Energy Phys.* **12**, 089 (2013), `arXiv:1307.3536 [hep-ph]`.
5. G. Aad *et al.* [ATLAS], The ATLAS Experiment at the CERN Large Hadron Collider, *JINST* **3**, S08003 (2008).
6. S. Chatrchyan *et al.* [CMS], The CMS experiment at the CERN LHC, *JINST* **3**, S08004 (2008).
7. D. Froidevaux and P. Sphicas, General-purpose detectors for the Large Hadron Collider, *Ann. Rev. Nucl. Part. Sci.* **56**, 375 (2006).
8. G. Aad *et al.* [ATLAS], Measurement of the Higgs boson mass from the $H \to \gamma\gamma$ and $H \to ZZ^* \to 4\ell$ channels with the ATLAS detector using 25 fb^{-1} of pp collision data, *Phys. Rev. D.* **90**, 052004 (2014), `arXiv:1406.3827 [hep-ex]`.
9. V. Khachatryan *et al.* [CMS], Precise determination of the mass of the Higgs boson and tests of compatibility of its couplings with the standard model predictions using proton collisions at 7 and 8 TeV, *Eur. Phys. J. C.* **75**, 212 (2015), `arXiv:1412.8662 [hep-ex]`.
10. G. Aad *et al.* [ATLAS, CMS], Combined measurement of the Higgs boson mass in pp collisions at $\sqrt{s} = 7$ and 8 TeV with the ATLAS and CMS Experiments, *Phys. Rev. Lett.* **114**, 191803 (2015), `arXiv:1503.07589 [hep-ex]`.
11. G. Aad *et al.* [ATLAS], Evidence for the spin-0 nature of the Higgs boson using ATLAS data, *Phys. Lett. B.* **726**, 120 (2013), `arXiv:1307.1432 [hep-ex]`.

12. G. Aad *et al.* [ATLAS], Study of the spin and parity of the Higgs boson in diboson decays with the ATLAS detector, *Eur. Phys. J. C.* **75**, 476 (2015), arXiv:1506.05669 [hep-ex].
13. V. Khachatryan *et al.* [CMS], Constraints on the spin-parity and anomalous HVV couplings of the Higgs boson in proton collisions at 7 and 8 TeV, *Phys. Rev. D.* **92**, 012004 (2015), arXiv:1411.3441 [hep-ex].
14. G. Aad *et al.* [ATLAS, CMS], Measurements of the Higgs boson production and decay rates and constraints on its couplings from a combined ATLAS and CMS analysis of the LHC pp collision data at $\sqrt{s} = 7$ and 8 TeV, Submitted to J. High Energy Phys. (2016), arXiv:1606.02266 [hep-ex].
15. G. Aad *et al.* [ATLAS], Measurements of fiducial and differential cross sections for Higgs boson production in the diphoton decay channel at $\sqrt{s} = 8$ TeV with ATLAS, *J. High Energy Phys.* **09**, 112 (2014), arXiv:1407.4222 [hep-ex].
16. G. Aad *et al.* [ATLAS], Fiducial and differential cross sections of Higgs boson production measured in the four-lepton decay channel in pp collisions at $\sqrt{s} = 8$ TeV with the ATLAS detector, *Phys. Lett. B.* **738**, 234 (2014), arXiv:1408.3226 [hep-ex].
17. G. Aad *et al.* [ATLAS], Measurements of the total and differential Higgs boson production cross sections combining the $H \to \gamma\gamma$ and $H \to ZZ^* \to 4\ell$ decay channels at $\sqrt{s} = 8\,\text{TeV}$ with the ATLAS detectors, *Phys. Rev. Lett.* **115**, 091801 (2015), arXiv:1504.05833 [hep-ex].
18. V. Khachatryan *et al.* [CMS], Measurement of differential cross sections for Higgs boson production in the diphoton decay channel in pp collisions at $\sqrt{s} = 8$ TeV, *Eur. Phys. J. C.* **76**, 13 (2016), arXiv:1508.07819 [hep-ex].
19. V. Khachatryan *et al.* [CMS], Measurement of differential and integrated fiducial cross sections for Higgs boson production in the four-lepton decay channel in pp collisions at $\sqrt{s} = 7$ and 8 TeV, *J. High Energy Phys.* **04**, 005 (2016), arXiv:1512.08377 [hep-ex].
20. G. Aad *et al.* [ATLAS], Measurement of fiducial differential cross sections of gluon-fusion production of Higgs bosons decaying to $WW^* \to e\nu\mu\nu$ with the ATLAS detector at $\sqrt{s} = 8$ TeV, Submitted to J. High Energy Phys. (2016), arXiv:1604.02997 [hep-ex].
21. V. Khachatryan *et al.* [CMS], Measurement of the transverse momentum spectrum of the Higgs boson produced in pp collisions at $\sqrt{s} = 7$ TeV using H→WW decays, Submitted to J. High Energy Phys. (2016), arXiv:1606.01522 [hep-ex].
22. G. Aad *et al.* [ATLAS], Measurement of Higgs boson production in the diphoton decay channel in pp collisions at center-of-mass energies of 7 and 8 TeV with the ATLAS detector, *Phys. Rev. D.* **90**, 112015 (2014), arXiv:1408.7084 [hep-ex].
23. S. Chatrchyan *et al.* [CMS], Observation of the diphoton decay of the 125 GeV Higgs boson and measurement of its properties, *Eur. Phys. J. C.* **74**, 3076 (2014), arXiv:1407.0558 [hep-ex].
24. G. Aad *et al.* [ATLAS], Measurements of Higgs boson production and couplings in the four-lepton channel in pp collisions at center-of-mass energies of 7 and 8 TeV with the ATLAS detector, *Phys. Rev. D.* **91**, 012006 (2015), arXiv:1408.5191 [hep-ex].
25. S. Chatrchyan *et al.* [CMS], Measurement of the properties of a Higgs boson in the four-lepton final state, *Phys. Rev. D.* **89**, 092007 (2014), arXiv:1312.5353 [hep-ex].
26. G. Aad *et al.* [ATLAS], Observation and measurement of Higgs boson decays to WW^* with the ATLAS detector, *Phys. Rev. D.* **92**, 012006 (2015), arXiv:1412.2641 [hep-ex].

27. S. Chatrchyan *et al.* [CMS], Measurement of Higgs boson production and properties in the WW decay channel with leptonic final states, *J. High Energy Phys.* **01**, 096 (2014), `arXiv:1312.1129 [hep-ex]`.
28. G. Aad *et al.* [ATLAS], Evidence for the Higgs-boson Yukawa coupling to tau leptons with the ATLAS detector, *J. High Energy Phys.* **04**, 117 (2015), `arXiv:1501.04943 [hep-ex]`.
29. S. Chatrchyan *et al.* [CMS], Evidence for the 125 GeV Higgs boson decaying to a pair of τ leptons, *J. High Energy Phys.* **05**, 104 (2014), `arXiv:1401.5041 [hep-ex]`.
30. G. Aad *et al.* [ATLAS], Search for the Standard Model Higgs boson produced by vector-boson fusion in 8 TeV pp collisions and decaying to bottom quarks with the ATLAS detector, Submitted to J. High Energy Phys. (2016), `arXiv:1606.02181 [hep-ex]`.
31. V. Khachatryan *et al.* [CMS], Search for the standard model Higgs boson produced through vector boson fusion and decaying to $b\bar{b}$, *Phys. Rev. D.* **92**, 032008 (2015), `arXiv:1506.01010 [hep-ex]`.
32. G. Aad *et al.* [ATLAS], Study of (W/Z)H production and Higgs boson couplings using $H \to WW^*$ decays with the ATLAS detector, *J. High Energy Phys.* **08**, 137 (2015), `arXiv:1506.06641 [hep-ex]`.
33. G. Aad *et al.* [ATLAS], Search for the Standard Model Higgs boson produced in association with a vector boson and decaying into a tau pair in pp collisions at $\sqrt{s} = 8$ TeV with the ATLAS detector, Submitted to Phys. Rev. D (2015), `arXiv:1511.08352 [hep-ex]`.
34. G. Aad *et al.* [ATLAS], Search for the $b\bar{b}$ decay of the Standard Model Higgs boson in associated $(W/Z)H$ production with the ATLAS detector, *J. High Energy Phys.* **01**, 069 (2015), `arXiv:1409.6212 [hep-ex]`.
35. S. Chatrchyan *et al.* [CMS], Search for the standard model Higgs boson produced in association with a W or a Z boson and decaying to bottom quarks, *Phys. Rev. D.* **89**, 012003 (2014), `arXiv:1310.3687 [hep-ex]`.
36. G. Aad *et al.* [ATLAS], Search for the associated production of the Higgs boson with a top quark pair in multilepton final states with the ATLAS detector, *Phys. Lett. B.* **749**, 519 (2015), `arXiv:1506.05988 [hep-ex]`.
37. V. Khachatryan *et al.* [CMS], Search for the associated production of the Higgs boson with a top-quark pair, *J. High Energy Phys.* **09**, 087 (2014), `arXiv:1408.1682 [hep-ex]`. [Erratum: *ibid.* **10**, 106 (2014)].
38. G. Aad *et al.* [ATLAS], Search for the Standard Model Higgs boson decaying into $b\bar{b}$ produced in association with top quarks decaying hadronically in pp collisions at $\sqrt{s} = 8$ TeV with the ATLAS detector, Submitted to J. High Energy Phys. (2016), `arXiv:1604.03812 [hep-ex]`.
39. G. Aad *et al.* [ATLAS], Search for the Standard Model Higgs boson produced in association with top quarks and decaying into $b\bar{b}$ in pp collisions at $\sqrt{s} = 8$ TeV with the ATLAS detector, *Eur. Phys. J. C.* **75**, 349 (2015), `arXiv:1503.05066 [hep-ex]`.
40. G. Aad *et al.* [ATLAS], Search for the Standard Model Higgs boson decay to $\mu^+\mu^-$ with the ATLAS detector, *Phys. Lett. B.* **738**, 68 (2014), `arXiv:1406.7663 [hep-ex]`.
41. V. Khachatryan *et al.* [CMS], Search for a standard model-like Higgs boson in the $\mu^+\mu^-$ and e^+e^- decay channels at the LHC, *Phys. Lett. B.* **744**, 184 (2015), `arXiv:1410.6679 [hep-ex]`.
42. S. Chatrchyan *et al.* [CMS], Search for a Higgs boson decaying into a Z and a photon in pp collisions at $\sqrt{s} = 7$ and 8 TeV, *Phys. Lett. B.* **726**, 587 (2013), `arXiv:1307.5515 [hep-ex]`.

43. G. Aad *et al.* [ATLAS], Search for Higgs boson decays to a photon and a Z boson in pp collisions at $\sqrt{s} = 7$ and 8 TeV with the ATLAS detector, *Phys. Lett. B.* **732**, 8 (2014), arXiv:1402.3051 [hep-ex].
44. V. Khachatryan *et al.* [CMS], Search for a Higgs boson decaying into $\gamma^*\gamma \to \ell\ell\gamma$ with low dilepton mass in pp collisions at $\sqrt{s} = 8$ TeV, *Phys. Lett. B.* **753**, 341 (2016), arXiv:1507.03031 [hep-ex].
45. G. Aad *et al.* [ATLAS], Search for Higgs and Z Boson Decays to $J/\psi\gamma$ and $\Upsilon(nS)\gamma$ with the ATLAS Detector, *Phys. Rev. Lett.* **114**, 121801 (2015), arXiv:1501.03276 [hep-ex].
46. S. Chatrchyan *et al.* [CMS], Search for invisible decays of Higgs bosons in the vector boson fusion and associated ZH production modes, *Eur. Phys. J. C.* **74**, 2980 (2014), arXiv:1404.1344 [hep-ex].
47. G. Aad *et al.* [ATLAS], Search for invisible decays of a Higgs boson using vector-boson fusion in *pp* collisions at $\sqrt{s} = 8$ TeV with the ATLAS detector, *J. High Energy Phys.* **01**, 172 (2016), arXiv:1508.07869 [hep-ex].
48. V. Khachatryan *et al.* [CMS], Search for Lepton-Flavour-Violating Decays of the Higgs Boson, *Phys. Lett. B.* **749**, 337 (2015), arXiv:1502.07400 [hep-ex].
49. G. Aad *et al.* [ATLAS], Search for lepton-flavour-violating decays of the Higgs and Z bosons with the ATLAS detector, Submitted to Eur. Phys. J. C (2016), arXiv:1604.07730 [hep-ex].
50. LEP Electroweak Working Group, Tevatron Electroweak Working Group, and SLD Electroweak and Heavy Flavour Groups [ALEPH, CDF, D0, DELPHI, L3, OPAL, SLD], Precision electroweak measurements and constraints on the standard model (2010), arXiv:1012.2367 [hep-ex].
51. D. Bardin *et al.*, Electroweak working group report (1997), arXiv:hep-ph/9709229 [hep-ph].
52. M. Baak, J. Cúth, J. Haller, A. Hoecker, R. Kogler, K. Mönig, M. Schott, and J. Stelzer [Gfitter Group], The global electroweak fit at NNLO and prospects for the LHC and ILC, *Eur. Phys. J. C.* **74**, 3046 (2014), arXiv:1407.3792 [hep-ph].
53. V. Khachatryan *et al.* [CMS], Performance of photon reconstruction and identification with the CMS detector in proton-proton collisions at $\sqrt{s} = 8$ TeV, *JINST* **10**, P08010 (2015), arXiv:1502.02702 [physics.ins-det].
54. V. Khachatryan *et al.* [CMS], Performance of electron reconstruction and selection with the CMS detector in proton-proton collisions at $\sqrt{s} = 8$ TeV, *JINST* **10**, P06005 (2015), arXiv:1502.02701 [physics.ins-det].
55. G. Aad *et al.* [ATLAS], Constraints on the off-shell Higgs boson signal strength in the high-mass ZZ and WW final states with the ATLAS detector, *Eur. Phys. J. C.* **75**, 335 (2015), arXiv:1503.01060 [hep-ex].
56. V. Khachatryan *et al.* [CMS], Search for Higgs boson off-shell production in proton-proton collisions at 7 and 8 TeV and derivation of constraints on its total decay width, (2016), Submitted to J. High Energy Phys., arXiv:1605.02329 [hep-ex].
57. V. Khachatryan *et al.* [CMS], Limits on the Higgs boson lifetime and width from its decay to four charged leptons, *Phys. Rev. D.* **92**, 072010 (2015), arXiv:1507.06656 [hep-ex].
58. D. de Florian, G. Ferrera, M. Grazzini, and D. Tommasini, Higgs boson production at the LHC: transverse momentum resummation effects in the $H \to 2\gamma$, $H \to WW \to l\nu l\nu$ and $H \to ZZ \to 4l$ decay modes, *J. High Energy Phys.* **06**, 132 (2012), arXiv:1203.6321 [hep-ph].
59. K. Hamilton, P. Nason, E. Re, and G. Zanderighi, NNLOPS simulation of Higgs boson production, *J. High Energy Phys.* **10**, 222 (2013), arXiv:1309.0017 [hep-ph].

60. L. Landau, On the angular momentum of a two-photon system, *Dokl. Akad. Nauk Ser. Fiz.* **60**, 207 (1948).
61. C.-N. Yang, Selection rules for the dematerialization of a particle into two photons, *Phys. Rev.* **77**, 242 (1950).
62. A. L. Read, Presentation of search results: The CL(s) technique, *J. Phys. G.* **28**, 2693 (2002).
63. J. R. Andersen *et al.* [LHC Higgs Cross Section Working Group], Handbook of LHC Higgs cross sections: 3. Higgs properties, (2013), `arXiv:1307.1347 [hep-ph]`.
64. G. Aad *et al.* [ATLAS], Test of CP Invariance in vector-boson fusion production of the Higgs boson using the optimal observable method in the ditau decay channel with the ATLAS detector, Submitted to Eur. Phys. J. C (2016), `arXiv:1602.04516 [hep-ex]`.
65. V. Khachatryan *et al.* [CMS], Combined search for anomalous pseudoscalar HVV couplings in VH production and H to VV decay, Accepted by Phys. Lett. B (2016), `arXiv:1602.04305 [hep-ex]`.
66. G. Aad *et al.* [ATLAS], Constraints on non-Standard Model Higgs boson interactions in an effective Lagrangian using differential cross sections measured in the $H \to \gamma\gamma$ decay channel at $\sqrt{s} = 8$ TeV with the ATLAS detector, *Phys. Lett. B.* **753**, 69 (2016), `arXiv:1508.02507 [hep-ex]`.
67. A. David, J. Heikkilä, and G. Petrucciani, Searching for degenerate Higgs bosons, *Eur. Phys. J. C.* **75**, 49 (2015), `arXiv:1409.6132 [hep-ph]`.
68. A. David, A. Denner, M. Duehrssen, M. Grazzini, C. Grojean, G. Passarino, M. Schumacher, M. Spira, G. Weiglein, and M. Zanetti [LHC Higgs Cross Section Working Group], LHC HXSWG interim recommendations to explore the coupling structure of a Higgs-like particle, (2012), `arXiv:1209.0040 [hep-ph]`.
69. M. Farina, C. Grojean, F. Maltoni, E. Salvioni, and A. Thamm, Lifting degeneracies in Higgs couplings using single top production in association with a Higgs boson, *J. High Energy Phys.* **05**, 022 (2013), `arXiv:1211.3736 [hep-ph]`.
70. G. Aad *et al.* [ATLAS], Search for $H \to \gamma\gamma$ produced in association with top quarks and constraints on the Yukawa coupling between the top quark and the Higgs boson using data taken at 7 TeV and 8 TeV with the ATLAS detector, *Phys. Lett. B.* **740**, 222 (2015), `arXiv:1409.3122 [hep-ex]`.
71. G. Aad *et al.* [ATLAS], Measurements of the Higgs boson production and decay rates and coupling strengths using pp collision data at $\sqrt{s} = 7$ and 8 TeV in the ATLAS experiment, *Eur. Phys. J. C.* **76**, 6 (2016), `arXiv:1507.04548 [hep-ex]`.
72. V. Khachatryan *et al.* [CMS], Search for the Associated Production of a Higgs Boson with a Single Top Quark in Proton-Proton Collisions at $\sqrt{s} = 8$ TeV, Submitted to J. High Energy Phys. (2015), `arXiv:1509.08159 [hep-ex]`.
73. C. Englert, M. McCullough, and M. Spannowsky, Gluon-initiated associated production boosts Higgs physics, *Phys. Rev. D.* **89**, 013013 (2014), `arXiv:1310.4828 [hep-ph]`.
74. J. Ellis and T. You, Updated global analysis of Higgs couplings, *J. High Energy Phys.* **06**, 103 (2013), `arXiv:1303.3879 [hep-ph]`.

Chapter 17

Flavour Physics and Implication for New Phenomena

Gino Isidori

Physik-Institut, Universität Zürich, CH-8057 Zürich, Switzerland
INFN, Laboratori Nazionali di Frascati, I-00044 Frascati, Italy

Flavour physics represents one of the most interesting and, at the same time, less understood sector of the Standard Theory. On the one hand, the peculiar pattern of quark and lepton masses, and their mixing angles, may be the clue to some new dynamics occurring at high-energy scales. On the other hand, the strong suppression of flavour-changing neutral-current processes, predicted by the Standard Theory and confirmed by experiments, represents a serious challenge to extend the Theory. This article reviews both these aspects of flavour physics from a theoretical perspective.

1. Introduction

The term *flavour* is used, in the jargon of particle physics, to characterise the different copies of fields with the same spin and gauge quantum numbers, and *flavour physics* refers to the study of the interactions that distinguish between these copies. Within the Standard Theory (ST) of fundamental interactions, as we know it now, all matter fields (quark, leptons, and neutrinos) appear in three flavours, and the only interaction that distinguish these three flavours is the Yukawa interaction, or the interaction of the matter fields with the Higgs boson.

The fact that flavour non-universality is generated only by Yukawa interaction is an unavoidable consequence in the Standard Theory, given its particle content. However, this structure was far from being obvious for decades: from the discovery of strange particles in the 1950s till the triumph of the ST predictions for quark-flavour mixing observed at the B-factories in the 2000s. During all these years the progress in understanding flavour physics has been intimately related to the overall progress in building and testing the ST of fundamental interactions.

At present we have a clear understanding of the underlying mechanism of flavour mixing and flavour non-universality within the ST, and this mechanism has been successfully verified in experiments. However, flavour physics still represents one of the most puzzling and, at the same time, interesting aspects of particle physics.

Our "ignorance" in this sector can be summarised by the following two open questions:

- What determines the observed pattern of masses and mixing angles of quarks and leptons?
- Which are the sources of flavour symmetry breaking accessible at low energies? Is there anything else beside the ST Yukawa couplings?

Answering these questions is a key part of the more general program of investigating the nature of physics beyond the ST. There are indeed convincing arguments, including the peculiar pattern of quark and lepton masses, which motivate us to consider the ST as the low-energy limit of a more complete theory.

The precise understanding of the mechanism of flavour mixing within the ST, summarised in Sections 2 and 3, is essential to formulate the above questions in a quantitative way. The present status of the partial answers obtained so far to the second question, and their implications for physics beyond the ST, are presented in Sections 4–6. Some of the theoretical ideas put forwards to address the first question are presented in Section 7.

2. Some Historical Remarks

The first building block of what we now call flavour physics was laid down by Cabibbo in 1963,[1] well before many of the ingredients of the Standard Theory were clear. The Cabibbo theory of semileptonic decays provided the first step toward a unified description of hadronic and leptonic weak interactions. Later on, the hypothesis of the existence of the charm quark, formulated by Glashow, Iliopoulos and Maiani,[2] represented a key ingredient both to understand the mechanism of quark flavour mixing within the ST and, at the same time, to understand how to extend the unified mechanism of weak and electromagnetic interactions from the lepton sector to the quark sector. Finally, the hypothesis formulated by Kobayashi and Maskawa[3] that quarks appear in three flavour turned out to be the correct explanation of the phenomenon of CP violation within the ST.

The theoretical foundations of the mechanism of flavour mixing within the ST were anticipated and followed by a long series of key experimental observations, starting from the discovery of CP violation in the neutral kaon system in 1964,[4] and culminated with the precise determination of all the elements of the Cabibbo–Kobayashi–Maskawa quark-flavour mixing matrix at the B-meson factories,[5] and at various dedicated K-decay experiments.[6] At the completion of the B-factory program, it has become clear that the ST provides a successful description of the mechanism of quark flavour mixing: possible contributions due to New Physics (NP), if any, can only be small corrections compared to the leading ST terms. The search for such tiny deviations is the main goal of present and future experimental efforts in flavour physics.[7,8]

The precise comparison between data and ST in flavour physics has been made possible by a significant amount of theoretical progress in understanding how QCD interactions modify weak interactions at low energies. This started with the pioneering work of Gaillard and Lee,[9] and Altarelli and Maiani,[10] further extended by Shifman, Vainshtein, and Zakharov,[11] and by Gilman and M. B. Wise.[12] A significant step forward was undertaken in the 1990s, where all the relevant flavour-changing processes have been computed at the next-to-leading order (NLO) accuracy, in particular by the Munich[13] and Rome[14] groups (see Ref. 15 for a complete list of NLO references). More recently specific processes, such as $B \to X_s\gamma$ and $B_{s,d} \to \ell^+\ell^-$, have been computed even at NNLO accuracy.[16,17]

3. The Flavour Sector of the Standard Theory

The ST Lagrangian can be divided into two main parts, the gauge and the Higgs (or symmetry breaking) sector. The gauge sector is extremely simple and highly symmetric: it is completely specified by the local symmetry $SU(3)_C \times SU(2)_L \times U(1)_Y$ and by the fermion content. This consists of five fields with different quantum numbers under the gauge group: the $SU(2)_L$ doublet of quarks (Q^i_L), the two right-handed quark singlets (U^i_R and D^i_R), the lepton doublet (Q^i_L), and the right-handed lepton singlet (E^i_R).

Each of these five different fields appears in three different replica or flavours ($i = 1, 2, 3$), giving rise to a large *global* flavour symmetry. Both the local and the global symmetries of the gauge sector of the ST are broken by the Higgs field. The local symmetry is spontaneously broken by the vacuum expectation value of the Higgs field, $\langle|\phi|\rangle = v = (2\sqrt{2}G_F)^{-1/2} \approx 174$ GeV, while the global flavour symmetry is *explicitly broken* by the Yukawa interaction of ϕ with the fermion fields:

$$-\mathcal{L}^{\rm ST}_{\rm Yukawa} = Y_d^{ij}\bar{Q}^i_L\phi D^j_R + Y_u^{ij}\bar{Q}^i_L\tilde{\phi}U^j_R + Y_e^{ij}\bar{L}^i_L\phi E^j_R + \text{h.c.} \quad (\tilde{\phi} = i\tau_2\phi^\dagger)\,. \tag{1}$$

The large global flavour symmetry of $\mathcal{L}^{\rm ST}_{\rm gauge}$, corresponding to the independent unitary rotations in flavour space of the five fermion fields, is a $U(3)^5$ group.[18] This can be decomposed as follows: $\mathcal{G}_{\rm flavour} = U(1)^5 \times \mathcal{G}_q \times \mathcal{G}_\ell$, where

$$\mathcal{G}_q = SU(3)_{Q_L} \times SU(3)_{U_R} \times SU(3)_{D_R}, \quad \mathcal{G}_\ell = SU(3)_{L_L} \otimes SU(3)_{E_R}\,. \tag{2}$$

Three of the five $U(1)$ subgroups can be identified with the total barion and lepton number, which are not broken by the Yukawa interaction, and the weak hypercharge, which is gauged and broken only spontaneously by $\langle\phi\rangle \neq 0$. The subgroups controlling flavour-changing dynamics and flavour non-universality are the non-Abelian groups $\mathcal{G}_q$ and $\mathcal{G}_\ell$, which are explicitly broken by $Y_{d,u,e}$ not being proportional to the identity matrix.

The diagonalisation of each Yukawa matrix requires, in general, two independent unitary matrices, $V_L Y V_R^\dagger = {\rm diag}(y_1, y_2, y_3)$. In the lepton sector we are free to choose the two matrices necessary to diagonalise Y_e without breaking gauge invariance. This is not the case in the quark sector, where we cannot diagonalise on the left both Y_d and Y_u at the same time. We are thus left with a non-trivial misalignment matrix V, between Y_d and Y_u, which is nothing but the Cabibbo–Kobayashi–Maskawa (CKM) mixing matrix:[1,3]

$$V = \begin{pmatrix} V_{ud} & V_{us} & V_{ub} \\ V_{cd} & V_{cs} & V_{cb} \\ V_{td} & V_{ts} & V_{tb} \end{pmatrix}. \tag{3}$$

For practical purposes it is often convenient to work in the mass eigenstate basis of both up- and down-type quarks. This can be achieved rotating independently the up and down components of the quark doublet Q_L, or moving the CKM matrix from the Yukawa sector to the charged weak current in $\mathcal{L}^{\rm ST}_{\rm gauge}$:

$$J^\mu_W|_{\rm quarks} = \bar{u}^i_L \gamma^\mu d^i_L \quad \overset{u,d \text{ mass-basis}}{\longrightarrow} \quad \bar{u}^i_L V_{ij} \gamma^\mu d^j_L \,. \tag{4}$$

However, it must be stressed that V originates from the Yukawa sector (in particular by the miss-alignment of Y_u and Y_d in the $SU(3)_{Q_L}$ subgroup of $\mathcal{G}_q$): in the absence of Yukawa couplings we can always set $V_{ij} = \delta_{ij}$.

To summarise, quark flavour physics within the ST is characterised by a large flavour symmetry, $\mathcal{G}_q$, defined by the gauge sector, whose only breaking sources are the two Yukawa couplings Y_d and Y_u. The CKM matrix arises by the mis-alignment of Y_u and Y_d in flavour space.

3.1. *The CKM matrix*

The residual invariance under the flavour group allows us to eliminate five of the six complex phases in V, that contains only four real physical parameters: three mixing angles and one CP-violating phase. The off-diagonal elements of the CKM matrix show a strongly hierarchical pattern: $|V_{us}|$ and $|V_{cd}|$ are close to 0.22, the elements $|V_{cb}|$ and $|V_{ts}|$ are of order 4×10^{-2} whereas $|V_{ub}|$ and $|V_{td}|$ are of $O(10^{-3})$.

The Wolfenstein parametrisation, namely the expansion of the CKM matrix elements in powers of the small parameter $\lambda \doteq |V_{us}| \approx 0.22$, is a convenient way to exhibit this hierarchy in a more explicit way:[19]

$$V = \begin{pmatrix} 1 - \frac{\lambda^2}{2} & \lambda & A\lambda^3(\varrho - i\eta) \\ -\lambda & 1 - \frac{\lambda^2}{2} & A\lambda^2 \\ A\lambda^3(1 - \varrho - i\eta) & -A\lambda^2 & 1 \end{pmatrix} + \mathcal{O}(\lambda^4). \tag{5}$$

Here A, ϱ, and η are three independent parameters of order 1. Because of the smallness of λ and the fact that for each element the expansion parameter is actually λ^2, this is a rapidly converging expansion.

The unitarity of the CKM matrix implies a series of relations of the type $\sum_{k=1\ldots3} V_{ki}^* V_{kj} = \delta_{ij}$. These relations are a distinctive feature of the ST, where the CKM matrix is the only source of quark flavour mixing. Their experimental verification is therefore a useful tool to set bounds on, or possibly reveal, new sources of flavour symmetry breaking. Among these relations, the one obtained for $i = 1$ and $j = 3$, namely

$$V_{ud}V_{ub}^* + V_{cd}V_{cb}^* + V_{td}V_{tb}^* = 0 \tag{6}$$

or

$$\frac{V_{ud}V_{ub}^*}{V_{cd}V_{cb}^*} + \frac{V_{td}V_{tb}^*}{V_{cd}V_{cb}^*} + 1 = 0 \leftrightarrow [\rho + i\eta] + [(1-\rho) - i\eta] + 1 = 0\ ,$$

is particularly interesting since it involves the sum of three terms all of the same order in λ and is usually represented as a unitarity triangle in the complex plane (see Fig. 1). We stress that Eq. (6) is invariant under any phase transformation of the quark fields. Under such transformations the unitarity triangle is rotated in the complex plane, but its angles and the sides remain unchanged. Both angles and sides of the unitary triangle are indeed observable quantities which can be measured in suitable experiments.

The values of $|V_{us}|$ and $|V_{cb}|$ (or λ and A), are determined with good accuracy from $K \to \pi\ell\nu$ and $B \to X_c\ell\nu$ decays, respectively. Using these inputs, all the other

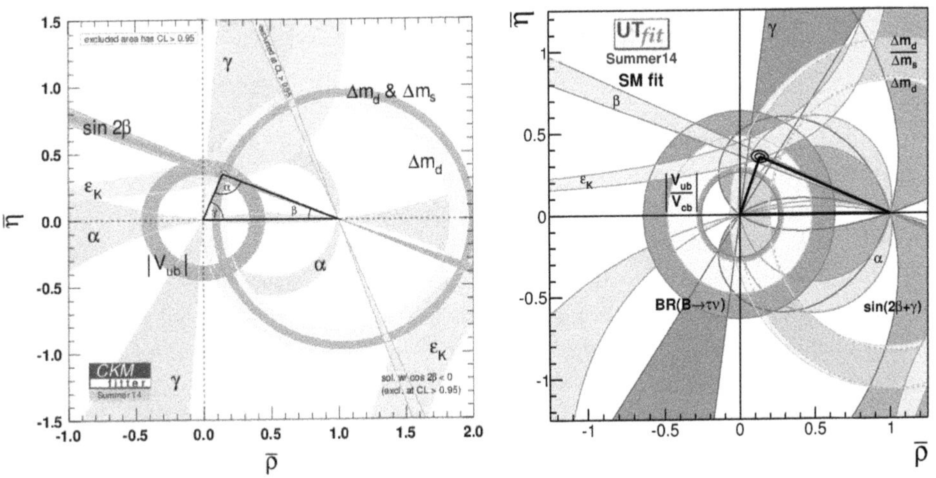

Fig. 1. Allowed region in the $\bar{\varrho}, \bar{\eta}$ plane as obtained by the CKMfitter[20] and UTfit[21] collaborations. Superimposed are the individual constraints from charmless semileptonic B decays ($|V_{ub}|$), mass differences in the B_d (Δm_d) and B_s (Δm_s) systems, CP violation in the neutral kaon (ε_K) and in the B_d systems ($\sin 2\beta$), the combined constrains on α and γ from various B decays.

constraints on the elements of the CKM matrix can be expressed as constraints on ρ and η. The list of the most sensitive observables used to (over) determine the CKM matrix elements include (see Fig. 1):

- The rates of inclusive and exclusive charmless semileptonic B decays, that depend on $|V_{ub}|$.
- The phase of the $B_d - \bar{B}_d$ mixing amplitude (measured from the time-dependent CP asymmetry in $B \to \psi K_S$ decays), that depends on $\sin 2\beta$.
- The rates of various $B \to DK$ decays constraining the angle γ.
- The rates of various $B \to \pi\pi, \rho\pi, \rho\rho$ decays constraining the combination $\alpha = \pi - \beta - \gamma$.
- The ratio between the mass splittings in the neutral B and B_s systems, that depends on $|V_{td}/V_{ts}|$.
- The indirect CP violating parameter of the kaon system (ϵ_K), that determines a hyperbola in the ρ–η plane.

The resulting constraints, as implemented by the CKMfitter and UTfit collaborations, are shown in Fig. 1. As can be seen, they are all consistent with a unique value of $\bar{\varrho} = \rho(1 - \frac{\lambda^2}{2})$ and $\bar{\eta} = \eta(1 - \frac{\lambda^2}{2})$.

The consistency of different constraints on the CKM unitarity triangle is a powerful consistency test of the ST in describing flavour-changing phenomena. From the plot in Fig. 1 it is quite clear, at least in a qualitative way, that there is little room for non-ST contributions in flavour changing transitions. A more quantitative evaluation of this statement is presented in the next section.

4. The Flavour Problem

As anticipated in the introduction, despite the impressive phenomenological success of the ST, there are various convincing arguments which motivate us to consider this model only as the low-energy limit of a more complete theory.

Assuming that the new degrees of freedom which complete the theory are heavier than the ST particles, we can integrate them out and describe physics beyond the ST in full generality by means of an effective field theory (EFT) approach. The ST Lagrangian becomes the renormalisable part of a more general local Lagrangian which includes an infinite tower of operators with dimension $d > 4$, constructed in terms of the ST fields and suppressed by inverse powers of an effective scale Λ. These operators are the residual effect of having integrated out the new heavy degrees of freedom, whose mass scale is parametrised by the effective scale $\Lambda > m_W$.

Integrating out heavy degrees of freedom is a procedure often adopted also within the ST: it allows us to simplify the evaluation of amplitudes which involve different energy scales. This approach is indeed a generalisation of the Fermi theory of weak interactions, where the dimension-six four-fermion operators describing weak decays are the results of having integrated out the W field. The only difference when applying this procedure to physics beyond the ST is that in this case, as also in the original work by Fermi, we don't know the nature of the degrees of freedom we are integrating out. This implies we are not able to determine a priori the values of the effective couplings of the higher-dimensional operators. The advantage of this approach is that it allows us to analyse all realistic extensions of the ST in terms of a limited number of free parameters.

The Lagrangian of the ST considered as an effective theory can be written as follows

$$\mathcal{L}_{\text{eff}} = \mathcal{L}^{\text{ST}}_{\text{gauge}} + \mathcal{L}^{\text{ST}}_{\text{Higgs}} + \mathcal{L}^{\text{ST}}_{\text{Yukawa}} + \Delta\mathcal{L}_{d>4} \ , \tag{7}$$

where $\Delta\mathcal{L}_{d>4}$ denotes the series of higher-dimensional operators invariant under the ST gauge group. The coefficients of these operators have the form $c_i/\Lambda^{(d_i-4)}$, where c_i is an adimensional coefficient and d_i denotes the canonical dimension of the effective operator. If the new dynamics appears at the TeV scale, as we expect from a natural stabilisation of the mechanism of electroweak symmetry breaking, the scale Λ cannot exceed a few TeV. Moreover, from naturalness arguments,[22] we should also expect that all the adimensional coefficients c_i are of $O(1)$ unless suppressed by some symmetry argument. The observation that this expectation is *not* fulfilled by several dimension-six operators contributing to flavour-changing processes is often denoted as the *flavour problem.*

If the ST Lagrangian were invariant under some flavour symmetry, this problem could easily be circumvented. For instance in the case of baryon- or lepton-number violating processes, which are exact symmetries of the ST Lagrangian, we can avoid the tight experimental bounds promoting B and L to be exact symmetries of the new dynamics at the TeV scale. The peculiar aspects of flavour physics is that there is no exact flavour symmetry in the low-energy theory. In this case it is not sufficient to invoke a flavour symmetry for the underlying dynamics. We also need to specify how this symmetry is broken in order to describe the observed low-energy spectrum and, at the same time, be in agreement with the precise experimental tests of flavour-changing processes.

The best way to quantify the flavour problem is obtained by looking at consistency of the tree- and loop-mediated constraints on the CKM matrix. In first approximation we can assume that New Physics (NP) effects are negligible in processes which are dominated by tree-level amplitudes. Following this assumption, the values of $|V_{us}|$, $|V_{cb}|$, and $|V_{ub}|$, as well as the constraints on α and γ can be considered as NP free. As can be seen in Fig. 1, this implies we can determine completely the CKM matrix assuming generic NP effects in loop-mediated amplitudes. We can then use the measurements of observables which are loop-mediated within the ST to bound the couplings of effective higher-dimensional operators which contribute to these observables at the tree level.

The loop-mediated constraints shown in Fig. 1 are those from the mixing of B_d, B_s, and K^0 with the corresponding anti-particles (generically denoted as $\Delta F = 2$

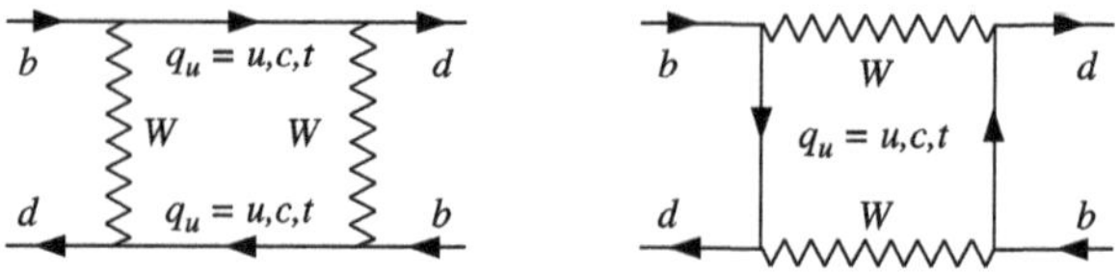

Fig. 2. Box diagrams contributing to B_d–$\bar{B}_d$ mixing in the unitary gauge.

amplitudes). Within the ST, these processes are generated by box amplitudes of the type in Fig. 2 (and similarly for B_s, and K^0) and are affected by small hadronic uncertainties. The leading contribution is obtained with the top-quark running inside the loop, giving rise to the highly suppressed result

$$\mathcal{M}^{\rm ST}_{\Delta F=2} \approx \frac{G_F^2 m_t^2}{16\pi^2} V^*_{3i} V_{3j} \langle \bar{M} | (\bar{d}^i_L \gamma^\mu d^j_L)^2 | M \rangle \times F\left(\frac{m_t^2}{m_W^2}\right) \quad [M = K^0, B_d, B_s], \tag{8}$$

where F is a loop function of $O(1)$ and i, j denote the flavour indexes of the meson valence quarks.

Magnitude and phase of all these mixing amplitudes have been determined with good accuracy from experiments and are consistent with the ST predictions. To translate this information into bounds on the scale of new physics, let's consider the following set of $\Delta F = 2$ dimension-six operators in $\Delta\mathcal{L}_{d>4}$:

$$\Delta\mathcal{L}_{d>4} \supset \sum \frac{c_{ij}}{\Lambda^2} \mathcal{O}^{ij}_{\Delta F=2} \quad \mathcal{O}^{ij}_{\Delta F=2} = (\bar{q}^i_L \gamma^\mu q^j_L)^2 \,. \tag{9}$$

These operators contribute at the tree-level to the meson–antimeson mixing amplitudes. The condition $|\mathcal{M}^{\rm NP}_{\Delta F=2}| < |\mathcal{M}^{\rm ST}_{\Delta F=2}|$ implies

$$\Lambda < \frac{3.4\,\text{TeV}}{|V^*_{3i} V_{3j}|/|c_{ij}|^{1/2}} < \begin{cases} 9 \times 10^3 \text{ TeV} \times |c_{21}|^{1/2} & \text{from} \quad K^0 - \bar{K}^0 \\ 4 \times 10^2 \text{ TeV} \times |c_{31}|^{1/2} & \text{from} \quad B_d - \bar{B}_d \\ 7 \times 10^1 \text{ TeV} \times |c_{32}|^{1/2} & \text{from} \quad B_s - \bar{B}_s \end{cases} . \tag{10}$$

A more refined analysis, with complete statistical treatment and separate bounds for the real and the imaginary parts of the various amplitudes, considering also operators with different Dirac structure, leads to the bounds reported in Table 1. The main message of these bounds is the following:

- New physics models with a generic flavour structure (c_{ij} of order 1) at the TeV scale are ruled out. If we want to keep Λ in the TeV range, physics beyond the ST must have a highly non-generic flavour structure.

In the specific case of the $\Delta F = 2$ operators in (9), in order to keep Λ in the TeV range, we must find a symmetry argument such that $|c_{ij}| \lesssim |V^*_{3i} V_{3j}|^2$. Reproducing a similar structure beyond the ST is a highly non-trivial task. However, as discussed below, it can be obtained under specific assumptions.

Table 1. Bounds on representative dimension-six $\Delta F = 2$ operators with effective coupling $c_{\rm NP}/\Lambda^2$. The bounds are quoted on Λ, setting $|c_{\rm NP}| = 1$, or on $c_{\rm NP}$, setting $\Lambda = 1$ TeV. The right column denotes the main observables used to derive these bounds.[23]

	Λ in TeV ($c_{\rm NP} = 1$)		Bounds on $c_{\rm NP}$ ($\Lambda = 1$ TeV)		
Operator	Re	Im	Re	Im	Observables
$(\bar{s}_L\gamma^\mu d_L)^2$	9.8×10^2	1.6×10^4	9.0×10^{-7}	3.4×10^{-9}	Δm_K; ϵ_K
$(\bar{s}_R\, d_L)(\bar{s}_L d_R)$	1.8×10^4	3.2×10^5	6.9×10^{-9}	2.6×10^{-11}	Δm_K; ϵ_K
$(\bar{c}_L\gamma^\mu u_L)^2$	1.2×10^3	2.9×10^3	5.6×10^{-7}	1.0×10^{-7}	Δm_D; $\lvert q/p\rvert, \phi_D$
$(\bar{c}_R\, u_L)(\bar{c}_L u_R)$	6.2×10^3	1.5×10^4	5.7×10^{-8}	1.1×10^{-8}	Δm_D; $\lvert q/p\rvert, \phi_D$
$(\bar{b}_L\gamma^\mu d_L)^2$	6.6×10^2	9.3×10^2	2.3×10^{-6}	1.1×10^{-6}	Δm_{B_d}; $S_{\psi K_S}$
$(\bar{b}_R\, d_L)(\bar{b}_L d_R)$	2.5×10^3	3.6×10^3	3.9×10^{-7}	1.9×10^{-7}	Δm_{B_d}; $S_{\psi K_S}$
$(\bar{b}_L\gamma^\mu s_L)^2$	1.4×10^2	2.5×10^2	5.0×10^{-5}	1.7×10^{-5}	Δm_{B_s}; $S_{\psi\phi}$
$(\bar{b}_R\, s_L)(\bar{b}_L s_R)$	4.8×10^2	8.3×10^2	8.8×10^{-6}	2.9×10^{-6}	Δm_{B_s}; $S_{\psi\phi}$

5. The Minimal Flavour Violation Hypothesis

The "protection" of $\Delta F = 2$ observables and, more generally, flavour-changing neutral-current (FCNC) processes occurring within the ST is a consequences of the specific symmetry and symmetry-breaking structure of the ST Lagrangian discussed in Section 3. In particular, the fact that the quark flavour group $\mathcal{G}_q$ is broken only by the two quark Yukawa couplings, and that the top-quark Yukawa coupling is the only $O(1)$ entry in $Y_{u,d}$, is the main reason why Eq. (8) is highly suppressed.

The strongest assumption we can make to suppress flavour-changing effects in generic extensions of the ST is the so-called Minimal Flavour Violation (MFV) hypothesis, namely the assumption that Y_u and Y_d are the only sources of flavour symmetry breaking also beyond the ST.[18,24,25] To implement and interpret this hypothesis in a consistent way, we can assume that $\mathcal{G}_q$ is a good symmetry and promote $Y_{u,d}$ to be non-dynamical fields (spurions) with non-trivial transformation properties under $\mathcal{G}_q$:

$$Y_u \sim (3, \bar{3}, 1), \quad Y_d \sim (3, 1, \bar{3}). \tag{11}$$

Employing the EFT language, an effective theory satisfies the MFV criterion in the quark sector if all higher-dimensional operators, constructed from ST fields and the $Y_{u,d}$ spurions, are formally invariant under the flavour group $\mathcal{G}_q$.[25] The dynamical idea behind this construction is the hypothesis that the breaking of the symmetry occurs at very high energy scales, and that $Y_{u,d}$ are the only independent combination of breaking terms (e.g. combination of appropriate vacuum expectation values) that survive at low energies.

According to the MFV criterion one should in principle consider operators with arbitrary powers of the (dimensionless) Yukawa fields. However, a strong simplification arises by the observation that all the eigenvalues of the Yukawa matrices are small, but for the top-quark one, and that the off-diagonal elements of the CKM matrix are very suppressed. This fact is enough to ensure that, even when including high powers of Y_u and Y_d, FCNC amplitudes get exactly the same CKM suppression as in the ST:

$$\mathcal{M}^{\rm MFV}_{\Delta F=1}(d^i \to d^j) \propto (V^*_{ti} V_{tj}), \quad \mathcal{M}^{\rm MFV}_{\Delta F=2}(d^i \bar{d}^j \to d^j \bar{d}^i) \propto (V^*_{ti} V_{tj})^2. \tag{12}$$

The proportionally constants in these relations are flavour universal, implying the same NP correction (relative to the ST) in $s \to d$, $b \to d$, and $b \to s$ transitions.

As a consequence of this structure, within the MFV framework several of the constraints used to determine the CKM matrix (and in particularly the unitarity triangle in Fig. 1) are not affected by NP.[26] For instance, the structure of the basic flavour-changing coupling in Eq. (12) implies that the weak CPV phase of B_d–$\bar{B}_d$ mixing is $\arg[(V_{td}V^*_{tb})^2]$, exactly as in the ST. This construction thus provides a natural (a posteriori) justification of why no NP effects have been observed in the quark sector. Moreover, the built-in CKM suppression leads to bounds on the effective scale of new physics in the few TeV domain . These bounds are very similar to the bounds on flavour-conserving operators derived by precision electroweak tests.

A few additional comments about the MFV hypothesis are listed below:

- Although MFV seems to be a natural solution to the flavour problem, we are still far from having proved the validity of this hypothesis from data. A proof of the MFV hypothesis can be achieved only with a positive evidence of NP exhibiting the flavour-universality pattern predicted by MFV (same relative correction in $s \to d$, $b \to d$, and $b \to s$ transitions of the same type). This could happen, for instance, via precise measurements of the rare decays $B_s \to \mu^+\mu^-$ and $B_d \to \mu^+\mu^-$.[27–29] Conversely, an evidence of NP in flavour-changing transitions not respecting the MFV pattern (e.g. an evidence of $\mathcal{B}(B_d \to \mu^+\mu^-)$ well above its ST prediction) would not only imply the existence of physics beyond the ST, but also the existence of new sources of flavour symmetry breaking beyond the Yukawa couplings.
- The MFV ansatz is quite successful on the phenomenological side; however, it is unlikely to be an exact property of the model valid to all energy scales. Despite some recent attempts to provide a dynamical justification of this symmetry-breaking ansatz, the most natural possibility is that MFV is only an accidental low-energy property of the theory. It could also well be that a less minimal connection between NP flavour-violating couplings and Yukawa couplings is at work. It is then very important to search for possible deviations (even if tiny) from the MFV predictions.
- Even if the MFV ansatz holds, it does not necessarily imply small deviations from the ST predictions in all flavour-changing phenomena. The MFV ansatz can be implemented in different ways. For instance, in models with two Higgs doublets we can change the relative normalisation of the two Yukawa couplings.[25] It is also possible to decouple the breaking of CP invariance from the breaking of the $\mathcal{G}_q$

quark-flavour group,[30] leaving more room for NP in CP-violating observables. All these variations lead to different and well defined patterns of possible deviations from the ST that we have only started to investigate and that represent one of the main goal of present and future experiments in flavour physics.[7,8]

- The usefulness of the MFV ansatz is closely linked to the theoretical expectation of NP in the TeV range. This expectation follows from a *natural* stabilisation of the Higgs sector, but it is in tension with the lack of any direct signal of NP at the LHC. The more the scale of NP is pushed up, the more it is possible to allow sizable deviations from the MFV ansatz.

6. Flavour Symmetry Breaking Beyond MFV

As anticipated, MFV is not the only option to "protect" flavour-changing transitions in extensions of the ST. A key feature common to most models able to accommodate NP not far from the TeV scale, ensuring a sufficient suppression of flavour-changing transitions, is some link between flavour-changing amplitudes and fermion masses. Indeed the strong phenomenological bounds on flavour-changing transitions always involve light quarks (or leptons) of the first two generations, and are particularly strong in the case of transitions among the first two families (see Table 1). Given the smallness of fermion masses of the first two generations, a link between flavour-changing amplitudes and fermion masses provides a good starting point for a natural suppression of flavour-changing transitions.

In the quark sector this link can be efficiently implemented considering only the $U(2)^3$ subgroup of the full quark flavour group ($\mathcal{G}_q$) that is obtained in the limit of vanishing Yukawa couplings for the first two generations of quarks.[30,31] This symmetry limit is a better approximation of the full ST Lagrangian, since top and bottom quarks are allowed to have a non-vanishing mass. The $U(2)^3$ subgroup is also sufficient to ensure enough protection from flavour-changing transitions beyond the ST, assuming the minimal breaking structure necessary to describe light fermion masses. The main difference of this ansatz compared to the MFV hypothesis is the breaking of the universal link between $s \to d$ transitions vs. transitions involving third generation quarks ($b \to d$ and $b \to s$).

So far we discussed mainly the quark sector, but a flavour problem exists also in the lepton sector. Similarly to the $\Delta F = 2$ bounds in Table 1, the strong experimental bounds on FCNC transitions of charged leptons ($\mu \to e\gamma$, $\mu \to 3e$, $\mu N \to eN$, $\tau \to \mu\gamma$, ...) can be translated into bounds on NP scales well above the TeV, for $O(1)$ flavour-changing coefficients. For instance the MEG[32] bound $\mathcal{B}(\mu \to e\gamma) < 5.7 \times 10^{-13}$ leads to an effective bound on Λ of the order of 10^5 TeV.

In order to allow TeV scale NP, some extension of the MFV hypothesis can be implemented also is the lepton sector. However, given there is not a unique way to accommodate non-vanishing neutrino masses, in this case there is more freedom to define the minimal sources of flavour symmetry breaking. Different versions of

Minimal Lepton Flavour Violation (MLFV) have been proposed in the literature, depending on how the irreducible breaking terms in the neutrino sector are identifed.[33–37] On general grounds, it is not difficult to provide a sufficient suppression of flavour-changing coefficients for TeV scale new physics, provided the (adimensional) flavour breaking terms associated to neutrino masses are sufficiently small. In the context of see-saw models, this imply masses for the heavy right-handed neutrinos typically around or below 10^{12} GeV.[33] As for the quark sector, the key tool to test these symmetry (and symmetry-breaking) assumptions relies on the observation of possible correlations in the rate of neutral-current LFV processes, such as $\tau \to \mu\gamma$ vs. $\mu \to e\gamma$.

7. Flavor Physics and Partial Compositeness

In the previous two sections we have discussed mechanisms to suppress flavour-changing transitions beyond the ST due to specific flavour symmetries and symmetry-breaking patterns. An interesting alternative is the possibility of a generic *dynamical suppression* of flavour-changing interactions, related to the weak mixing of the light ST fermions with some new dynamics occurring at the TeV scale. This is what happens in the so-called models with partial compositeness,[38,39] where the hierarchy of fermion masses is attributed to the hierarchical mixing of the ST fermions with the heavier (composite) states of the theory.

Also the general features of this class of models can be described by means of an effective theory approach.[40,41] The two main assumptions of this EFT approach are the following:

- There exists a (non-canonical) basis for the ST fermions where their kinetic terms exhibit a rather hierarchical form:

$$\mathcal{L}_{\rm kin}^{\rm quarks} = \sum_{\Psi=Q_L,U_R,D_R} \overline{\Psi} Z_\psi^{-2} \not{D} \Psi,$$

$$Z_\psi = {\rm diag}(z_\psi^{(1)}, z_\psi^{(2)}, z_\psi^{(3)}), \quad z_\psi^{(1)} \ll z_\psi^{(2)} \ll z_\psi^{(3)} \lesssim 1. \tag{13}$$

- In such basis there is no flavour symmetry and all the flavour-violating interactions, including the Yukawa couplings, are $\mathcal{O}(1)$.

Once the fields are transformed into the canonical basis, the hierarchical kinetic terms act as a distorting lens, through which all interactions are seen as approximately aligned on the magnification axes of the lens. The hierarchical $z_\psi^{(i)}$ can be interpreted as the effect of the mixing of an elementary (ST-like) sector of massless fermions with a corresponding set of heavy composite fermions: the elementary fermions feel the breaking of the electroweak (and flavour) symmetry only via this mixing.

The values of the $z_\psi^{(i)}$ can be deduced, up to an overall normalisation, from the know structure of the Yukawa couplings, that can be decomposed as follows

$$Y_u^{ij} \sim z_Q^{(i)} z_U^{(j)}, \quad Y_d^{ij} \sim z_Q^{(i)} z_D^{(j)}. \tag{14}$$

Inverting such relations we can express the $z_\psi^{(i)}$ combinations appearing in the effective couplings of dimension-six operators involving ST fields [e.g. the combination $(z_Q^{(1)} z_Q^{(2)})^2$ for the operator $(\bar{s}_L \gamma_\mu d_L)^2$, etc.] into appropriate powers of quark masses and CKM angles. The resulting suppression of FCNC amplitudes turns out to be quite effective being linked to the hierarchical structure of the Yukawa couplings.

As shown in a recent analysis,[41] this framework is compatible with the strong flavour bounds in kaon sector for scales of the composite states (vector resonances) around 10 TeV. In this case one can expect deviations from the ST at the present level of experimental sensitivity in the electric dipole moment (EDM) of the neutron (where there is actually a significant tension with the present bound), CP-violating observables in the kaon system (ϵ'/ϵ and ϵ_K), and $b \to s$ FCNC transitions. However, in the lepton sector the minimal framework is not satisfactory (a severe fine-tuning is needed to satisfy current bounds on lepton-flavour violating processes).

It should be stressed that also in partial-compositeness models is possible to postulate the existence of additional protective flavour symmetries (as discussed e.g. in Refs. 43–45) and, for instance, recover a MFV structure. In this case the bounds on the composite states turn out to be well below 10 TeV.

8. Dynamical Yukawa Couplings

The MFV principle does not provide an explanation for the observed pattern of masses and mixings of quarks and leptons: the Yukawa couplings are simply treated as inputs, as in the ST. To a large extent, also the mechanism of partial compositeness does not explain the observed pattern of quark and lepton Yukawa couplings: the hierarchal mixing between elementary and composite fermions is an input of the construction.

A more ambitious goal is that of deriving the observed structure of the Yukawa couplings from some fundamental principle. The simplest realisation of the idea of a dynamical character for the Yukawa couplings is to assume that

$$Y = \frac{\langle 0|\Phi|0\rangle}{\Lambda} \tag{15}$$

with Λ being some high energy scale and Φ a set of scalar fields (or composite operators) with transformation properties such as to make invariant the effective potential $V(Y)$ under the flavour group $\mathcal{G}_{\text{flavour}}$ (or some of its subgroups). A general problem that one encounters along this line is the unwanted appearance of a large number of Goldstone bosons, associated to the spontaneous breaking of the large global continuous flavour symmetry. This problem could be avoided assuming that the flavour symmetry is gauged at some high energy scale.[46]

An interesting alternative to continuous flavour symmetries, that naturally avoids the problem of Goldstone bosons, is the possibility that the fundamental flavour symmetry is a suitable discrete subgroup of $\mathcal{G}_{\rm flavour}$. This option has received a lot of attention in the recent past, mainly because of neutrino physics:[47] the neutrino mixing matrix exhibits an almost tri-bimaximal structure and the latter is naturally expected in the context of discrete flavour symmetries. However, the description of both quark and lepton sectors in terms of a unique discrete flavour symmetry is less trivial and significantly more complicated.[47–49] Moreover, this option has become less appealing also in the pure neutrino sector after the observation of a sizable 1–3 neutrino mixing angle,[50,51] that implies sizable deviations from the tri-bimaximal mixing structure.

The idea that quark masses and, more generally, the Yukawa couplings, could arise from the minimisation of a potential invariant under some continuous flavour symmetry is an old idea. Earlier attempts dates back to the sixties, when Michel and Radicati,[52] and Cabibbo and Maiani[53] developed generic group-theoretical methods to identify the *natural extrema* of $SU(3)_L \times SU(3)_R$ invariant potentials. Several further attempts towards a dynamical origin of the Yukawa couplings, employing various subgroups of $\mathcal{G}_{\rm flavour}$ have been discussed in the literature.[54–64] In models based on small symmetry groups, such as the $U(1)$ horizontal symmetry originally proposed by Froggat and Nielsen,[54] it is quite easy to reproduce the observed mass matrices in terms of a reduced number of free parameters, while it is difficult to avoid problems with FCNCs, unless some amount of fine-tuning is introduced.

In models based on large (MFV-like) symmetry groups, it is difficult to explain the full pattern of quark and lepton masses in absence of significant fine-tuning among the coefficients of the potential.[61] In this context, an interesting recent development has been presented in Ref. 65. There it has been shown that, among the most stable solution of the general minimisation problem of $V(Y)$, corresponding to maximally unbroken subgroups[52,53] of $\mathcal{G}_{\rm flavour}$, there exists a class of solutions quite close to a realistic spectrum. In the quark sector, this corresponds to a hierarchical mass pattern of the third vs. the first two generations, with unity CKM matrix. In the lepton sector, it implies hierarchical masses for charged leptons and degenerate Majorana neutrinos, with one maximal, one large, and one vanishing mixing angle. Both these textures are close to the real situation, and can be brought in full agreement with data adding small perturbations. In the neutrino sector, this implies a firm prediction that can be tested in the near future, namely an almost degenerate spectrum with an average neutrino mass close to $m_\nu \approx 0.1$ eV.

The radical alternative to predictions of quark and lepton masses based on continuous or discrete symmetries is the idea that they are simply random variables, possibly selected by anthropic arguments. The latter option has recently gained consensus, given the lack of deviations from the ST after the first run of the LHC.[66] Drawing any firm conclusion in this respect is very difficult, and it will remain so also in the future. However, it is worth to stress that the measurement of the absolute value of neutrino masses could provide a significant additional piece of this fascinating puzzle: a value close to the present bounds, compatible with the hypothesis of a degenerate spectrum, would certainly speak in favour of some underlying large and mildly broken flavour symmetry.[65,67]

9. Conclusions

Flavour physics has a twofold role in investigating the nature of physics beyond the ST. On the one hand, for NP models with new particles close to the TeV scale, existing low-energy flavour-physics measurements put very stringent limits on the flavour structure of the model. As illustrated in general terms and with a few specific examples, for such models present data tell us that the new degrees of freedom must have a highly non-generic flavour structure. However, this structure has not been clearly identified yet. In this perspective, if direct signals of NP will appear during the next LHC run, future progress in flavour physics will be an essential tool to investigate the peculiar flavour structure of the new degrees of freedom.

On the other hand, the paradigm of NP at the TeV scale is seriously challenged by the absence of deviations from the SM at the high-energy frontier. In this perspective, flavour physics remains a very powerful tool to search for physics beyond the ST, being potentially sensitive to NP scales much higher than those directly accessible at present and near-future high-energy facilities.

References

1. N. Cabibbo, *Phys. Rev. Lett.* **10**, 531 (1963).
2. S. L. Glashow, J. Iliopoulos, L. Maiani, *Phys. Rev. D* **2**, 1285 (1970).
3. M. Kobayashi, T. Maskawa, *Prog. Theor. Phys.* **49**, 652 (1973).
4. J. H. Christenson, J. W. Cronin, V. L. Fitch, R. Turlay, *Phys. Rev. Lett.* **13**, 138 (1964).
5. A. J. Bevan *et al.* [BaBar & Belle Collab.], *Eur. Phys. J. C* **74**, 11, 3026 (2014).
6. M. Antonelli *et al.* [FlaviaNet Kaon Working Group], *Eur. Phys. J. C* **69**, 399 (2010).
7. R. Aaij *et al.* [LHCb Collaboration], *Eur. Phys. J. C* **73**, 4, 2373 (2013).
8. T. Abe *et al.* [Belle-II Collaboration], arXiv:1011.0352.
9. M. K. Gaillard, B. W. Lee, *Phys. Rev. D* **10**, 897 (1974); *Phys. Rev. Lett.* **33**, 108 (1974).
10. G. Altarelli, L. Maiani, *Phys. Lett. B* **52**, 351 (1974).
11. M. A. Shifman, A. I. Vainshtein, V. I. Zakharov, *Nucl. Phys. B* **120**, 316 (1977).
12. F. J. Gilman, M. B. Wise, *Phys. Rev. D* **20**, 2392 (1979).
13. G. Buchalla, A. J. Buras, *Nucl. Phys. B* **400**, 225 (1993).
14. M. Ciuchini, E. Franco, G. Martinelli, L. Reina, *Nucl. Phys. B* **415**, 403 (1994).
15. G. Buchalla, A. J. Buras, M. E. Lautenbacher, *Rev. Mod. Phys.* **68**, 1125 (1996).
16. M. Misiak *et al.*, *Phys. Rev. Lett.* **98**, 022002 (2007); *Phys. Rev. Lett.* **114**, 221801 (2015).
17. C. Bobeth *et al.*, *Phys. Rev. Lett.* **112**, 101801 (2014).
18. R. S. Chivukula, H. Georgi, *Phys. Lett. B* **188**, 99 (1987).
19. L. Wolfenstein, *Phys. Rev. Lett.* **51**, 1945 (1983).

20. J. Charles *et al.* [CKMfitter Collaboration], *Eur. Phys. J. C* **41**, 1 (2005), online update at http://www.slac.stanford.edu/xorg/ckmfitter/
21. M. Bona *et al.* [UTfit Collaboration], *JHEP* **0610**, 081 (2006), online update at http://www.utfit.org/
22. G. 't Hooft, *NATO Sci. Ser. B* **59**, 135 (1980).
23. G. Isidori, Y. Nir, G. Perez, *Ann. Rev. Nucl. Part. Sci.* **60**, 355 (2010).
24. L. J. Hall, L. Randall, *Phys. Rev. Lett.* **65**, 2939 (1990).
25. G. D'Ambrosio, G. F. Giudice, G. Isidori, A. Strumia, *Nucl. Phys. B* **645**, 155 (2002).
26. A. J. Buras *et al.*, *Phys. Lett. B* **500**, 161 (2001).
27. R. Aaij *et al.* [LHCb Collaboration], *Phys. Rev. Lett.* **111**, 101805 (2013).
28. S. Chatrchyan *et al.* [CMS Collaboration], *Phys. Rev. Lett.* **111**, 101804 (2013).
29. The CMS and LHCb Collaborations, CMS-PAS-BPH-13-007, LHCb-CONF-2013-012.
30. A. L. Kagan, G. Perez, T. Volansky and J. Zupan, *Phys. Rev. D* **80**, 076002 (2009).
31. R. Barbieri, G. Isidori, J. Jones-Perez, P. Lodone and D. M. Straub, *Eur. Phys. J. C* **71**, 1725 (2011).
32. J. Adam *et al.* [MEG Collaboration], *Phys. Rev. Lett.* **110**, 201801 (2013).
33. V. Cirigliano, B. Grinstein, G. Isidori and M. B. Wise, *Nucl. Phys. B* **728**, 121 (2005).
34. B. Grinstein, V. Cirigliano, G. Isidori and M. B. Wise, *Nucl. Phys. B* **763**, 35 (2007).
35. S. Davidson and F. Palorini, *Phys. Lett. B* **642**, 72 (2006).
36. M. B. Gavela, T. Hambye, D. Hernandez and P. Hernandez, *JHEP* **0909**, 038 (2009).
37. R. Alonso, G. Isidori, L. Merlo, L. A. Munoz and E. Nardi, *JHEP* **1106**, 037 (2011).
38. D. B. Kaplan, *Nucl. Phys. B* **365**, 259 (1991).
39. K. Agashe, G. Perez and A. Soni, *Phys. Rev. D* **71**, 016002 (2005).
40. S. Davidson, G. Isidori and S. Uhlig, *Phys. Lett. B* **663**, 73 (2008).
41. B. Keren-Zur, P. Lodone, M. Nardecchia, D. Pappadopulo, R. Rattazzi and L. Vecchi, *Nucl. Phys. B* **867**, 429 (2013).
42. R. Barbieri, G. Isidori and D. Pappadopulo, *JHEP* **0902**, 029 (2009).
43. M. Redi and A. Weiler, *JHEP* **1111**, 108 (2011).
44. R. Barbieri, D. Buttazzo, F. Sala and D. M. Straub, *JHEP* **1210**, 040 (2012).
45. R. Barbieri, D. Buttazzo, F. Sala, D. M. Straub and A. Tesi, *JHEP* **1305**, 069 (2013).
46. B. Grinstein, M. Redi and G. Villadoro, *JHEP* **1011**, 067 (2010).
47. G. Altarelli and F. Feruglio, *Rev. Mod. Phys.* **82**, 2701 (2010).
48. F. Feruglio, *Eur. Phys. J. C.* **75**, 8, 373 (2015).
49. R. Zwicky and T. Fischbacher, *Phys. Rev. D* **80**, 076009 (2009).
50. F. P. An *et al.* [Daya Bay Collaboration], *Phys. Rev. Lett.* **108**, 171803 (2012).
51. J. K. Ahn *et al.* [RENO Collaboration], *Phys. Rev. Lett.* **108**, 191802 (2012).
52. L. Michel and L. A. Radicati, in *Proc. of the Fifth Coral Gables Conference on Symmetry principles at High Energy*, ed. by B. Kursunoglu *et al.*, (W. H. Benjamin, Inc. New York, 1965); *Annals Phys.* **66**, (1971) 758.
53. N. Cabibbo and L. Maiani, in *Evolution of Particle Physics* (Academic Press, 1970), 50, App. I.
54. C. D. Froggatt, H. B. Nielsen, *Nucl. Phys. B* **147**, 277 (1979).
55. L. E. Ibanez and G. G. Ross, *Phys. Lett. B* **332**, 100 (1994).
56. A. Anselm and Z. Berezhiani, *Nucl. Phys. B* **484**, 97 (1997).
57. R. Barbieri, L. J. Hall, G. L. Kane and G. G. Ross, hep-ph/9901228.
58. Z. Berezhiani and A. Rossi, *Nucl. Phys. Proc. Suppl.* **101**, 410 (2001).
59. P. F. Harrison and W. G. Scott, *Phys. Lett. B* **628**, 93 (2005).
60. T. Feldmann, M. Jung and T. Mannel, *Phys. Rev. D* **80**, 033003 (2009).
61. R. Alonso, M. B. Gavela, L. Merlo and S. Rigolin, *JHEP* **1107**, 012 (2011).
62. E. Nardi, *Phys. Rev. D* **84**, 036008 (2011).

63. R. Alonso, M. B. Gavela, D. Hernandez and L. Merlo, *Phys. Lett. B* **715**, 194 (2012).
64. J. R. Espinosa, C. S. Fong and E. Nardi, *JHEP* **1302**, 137 (2013).
65. R. Alonso, M. B. Gavela, G. Isidori and L. Maiani, *JHEP* **1311**, 187 (2013).
66. G. F. Giudice, PoS **EPS-HEP2013**, 163 (2013).
67. G. Blankenburg, G. Isidori and J. Jones-Perez, *Eur. Phys. J. C* **72**, 2126 (2012).

Chapter 18

Rare Decays Probing Physics Beyond the Standard Theory

Frederic Teubert

CERN, Physics Department, Geneva, 23, CH-1211, Switzerland

In the last 50 years we have seen how an initially *ad hoc* and not widely accepted theory of the strong and electroweak interactions (Standard Theory: ST) has correctly predicted the entire accelerator based experimental observations with incredible accuracy. Decays of the ST particles (quarks and leptons), which are rare due to some symmetry of the theory, have played an important role in the making of the ST. These rare decays have been powerful tools to search for interactions of the ST particles with new particles which not necessarily have the same symmetries. In this article, I will describe the indirect search for evidence of new physics (NP) using quark and lepton flavour changing neutral decays, which are highly suppressed within the ST and constitute strong probes of potential new flavour structures.

1. Historical Role of Rare Decays

Rare decays of mesons have played a very important role in the making of the ST of particle physics. In particular, they have been relevant in the development of the Higgs mechanism, which generates fermion masses and quark mixing, and in the establishment of the symmetries of the theory.

A very well known example is the discovery[1] in 1964 that CP was not a symmetry conserved in K decays. A decade before it had been proposed[2] and then experimentally verified[3] that parity (P) was not conserved in weak decays, however nobody imagined that this was also the case for the combination of charge conjugation (C) and P symmetries. It was known that flavour was not conserved in weak interactions, and indeed the neutral K^0 could mix with its anti-particle.[4] If CP is conserved it is possible to define two states (K_1^0 and K_2^0) that are eigenstates of both the weak and CP operators. Then K_1^0 decays into the pure CP-odd state of two π but K_2^0 cannot. All-possible decay channels for K_2^0 are suppressed by parity violation (semileptonic) or by phase-space. Consequently, K_2^0 has a much longer lifetime than K_1^0 by a factor about 500. Experimenters in Ref. 1 shot protons at a target to produce K^0 and after a long enough trip in a vacuum pipe they obtained a pure K_2^0 beam. Amazingly, it was measured that very rarely, once in every 500 decays, the K_2^0 decays into two π rather than three. At that time only three quark were known and the observation of this rare CP violating K_2^0 decay could not be explained within the theoretical framework currently being used.

Another good example of the importance of rare decays is from around the same times, at the birth of the ST. The weak coupling did not seem to be universal: the observed decay probability of the semileptonic $\pi^+ \to \mu^+\nu$ decay, after correcting for the different phase space, was about 20 times larger than the very similar $K^+ \to \mu^+\nu$ decay. In 1963 Cabibbo explained[5] these observations by introducing the 'Cabibbo angle' (θ_c), such that the weak bosons couple to a linear combination of the 'd' and 's' quarks (using the modern language as in 1963 quarks had not yet been proposed). The suppression of the semileptonic K^+ decay with respect to the π^+ decay arises from the square of the ratio of the linear coefficients, $(\sin\theta_c/\cos\theta_c)^2$, in the decay amplitudes. One would then expect large flavour changing neutral currents (FCNC) in the equivalent case of leptonic K^0 decays. However, the probability of the process $K_2^0 \to \mu^+\mu^-$ was eventually measured[6] to be 7×10^{-9}, a very rare decay indeed. This large suppression of the observed FCNC motivated Glashow, Ilioupoulos and Maiani[7] in 1970 to predict the existence of an unobserved fourth quark (c-quark), which forms an SU(2) doublet with the known s-quark. In this model, that later developed into the ST, the existence of two SU(2) doublets, $((u,d)$ and $(c,s))$, implies a series of cancellations which result in a strong suppression of FCNC (GIM mechanism), in agreement with experimental observations. There are many other examples of how rare decays of known existing particles have helped to shape the ST. I have taken these two examples as they demonstrate two important and complementary cases:

- Processes that are protected by some symmetry of the current theory that is not necessarily conserved in the next version of the theory.
- Processes that should not be rare in the current theory, but contradictory observations suggest a new mechanism, to be included in the next version of the theory.

We will see in the next sections how this historically successful strategy is being pursued today and some experimental processes that can reveal properties of a new theory to supersede the ST. Before that I will introduce the flavour structure of the ST and which type of rare decays are *a priori* more interesting probes of NP.

2. Flavour Structure and Symmetries in the ST

According to the ST, the basic constituents of matter can be grouped into three families, or flavours, of quarks and leptons. The four fermions within each family have different combinations of strong, weak, and electromagnetic charges, which determine completely their fundamental interactions with the exception of gravity, which is not included in the ST. As far as we know, quarks and leptons of the second and third family are identical copies of those in the first family but with different, heavier, masses. One of the biggest questions that is not answered by the ST is why there are three almost identical replicas of quarks and leptons, and what is the origin of their different masses.

In the limit of unbroken electroweak symmetry all basic constituents of matter have a vanishing mass. The problem of quark and lepton masses is therefore intimately related to another open key questions in particle physics: what is the mechanism behind the breaking of the electroweak symmetry? Within the ST these two problems are both addressed by the Higgs mechanism: the masses of quarks and leptons, as well as the masses of the W and Z bosons, are the result of the interaction of these basic fields with a new type of field, the Higgs scalar field, whose ground state spontaneously breaks the electroweak symmetry.

The recent observation by the ATLAS[8] and CMS[9] experiments of a new state compatible with the properties of the Higgs boson (or the spin-0 excitation of the Higgs field) has significantly reinforced the evidence in favour of the Higgs mechanism and the validity of the ST. However, we also have clear empirical indications that this theory is not complete: the phenomenon of neutrino oscillations[10] and the evidence for dark matter[11] cannot be explained within the ST.

The description of quark and lepton masses in terms of the Higgs mechanism is particularly unsatisfactory since the corresponding interactions between fermions and Higgs field are not controlled by any symmetry principle, contrary to all other known interactions, resulting in a large number of free parameters. Besides determining quark masses, the interaction of the quarks with the Higgs is responsible for the peculiar pattern of mixing of the various families of quarks under the weak interactions, and of the violation of the CP symmetry in the ST.

In particular, the interplay of weak and Higgs interactions implies that FCNC processes can occur only at higher orders in the electroweak interactions and are strongly suppressed, in accordance with the GIM mechanism described in Section 1. This strong suppression makes FCNC processes natural candidates to search for physics beyond the ST. If the new degrees of freedom do not have the same flavour structure of the quark/lepton–Higgs interaction present in the ST, they may contribute to FCNC processes at a level comparable to the ST amplitudes. Even if their masses are well above the electroweak scale, they can produce sizable deviations from the ST predictions for these rare processes.

Not all processes that are rare and occur only at higher orders are necessarily powerful tools to search for NP. Within the ST, the interactions between photons and leptons, and between quarks and gluons, are governed by gauge theories (QED and QCD) with non-broken gauge symmetries. The decoupling theorem[12] implies that contributions of heavy particles (with masses much larger than the momentum transfer of the process) are irrelevant. On the other hand, the weak interactions are described by a gauge theory with broken symmetry. Therefore, higher order corrections are sensitive to the size of the squared mass difference within an isospin doublet, which is a measure of how badly the isospin symmetry is broken. This is the reason why the dominant effect from weak higher order corrections within the ST is proportional to $(m_{\rm top}^2 - m_b^2)$.

Therefore a rare process that occurs at higher orders in the ST is a good tool to search for new heavier particles that can modify the loop contribution if the process is protected by some symmetry in the ST that does not necessarily hold for the new theory. One example has already been given above, FCNC in quark transitions. Another excellent example is lepton flavour violating decays (LFV). Within the ST neutrinos are massless and the lepton Yukawa matrices can be diagonalised independently. Therefore, there are no FCNC in lepton decays and the lepton flavour is conserved. However, the discovery that different neutrino flavours can mix, and therefore that neutrinos have non-zero masses cannot be explained by the ST. Depending on what is the mechanism to generate neutrino masses and what is their nature (Majorana or Dirac) one can expect different levels of charged LFV decays. Searches for LFV decays are therefore extremely interesting, not only as evidence for NP, but as critical data to constrain the mass generation mechanism in the lepton sector, which may or may not be related to the ST Higgs mechanism.

In the next sections I will discuss the status of some of the most interesting examples of rare decays in quark and lepton FCNC. These correspond to the first class of interesting processes described in Section 1. FCNC processes are analogous to the discovery of CP violation observed in the decay $K_2^0 \to \pi^+\pi^-$, which occurs because the symmetry protecting it is not present in the updated theory.

I also mentioned in Section 1 a historical example of a second class of interesting rare processes that should not have been rare: $K_2 \to \mu^+\mu^-$. I cannot resist the temptation to just briefly mention an excellent example of this kind of process that is currently being pursued: the search for nucleon electric dipole moment (EDM). There is no good reason why within the ST there is only CP violation in the weak interactions. In general one could add a CP violating term in the Lagrangian describing strong interactions and remain consistent with the ST gauge symmetries. Either there is a new symmetry[13] involving new particles (axion-like) or a precise measurement of the neutron and proton EDMs will reveal that CP violation is also present in strong interactions at some level.

3. Quark Flavour Changing Neutral Decays

Within the ST quarks can change flavour in electrically neutral processes via higher order loops. The quark transitions $b \to s$ and $b \to d$ have been measured in B-meson decays, $c \to u$ has been measured in D-mesons decays and $s \to d$ in K-mesons decays. At first order these transitions occur through two kinds of Feynman diagrams shown in Fig. 1. The first corresponds to the so-called "box" diagram and is in particular relevant to describe the mixing between neutral mesons: the example of Fig. 1(a) shows B_s^0 mixing. The second kind of diagram, the so-called "penguin"

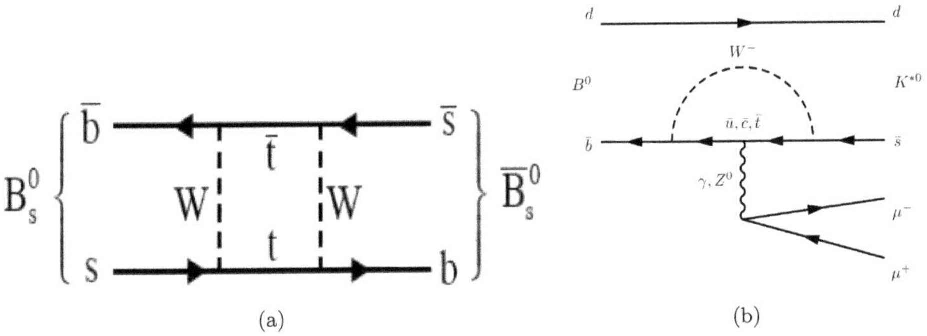

Fig. 1. Examples of loop processes within the ST that allow the FCNC $b \to s$ quark transition. (a) is an example of a box diagram and (b) is an example of a penguin diagram.

diagram, is responsible for a large variety of FCNC rare decays. The example shown in Fig. 1(b) is a $b \to sll$ transition. In particular, if the bosons radiated are the electroweak bosons (Z, W or γ like in Fig. 1), the uncertainties in the calculation of the rates due to the less well known non-perturbative QCD effects are drastically reduced as compared to the case when a gluon is radiated. These "electroweak penguins" are particularly interesting for the discussion in this chapter, as will be shown in a few interesting examples in the next sections.

3.1. $K^+ \to \pi^+\nu\nu$, $K_L^0 \to \pi^0\nu\nu$

One of the strongest constraints on the possible size of NP contributions comes from Kaon physics, in particular the precise measurement of the mass difference ($\Delta m_K = m(\mathrm{K}_L) - m(\mathrm{K}_S)$) of the neutral kaon weak eigenstates and the CP-violating quantities c_K and ϵ'. This is because the ST suppression factors are bigger in the Kaon sector, since the u and c-quark contributions to FCNC processes are very strongly suppressed by the GIM mechanism, while the contribution of the t-quark is strongly suppressed by the Cabbibo–Kobayashi–Maskawa (CKM) matrix elements. Progress in this area is limited by theoretical uncertainties affecting, in particular, the ST prediction of ϵ'. The situation is better when the process occurs through an 'electroweak penguins' with a charged lepton pair in the final state. However, there is still a limitation due to the long distance contributions via one or two photon conversions. This is the reason why there is great interest in decays with a neutrino pair in the final state. The $K^+ \to \pi^+\nu\nu$ and $K_L^0 \to \pi^0\nu\nu$ decays are determined by short distance physics. There is a single operator that determines the decay rates within the ST and in most NP scenarios. In Fig. 2 one can see the leading ST Feynman diagrams that contribute to these processes.

Within the ST, these decays are predicted with good precision:[14]

$$\mathrm{BR}(K_L^0 \to \pi^0\nu\nu)_{ST} = (2.43 \pm 0.39 \pm 0.06) \times 10^{-11}$$

$$\mathrm{BR}(K^+ \to \pi^+\nu\nu)_{ST} = (7.81 \pm 0.75 \pm 0.29) \times 10^{-11}$$

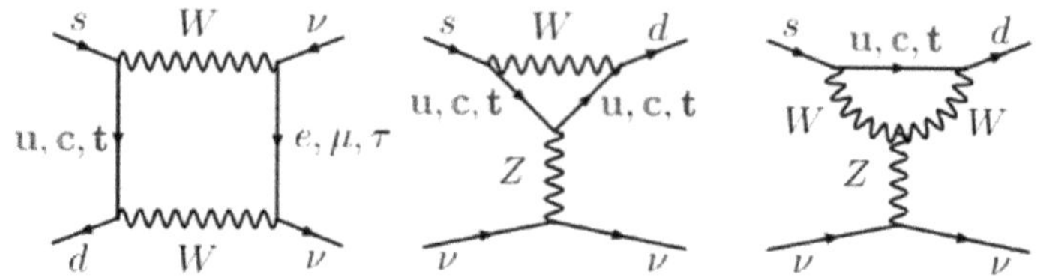

Fig. 2. Box and penguin Feynman diagrams representing the lowest order contributions to the FCNC $s \to d$ quark transition.

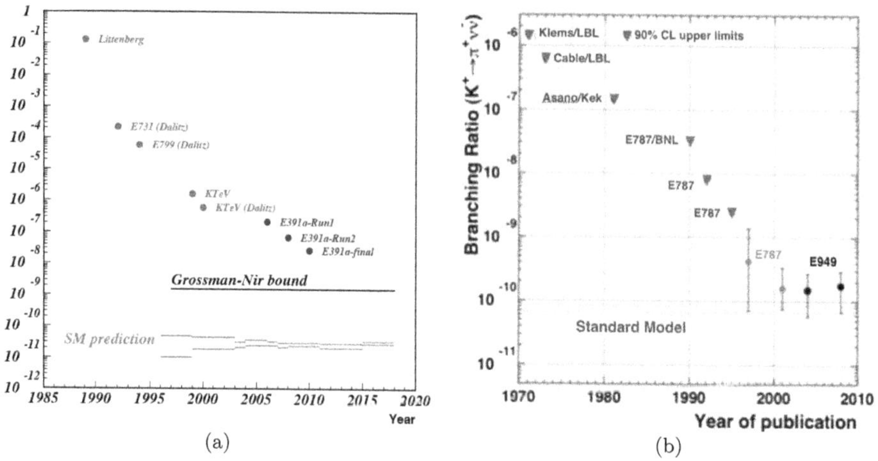

Fig. 3. Historical evolution of the 90% C.L. limits and measurements of the branching ratio of the decays $K_L^0 \to \pi^0\nu\nu$ (a) and $K^+ \to \pi^+\nu\nu$ (b). The Grossman–Nir bound[18] corresponds to the limit on the branching ratio $K_L^0 \to \pi^0\nu\nu$ inferred from the measured branching ratio $K^+ \to \pi^+\nu\nu$ assuming lepton flavour conservation.

where the first uncertainty is due to the experimental uncertainty on the input parameters while the second is due to the intrinsic theoretical limitations. In Fig. 3 one can see the results of the experimental searches for these very rare decays in the last 50 years. The upgraded E949 experiment at the Brookhaven National Laboratory Alternating Gradient Synchrotron (AGS) provided the first evidence[15] for the charged K decay: $\mathrm{BR}(K^+ \to \pi^+\nu\nu) = (17 \pm 11) \times 10^{-11}$ while the E391a experiment at KEK in Japan measured a limit[16] of $\mathrm{BR}(K_L^0 \to \pi^0\nu\nu) < 2.6 \times 10^{-8}$ at 90% C.L. The accelerator and detector technology has advanced and the sensitivity of rare K decay experiments has also improved. Today there are two experiments taking data and aiming for a precise measurement of these decays: KOTO at J-PARC in Japan, and NA62 at the SPS at CERN.

The NA62 experiment[17] has the potential to measure the $\mathrm{BR}(K^+ \to \pi^+\nu\nu)$ with at least a 10% precision. With an expected signal acceptance of 10% and Signal over Background (S/B) larger than 4.5, the experiment requires $\sim 10^{13}$ K decays to achieve such a goal. The CERN SPS provides 10^{12} 400 GeV protons on target per second, which produces a very high intensity K beam, resulting in 5 million K

decays per second in a 60 m long vacuum chamber. The sample available to the NA62 experiment corresponds to $\sim 4.5 \times 10^{12}$ K decays whose flight path is in their acceptance per year ($\sim 10^7$ sec). Assuming the branching ratio of the ST they expect to see 45 signal candidates per year with <10 background events.

The key for the experiment's success is the background rejection and the uncertainty in the background estimation. The background rejection requires a precise measurement of the incoming K (achieved with a silicon pixel tracker operating at the secondary beam rate of 750 MHz, so-called Gigatracker) and a precise measurement of the outgoing π^+ as well as a very efficient π^0 veto. NA62 started commissioning parts of the detector at the end of 2014 and expects to start data taking in 2015.

The KOTO experiment[19] has the potential to reach a first observation of the decay $K_L^0 \to \pi^0 \nu\nu$, at the level of the ST prediction, and has plans for upgrades that could allow for a ~10% measurement of the branching fraction. The J-PARC accelerator can provide 2×10^{14} 30 GeV protons on target every three seconds. Moreover, the neutral K beam is highly collimated ('pencil beam') so that the reconstructed π^0 momentum component transverse to the beam direction can be used as a constraint. The KOTO detector situated ~20 m from the target at an angle of 16° from the incident proton beam, has taken data in 2013 for a short period of time (only 100 hours) at 10% of the nominal intensity. KOTO has been able to observe[20] $\sim 8 \times 10^7$ K decays reaching a single event sensitivity of 1.3×10^{-8} very similar to the previous best experiment.[16] On the other hand the level of background extrapolated to the signal region is higher than originally anticipated due to a significantly higher contribution from '*halo neutrons*'. One signal candidate event is observed compatible with the 0.4 events expected from background. KOTO is expected to resume data taking in 2015.

The $K^+ \to \pi^+ \nu\nu$ and $K_L^0 \to \pi^0 \nu\nu$ decays are indeed extremely interesting measurements in terms of sensitivity to NP. Deviations from the ST at the (10–20)% level[21] appear in many reasonable NP models, compatible with all existing experimental constraints. The next decade promises to be very interesting in terms of the potential experimental sensitivity.

3.2. $B_d^0 \to K^{*0} \mu^+ \mu^-$

An example of the simplest realization of the EW penguin in $b \to s$ FCNC can be seen in Fig. 4(a) and consists of a photon emitted from the internal loop. The inclusive process $b \to s\gamma$ has been measured[22] precisely at experiments at KEKB and SLAC, and experiments at CESR and LEP with an uncertainty of ~7%, in agreement with the ST prediction.[23] This agreement is one of the strongest constraints in NP models, in particular supersymmetric models.

If the photon emitted from the internal loop decays into a lepton pair (hence the amplitude is further suppressed by a factor $\alpha_{\rm QED}$) or it is replaced by a Z boson, the process can provide a rich laboratory to test NP models. An example is shown

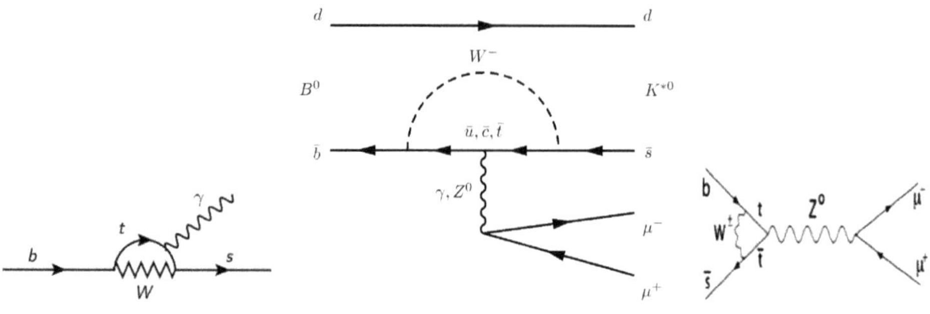

(a) (b) (c)

Fig. 4. Three realizations of the EW penguin in $b \to s$ FCNC. (a) One of the simplest form of radiative decays. (b) The dominant contribution within the ST to the decays $B_d^0 \to K^{*0}\mu^+\mu^-$. (c) The dominant contribution within the ST to the very rare decay $B_s^0 \to \mu^+\mu^-$.

in Fig. 4(b). If we use the language of effective field theories to parameterize NP contributions in terms of a sum of local four-fermion operators (Q_i which depend only on ST fermions) modulated by Wilson coefficients (C_i which depend of the heavy degrees of freedom, i.e. NP particles), then $B_d^0 \to K^{*0}\mu^+\mu^-$ is the "golden mode" to test new vector (axial-vector) couplings (i.e. C_9 and C_{10} in terms of Wilson coefficients contributing to the $b \to s$ transition). Incidentally, $B_d^0 \to K^{*0}\mu^+\mu^-$ complements the $b \to s\ \gamma$ decay which is mostly sensitive to NP dipole operators (i.e. C_7) and the $B^0_{(d,s)} \to \mu^+\mu^-$ decays mostly sensitive to NP (pseudo-)scalar operators (i.e. C_S, C_P). The charge of the pion in the decay $K^* \to K\pi$ defines the flavour of the B meson and an angular analysis can be performed unambiguously to test the helicity structure of the 'electroweak penguins'.

The system is completely defined by four variables: q^2, the square of the invariant mass of the dimuon system, θ_1, the angle between the positive lepton and the direction opposite the B-meson in the dimuon rest frame, θ_K, the equivalent angle of the K^+ in the K^* rest frame and ϕ the angle between the two planes defined by (K,π) and (μ^+,μ^-) in the B-meson rest frame. The four fold differential distribution contains a total of eleven angular terms that can be written in terms of seven q^2 dependent complex decay amplitudes. These amplitudes can be expressed in terms of five complex Wilson coefficients (C_S, C_P, C_7, C_9 and C_{10}), their five helicity counter-parts and six form-factors, which play the role of nuisance parameters in the fit.

The LHCb experiment at the LHC is designed to profit from the enormous production rate of b-quarks ($\sim 3 \times 10^{11}$ per fb^{-1}) and c-quarks ($\sim 6 \times 10^{12}$ per fb^{-1}) in proton–proton collisions at the LHC energies (7–8 TeV of Run-I). About 40% of these b-quarks hadronise to form a B_d^0 meson and about one in 10^6 decays into $B_d^0 \to K^{*0}\mu^+\mu^-$. With ~ 3 fb^{-1} of data collected in 2011–2012, LHCb has been able to trigger and select ~ 2400 candidates in the range 0.1 GeV$^2 < q^2 < 19$ GeV^2 with S/B > 5. This is about one order of magnitude larger than the samples available at previous experiments (BaBar, Belle and CDF) and similar to what the experiments

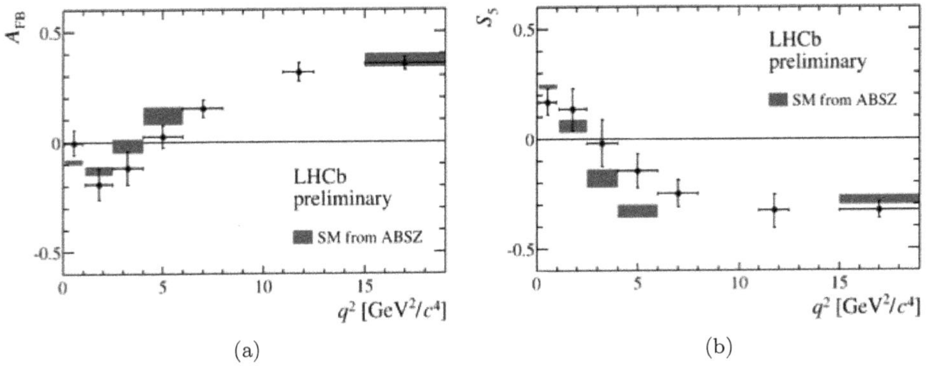

Fig. 5. Two examples of the CP-averaged coefficients in the angular terms as a function of q^2. The shaded boxes show the ST prediction taken from Ref. 25.

ATLAS and CMS, also at the LHC, have collected with ten times more luminosity (however ATLAS and CMS with significantly worse S/B).

The statistics and the quality of the data accumulated by the LHCb experiment allows for the first time to perform a full angular analysis of these decays. The preliminary results[24] of this "tour de force" analysis have been recently shown at conferences. In Fig. 5 two examples of the CP-averaged (i.e. the average of the coefficients measured with B_d^0 and anti-B_d^0 decays) angular coefficients measured by LHCb are shown as a function of q^2. While most of the measurements agree reasonably well with the ST predictions,[25] for these two examples: A_{FB} (modulating the $\sin^2\theta_K \times \cos\theta_1$ angular term) and S_5 (modulating the $\sin(2\theta_K) \times \sin\theta_1 \times \cos\phi$ angular term) there seems to be a hint of disagreement. These are early times and more data and a careful reassessment of the ST prediction uncertainties are needed before reaching a conclusion.

Nevertheless, several authors have already attempted to see if the overall pattern of the measurements is consistent with a given value for the relevant Wilson coefficients. As mentioned before the inclusive $b \to s\gamma$ measurements strongly constrain non-ST values for C_7. The scalar C_S and pseudo-scalar C_P coefficients are constrained, for example, by the measurement of the branching ratio of the very rare decays $B^0_{(d,s)} \to \mu^+\mu^-$. Therefore, the small disagreements observed in the full angular analysis of the decay $B_d^0 \to K^{*0}\mu^+\mu^-$ and also other similar decays like $B_s^0 \to \phi\mu^+\mu^-$ or $B^+ \to K^+\mu^+\mu^-$ seem to be consistent with a non-ST value of the C_9 Wilson coefficient, as can be seen in Fig. 6 taken from Ref. 26. All these measurements will improve significantly in the next decade. Therefore if NP is the responsible for these early hints, we should get a definitive answer in the near future.

3.3. $B^0_{(d,s)} \to \mu^+\mu^-$

The pure leptonic decays of K, D and B-mesons are a particular interesting case of 'electroweak penguins', see Fig. 4(c). Compared with the decays described in Section 3.2, the helicity configuration of the final state suppresses

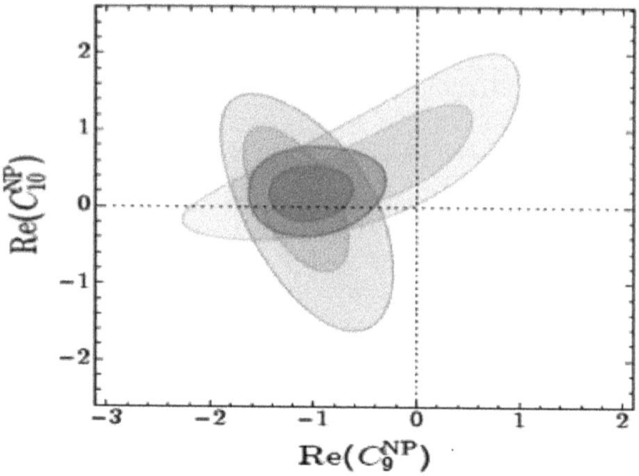

Fig. 6. Constrains on the contribution of NP to the real part of C9 and C10 at the 1σ and 2σ level taken from Ref. 26. The darkest contours correspond to the global fit. The contours with negative correlation correspond to the results using only the full angular analysis of the decay $B_d^0 \to K^{*0}\mu^+\mu^-$. The contours with positive correlation are from other measurements. By definition the ST prediction is (0,0).

the vector (axial-vector) contribution by a factor proportional to $(M_\mu/M_{K,D,B})^2$. Therefore, these decays are particularly sensitive to new (pseudo-)scalar interactions. In the case of B_d^0 and B_s^0 meson decays the contribution of the absorptive part can be safely neglected. As a consequence, the rate is well predicted[27] theoretically: $\mathrm{BR}(B_s^0 \to \mu^+\mu^-) = (3.65 \pm 0.23) \pm 10^{-9}$ and $\mathrm{BR}(B_d^0 \to \mu^+\mu^-) = (1.06 \pm 0.09) \pm 10^{-10}$. In the B_s^0 case, this prediction corresponds to a flavour-averaged time-integrated measurement, taking into account the correction due to the non-vanishing width difference.

The experimental signature is sufficiently clean to reach an expected S/B $\sim$3 for the B_s^0 decay, assuming the ST branching fraction. The main background in the invariant mass region around the B_s^0 mass is due to combinations of uncorrelated muons and can be estimated from the mass sidebands. The most important handle to reduce this combinatorial background is the invariant mass resolution of the experiment, which is also crucial to differentiate between the B_d^0 and B_s^0 decays ($\Delta m \sim 87$ MeV). Moreover, the large fraction of $B^0_{(d,s)} \to hh$ decays is an important source of background due to misidentified hadrons in the region around the B_d^0 mass (this background is very small in the B_s^0 mass region). Given the experimental detector resolution and trigger acceptance, the CMS experiment with 25 fb^{-1} of data accumulated in Run-I has similar sensitivity to the LHCb experiment with 3 fb^{-1} collected in the same period. Both experiments have provided a clear observation[28] of the decay $B_s^0 \to \mu^+\mu^-$ (CMS and LHCb observe 28 and 11 B_s^0 signal candidates with a background of 10 and 3.6 events respectively) but not yet a significant observation for the decay $B_d^0 \to \mu^+\mu^-$. The combined analysis of the two experiments has been published recently in Nature.[29] The invariant mass

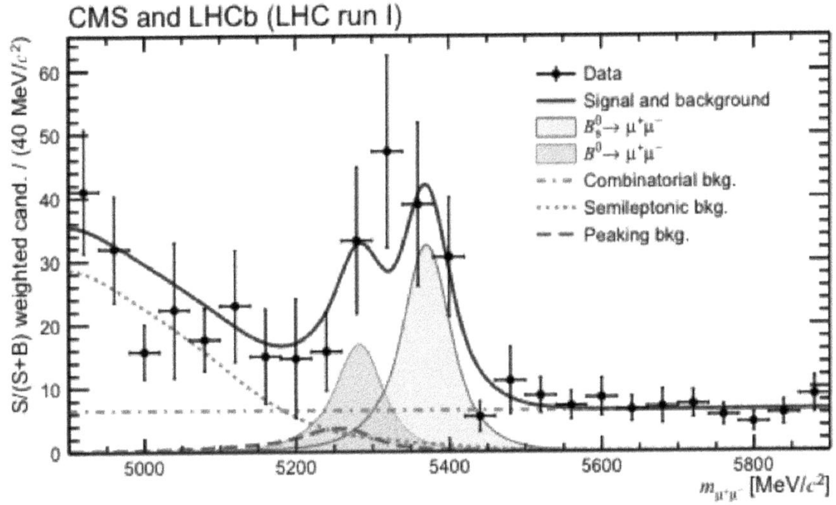

Fig. 7. Weighted distribution of the dimuon invariant mass from Ref. 29. Superimposed on the data points in black are the combined fit (solid line) and its components: the B_s^0 (light grey) and B_d^0 (gray) signal components: the combinatorial background (dash-dotted line); exclusive backgrounds (dotted line); and the exclusive peaking backgrounds (dashed line).

distribution is shown in Fig. 7 and the combined value for the branching ratio is $\mathrm{BR}(B_s^0 \to \mu^+\mu^-) = (2.8^{+0.7}_{-0.6}) \times 10^{-9}$ in agreement with the ST prediction within the present uncertainties.

This very rare decay has been searched since the discovery of B mesons 30 years ago. Thanks to the ingenuity and persistence of the experimenters it has been eventually measured at the LHC and found to be in agreement with the ST within the current uncertainties. In the next decade it will be very interesting to see how it turns out the measurement of $\mathrm{BR}(B_d^0 \to \mu^+\mu^-)$.

4. Lepton Flavour Changing Neutral Currents

The search for FCNC in charged lepton decays has been unsuccessful so far.[30] Nevertheless, historically these limits have played a very important role. By the end of the 50s it was known[31] that $\mathrm{BR}(\mu \to e\gamma)$ had to be below 10^{-5} and this limit was used to argue the existence of a second neutrino,[32] needed to cancel possible large loop-induced neutral currents, in analogy to the GIM mechanism described in Section 1. Figure 8 shows the status of the searches for $\mu \to e\gamma$, $\mu \to eee$, $\tau \to \mu\gamma$, $\tau \to \mu\mu\mu$ and $\mu \to e$ conversions in the presence of the field of nucleus, as they were by the end of 2008. There have been significant improvements of this picture in recent years by experiments like MEG at PSI, Belle at KEKB, BaBar at SLAC and LHCb at LHC.

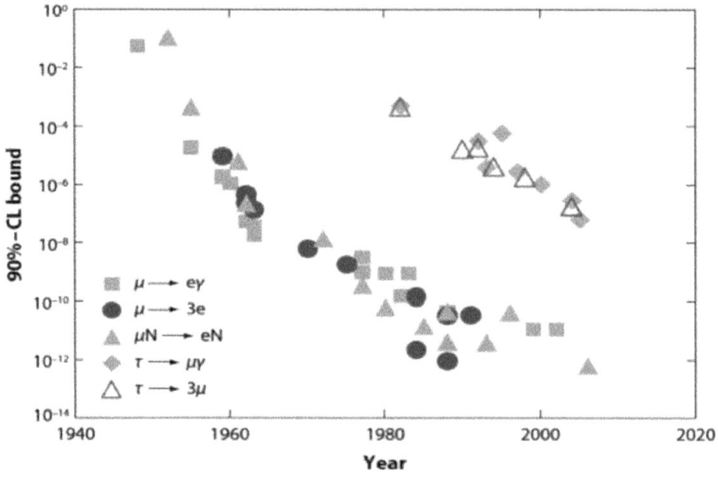

Fig. 8. Historical evolution of the 90% C.L. limits on μ and τ flavour violation decays as it was in 2008 taken from Ref. 30.

The MEG experiment collected stopped μ^+ at the PSI facilities between 2009 and 2013 to search for the process $\mu^+ \to e^+\gamma$. The experimental signature consists of a monochromatic positron and photon back-to-back in the μ^+ rest frame. Therefore, the experiment needs an excellent energy and tracking resolutions as well as a precise measurement of the time coincidence between positron and photon. The MEG collaboration has recently published[33] the results of the analysis using 3.6×10^{14} stopped μ^+ collected up to 2011. The very few events seen in the expected signal region are compatible and slightly below the expected background, reaching an observed limit of $\mathrm{BR}(\mu^+ \to e^+\gamma) < 5.7 \times 10^{-13}$ at 90% C.L. The expected sensitivity will improve when adding the rest of the data available, and should improve by a factor ten with the upgrade of the experiment.

Large samples of τ leptons are produced in high luminosity e+e− colliders through the process $e^+e^- \to \tau^+\tau^-$. The analysis of the whole data sample collected by BaBar[34] and Belle[35] (500–800 fb^{-1} which corresponds to 10^9 $\tau^+\tau^-$ pairs) shows no evidence of LFV τ decays and sets the limits on $\mathrm{BR}(\tau \to \mu\gamma)$ and $\mathrm{BR}(\tau \to \mu\mu\mu)$ to a few 10^{-8}.

The start of the LHC proton–proton collider and the enormous production of charm and beauty mesons, and of τ leptons from their decays, has opened a new window of opportunity increasing the τ production rate by five orders of magnitude compared to existing e^+e^- colliders. The LHCb experiment has collected 3 fb^{-1} of proton–proton collisions at 7–8 TeV centre-of-mass energies by the end of 2012, which correspond to 10^{11} τ leptons. The LHCb analysis[36] has less efficiency and purity than those done at the e^+e^- colliders, but nevertheless reaches similar sensitivities for the decay $\tau \to \mu\mu\mu$. Using the initial sample from LHCb, the combined values of the limits computed by the HFAG[37] improve the previous limits to: BR($\tau \to \mu\mu\mu$)$< 1.2 \times 10^{-8}$ and BR($\tau \to \mu\gamma$) $< 5 \times 10^{-8}$ at 90% C.L.

Improvements in the forthcoming decade are expected from the upgrade of MEG and Belle experiments and from the data accumulated at the LHC and the upgrade of the LHC experiments. Moreover new experiments are being built: COMET[38] at J-PARC and Mu2e[39] at the Fermilab booster plan to use very large samples of μ, collected at a rate of about 10^{10} per second, to search for $\mu \to e$ conversions in the presence of the field of a nucleus. They both expect to reach sensitivities below 10^{-16} on this process, i.e. a four order of magnitude improvement. It is clear that Fig. 8 will look very different in ten years from now.

5. Final Remarks

The concept of symmetry has played a fundamental role in the making of the ST. Because of these symmetries some processes are highly suppressed in the ST, but may receive important NP contributions if these symmetries are not respected by the theory that should supersede the ST. Flavour is one of the properties of the ST that cannot be explained by fundamental symmetries: why do we have three almost identical replicas of quarks and leptons, and what is the origin of their different masses? A new theory of flavour should be able to answer these questions, and naturally include new flavour transitions. This is the reason why this chapter has been devoted to describe a few of the most interesting attempts to search for FCNC in quark and lepton decays. The field is very active today: running experiments together with new experiments scheduled to start data taking in the next decade will provide new hope for a deeper understanding of nature.

Acknowledgments

I would like to thank L. Maiani and G. Rolandi for their invitation to contribute to this interesting book. A. Ceccucci, S. Kettell and T. Komatsubara for their help with the text and figures in Section 3.2, and M. Kenzie for his very useful comments to the manuscript.

References

1. J. H. Christenson, J. W. Cronin, V. L. Fitch, and R. Turlay, *Phys. Rev. Lett.* **13**, 138 (1964).
2. T. D. Lee and C. N. Yang, *Phys. Rev.* **104**, 254 (1956) [Erratum *ibid.* **106**, 1371 (1957).
3. C. S. Wu, E. Ambler, R. W. Hayward, D. D. Hoppes and R. P. Hudson, *Phys. Rev.* **105**, 1413 (1957).
4. M. Gell-Mann and A. Pais, *Phys. Rev.* **97**, 1387 (1955).
5. N. Cabibbo, *Phys. Rev. Lett.* **10**, 531 (1963).
6. A. P. Heinson *et al.* (BNL E791), *Phys. Rev. D* **51**, 985 (1995).
7. S. L. Glashow, J. Iliopoulos and L. Maiani, *Phys. Rev. D* **2**, 1285 (1970).

8. ATLAS Collaboration, *Phys. Lett. B* **716**, 1 (2012).
9. CMS Collaboration, *Phys. Lett. B* **716**, 30 (2012).
10. R. Davis, D. Harmer and K. C. Hoffman, *Phys. Rev. Lett.* **20**, 1205 (1968).
    Super-Kamiokande Collaboration, *Phys. Rev. Lett.* **81**, 1562 (1998).
11. V. C. Rubin and W. K. Ford, *Astrophysical Journal* **159**, 379 (1970).
12. T. Appelquist and J. Carazzone, *Phys. Rev. D* **11**, 2856 (1975).
13. R. D. Peccei and H. R. Quinn, *Phys. Rev. D* **16**, 1440 (1977).
14. G. Buchalla and A. J. Buras, *Nucl. Phys. B* **412**, 106 (1994).
    J. Brod, M. Gorbahn and E. Stamou, *Phys. Rev. D* **83**, 034030 (2011).
15. E949 and E787 Collaborations, *Phys. Rev. D* **77**, 052003 (2008).
    BNL-E949 Collaboration, *Phys. Rev. D* **79**, 092004 (2009).
16. E39a Collaboration, *Phys. Rev. D* **81**, 072004 (2010).
17. NA62 Collaboration, Proposal to measure the Rare Decay $K^+ \to \pi^+\nu\nu$ at the CERN SPS, CERN-SPSC-2005-013/SPSC-P-326 (2005).
18. Y. Grossman and Y. Nir, *Phys. Lett. B* **398**, 163 (1997).
19. KOTO Collaboration, Proposal for a $K_L^0 \to \pi^0\nu\nu$ Experiment at J-PARC, J-PARC Experimental Proposal (2006).
20. KOTO collaboration, $K_L^0 \to \pi^0\nu\nu$ at KOTO, [arXiv:1411.4250 [hep-ex]].
21. G. Isidori and F. Teubert, *Eur. Phys. J. Plus* **129**, 40 (2014).
22. K. A. Olive *et al.* (Particle Data Group), *Chin. Phys. C* **38** (2014) 090001.
23. M. Misiak *et al.*, *Phys. Rev. Lett.* **98**, 022002 (2007).
24. LHCb Collaboration, LHCb-CONF-2015-002 (2015).
25. A. Bharucha, D. Straub and R. Zwicky, $B \to V\ l^+l^-$ in the Standard Model from Light-Cone Sum Rules, [arXiv:1503.05534 [hep-ph]].
26. W. Altmannshofer and D. M. Straub, Implications of $b \to s$ measurements, [arXiv:1503.06199 [hep-ph]].
27. C. Bobeth, M. Gorbahn, T. Hermann, M. Misiak, E. Stamou and M. Steinhauser, *Phys. Rev. Lett.* **112**, 101801 (2014).
28. CMS Collaboration, *Phys. Rev. Lett.* **111**, 101804 (2013).
    LHCb Collaboration, *Phys. Rev. Lett.* **111**, 101805 (2013).
29. CMS and LHCb Collaborations, *Nature* **522**, 68 (2015).
30. W. J. Marciano, T. Mori and J. Michael Roney, *Annu. Rev. Nucl. Part. Sci.* **58**, 315 (2008).
31. S. Lokanathan and J. Steinberger, *Phys. Rev.* **98**, 240(A) (1955).
32. G. Feinberg, *Phys. Rev.* **110**, 1482 (1958).
33. MEG Collaboration, *Phys. Rev. Lett.* **110**, 201801 (2013).
34. BaBar Collaboration, *Phys. Rev. Lett.* **104**, 021802 (2010); BaBar Collaboration, *Phys. Rev. D* **81**, 111101(R) (2010).
35. Belle Collaboration, *Phys. Lett. B* **666**, 16 (2008).
36. LHCb Collaboration, *JHEP* **1502**, 121 (2015).
37. Y. Amhis *et al.* (HFAG), Averages of b-hadron, c-hadron, and τ-lepton properties as of summer 2014, [arXiv:1412.7515 [hep-ex]].
38. COMET Collaboration, Conceptual Design Report for Experimental Search for Lepton Flavour Violating $\mu \to e$ Conversion at Sensitivity of 10^{-6} with a Slow-Extracted Bunched Proton Beam, J-PARC P21 (2009).
39. Mu2e Collaboration, Mu2e Technical Design Report, October 2014, Fermi National Accelerator Laboratory.

Chapter 19

Neutrino Masses and Flavor Oscillations

Yifang Wang and Zhi-zhong Xing*
Institute of High Energy Physics, Chinese Academy of Sciences, P.O. Box 918, Beijing 100049, China
xingzz@ihep.ac.cn

This essay is intended to provide a brief description of the peculiar properties of neutrinos within and beyond the standard theory of weak interactions. The focus is on the flavor oscillations of massive neutrinos, from which one has achieved some striking knowledge about their mass spectrum and flavor mixing pattern. The experimental prospects towards probing the absolute neutrino mass scale, possible Majorana nature and CP-violating effects, will also be addressed.

1. Neutrinos and Their Sources

1.1. *From Pauli's hypothesis to the discoveries of neutrinos*

Soon after Henri Becquerel discovered the radioactivity of uranium in 1896,[1] many nuclear physicists started to pay attention to the beta decays $(A, Z) \to (A, Z+1) + e^-$, in which the energy spectrum of electrons was expected to be *discrete* thanks to the laws of energy and momentum conservations. However, James Chadwick observed a *continuous* electron energy spectrum of the beta decay in 1914,[2] and such a result was firmly confirmed by Charles Ellis and his colleagues in the 1920s.[3] At that time there were two different ideas to resolve this "new physics" phenomenon (i.e., the discrepancy between *observed* and *expected* energy spectra of electrons): one was to give up the energy conservation law and the other was to add in a new particle. Niels Bohr was the representative of the former idea, which turned out to be wrong. Wolfgang Pauli conjectured that an unobservable, light, spin-1/2 and neutral particle — known as the electron antineutrino later — appeared in the beta decay and carried away some energy and momentum, and thus the energy spectrum of electrons in the process $(A, Z) \to (A, Z+1) + e^- + \overline{\nu}_e$ was continuous. Pauli first put forward the concept of neutrinos in his famous letter to the "Dear radioactive ladies and gentlemen" who had gathered in Tübingen on 4 December 1930.[4] Three years later he gave a talk on his neutrino hypothesis in the renowned Solvay Conference, where Enrico Fermi was in the audience and took this hypothesis seriously. In the end of 1933, Fermi published his most important theoretical work, an effective theory of the beta decay,[5] which is actually a low-energy version of today's standard picture of weak charged-current interactions. Fermi's seminal work made it possible to calculate the reaction rates of nucleons and electrons (or positrons) interacting with neutrinos (or antineutrinos).

In 1936, Hans Bethe pointed out that an inverse beta decay mode of the type $\overline{\nu}_e + p \to n + e^+$ (or more general, $\overline{\nu}_e + (A, Z) \to (A, Z-1) + e^+$) could be a possible way to verify the existence of electron antineutrinos produced from either fission bombs or fission reactors.[6] This preliminary idea was elaborated by Bruno Pontecorvo in 1946,[7] and it became feasible with the development of the liquid scintillation counting techniques in the 1950s. Although the incident $\overline{\nu}_e$ is invisible, it can trigger the inverse beta decay where the emitted positron annihilates with an electron and the daughter nucleus is captured in the detector. Both events are observable because they emit gamma rays, and the corresponding flashes in the liquid scintillator are separated by some microseconds. Frederick Reines and Clyde Cowan did the first reactor antineutrino experiment and obtained a positive result in 1956,[8] and they reported a new result consistent with the parity-violating theory of weak interactions in 1960. The Nobel Prize finally came to Reines in 1995, when Cowan had passed away 21 years before.

The discovery of electron antineutrinos motivated Pontecorvo to speculate on the possibility of lepton number violation and neutrino–antineutrino transitions in 1957.[9] His argument was actually based on a striking conjecture made by Ettore Majorana in 1937: a massive neutrino could be its own antiparticle.[10]

In 1962, the muon neutrino — a sister of the electron neutrino — was discovered by Leon Lederman, Melvin Schwartz and Jack Steinberger in an accelerator-based experiment.[11] This discovery, which immediately motivated Ziro Maki, Masami Nakagawa and Shoichi Sakata to conjecture the $\nu_e \leftrightarrow \nu_\mu$ conversion,[12] was also recognized by the Nobel Prize in 1988. The tau neutrino, another sister of the electron neutrino, was finally observed at the Fermilab in the end of 2000.[13] Within the standard model the complete lepton family consists of three charged members (e, μ, τ) and three neutral members (ν_e, ν_μ, ν_τ), and their corresponding antiparticles.

1.2. *Where do neutrinos come from?*

Neutrinos and antineutrinos may originate from many physical and astrophysical processes via weak interactions. Figure 1 illustrates some typical examples of neutrino or antineutrino sources in the Universe.

Example (1): Neutrinos and antineutrinos from the Big Bang. The standard cosmology predicts the existence of a cosmic neutrino (or antineutrino) background in the Universe. Today such *relic* neutrinos and antineutrinos should have an overall number density around 330 cm^{-3}, but their temperature is so low (only about 1.9 K, or roughly 1.6×10^{-4} eV) that there is no way to detect them. In the long run

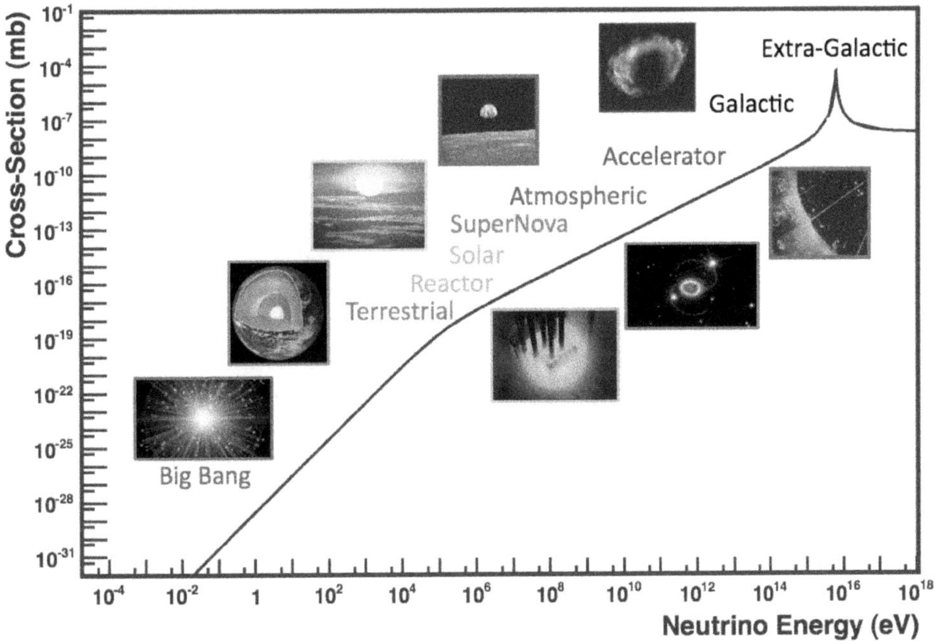

Fig. 1. Some representative sources of neutrinos and (or) antineutrinos and their corresponding energies.[14] The cross sections of $\overline{\nu}_e + e^- \to \overline{\nu}_e + e^-$ scattering associated with different sources are also shown for comparison, where the peak around 6.3 PeV is related to the Glashow resonance.[15]

it might be possible to capture the relic electron neutrinos on some beta-decaying nuclei,[16] as the PTOLEMY project is trying.[17]

Example (2): Electron antineutrinos from the Earth. Since its birth, the Earth's interior has kept a number of radioactive nuclei (e.g., ^{40}K, ^{238}U and ^{232}Th). That is why numerous electron antineutrinos can be produced from terrestrial "natural radioactivity" (i.e., the beta decays), at a rate of several millions per square centimeter per second. So far such interesting geo-$\overline{\nu}_e$ events have been observed at the 3σ level in the KamLAND[18] and Borexino[19] experiments.

Example (3): Electron neutrinos from the Sun. Solar electron neutrinos come along with a number of thermonuclear fusion reactions inside the Sun. One may understand why the Sun shines with the help of $4p \to {}^4\text{He} + 2e^+ + 2\nu_e + 26.7$ MeV: about 98% of the energy radiates in the form of light and only 2% of the energy is taken away by neutrinos.[20] The only way to verify such a picture on the Earth is to detect the electron neutrinos emitted from the core of the Sun. In 1968 solar neutrinos were first observed by Raymond Davis in his radiochemical experiment (see Section 4.1 for a more detailed description).[21]

Example (4): Neutrinos and antineutrinos from supernovae. The explosion of a supernova may release the gravitational binding energy of $\mathcal{O}(10^{53})$ erg in the form of neutrinos and antineutrinos.[22] On 23 February 1987 the ν_e and $\overline{\nu}_e$ events from the Supernova 1987A explosion were observed by the Kamiokande-II,[23] IMB[24] and

Baksan[25] detectors. This observation was a great milestone in neutrino astronomy. Davis and Masatoshi Koshiba received the Nobel Prize in 2002 for their pioneering detections of solar and supernova neutrinos, respectively.

Example (5): Neutrinos and antineutrinos from the Earth's atmosphere. When a cosmic ray (which is mainly composed of high-energy protons coming from somewhere in the galactic or extragalactic space) penetrates the atmosphere around the Earth, it may interact with the ambient nuclei and generate a particle shower containing charged pions and muons. The decays of $\pi^{\pm}$ and $\mu^{\pm}$ can therefore produce atmospheric ν_{μ}, $\overline{\nu}_{\mu}$, ν_e and $\overline{\nu}_e$ events, which have been observed in several experiments.[26] In particular, the phenomenon of atmospheric neutrino oscillations was firmly established by the Super-Kamiokande (SK) Collaboration in 1998.[27]

Example (6): Ultrahigh-energy (UHE) cosmic neutrinos and antineutrinos from distant astrophysical sources, including the expected active galactic nuclei, gamma ray bursts, supernova remnants and the Greisen–Zatsepin–Kuzmin cut-off of cosmic rays.[29] The UHE ν_{μ}, $\overline{\nu}_{\mu}$, ν_e and $\overline{\nu}_e$ events can be produced from UHE $p\gamma$ or pp collisions via $\pi^{\pm}$ and $\mu^{\pm}$ decays, and thus they may serve as a unique cosmic messenger and provide us with useful information about the cosmos that cannot be extracted from the measurements of cosmic rays and gamma rays. So far the IceCube detector at the South Pole has observed 37 extraterrestrial neutrino candidate events with deposited energies ranging from 30 TeV to 2 PeV.[28] Among them, the three PeV events represent the highest-energy neutrino interactions ever observed, but their astrophysical origin remains mysterious.

Of course, neutrinos and (or) antineutrinos can also be produced from some man-made facilities, especially the nuclear reactors and particle accelerators. They also play a crucial role in discovering neutrinos, observing flavor oscillations and measuring fundamental parameters, as one will see in sections 3–5.

2. Weak Interactions of Neutrinos in the Standard Theory

As an important part of the matter content in the standard electroweak model based on the $SU(2)_{\rm L} \times U(1)_{\rm Y}$ gauge group, neutrinos are assumed to be the *massless* Weyl particles. Hence only the left-handed neutrinos and right-handed antineutrinos exist, and they take part in weak charged- and neutral-current interactions via

$$-\mathcal{L}_{\rm cc} = \frac{g}{2\sqrt{2}} \sum_{\alpha} \left[\overline{\alpha}\, \gamma^{\mu}(1-\gamma_5)\nu_{\alpha} W^{-}_{\mu} + {\rm h.c.}\right] ,$$
$$-\mathcal{L}_{\rm nc} = \frac{g}{4\cos\theta_{\rm w}} \sum_{\alpha} \left[\overline{\nu_{\alpha}}\, \gamma^{\mu}(1-\gamma_5)\nu_{\alpha}\right] Z_{\mu} , \qquad (1)$$

where $\alpha = e, \mu, \tau$. Eq. (1) allows one to calculate the cross sections of neutrino–electron, neutrino–neutrino and neutrino–nucleon scattering processes.[29] Note that the reactions $\nu_e + e^- \to \nu_e + e^-$ and $\overline{\nu}_e + e^- \to \overline{\nu}_e + e^-$ can happen via both charged- and neutral-current interactions, but $\nu_{\mu} + e^- \to \nu_{\mu} + e^-$ (or $\nu_{\tau} + e^- \to \nu_{\tau} + e^-$) and $\overline{\nu}_{\mu} + e^- \to \overline{\nu}_{\mu} + e^-$ (or $\overline{\nu}_{\tau} + e^- \to \overline{\nu}_{\tau} + e^-$) can only occur via the neutral-current interactions. That is why the behavior of neutrino flavor conversion in a dense medium may be modified by the coherent forward $\nu_e e^-$ or $\overline{\nu}_e e^-$ scattering. This effect is referred to as the Wolfenstein–Mikheyev–Smirnov (MSW) matter effect.[30]

The simplest quasi-elastic neutrino–nucleon scattering processes are the inverse beta decays $\overline{\nu}_e + p \to e^+ + n$ and $\nu_e + n \to e^- + p$, which take place via the charged-current weak interactions. Their cross sections can be approximately expressed as $\sigma(\overline{\nu}_e p) = \sigma(\nu_e n) \simeq 9.1 \times 10^{-44} (E_\nu/\text{MeV})^2 \text{cm}^2$. In comparison, the elastic neutrino–nucleon scattering reaction $\nu_\alpha + N \to \nu_\alpha + N$ (for $\alpha = e, \mu, \tau$) is mediated by the neutral-current weak interactions.

Historically, the existence of weak neutral currents was first established in the Gargamelle bubble chamber at CERN in 1973.[31] This experiment, which observed the highly expected events of $\nu_\mu + N \to \nu_\mu + \text{hadrons}$ and $\overline{\nu}_\mu + N \to \overline{\nu}_\mu + \text{hadrons}$, crowned the long-range neutrino program initiated by CERN at that time and brought CERN a leading role in the field of high energy physics. It also provided an unprecedentedly strong support to the standard electroweak model formulated by Sheldon Glashow, Steven Weinberg and Abdus Salam in the 1960s.[32] These three theorists received the Nobel Prize in 1979 for their contributions to the electroweak theory and especially for their prediction of the weak neutral current. Four years later, the three mediators of the weak force (i.e., the $W^\pm$ and Z^0 bosons) were finally discovered by Carlo Rubbia and his colleagues at CERN.[33]

The standard theory was thoroughly tested in the 1990s with the help of the Large Electron–Positron Collider (LEP) running on the Z^0 resonance at CERN. In particular, the number of neutrino species was determined to be $N_\nu = 2.984 \pm 0.008$ via the decay $Z^0 \to \nu_\alpha + \overline{\nu}_\alpha$.[26] Such a result is consistent very well with 3 as required in the theory. Extra light neutrino species are not impossible, but they must be "sterile" — in the sense that they do not directly take part in the standard weak interactions, and hence their existence is not subject to the LEP measurement.

Note that the structure of the standard theory itself is too economical to allow the neutrinos to be massive. On the one hand, the particle content of the model is so limited that there are neither right-handed neutrinos nor any Higgs triplets. Hence a normal Dirac neutrino mass term is not allowed, nor a gauge-invariant Majorana mass term. On the other hand, the model is a renormalizable quantum field theory. The renormalizability implies that an effective dimension-5 operator, which can give each neutrino a Majorana mass, is also forbidden.

3. Neutrino Masses, Flavor Mixing and Oscillations

3.1. *Massive neutrinos and their electromagnetic properties*

There are several ways to slightly extend the standard theory such that the neutrinos can acquire their masses with little influence on the great success of the theory itself.[34] Here let us take two typical examples for illustration.

(1) If the renormalizability of the standard theory is relaxed, then the lowest-dimension operator that violates lepton number and generates neutrino masses must be the unique dimension-5 Weinberg operator $HH\ell\ell/\Lambda$, where Λ denotes the cut-off

energy scale in such an effective field theory, H and ℓ are the Higgs and lepton doublets, respectively.[35] After spontaneous gauge symmetry breaking, this operator yields the neutrino masses $m_i \sim \langle H\rangle^2/\Lambda$ (for $i = 1, 2, 3$), which can be sufficiently small ($\lesssim$1 eV) provided $\Lambda \gtrsim 10^{13}$ GeV and $\langle H\rangle \sim 10^2$ GeV. In this sense the study of neutrino mass generation can serve as a striking low-energy window onto new physics at superhigh energy scales.

(2) If two or more heavy right-handed neutrinos are added into the standard theory and lepton number is violated by their Majorana mass term, then the Lagrangian responsible for neutrino masses can be written as

$$-\mathcal{L}_{\rm mass} = \overline{\ell_{\rm L}} Y_\nu \tilde{H} N_{\rm R} + \frac{1}{2}\overline{N^c_{\rm R}} M_{\rm R} N_{\rm R} + {\rm h.c.}, \tag{2}$$

in which the first term stands for the neutrino Yukawa interactions, and the second term is lepton-number-violating. After the $SU(2)_{\rm L} \times U(1)_{\rm Y}$ gauge symmetry is spontaneously broken to $U(1)_{\rm em}$, one is left with the effective Majorana neutrino mass matrix $M_\nu \simeq -\langle H\rangle^2 Y_\nu M_{\rm R}^{-1} Y_\nu^T$, which is often referred to as the canonical *seesaw* formula.[36] Because $N_{\rm R}$ is the $SU(2)_{\rm L}$ singlet, the mass scale of $M_{\rm R}$ can be greatly higher than the electroweak scale $\langle H\rangle$. Hence the mass scale of M_ν is highly suppressed, providing a natural explanation of the smallness of neutrino masses.

Instead of introducing the heavy right-handed neutrinos, one may also introduce a Higgs triplet or a few triplet fermions into the standard theory so as to explain why the three active neutrinos should have naturally small masses.[29] Such seesaw mechanisms essentially have the same spirit, which attributes the smallness of neutrino masses to the largeness of new degrees of freedom. Furthermore, they require massive neutrinos to be the Majorana particles and thus allow some lepton-number-violating processes to happen.

It is worth pointing out that a pure Dirac neutrino mass term, originating from the neutrino Yukawa interactions on the right-hand side of Eq. (2), is less convincing and less interesting from a theoretical point of view. The reason for this argument is two-fold: (a) such a scenario cannot explain why the neutrino masses are so small as compared with the charged lepton masses; (b) given $N_{\rm R}$, the lepton-number-violating term $\overline{N^c_{\rm R}} M_{\rm R} N_{\rm R}$ should not be absent because it is not forbidden by gauge symmetry and Lorentz invariance. If massive neutrinos really have the Majorana nature, they can trigger the neutrinoless double-beta ($0\nu\beta\beta$) decays and some other lepton-number-violating processes. In particular, they are likely to have something to do with the observed asymmetry of matter and antimatter in the Universe via the seesaw and leptogenesis[37] mechanisms. Hence the phenomenology of Majorana neutrinos is much richer and more interesting than that of Dirac neutrinos.

Although a massive neutrino does not possess any electric charge, it can have electromagnetic interactions via quantum loops.[38] Now that Dirac and Majorana neutrinos couple to the photon in different ways, their corresponding electromagnetic form factors must be different. Given the standard weak interactions, one finds that a massive Dirac neutrino has no electric dipole moment and its magnetic dipole moment is finite but extremely small: $\mu_\nu \sim 3 \times 10^{-20}(m_\nu/0.1\ \text{eV})\mu_\text{B}$ with μ_B being the Bohr magneton. In contrast, a massive Majorana neutrino has neither electric nor magnetic dipole moments, simply because its antiparticle is just itself.

But both Dirac and Majorana neutrinos can have the *transition* dipole moments (i.e., from one mass eigenstate to another mass eigenstate), which may result in neutrino decays, neutrino–electron scattering, neutrino interactions with external magnetic fields, etc.[39] In a realistic neutrino–electron scattering experiment, what can be constrained is actually an effective transition dipole moment μ_eff consisting of both electric and magnetic components. Hence it is practically impossible to distinguish between Dirac and Majorana neutrinos in such measurements. Current experimental upper bounds on μ_eff are at the level of $10^{-11}\mu_\text{B}$,[39] far above the afore-mentioned theoretical expectation $\mu_\nu \sim 10^{-20}\mu_\text{B}$.

3.2. *Lepton flavor mixing and neutrino oscillations*

In the basis where the flavor eigenstates of three charged leptons are identified with their mass eigenstates, one may diagonalize the Majorana neutrino mass matrix M_ν by means of a unitary transformation. Then the leptonic charged-current interactions in Eq. (1) can be reexpressed in terms of the mass eigenstates:

$$-\mathcal{L}_\text{cc} = \frac{g}{\sqrt{2}}\, \overline{(e\ \mu\ \tau)_\text{L}}\ \gamma^\mu\ U \begin{pmatrix} \nu_1 \\ \nu_2 \\ \nu_3 \end{pmatrix}_\text{L} W^-_\mu + \text{h.c.}, \tag{3}$$

where the 3×3 unitary matrix U describes the strength of lepton flavor mixing and can be parameterized by using three rotation angles and three CP-violating phases:

$$U = \begin{pmatrix} c_{12}c_{13} & s_{12}c_{13} & s_{13}e^{-\text{i}\delta} \\ -s_{12}c_{23} - c_{12}s_{13}s_{23}e^{\text{i}\delta} & c_{12}c_{23} - s_{12}s_{13}s_{23}e^{\text{i}\delta} & c_{13}s_{23} \\ s_{12}s_{23} - c_{12}s_{13}c_{23}e^{\text{i}\delta} & -c_{12}s_{23} - s_{12}s_{13}c_{23}e^{\text{i}\delta} & c_{13}c_{23} \end{pmatrix} P_\nu, \tag{4}$$

where $c_{ij} \equiv \cos\theta_{ij}$, $s_{ij} \equiv \sin\theta_{ij}$ (for $ij = 12, 13, 23$), δ is referred to as the Dirac CP-violating phase, and $P_\nu = \text{Diag}\left\{e^{\text{i}\rho}, e^{\text{i}\sigma}, 1\right\}$ contains two extra phase parameters of the Majorana nature. The matrix U is often called the Pontecorvo–Maki–Nakagawa–Sakata (PMNS) matrix, and its unitarity has been tested at the percent level.[40,a]

Equation (3) tells us that a ν_α neutrino can be produced from the $W^+ + \alpha^- \to \nu_\alpha$ interaction, and a ν_β neutrino can be detected through the $\nu_\beta + W^- \to \beta^-$ interaction (for $\alpha, \beta = e, \mu, \tau$). The $\nu_\alpha \to \nu_\beta$ oscillation may happen if the ν_i beam

[a]Note that whether U is unitary or not depends on the mechanism of neutrino mass generation. In the canonical seesaw mechanism,[36] for instance, the mixing between light and heavy Majorana neutrinos may lead to tiny unitarity-violating effects for the PMNS matrix U itself.

with energy $E \gg m_i$ travels a proper distance L in vacuum. The probability of such a flavor oscillation is given by[29]

$$P(\nu_\alpha \to \nu_\beta) = \delta_{\alpha\beta} - 4\sum_{i<j}\left(\text{Re}\lozenge^{ij}_{\alpha\beta}\sin^2\Delta_{ji}\right) + 8\text{Im}\lozenge^{ij}_{\alpha\beta}\prod_{i<j}\sin\Delta_{ji}, \quad (5)$$

in which $\Delta_{ji} \equiv \Delta m^2_{ji}L/(4E)$ and $\lozenge^{ij}_{\alpha\beta} \equiv U_{\alpha i}U_{\beta j}U^*_{\alpha j}U^*_{\beta i}$ (for $i, j = 1, 2, 3$ and $\alpha, \beta = e, \mu, \tau$). The probability of the $\overline{\nu}_\alpha \to \overline{\nu}_\beta$ oscillation can easily be read off from Eq. (5) by making the replacement $U \to U^*$. There are two types of neutrino oscillation experiments: the "appearance" one ($\alpha \neq \beta$) and the "disappearance" one ($\alpha = \beta$). Both solar neutrino oscillations ($\nu_e \to \nu_e$) and reactor antineutrino oscillations ($\overline{\nu}_e \to \overline{\nu}_e$) are of the disappearance type. The atmospheric muon-neutrino (or muon-antineutrino) oscillations essentially belong to the disappearance type, and the accelerator neutrino oscillations can be of either type.

At this point let us explain why it is extremely difficult to do a realistic neutrino–antineutrino oscillation experiment. We consider an $\overline{\nu}_\alpha$ beam produced from the standard charged-current interactions $\alpha^+ + W^- \to \overline{\nu}_\alpha$. After traveling a distance L this beam will be detected at a detector through the standard charged-current interactions $\nu_\beta \to \beta^- + W^+$. Different from the normal $\nu_\alpha \to \nu_\beta$ or $\overline{\nu}_\alpha \to \overline{\nu}_\beta$ oscillations, the $\overline{\nu}_\alpha \to \nu_\beta$ oscillation involves a suppression factor m_i/E in its amplitude. This factor reflects the fact that the incoming α^+ leads to an antineutrino $\overline{\nu}_\alpha$ in a dominantly right-handed helicity state, whereas the standard charged-current interactions that produce the outgoing β^- would prefer the incident neutrino ν_β being in a left-handed state.[41] Because of $m_i \lesssim 1$ eV and $E \gtrsim 1$ MeV in a realistic experiment, this helicity suppression factor (i.e., $m_i/E \lesssim 10^{-6}$) makes it impossible to observe the phenomenon of neutrino–antineutrino oscillations.

4. Observations of Neutrino Oscillations

4.1. *Solar neutrino oscillations*

In 1946 Pontecorvo put forward a radiochemical technique which can be used to measure solar electron neutrinos via the reaction $^{37}\text{Cl} + \nu_e \to {}^{37}\text{Ar} + e^-$.[7] The incident neutrino's energy threshold for this reaction to happen is 0.814 MeV, low enough to make it sensitive to solar ^{8}B neutrinos. In 1964 John Bahcall carefully calculated the solar neutrino flux and the capture rate of ^{8}B neutrinos, demonstrating the experimental feasibility of Pontecorvo's idea.[42] This motivated Davis to build a 10^5-gallon Chlorine-Argon neutrino detector in the Homestake Gold Mine in the middle of the 1960s. The final result of this experiment was published in 1968 and caused a big puzzle: the measured flux of solar ^{8}B neutrinos was only about one third of the value predicted by the standard solar model (SSM).[21] Such a deficit was later confirmed in a number of solar neutrino experiments, including the Homestake,[43] GALLEX/GNO,[44] SAGE,[45] SK[46] and SNO[47] experiments. Among them, the SNO experiment was especially crucial because it model-independently demonstrated the flavor conversion of solar ν_e neutrinos into ν_μ and ν_τ neutrinos.

Given heavy water as the target material of the SNO detector, the solar ^{8}B neutrinos were measured via the charged-current (CC) reaction $\nu_e + \mathrm{D} \to e^- + p + p$, the neutral-current (NC) reaction $\nu_\alpha + \mathrm{D} \to \nu_\alpha + p + n$ and the elastic-scattering process $\nu_\alpha + e^- \to \nu_\alpha + e^-$ (for $\alpha = e, \mu, \tau$).[47] The observed neutrino fluxes in these three different channels are expected to satisfy $\phi_{\mathrm{CC}} = \phi_e$, $\phi_{\mathrm{NC}} = \phi_e + \phi_{\mu\tau}$ and $\phi_{\mathrm{ES}} = \phi_e + 0.155\phi_{\mu\tau}$, where $\phi_{\mu\tau}$ denotes a sum of the fluxes of ν_μ and ν_τ neutrinos. So $\phi_{\mathrm{CC}} = \phi_{\mathrm{NC}} = \phi_{\mathrm{ES}}$ would hold if there were no flavor conversion (i.e., $\phi_{\mu\tau} = 0$). The SNO data $\phi_{\mathrm{CC}} = 1.68^{+0.06}_{-0.06}(\mathrm{stat})^{+0.08}_{-0.09}(\mathrm{syst})$, $\phi_{\mathrm{NC}} = 4.94^{+0.21}_{-0.21}(\mathrm{stat})^{+0.38}_{-0.34}(\mathrm{syst})$ and $\phi_{\mathrm{ES}} = 2.35^{+0.22}_{-0.22}(\mathrm{stat})^{+0.15}_{-0.15}(\mathrm{syst})$ as illustrated in Fig. 2 from Ref. 48 definitely demonstrated $\phi_{\mu\tau} \neq 0$. Now we are sure that the deficit of solar ^{8}B neutrinos, whose typical energies are about 6 MeV to 7 MeV, is due to $\nu_e \to \nu_\mu$ and $\nu_e \to \nu_\tau$ oscillations modified by significant MSW matter effects in the Sun. A careful analysis shows that the observed survival probability of solar ^{8}B neutrino oscillations can approximate to $P(\nu_e \to \nu_e) \simeq \sin^2\theta_{12} \simeq 0.32$,[49] leading us to $\theta_{12} \simeq 34°$.

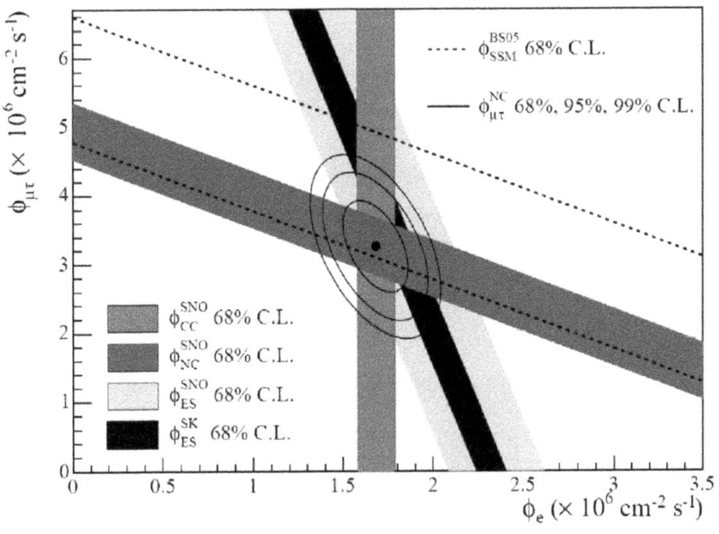

Fig. 2. The $\nu_\mu + \nu_\tau$ flux versus the ν_e flux determined from the SNO data. The total solar ^{8}B neutrino flux predicted by the SSM is shown as dashed lines, parallel to the NC measurement. The narrowed band parallel to the SNO's ES measurement corresponds to the SK's ES result. The best-fit point is obtained by using only the SNO data.[48]

Moreover, the Borexino experiment has accomplished a real-time measurement of the mono-energetic solar ^{7}Be neutrinos with $E = 0.862$ MeV and observed a remarkable deficit corresponding to $P(\nu_e \to \nu_e) = 0.56 \pm 0.1$.[50] Such a result can roughly be explained as a vacuum oscillation effect, because the low-energy ^{7}Be neutrino oscillation is not very sensitive to matter effects.[49] In this case we are left with the averaged survival probability $P(\nu_e \to \nu_e) \simeq 1 - \sin^2 2\theta_{12}/2 \simeq 0.56$ as a reasonable approximation for solar ^{7}Be neutrinos, and thus obtain $\theta_{12} \simeq 35^\circ$. This result is essentially consistent with the one extracted from solar ^{8}B neutrinos.

4.2. *Atmospheric neutrino oscillations*

The atmospheric ν_μ, $\overline{\nu}_\mu$, ν_e and $\overline{\nu}_e$ events are produced in the Earth's atmosphere by cosmic rays, mainly via the decays $\pi^+ \to \mu^+ + \nu_\mu$ with $\mu^+ \to e^+ + \nu_e + \overline{\nu}_\mu$ and $\pi^- \to \mu^- + \overline{\nu}_\mu$ with $\mu^- \to e^- + \overline{\nu}_e + \nu_\mu$. So the ratio of ν_μ and $\overline{\nu}_\mu$ events to ν_e and $\overline{\nu}_e$ events is expected to be nearly 2 : 1 at low energies ($\lesssim$1 GeV). But a smaller ratio was observed at the Kamiokande[51] and IMB[52] detectors in the late 1980s and early 1990s, indicating a preliminary deficit of atmospheric muon neutrinos and muon antineutrinos. If there were no neutrino oscillation, the atmospheric neutrinos that enter and excite an underground detector would have an almost perfect spherical symmetry. Namely, the downward-going and upward-going neutrino fluxes should be equal to each other, or equivalently $\Phi_e(\theta_z) = \Phi_e(\pi - \theta_z)$ and $\Phi_\mu(\theta_z) = \Phi_\mu(\pi - \theta_z)$ for the zenith angle θ_z. In 1998 the SK Collaboration observed an approximate up-down flux symmetry for atmospheric ν_e and $\overline{\nu}_e$ events and a significant up-down flux asymmetry for atmospheric ν_μ and $\overline{\nu}_\mu$ events.[27]

The SK detector is a 5×10^4-ton tank of ultra-pure water, located approximately 1 km underground in the Mozumi Mine in Kamioka. As illustrated in Fig. 3, the inside surface of the tank is lined with more than 1.1×10^4 photo-multiplier tubes (PMTs). An additional layer of water called the outer detector is also instrumented PMTs to detect any charged particles entering the central volume and to shield the inner detector by absorbing any neutrons produced in the nearby rock. A neutrino

Fig. 3. A brief view from inside the SK detector's water tank during filling.[27] SK image copyright: Kamioka Observatory, ICRR (Institute for Cosmic Ray Research), The University of Tokyo.

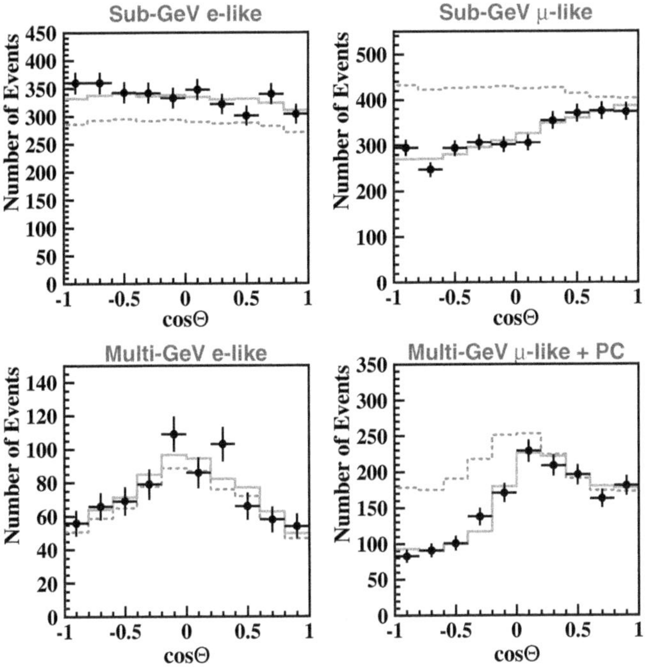

Fig. 4. The SK zenith-angle distributions for fully contained 1-ring e-like and μ-like events with visible energy <1.33 GeV (sub-GeV) and >1.33 GeV (multi-GeV). For multi-GeV μ-like events, a combined distribution with partially contained events is illustrated. The dotted histograms show the non-oscillation Monte Carlo events, and the solid histograms show the best-fit expectations for atmospheric $\nu_\mu \to \nu_\mu$ oscillations.[26]

interacting with the electrons or nuclei of water can produce a charged particle that moves faster than the speed of light in water, creating a cone of light known as Cherenkov radiation. The Cherenkov light is projected as a ring on the wall of the detector and recorded by the PMTs. Hence the direction and flavor of an incident neutrino can be identified by using the details of the ring pattern.

As shown in Fig. 4, the observed deficit of atmospheric upward-going ν_μ and $\overline{\nu}_\mu$ events at SK could naturally be attributed to $\nu_\mu \to \nu_\tau$ and $\overline{\nu}_\mu \to \overline{\nu}_\tau$ oscillations, because the detector itself was insensitive to ν_τ and $\overline{\nu}_\tau$ events. This was actually the first *model-independent* evidence for neutrino oscillations, and it marked the threshold of a new era in particle physics. Since 1998 a number of breakthroughs have been made in experimental neutrino physics.

In 2004 the SK Collaboration carried out a careful analysis of the ν_μ (or $\overline{\nu}_\mu$) disappearance probability as a function of the neutrino flight length L over the neutrino energy E, and observed a dip in the L/E distribution as the first *direct* evidence for atmospheric neutrino oscillations.[53] This dip was consistent with the prediction from the sinusoidal flavor transition probability of neutrino oscillations, but inconsistent with the exotic neutrino decay and neutrino decoherence scenarios.

To directly observe the atmospheric $\nu_\mu \to \nu_\tau$ oscillation is quite difficult because it requires the neutrino beam energy greater than a threshold of 3.5 GeV, such that a tau lepton can be produced via the charged-current interaction of incident ν_τ with the target nuclei in the detector. But the SK data are found to be best described by neutrino oscillations that include the ν_τ appearance in addition to the overwhelming signature of the ν_μ disappearance. A neural network analysis of the zenith-angle distribution of multi-GeV contained events has recently demonstrated this observation at the 3.8σ level.[54]

4.3. *Accelerator neutrino oscillations*

If the observed deficit of atmospheric ν_μ and $\overline{\nu}_\mu$ events is ascribed to neutrino oscillations, then a fraction of the accelerator-produced ν_μ and $\overline{\nu}_\mu$ events should also disappear on their way to a remote detector. This expectation has definitely been confirmed by two long-baseline neutrino oscillation experiments: K2K[55] and MINOS.[56] The K2K experiment was designed in such a way that the ν_μ beam was produced at the KEK accelerator and measured 250 km away at the SK detector in Kamioka. In comparison, the baseline length of the MINOS experiment is 735 km, from the source of ν_μ neutrinos at Fermilab to the far detector in northern Minnesota. Both of them have observed a reduction of the ν_μ flux and a distortion of the ν_μ energy spectrum, implying $\nu_\mu \to \nu_\mu$ oscillations. The most striking result obtained from the atmospheric and accelerator neutrino oscillation experiments is $\sin^2 2\theta_{23} \simeq 1$ or $\theta_{23} \simeq 45^\circ$, which might hint at a special flavor structure or a certain flavor symmetry in the neutrino sector.[57]

An especially important accelerator neutrino oscillation experiment is the T2K experiment with a ν_μ beam produced from the J-PARC Main Ring in Tokai and pointing to the SK detector at a distance of 295 km. Its main goal is to discover $\nu_\mu \to \nu_e$ appearance oscillations and perform a precision measurement of $\nu_\mu \to \nu_\mu$ disappearance oscillations. Since its preliminary data were first released in June 2011, the T2K experiment has proved to be very successful in establishing the ν_e appearance out of a ν_μ beam at the 7.3σ level and constraining the neutrino mixing parameters θ_{13}, θ_{23} and δ.[58] The point is that the leading term of $P(\nu_\mu \to \nu_e)$ is sensitive to $\sin^2 2\theta_{13} \sin^2 \theta_{23}$, and its sub-leading term is sensitive to δ and terrestrial matter effects.[59] Figure 5 shows the allowed region of $\sin^2 2\theta_{13}$ changing with the CP-violating phase δ as constrained by the T2K data,[58] from which one can see an unsuppressed value of θ_{13} together with a preliminary hint $\delta \sim -\pi/2$ even though the neutrino mass ordering (i.e., the sign of Δm^2_{32}) remains undetermined.

Different from the K2K, MINOS and T2K experiments, the OPERA experiment was designed to search for the ν_τ appearance in a ν_μ beam traveling from CERN to Gran Sasso at a distance of 730 km. After several years of data taking, the OPERA Collaboration reported four ν_τ candidate events in 2014. These events are consistent with $\nu_\mu \to \nu_\tau$ oscillations with the 4.2σ significance.[60]

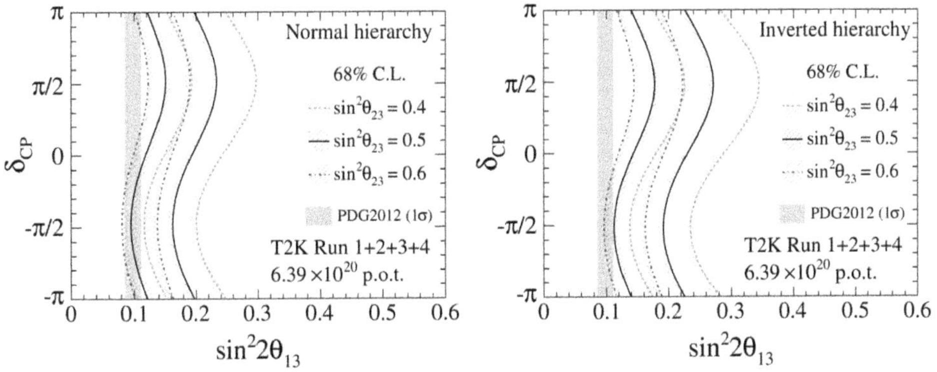

Fig. 5. The allowed region of $\sin^2 2\theta_{13}$ as a function of the CP-violating phase δ, constrained by the present T2K neutrino oscillation data.[58]

4.4. *Reactor antineutrino oscillations*

Since the first discovery of electron antineutrinos with the help of the Savannah River reactor in 1956,[8] reactors have been playing an important role in neutrino physics. In particular, two of the three neutrino mixing angles (θ_{12} and θ_{13}) have been measured in the KamLAND[61] and Daya Bay[62] reactor antineutrino oscillation experiments to an unprecedentedly good degree of accuracy.

The average baseline length of the KamLAND experiment was $L = 180$ km, and hence it was sensitive to the Δm^2_{21}-driven $\overline{\nu}_e \to \overline{\nu}_e$ oscillation and allowed a terrestrial test of the large-mixing-angle (LMA) MSW solution to the solar neutrino problem. Under CPT invariance the KamLAND measurement[61] firmly established the LMA solution for the first time, and pinned down the correct parameter space of solar $\nu_e \to \nu_e$ oscillations constrained by the SNO and SK experiments, as shown in Fig. 6 in the two-flavor scheme.[63] A striking sinusoidal behavior of $P(\overline{\nu}_e \to \overline{\nu}_e)$ against L/E was also demonstrated in the KamLAND experiment.[63]

While the CHOOZ[64] and Palo Verde[65] reactor antineutrino experiments tried to search for the Δm^2_{31}-driven $\overline{\nu}_e \to \overline{\nu}_e$ oscillations at the end of the 20th century, they found no indication in favor of such oscillations and thus set an upper bound on the smallest neutrino mixing angle θ_{13}. This situation has been changed by the Daya Bay,[62] RENO[66] and Double Chooz[67] experiments in the past few years.

The Daya Bay experiment was designed to probe the smallest neutrino mixing angle θ_{13} with an unprecedented sensitivity $\sin^2 2\theta_{13} \sim 1\%$ by measuring the Δm^2_{31}-driven $\overline{\nu}_e \to \overline{\nu}_e$ oscillation with a baseline length $L \simeq 2$ km. In this experiment the electron antineutrino beam takes its source at the Daya Bay nuclear power complex located in Shenzhen, as shown in Fig. 7. The eight antineutrino detectors deployed at the near (two plus two) and far (four) sites are all the liquid scintillator detectors.

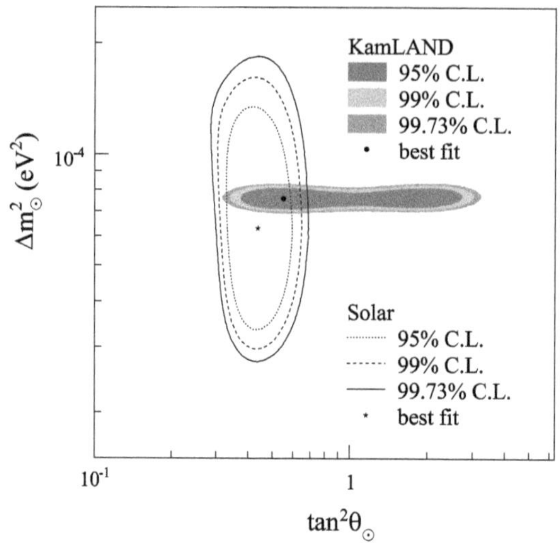

Fig. 6. The allowed region for two-flavor neutrino oscillation parameters from the KamLAND and solar neutrino experiments, where $\Delta m^2_\odot \simeq \Delta m^2_{21}$ and $\tan^2\theta_\odot \simeq \tan^2\theta_{12}$ hold.[63]

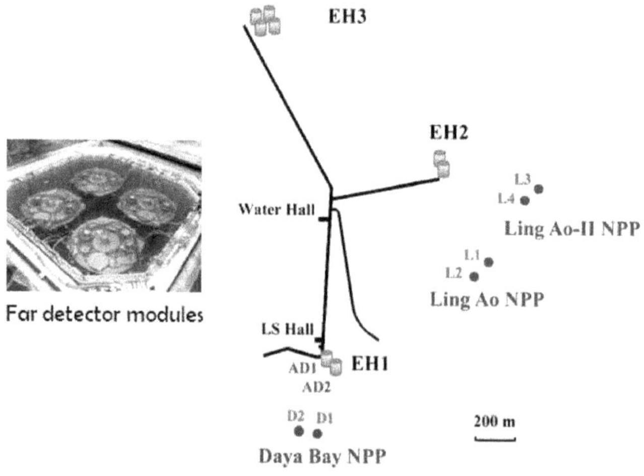

Fig. 7. The layout of the Daya Bay reactor antineutrino experiment with three pairs of reactor cores (Daya Bay, Ling Ao I and Ling Ao II). Four detector modules are deployed at the far site, and two detector modules are deployed at each of the two near sites.[62]

In March 2012 the Daya Bay Collaboration announced a 5.2σ discovery of $\theta_{13} \neq 0$, with $\sin^2 2\theta_{13} = 0.092 \pm 0.016(\text{stat}) \pm 0.005(\text{syst})$ (see Fig. 8 for illustration).[62] A similar but slightly less significant result was later achieved in the RENO[66] and Double Chooz[67] reactor antineutrino experiments.

The Daya Bay Collaboration has also measured the energy dependence of $\overline{\nu}_e$ disappearance and observed a nearly full oscillation cycle against L/E.[68] An improved result of the oscillation amplitude $\sin^2 2\theta_{13} = 0.090^{+0.008}_{-0.009}$ has recently been obtained

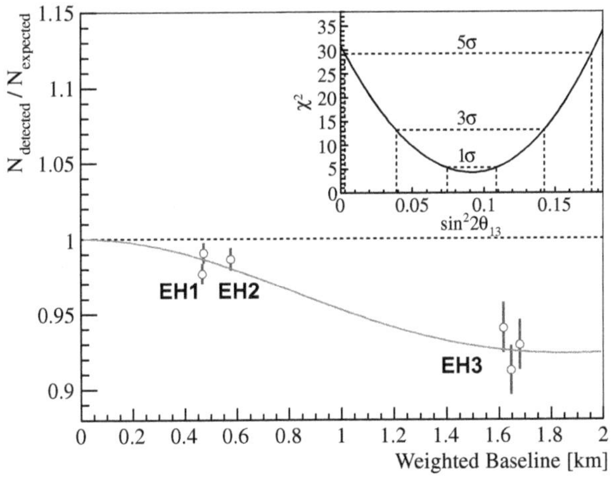

Fig. 8. The survival probability of $\overline{\nu}_e \to \overline{\nu}_e$ oscillations observed at the near and far experimental halls (i.e., EH1, EH2 and EH3) in the Daya Bay experiment.[62]

by using the observed $\overline{\nu}_e$ rate and the observed energy spectrum in the three-flavor framework.[68] The relative large value of θ_{13} is very encouraging for the next-generation precision neutrino experiments, which aim to determine the neutrino mass ordering and probe leptonic CP violation in the foreseeable future.

4.5. *Determination of oscillation parameters*

The aforementioned neutrino or antineutrino oscillation experiments involve different sources, different flavors, different energies and different baseline lengths. But the relevant experimental data can all be explained in the scheme of three-flavor oscillations, which depend on two independent neutrino mass-squared differences (Δm^2_{21}, Δm^2_{32}), three flavor mixing angles (θ_{12}, θ_{13}, θ_{23}) and one CP-violating phase (δ). A global fit of all the available experimental data is therefore needed in order to determine or constrain the six oscillation parameters.

A global three-flavor analysis of current experimental data on solar (SNO, SK, Borexino), atmospheric (SK), accelerator (MINOS, T2K) and reactor (KamLAND, Daya Bay, RENO) neutrino or antineutrino oscillations has recently been done by several groups.[69–71] For the sake of simplicity, here we only quote the main results obtained by the Italian group,[69,b] as listed in Table 1.

Table 1 shows that the output values of θ_{13}, θ_{23} and δ in such a global fit are sensitive to the sign of Δm^2_{31}. That is why it is crucial to determine the neutrino mass ordering in the upcoming neutrino oscillation experiments. The hint $\delta \neq 0^\circ$ (or 180°) at the 1σ level is still preliminary but quite encouraging, because

[b]In this reference the notations $\delta m^2 \equiv m_2^2 - m_1^2$ and $\Delta m^2 \equiv m_3^2 - (m_1^2 + m_2^2)/2$ are used. Their relations with Δm^2_{21} and Δm^2_{31} are rather simple: $\Delta m^2_{21} = \delta m^2$ and $\Delta m^2_{31} = \Delta m^2 + \delta m^2/2$.

Table 1. The three-flavor neutrino oscillation parameters determined or constrained from a global analysis of current experimental data.[69]

Parameter	Best fit	1σ range	2σ range	3σ range
		Normal neutrino mass ordering ($m_1 < m_2 < m_3$)		
$\Delta m^2_{21}/10^{-5}$ eV^2	7.54	7.32 — 7.80	7.15 — 8.00	6.99 — 8.18
$\Delta m^2_{31}/10^{-3}$ eV^2	2.47	2.41 — 2.53	2.34 — 2.59	2.26 — 2.65
$\sin^2\theta_{12}/10^{-1}$	3.08	2.91 — 3.25	2.75 — 3.42	2.59 — 3.59
$\sin^2\theta_{13}/10^{-2}$	2.34	2.15 — 2.54	1.95 — 2.74	1.76 — 2.95
$\sin^2\theta_{23}/10^{-1}$	4.37	4.14 — 4.70	3.93 — 5.52	3.74 — 6.26
$\delta/180°$	1.39	1.12 — 1.77	0.00 — 0.16 $\oplus$ 0.86 — 2.00	0.00 — 2.00
		Inverted neutrino mass ordering ($m_3 < m_1 < m_2$)		
$\Delta m^2_{21}/10^{-5}$ eV^2	7.54	7.32 — 7.80	7.15 — 8.00	6.99 — 8.18
$\Delta m^2_{13}/10^{-3}$ eV^2	2.42	2.36 — 2.48	2.29 — 2.54	2.22 — 2.60
$\sin^2\theta_{12}/10^{-1}$	3.08	2.91 — 3.25	2.75 — 3.42	2.59 — 3.59
$\sin^2\theta_{13}/10^{-2}$	2.40	2.18 — 2.59	1.98 — 2.79	1.78 — 2.98
$\sin^2\theta_{23}/10^{-1}$	4.55	4.24 — 5.94	4.00 — 6.20	3.80 — 6.41
$\delta/180°$	1.31	0.98 — 1.60	0.00 — 0.02 $\oplus$ 0.70 — 2.00	0.00 — 2.00

it implies a potential effect of leptonic CP violation which is likely to show up in some long-baseline neutrino oscillation experiments in the foreseeable future. The possibility $\theta_{23} = 45°$ cannot be ruled out at the 2σ level, and thus a more precise determination of θ_{23} is required in order to resolve its octant.

It is worth pointing out that $|U_{\mu i}| = |U_{\tau i}|$ (for $i = 1, 2, 3$), the so-called μ-τ permutation symmetry of the PMNS matrix U itself, holds if either the conditions $\theta_{13} = 0°$ and $\theta_{23} = 45°$ or the conditions $\delta = 90°$ (or $270°$) and $\theta_{23} = 45°$ are satisfied.[72] Now that $\theta_{13} = 0°$ has definitely been excluded, it is imperative to know the values of θ_{23} and δ as accurately as possible, so as to fix the strength of μ-τ symmetry breaking associated with the structure of U.

5. Neutrino Mass Ordering and CP Violation

The neutrino mass ordering can be explored with either reactor electron antineutrinos or atmospheric muon neutrinos in the "disappearance" oscillation experiments, or with accelerator muon neutrinos in the "appearance" oscillation experiments. Let us take the JUNO,[73] PINGU[74] and LBNE[75] experiments for example to illustrate the future prospects in this regard.

The JUNO electron antineutrino detector is expected to be a 20-kiloton liquid-scintillator detector located in the Jiangmen city of Guangdong province in southern China, about 53 km away from the Yangjiang (17.4 GW$_{\rm th}$) and Taishan (18.4 GW$_{\rm th}$) reactor facilities which serve as the $\overline{\nu}_e$ source. Given Eq. (5), the survival probability of $\overline{\nu}_e \to \overline{\nu}_e$ oscillations can be explicitly expressed as

$$P(\overline{\nu}_e \to \overline{\nu}_e) = 1 - \sin^2 2\theta_{12} \cos^4 \theta_{13} \sin^2 \Delta_{21} - \frac{1}{2} \sin^2 2\theta_{13} \, [1 - \cos \Delta_* \cos \Delta_{21} + \cos 2\theta_{12} \sin \Delta_* \sin \Delta_{21}] \, , \tag{6}$$

where $\Delta_* \equiv \Delta_{31} + \Delta_{32}$. In Eq. (6) the oscillating argument Δ_{21} is unambiguous, and the neutrino mass ordering is determined by the sign of Δ_* (normal: positive; inverted: negative). To distinguish the inverted neutrino mass hierarchy from the normal one, it is necessary to measure the Δ_*-driven oscillations over many cycles on condition that $\Delta_{21} \sim \pi/2$ is satisfied for $L \sim 53$ km as taken in the JUNO experiment.[76] Figure 9 illustrates why this idea works.

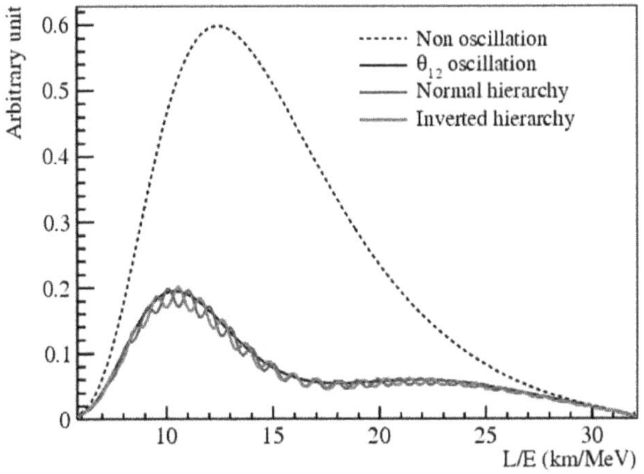

Fig. 9. The reactor antineutrino spectrum changing with L/E at a baseline $L \sim 53$ km, where the blue (normal) or red (inverted) fine structure can tell the neutrino mass hierarchy after a Fourier transformation of the spectrum.[76]

Now the JUNO experiment's civil construction is underway, and its detector assembly is planned for 2018 to 2019. Data taking will commence in 2020, with a target of about six years of operation to pin down the neutrino mass ordering at the 3σ or 4σ level.[73] The challenges for this experiment, which must be met successfully, are mainly technological, such as how to improve the scintillator light yield, attenuation length and PMT quantum efficiency.[77]

The PINGU experiment is a proposed low-energy infill extension of the IceCube experiment at the South Pole.[74] Its design closely follows the one used for IceCube and DeepCore. The idea is to further infill the central DeepCore volume with 40 new strings of 60 optical modules each, so that the neutrino trigger energy threshold can be lowered to a few GeV and thus high-quality reconstructions for neutrino events can be achieved between 5 and 15 GeV. Such a detector geometry will be able to distinguish between the normal and inverted neutrino mass hierarchies at the 3σ significance with an estimated 3.5 years of data taking.

The survival probability of atmospheric muon neutrinos that reach the PINGU detector after propagation through the Earth (i.e., from below) depends on their beam energy E and propagation length L. Thanks to interactions with electrons within the Earth, a resonant flavor conversion can happen at a specific pattern of neutrino energies and Earth-crossing paths. This matter-induced resonant conversion occurs only for neutrinos in the normal mass ordering or only for antineutrinos in the inverted mass ordering, as the behaviors of $\nu_\mu \to \nu_\mu$ and $\overline{\nu}_\mu \to \overline{\nu}_\mu$ oscillations depend respectively on $\Delta m^2_{31} \mp 2\sqrt{2}G_{\rm F}N_e E$, where N_e is the number density of electrons in matter and E denotes the neutrino beam energy. The PINGU detector is capable of discriminating the cross sections and kinematics of neutrino and antineutrino interactions with nuclei, so it is capable of identifying different detected event rates which depend on different neutrino mass orderings.

Given an accelerator-driven neutrino beam, the long-baseline oscillation experiments are also sensitive to the neutrino mass ordering. Because of the interaction of neutrinos with terrestrial matter as they pass through the Earth, the probability of $\nu_\mu \to \nu_e$ oscillations can be approximately expressed as[59]

$$\begin{aligned} P(\nu_\mu \to \nu_e) \simeq\ & \sin^2 2\theta_{13} \sin^2\theta_{23} \frac{\sin^2(x-1)\Delta_{31}}{(x-1)^2} + \alpha \sin 2\theta_{12} \sin 2\theta_{13} \sin 2\theta_{23} \\ & \times \cos(\Delta_{31}+\delta) \frac{\sin x\Delta_{31} \sin(x-1)\Delta_{31}}{x(x-1)} \\ & + \alpha^2 \sin^2 2\theta_{12} \cos^2\theta_{23} \frac{\sin^2 x\Delta_{31}}{x^2} , \end{aligned} \tag{7}$$

where $x \equiv 2\sqrt{2}G_{\rm F}N_e E/\Delta m^2_{31}$ and $\alpha \equiv \Delta m^2_{21}/\Delta m^2_{31}$. One may easily obtain the expression of $P(\overline{\nu}_\mu \to \overline{\nu}_e)$ from Eq. (7) with the replacements $\delta \to -\delta$ and $x \to -x$. So the sign of Δm^2_{31} affects the behaviors of neutrino oscillations via the signs of x and α. That is why the matter-induced resonant conversion can only occur for neutrinos in the normal mass hierarchy ($x > 0$) or for antineutrinos in the inverted mass hierarchy ($x < 0$), similar to the case of atmospheric neutrino or antineutrino oscillations. In practice the baseline length L of an experiment is crucial for its sensitivity to the mass hierarchy. The LBNE experiment[75] with $L \simeq 1300$ km is therefore expected to be more promising than the T2K experiment[58] with $L \simeq 295$ km and the NOνA experiment[78] with $L \simeq 810$ km in this respect. But the undetermined CP-violating phase δ may in general give rise to some uncertainties associated with a determination of the neutrino mass hierarchy in the long-baseline experiments. In particular, a careful analysis shows that the mass hierarchy sensitivity is most optimistic (or pessimistic) for $\delta \simeq -\pi/2$ in the normal (or inverted) hierarchy case, or for $\delta \simeq +\pi/2$ in the inverted (or normal) hierarchy case.[75] Regardless of possible values of δ, LBNE in combination with T2K and NOνA promises to resolve the neutrino mass hierarchy with a significance of more than 3σ by 2030.[77]

In addition, the proposed Hyper-Kamiokande (HK) detector will be a next-generation underground water Cherenkov detector serving as the far detector of the 295 km-baseline neutrino oscillation experiment for the J-PARC neutrino beam.[79] It is expected to be ten times larger than the SK detector and capable of probing the neutrino mass ordering, resolving the octant of the largest flavor mixing angle θ_{23} and observing leptonic CP violation as well as proton decays and extraterrestrial neutrinos from distant astrophysical sources.

CP violation in the lepton sector may have far-reaching impacts on our understanding of the origin of matter–antimatter asymmetries at both microscales and macroscales. The LBNE and HK experiments, together with other next-generation long-baseline neutrino oscillation experiments, are aiming at a determination of the CP-violating phase δ. The latter can be extracted from comparing between the probabilities of $\nu_\mu \to \nu_e$ and $\overline{\nu}_\mu \to \overline{\nu}_e$ oscillations, but it is in general contaminated by terrestrial matter effects. In the leading-order approximation,

$$\mathcal{A}_{\rm CP} \equiv \frac{P(\nu_\mu \to \nu_e) - P(\overline{\nu}_\mu \to \overline{\nu}_e)}{P(\nu_\mu \to \nu_e) + P(\overline{\nu}_\mu \to \overline{\nu}_e)} \simeq -\frac{\sin 2\theta_{12} \sin\delta}{\sin\theta_{13} \tan\theta_{23}} \Delta_{21} + \text{matter effects}, \quad (8)$$

where the term of matter effects should more or less be correlated with the neutrino mass ordering. To lower the matter contamination, one may therefore consider a low-energy neutrino (or antineutrino) beam with a much shorter baseline length.[80] A proposal of this kind is the MOMENT project with a neutrino beam energy $E \sim 300$ MeV and a baseline length $L \sim 120$ km,[81] towards probing leptonic CP violation before a more powerful neutrino factory is built.

6. Two Non-Oscillation Aspects

6.1. *Neutrinoless double-beta decays*

Soon after Fermi developed an effective beta decay theory,[5] Maria Goeppert-Mayer pointed out that certain even-even nuclei should have a chance to decay into the second nearest neighbors via two simultaneous beta decays:[82] $(A, Z) \to (A, Z+2) + 2e^- + 2\overline{\nu}_e$, where the kinematic conditions $m(A, Z) > m(A, Z+2)$ and $m(A, Z) < m(A, Z+1)$ must be satisfied. In 1939 Wendell Furry further pointed out that the $0\nu\beta\beta$ decays $(A, Z) \to (A, Z+2) + 2e^-$ could happen via an exchange of the *virtual* neutrinos between two associated beta decays,[83] provided the neutrinos are massive and have the Majorana nature.[10] If such a $0\nu\beta\beta$ process is measured, does it definitely imply the existence of a Majorana mass term for neutrinos? The answer is affirmative according to the Schechter–Valle theorem,[84] no matter whether there are new physics contributions to the $0\nu\beta\beta$ decays. Hence the $0\nu\beta\beta$ transitions can serve for an experimentally feasible probe towards identifying the Majorana nature of massive neutrinos at low energies.

The half-life of a $0\nu\beta\beta$-decaying nuclide can be expressed as follows:

$$T^{0\nu}_{1/2} = \left(G^{0\nu}\right)^{-1} \left|M^{0\nu}\right|^{-2} \left|\langle m\rangle_{ee}\right|^{-2} , \qquad \langle m\rangle_{ee} \equiv \sum_i \left(m_i U^2_{ei}\right) , \tag{9}$$

where $G^{0\nu}$ is the phase-space factor, $M^{0\nu}$ stands for the relevant nuclear matrix element, and $\langle m\rangle_{ee}$ denotes the effective Majorana neutrino mass in the absence of new physics contributions. Among them, the calculation of $|M^{0\nu}|$ relies on the chosen nuclear models which are only able to approximately describe the many-body interactions of nucleons in nuclei, and thus it involves the largest theoretical uncertainty (e.g., a factor of two or three for some typical nuclei).[85] This causes quite a big uncertainty associated with the determination of $|\langle m\rangle_{ee}|$.

So far no convincing evidence for an occurrence of the $0\nu\beta\beta$ decay has been established, although a lot of experimental efforts have been made in the past few decades. Such an experiment is designed to observe the two electrons emitted in a given $0\nu\beta\beta$ decay, and its signature is based on the fact that the sum of the energies of the two emitted electrons is equal to the Q-value of this process. In contrast, the energy spectrum of the two emitted electrons in a normal double-beta decay must be continuous. At present the strongest upper bound on the effective mass term $|\langle m\rangle_{ee}|$ can be set by the ${}^{76}_{32}\mathrm{Ge} \to {}^{76}_{34}\mathrm{Se} + 2e^-$ and ${}^{136}_{54}\mathrm{Xe} \to {}^{136}_{56}\mathrm{Ba} + 2e^-$ experiments.[85] In particular, the GERDA,[86] EXO-200[87] and KamLAND-Zen[88] experiments have obtained $T^{0\nu}_{1/2} > 2.1 \times 10^{25}$ yr, 1.1×10^{25} yr and 1.9×10^{25} yr at the 90% confidence level, respectively. These results lead to the constraints $|\langle m\rangle_{ee}| < 0.22$–$0.64$ eV, 0.2–0.69 eV and 0.15–0.52 eV at the same confidence level, respectively, after the relevant uncertainties of nuclear matrix elements are taken into account.[85]

The expected magnitude of $|\langle m\rangle_{ee}|$ in the standard three-flavor case is illustrated in Fig. 10, where current neutrino oscillation data have been input and arbitrary values of the CP-violating phases have been taken.[89] It is clear that the inverted neutrino mass ordering or a near neutrino mass degeneracy may allow $|\langle m\rangle_{ee}| \geq 0.01$ eV, which should be accessible in the next-generation $0\nu\beta\beta$-decay experiments. If the neutrino mass spectrum is normal and hierarchical, however, there will be little prospect of observing any $0\nu\beta\beta$ decays in the foreseeable future, simply because of $|\langle m\rangle_{ee}| \sim \mathcal{O}(10^{-3})$ eV in this unfortunate case.

6.2. *The absolute neutrino mass scale*

Since the flavor oscillations of massive neutrinos are only sensitive to the neutrino mass-squared differences, a determination of the absolute neutrino mass scale has to rely on some non-oscillation experiments. Searching for the $0\nu\beta\beta$ decay is one of the feasible ways for this purpose if massive neutrinos are the Majorana particles, because the magnitude of its effective mass term $\langle m\rangle_{ee}$ is associated with m_i as shown in Eq. (9) and Fig. 10. Another way is to detect the beta decays, such as

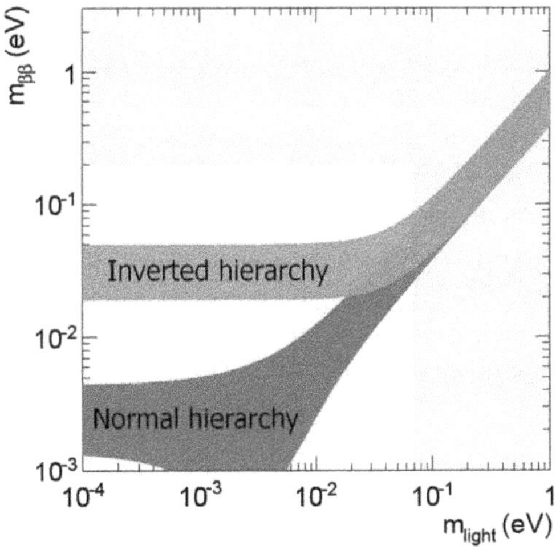

Fig. 10. The effective Majorana neutrino mass $m_{\beta\beta} \equiv |\langle m \rangle_{ee}|$ as a function of the lightest neutrino mass $m_{\rm light} \equiv m_1$ (normal hierarchy, red band) or m_3 (inverted hierarchy, green band).[89] Here the horizontally-excluded region comes from the $0\nu\beta\beta$ experiments,[86–88] and the vertically-excluded region is due to the cosmological bound.[90]

${}^3_1{\rm H} \to {}^3_2{\rm He} + e^- + \overline{\nu}_e$, whose effective neutrino mass term $\langle m \rangle_e$ is defined via

$$\left(\langle m \rangle_e\right)^2 \equiv \sum_i \left(m_i^2 |U_{ei}|^2\right) \, . \tag{10}$$

The most promising experiment of this kind is the KATRIN experiment,[91] which may hopefully probe $\langle m \rangle_e$ with a sensitivity of about 0.2 eV in the near future. But up to now only $\langle m \rangle_e < 2.05$ eV has been obtained at the 95% confidence level from the Troitzk beta-decay experiment.[92]

Furthermore, one may get useful information on the mass scale of light neutrinos from cosmology. Based on the standard ΛCDM model, a global analysis of current cosmological data (especially those on the cosmic microwave background (CMB) radiation and large-scale structure (LSS) formation) can provide us with the most powerful sensitivity to the sum of light neutrino masses via the relation

$$\Omega_\nu h^2 = \frac{1}{93\ {\rm eV}} \Sigma_\nu , \qquad \Sigma_\nu \equiv \sum_i m_i , \tag{11}$$

in which Ω_ν denotes the light neutrino contribution to today's energy density of the Universe, and h is the Hubble constant. For example, $\Sigma_\nu < 0.23$ eV has recently been reported by the Planck Collaboration at the 95% confidence level.[90] If a combination of the next-generation CMB and LSS measurements can reach a sensitivity of about 0.02 eV for the sum of three neutrino masses,[93] then it will be possible to determine the absolute neutrino mass scale via a definite determination of Σ_ν even though the neutrino mass ordering is normal.

Note that it is also possible to determine or constrain the absolute neutrino mass scale m_ν through the study of kinematic effects of supernova neutrinos, because their flight time from a supernova's core to a terrestrial detector will be more or less delayed as compared with the massless particles.[94] A careful analysis of the $\overline{\nu}_e$ events from the Supernova 1987A explosion led us to an upper bound of about 6 eV on m_ν.[95] The prospects of this astrophysical approach depend on the emergence of new neutrino detectors or the existence of antineutrino pulses in the first instants of a supernova explosion.[96] Given the JUNO liquid scintillator detector as an example, $m_\nu < 0.83 \pm 0.24$ eV is expected to be achievable at the 95% confidence level for a typical galactic supernova at a distance of 10 kpc from the Earth.[97]

7. Summary and Outlook

Since 1998, quite a lot of significant breakthroughs have been made in experimental neutrino physics. On the one hand, the exciting phenomena of atmospheric, solar, reactor and accelerator neutrino or antineutrino oscillations have all been observed, and the oscillation parameters Δm^2_{21}, $|\Delta m^2_{31}|$, θ_{12}, θ_{13} and θ_{23} have been determined to an impressive degree of accuracy. On the other hand, the geo-antineutrino events and extraterrestrial PeV neutrino events have been observed, and the sensitivities to neutrino masses in the beta decays, $0\nu\beta\beta$ decays and cosmology have been improved to a great extent. Furthermore, a lot of theoretical efforts have also been made towards understanding the origin of tiny neutrino masses and the flavor structure behind the observed neutrino mixing pattern, and towards studying possible implications of massive neutrinos on the cosmological matter–antimatter asymmetry, warm dark matter and many violent astrophysical processes.[29,98] All these have demonstrated neutrino physics to be one of the most important frontiers of particle physics, astrophysics and cosmology.

But a number of fundamental questions about massive neutrinos remain open. The burning ones include how small the absolute neutrino mass scale is, whether the neutrino mass spectrum is normal or inverted, whether massive neutrinos are the Majorana particles, how large the CP-violating phase δ is, which octant the largest flavor mixing angle θ_{23} belongs to, whether there are light and (or) heavy sterile neutrinos, what the role of neutrinos is in dark matter, whether the observed matter-antimatter asymmetry of the Universe is related to CP violation in neutrino oscillations, etc. Motivated by so many questions, we are trying to discover a new physics world with the help of massive neutrinos in the coming decades.

Acknowledgments

We would like to thank Luciano Maiani and Gigi Rolandi for inviting us to contribute to this book. We are also grateful to Yu-Feng Li, Jue Zhang, Zhen-hua Zhao, Shun Zhou and Ye-Ling Zhou for their helpful comments on this essay. This work is supported in part by the National Natural Science Foundation of China under grant No. 11135009 and 11390380; by the National Basic Research Program of China under grant No. 2013CB834300; by the Strategic Priority Research Program of the Chinese Academy of Sciences (CAS) under grant No. XDA10000000; and by the CAS Center for Excellence in Particle Physics.

References

1. A. H. Becquerel, *Compt. Rend. Math.* **122**, 420 (1896).
2. J. Chadwick, *Verhandle. Deut. Phys.* **16**, 383 (1914).
3. C. D. Ellis and W. A. Wooster, *Proc. Roy. Soc. London A* **117**, 109 (1927).
4. W. Pauli, Lecture given in Zürich in 1957, published in *Physik und Erkenntnistheorie* (Friedr, Vieweg, & Sohn, Braunschweig/Wiesbaden, 1984), p. 156.
5. E. Fermi, *La Ricerca Scientifica* **2**, 12 (1933); *Z. Phys.* **88**, 161 (1934).
6. H. A. Bethe and R. F. Bacher, *Rev. Mod. Phys.* **8**, 184 (1936).
7. B. Pontecorvo, Chalk River Lab. Report PD-205 (1946).
8. C. L. Cowan, F. Reines, F. B. Harrison, H. W. Kruse, and A. D. McGuire, *Science* **124**, 103 (1956).
9. B. Pontecorvo, *Sov. Phys. JETP* **6**, 429 (1957).
10. E. Majorana, *Nuovo Cim.* **14**, 171 (1937).
11. G. Danby *et al.*, *Phys. Rev. Lett.* **9**, 36 (1962).
12. Z. Maki, M. Nakagawa, and S. Sakata, *Prog. Theor. Phys.* **28**, 870 (1962).
13. K. Kodama *et al.* (DONUT Collaboration), *Phys. Rev. Lett.* **504**, 218 (2001).
14. J. A. Formaggio and G. P. Zeller, *Rev. Mod. Phys.* **84**, 1307 (2012).
15. S. L. Glashow, *Phys. Rev.* **118**, 316 (1960).
16. S. Weinberg, *Phys. Rev.* **128**, 1457 (1962); J. M. Irvine and R. Humphreys, *J. Phys. G.* **9**, 847 (1983); A. Cocco, G. Mangano, and M. Messina, *JCAP* **0706**, 015 (2007); Y. F. Li, Z. Z. Xing, and S. Luo, *Phys. Lett. B* **692**, 261 (2010).
17. S. Betts *et al.*, arXiv:1307.4738 (2013).
18. T. Araki *et al.* (KamLAND Collaboration), *Nature* **436**, 499 (2005); S. Abe *et al.* (KamLAND Collaboration), *Phys. Rev. Lett.* **100**, 221803 (2008).
19. G. Bellini *et al.* (Borexino Collaboration), *Phys. Lett. B* **687**, 299 (2010).
20. H. A. Bethe, *Phys. Rev.* **55**, 434 (1939); G. Gamow and M. Schönberg, *Phys. Rev.* **58**, 1117 (1940); *Phys. Rev.* **59**, 539 (1941).
21. R. Davis, D. S. Harmer, and K. C. Hoffman, *Phys. Rev. Lett.* **20**, 1205 (1968); J. N. Bahcall and G. Shaviv, *Phys. Rev. Lett.* **20**, 1209 (1968).
22. H. A. Bethe and J. R. Wilson, *Astrophys. J.* **295**, 14 (1985); H. A. Bethe, *Rev. Mod. Phys.* **62**, 801 (1990).
23. K. Hirata *et al.* (Kamiokande Collaboration), *Phys. Rev. Lett.* **58**, 1490 (1987).
24. R. M. Bionta *et al.* (IMB Collaboration), *Phys. Rev. Lett.* **58**, 1494 (1987).
25. E. N. Alekseev *et al.*, *JETP Lett.* **45**, 589 (1987).
26. K. A. Olive *et al.* (Particle Data Group), *Chin. Phys. C* **38**, 090001 (2014).
27. Y. Fukuda *et al.* (Super-Kamiokande Collaboration), *Phys. Rev. Lett.* **81**, 1562 (1998).
28. M. G. Aartsen *et al.* (IceCube Collaboration), *Science* **342**, 1242856 (2013); *Phys. Rev. Lett.* **113**, 101101 (2014).
29. Z. Z. Xing and S. Zhou, *Neutrinos in Particle Physics, Astronomy and Cosmology* (Zhejiang University Press and Springer-Verlag, 2011).
30. L. Wolfenstein, *Phys. Rev. D* **17**, 2369 (1978); S. P. Mikheyev and A. Yu. Smirnov, *Sov. J. Nucl. Phys.* **42**, 913 (1985).

31. F. J. Hasert *et al.*, *Phys. Lett. B* **46**, 138 (1973).
32. S. L. Glashow, *Nucl. Phys.* **22**, 579 (1961); S. Weinberg, *Phys. Rev. Lett.* **19**, 1264 (1967); A. Salam, in *Elementary Particle Physics* (*Nobel Symposium No. 8*), edited by N. Svartholm (Almqvist and Wilsell, 1968), p. 367.
33. Arnison *et al.* (UA1 Collaboration), *Phys. Lett. B* **122**, 103 (1983); *Phys. Lett. B* **126**, 398 (1983).
34. Z. Z. Xing, *Prog. Theor. Phys. Suppl.* **180**, 112 (2009).
35. S. Weinberg, *Phys. Rev. Lett.* **43**, 1566 (1979).
36. P. Minkowski, *Phys. Lett. B* **67**, 421 (1977); T. Yanagida, in *Proceedings of the Workshop on Unified Theory and the Baryon Number of the Universe*, edited by O. Sawada and A. Sugamoto (KEK, Tsukuba, 1979), p. 95; M. Gell-Mann, P. Ramond, and R. Slansky, in *Supergravity*, edited by P. van Nieuwenhuizen and D. Freedman (North Holland, Amsterdam, 1979), p. 315; S. L. Glashow, in *Quarks and Leptons*, edited by M. Lévy *et al.* (Plenum, New York, 1980), p. 707; R. N. Mohapatra and G. Senjanovic, *Phys. Rev. Lett.* **44**, 912 (1980).
37. M. Fukugita and T. Yanagida, *Phys. Lett. B* **174**, 45 (1986).
38. W. J. Marciano and A. I. Sanda, *Phys. Lett. B* **67**, 303 (1977); B. W. Lee and R. Shrock, *Phys. Rev. D* **16**, 1444 (1977); K. Fujikawa and R. Shrock, *Phys. Rev. Lett.* **45**, 963 (1980); R. Shrock, *Nucl. Phys. B* **206**, 359 (1982); P. Pal and L. Wolfenstein, *Phys. Rev. D* **25**, 766 (1982).
39. For a review, see: C. Giunti and A. Studenikin, *Phys. Atom. Nucl.* **72**, 2089 (2009).
40. S. Antusch, C. Biggio, E. Fernandez-Martinez, M. B. Gavela, and J. Lopez-Pavon, *JHEP* **0610**, 084 (2006); S. Antusch and O. Fischer, *JHEP* **1410**, 94 (2014).
41. J. Schechter and J. W. F. Valle, *Phys. Rev. D* **23**, 1666 (1981); Z. Z. Xing, *Phys. Rev. D* **87**, 053019 (2013); Z. Z. Xing and Y. L. Zhou, *Phys. Rev. D* **88**, 033002 (2013).
42. J. N. Bahcall, *Phys. Rev. Lett.* **12**, 300 (1964).
43. B. T. Cleveland *et al.* (Homestake Collaboration), *Astrophys. J.* **496**, 505 (1998).
44. W. Hampel *et al.* (GALLEX Collaboration), *Phys. Lett. B* **447**, 127 (1999); M. Altmann *et al.* (GNO Collaboration), *Phys. Lett. B* **490**, 16 (2000).
45. J. N. Abdurashitov *et al.* (SAGE Collaboration), *JETP* **95**, 181 (2002).
46. Y. Fukuda *et al.* (Super-Kamiokande Collaboration), *Phys. Rev. Lett.* **81**, 1158 (1998).
47. Q. R. Ahmad *et al.* (SNO Collaboration), *Phys. Rev. Lett.* **89**, 011301 (2002).
48. B. Aharmim *et al.* (SNO Collaboration), *Phys. Rev. C* **72**, 055502 (2005).
49. B. Kayser, arXiv:0804.1497 (2008).
50. C. Arpesella *et al.* (Borexino Collaboration), *Phys. Lett. B* **658**, 101 (2008); *Phys. Rev. Lett.* **101**, 091302 (2008).
51. K. S. Hirata *et al.* (Kamiokande-II Collaboration), *Phys. Lett. B* **205**, 416 (1988); *Phys. Lett. B* **280**, 146 (1992).
52. D. Casper *et al.*, *Phys. Rev. Lett.* **66**, 2561 (1991); R. Becker-Szendy *et al.*, *Phys. Rev. D* **46**, 3720 (1992).
53. Y. Ashie *et al.* (Super-Kamiokande Collaboration), *Phys. Rev. Lett.* **93**, 101801 (2004).
54. K. Abe *et al.* (Super-Kamiokande Collaboration), *Phys. Rev. Lett.* **110**, 181802 (2013).
55. M. H. Ahn *et al.* (K2K Collaboration), *Phys. Rev. Lett.* **90**, 041801 (2003).
56. D. Michael *et al.* (MINOS Collaboration), *Phys. Rev. Lett.* **97**, 191801 (2006).
57. A. Strumia and F. Vissani, arXiv:hep-ph/0606054 (2006).
58. K. Abe *et al.* (T2K Collaboration), *Phys. Rev. Lett.* **107**, 041801 (2011); *Phys. Rev. Lett.* **111**, 211803 (2013); *Phys. Rev. Lett.* **112**, 061802 (2014); *Phys. Rev. Lett.* **112**, 181801 (2014).
59. See, e.g., M. Freund, *Phys. Rev. D* **64**, 053003 (2001).
60. N. Agafonova *et al.* (OPERA Collaboration), *PTEP* **2014**, 101C01 (2014).

61. K. Eguchi *et al.* (KamLAND Collaboration), *Phys. Rev. Lett.* **90**, 021802 (2003).
62. F. P. An *et al.* (Daya Bay Collaboration), *Phys. Rev. Lett.* **108**, 171803 (2012); *Chin. Phys. C* **37**, 011001 (2013).
63. S. Abe *et al.* (KamLAND Collaboration), *Phys. Rev. Lett.* **100**, 221803 (2008).
64. M. Apollonio *et al.* (CHOOZ Collaboration), *Phys. Lett. B* **420**, 397 (1998).
65. F. Boehm *et al.* (Palo Verde Collaboration), *Phys. Rev. Lett.* **84**, 3764 (2000).
66. J. K. Ahn *et al.* (RENO Collaboration), *Phys. Rev. Lett.* **108**, 191802 (2012).
67. Y. Abe *et al.* (Double Chooz Collaboration), *Phys. Rev. D* **86**, 052008 (2012); *Phys. Lett. B* **723**, 66 (2013).
68. F. P. An *et al.* (Daya Bay Collaboration), *Phys. Rev. Lett.* **112**, 061801 (2014).
69. F. Capozzi, G. L. Fogli, E. Lisi, A. Marrone, D. Montanino, and A. Palazzo, *Phys. Rev. D* **89**, 093018 (2014).
70. D. V. Forero, M. Tortola, and J. W. F. Valle, *Phys. Rev. D* **90**, 093006 (2014).
71. M. C. Gonzalez-Garcia, M. Maltoni, and T. Schwetz, *JHEP* **1411**, 052 (2014).
72. Z. Z. Xing and S. Zhou, *Phys. Lett. B* **666**, 166 (2008).
73. Y. F. Li, J. Cao, Y. Wang, and L. Zhan, *Phys. Rev. D* **88**, 013008 (2013).
74. M. G. Aartsen *et al.* (IceCube-PINGU Collaboration), arXiv:1401.2046 (2014).
75. C. Adams *et al.* (LBNE Collaboration), arXiv:1307.7335 (2013).
76. L. Zhan, Y. Wang, J. Cao, and L. Wen, *Phys. Rev. D* **78**, 111103 (2008); *Phys. Rev. D* **79**, 073007 (2009).
77. R. N. Cahn *et al.*, arXiv:1307.5487 (2013).
78. D. S. Ayres *et al.* (NOνA Collaboration), arXiv:hep-ex/0503053 (2005).
79. K. Abe *et al.* (Hyper-Kamiokande Working Group), arXiv:1109.3262 (2011).
80. See, e.g., M. Koike and J. Sato, *Phys. Rev. D* **61**, 073012 (2000); H. Minakata and H. Nunokawa, *Phys. Lett. B* **495**, 369 (2000).
81. J. Cao *et al.*, *Phys. Rev. S. T. A. B.* **17**, 090101 (2014).
82. M. Goeppert-Mayer, *Phys. Rev.* **48**, 512 (1935).
83. W. H. Furry, *Phys. Rev.* **56**, 1184 (1939).
84. J. Schechter and J. W. F. Valle, *Phys. Rev. D* **25**, 2951 (1982).
85. S. M. Bilenky and C. Giunti, *Int. J. Mod. Phys. A* **30**, 1530001 (2015).
86. M. Agostini *et al.* (GERDA Collaboration), *Phys. Rev. Lett.* **111**, 122503 (2013).
87. J. Albert *et al.* (EXO-200 Collaboration), *Nature* **510**, 229 (2014).
88. A. Gando *et al.* (KamLAND-Zen Collaboration), *Phys. Rev. Lett.* **110**, 062502 (2013).
89. J. J. Gomez-Cadenas and J. Martin-Albo, arXiv:1502.00581 (2015).
90. P. A. R. Ade *et al.* (Planck Collaboration), *Astron. Astrophys.* **571**, A16 (2014).
91. L. Bornschein *et al.* (KATRIN Collaboration), hep-ex/0309007 (2003).
92. V. N. Aseev *et al.*, *Phys. Rev. D* **84**, 112003 (2011).
93. K. N. Abazajian *et al.*, *Astropart. Phys.* **63**, 66 (2015).
94. G. T. Zatsepin, *Pisma Zh. Eksp. Teor. Fiz.* **8**, 333 (1968).
95. T. J. Loredo and D. Q. Lamb, *Phys. Rev. D* **65**, 063002 (2002); G. Pagliaroli, F. Rossi-Torres, and F. Vissani, *Astropart. Phys.* **33**, 287 (2010).
96. S. Dell'Oro, S. Marcocci, and F. Vissani, *Phys. Rev. D* **90**, 033005 (2014).
97. J. S. Lu, J. Cao, Y. F. Li, and S. Zhou, arXiv:1412.7418 (2014).
98. See, e.g., H. Fritzsch and Z. Z. Xing, *Prog. Part. Nucl. Phys.* **45**, 1 (2000); Z. Z. Xing, *Int. J. Mod. Phys. A* **19**, 1 (2004); G. Altarelli and F. Feruglio, *Rev. Mod. Phys.* **82**, 2701 (2010); S. F. King and C. Luhn, *Rept. Prog. Phys.* **76**, 056201 (2013).

Chapter 20

The Supersymmetric Standard Model

Pierre Fayet

Laboratoire de Physique Théorique de l'École Normale Supérieure,
UMR 8549, CNRS-ENS, associée à l'Université Paris 6 UPMC,
24 rue Lhomond, 75231 Paris Cedex 05, France
and
Département de Physique, École Polytechnique,
91128 Palaiseau Cedex, France

The Standard Model may be included within a supersymmetric theory, postulating new *sparticles* that differ by half-a-unit of spin from their standard model partners, and by a new quantum number called R-parity. The lightest one, usually a neutralino, is expected to be stable and a possible candidate for dark matter.

The electroweak breaking requires two doublets, leading to several charged and neutral Brout–Englert–Higgs bosons. This also leads to gauge/Higgs unification by providing extra spin-0 partners for the spin-1 $W^{\pm}$ and Z. It offers the possibility to view, up to a mixing angle, the new 125 GeV boson as the spin-0 partner of the Z under two supersymmetry transformations, i.e. as a Z that would be deprived of its spin. Supersymmetry then relates *two existing particles of different spins*, in spite of their different gauge symmetry properties, through supersymmetry transformations acting on physical fields in a non-polynomial way.

We also discuss how the compactification of extra dimensions, relying on R-parity and other discrete symmetries, may determine both the supersymmetry-breaking and grand-unification scales.

Is there a superworld of new particles? Could half of the particles at least have escaped our observations? Do new states of matter exist? After the prediction of antimatter by Dirac, supersymmetric extensions of the standard model lead to anticipate the possible existence of spin-0 squarks and sleptons, with the gluons, $W^{\pm}$, Z and photon associated with gluinos, charginos and neutralinos.[1–4] These new states differ from ordinary particles by half-a-unit of spin and are distinguished by a R-parity quantum number related to baryon and lepton numbers, making the lightest superpartner stable, and a possible candidate for the dark matter of

Table 1. Particle content of the standard model.

spin-1 gauge bosons:	gluons, W^+, W^-, Z, photon
spin-$\frac{1}{2}$ fermions:	6 quarks: $\binom{u}{d}\binom{c}{s}\binom{t}{b}$ 6 leptons: $\binom{\nu_e}{e^-}\binom{\nu_\mu}{\mu^-}\binom{\nu_\tau}{\tau^-}$
spin-0	scalar BEH boson

the Universe. Spontaneous electroweak breaking is induced by two spin-0 doublets instead of one in the standard model, leading to several charged and neutral spin-0 BEH bosons. These may even be related to the massive gauge bosons, with the possibility that the new 125 GeV boson recently discovered at CERN[5,6] be a spin-0 partner of the Z under *two* supersymmetry transformations.[1,7,8] But, where is all this coming from?

1. Fundamental Interactions, Symmetry Breaking and the New Spin-0 Boson

Special relativity and quantum mechanics, operating within quantum field theory, led to the Standard Model of particles and interactions (SM). It has met a long series of successes with the discoveries of weak neutral currents (1973), charmed particles (1974–76), gluons mediators of strong interactions (1979), $W^\pm$ and Z's mediators of weak interactions (1983), and the sixth quark known as the top quark (1995). Weak, electromagnetic and strong interactions are well understood from the exchanges of spin-1 mediators between spin-$\frac{1}{2}$ quarks and leptons, generically referred to as the constituents of matter (cf. Table 1).

The eight gluons mediate the strong interactions, invariant under the color $SU(3)$ gauge group. The $W^\pm$, Z and photon are associated with the $SU(2) \times U(1)$ electroweak gauge group.[9–12] The $W^\pm$ and Z masses, $m_W \simeq 80$ GeV/c^2 and $m_Z \simeq 91$ GeV/c^2, are generated through the spontaneous breaking of the electroweak symmetry, induced in the standard model by a doublet of spin-0 fields φ.[11,12] Three of its four real components, instead of being associated with unwanted massless Goldstone bosons,[13] are eliminated by the Brout–Englert–Higgs mechanism[14–16] to provide the extra degrees of freedom for the massive $W^\pm$ and Z. The fourth component, taken as $\phi = \sqrt{2\,\varphi^\dagger\varphi}$, adjusts so that the potential

$$V(\varphi) \;=\; \lambda_{\rm SM}\,(\varphi^\dagger\varphi)^2 \,-\, \mu^2_{\rm SM}\,\varphi^\dagger\varphi \tag{1}$$

is minimum, for $\phi = v = \sqrt{\mu^2_{\rm SM}/\lambda_{\rm SM}}$.[11–15]

The electroweak symmetry, said to be "spontaneously broken", is in fact simply hidden, with ϕ being gauge-invariant. The $W^\pm$ and Z acquire masses $m_W = gv/2$, $m_Z = \sqrt{g^2+g'^2}\; v/2 = m_W/\cos\theta$, with $\tan\theta = g'/g$. The elementary charge and the Fermi coupling of weak interactions are given by $e = g\,\sin\theta$ and $G_F/\sqrt{2} = g^2/8m_W^2 = 1/2v^2$, so that $v = (G_F\sqrt{2})^{-1/2} \simeq 246$ GeV. Charged lepton and quark fields interact with φ with coupling constants $\lambda_{l,q}$, so that the corresponding particles, sensitive to the physical BEH field $\phi = \sqrt{2\,\varphi^\dagger\varphi}$ with $\langle\phi\rangle = v$, acquire masses $m_{l,q} = \lambda_{l,q}\, v/\sqrt{2}$, neutrinos remaining massless at this stage. The waves corresponding to the space-time variations of $\phi = \sqrt{2\,\varphi^\dagger\varphi}$, when quantized, are associated with spin-0 Brout–Englert–Higgs bosons, commonly referred to as Higgs bosons. Their mass,

$$m_h = \sqrt{2\mu_{\rm SM}^2} = \sqrt{2\lambda_{\rm SM}v^2}, \tag{2}$$

is fixed by the quartic coupling $\lambda_{\rm SM}$ in the scalar potential $V(\varphi)$ in (1), a mass of 125 GeV/c^2 corresponding to a coupling

$$\lambda_{\rm SM} = \frac{m_h^2}{2v^2} = \frac{g^2+g'^2}{8}\,\frac{m_h^2}{m_Z^2} = \frac{G_F\, m_h^2}{\sqrt{2}} \simeq 0.13. \tag{3}$$

The possible origin of this coupling will be discussed later, within supersymmetric theories. They lead to consider several BEH bosons originating from the two spin-0 doublets

$$h_1 = \begin{pmatrix} h_1^0 \\ h_1^- \end{pmatrix}, \quad h_2 = \begin{pmatrix} h_2^+ \\ h_2^0 \end{pmatrix}, \tag{4}$$

relating their quartic couplings to the squares of the electroweak gauge couplings, in particular through

$$\textit{Supersymmetry} \quad \Rightarrow \quad \lambda_{\rm SM} \;\rightarrow\; \frac{g^2+g'^2}{8}\,, \tag{5}$$

with $(g^2+g'^2)/8 = G_F m_Z^2/\sqrt{2} \simeq .069$. Here we first focus for simplicity on h_2 and on the "large $\tan\beta$ limit", for which h_2 acquires a non-vanishing v.e.v. much larger than for h_1. We then get a neutral BEH boson that would have *the same mass as the Z*,[1] according to

$$\boxed{\begin{array}{l} \textit{Supersymmetry} \quad \Rightarrow \\[1ex] m_h = \sqrt{2\lambda_{\rm SM}v^2} = \dfrac{\sqrt{g^2+g'^2}\;v}{2} = m_Z \simeq 91\ \text{GeV}/c^2, \end{array}} \tag{6}$$

up to supersymmetry-breaking effects.

This mass equality results from an unbroken supersymmetry in the sector of neutral particles, with the spin-1 Z and the spin-0 h in the same massive multiplet of supersymmetry. It remains valid even *independently of the value of the mixing angle* β defined from the ratio of the two doublet v.e.v.'s by

$$\tan\beta = \frac{v_2}{v_1}, \tag{7}$$

with $\langle h_i^0 \rangle = v_i/\sqrt{2}$, as long as supersymmetry remains unbroken in this sector.[1] The corresponding spin-0 boson then appears as *the spin-0 partner of the Z under two supersymmetry transformations.*[7,8] It was even originally denoted by z to make this association explicit.

Finding such a spin-0 boson with a mass of 125 GeV/c^2,[5,6] not much higher than the Z mass, may thus be considered, at least, as a very encouraging sign for supersymmetry. This is especially true as the value $m_h \simeq m_Z$ required by unbroken supersymmetry may be increased up to 125 GeV/c^2 by supersymmetry-breaking effects. This is the case, most notably, in models such as the N/nMSSM or USSM, that include an extra singlet next to the two doublets in (4), with a trilinear $\lambda\, H_2 H_1 S$ superpotential coupling.[1]

The lightest spin-0 mass may then easily reach 125 GeV, without having to rely on very large effects from radiative corrections. This is a much better situation than in the usual MSSM for which no electroweak breaking is obtained in the absence of the supersymmetry-breaking terms, m_h is required to be less than m_Z at the classical level, and it is difficult to obtain such a 125 GeV spin-0 boson from sufficiently large radiative corrections involving very heavy stop quarks.

The scalar boson of the standard model has long remained its last missing particle after the discovery of the top quark in 1995. The new boson found at CERN in 2012 shows the properties expected from a scalar boson associated with the differentiation between electromagnetic and weak interactions, and the generation of masses. It may well be identified with the one of the standard model, which may then be considered as complete.

Still, it would be presumptuous to imagine that our knowledge of particles and interactions is now complete, without new particles or interactions remaining to be discovered. The standard model does not answer many fundamental questions, concerning the origin of symmetries and symmetry breaking, the quark and lepton mass spectrum and mixing angles, etc. Gravitation, classically described by general relativity, cannot easily be cast into a consistent quantum theory. This is why string theories were developed, which seem to require supersymmetry for consistency.

The nature of dark matter and dark energy which govern the evolution of the Universe and its accelerated expansion remains unknown, as the origin of the predominance of matter over antimatter. Dark matter may be composed, for its main part, non-baryonic, of new particles, such as the neutralinos of supersymmetric theories. There may also be new forces or interactions beyond the four known ones. And maybe, beyond space and time, new hidden dimensions, extremely small or even stranger, like the anticommuting quantum dimensions of supersymmetry.

2. Introducing Supersymmetry

In contrast with pions, kaons and other spin-0 mesons, composed of quarks and antiquarks, the new 125 GeV boson presents at this stage all the characteristics of an elementary spin-0 particle, the first one of its kind. The possible existence of such a scalar has long been questioned, many physicists having serious doubts about the very existence of fundamental spin-0 fields. More specifically in a theory involving very high mass or energy scales much larger than the electroweak scale, such as a grand-unification scale[17,18] (now usually believed to be of the order of 10^{16} GeV), or the Planck scale $\simeq 10^{19}$ GeV possibly associated with quantum gravity, such spin-0 fields tend to acquire very large mass terms. They would then disappear from the low-energy theory, no longer being available to provide an appropriate breaking of the electroweak symmetry.

Many efforts were thus devoted to replace fundamental spin-0 fields by composite fields built from spin-$\frac{1}{2}$ ones, without however much success at this point. These spin-$\frac{1}{2}$ subconstituent fields could have been, for example, techniquark fields interacting through a new interaction specially introduced for this purpose,[19–22] in view of ultimately avoiding fundamental spin-0 fields and particles associated with the electroweak breaking, like the one discovered recently. Furthermore it would still remain difficult to completely avoid considering fundamental spin-0 fields, e.g. to trigger the breaking of the initial extended technicolor gauge group.

In the meantime however, and even before these increased questionings about fundamental spin-0 bosons, the situation concerning our view of spin-0 fields had already changed considerably with the introduction of supersymmetry, in the early 1970's. This one provides a natural framework for fundamental spin-0 fields. They may now be treated on the same footing as spin-$\frac{1}{2}$ ones, also benefiting from the same mass terms when supersymmetry is unbroken; and of mass terms which may remain moderate as compared to very large scales if supersymmetry is not too badly broken, then remaining available to trigger the electroweak breaking.

The supersymmetry algebra involves a self-conjugate (Majorana) spin-$\frac{1}{2}$ generator Q satisfying the anticommutation and commutation relations[23–28]

$$\left\{ \begin{array}{lcl} \{\, Q \,,\, \bar{Q} \,\} &=& -\,2\,\gamma_\mu P^\mu , \\ [\, Q,\, P^\mu \,] &=& 0. \end{array} \right. \tag{8}$$

They express that supersymmetry transformations may be combined to generate translations, and commute with them. This algebra was originally introduced as a parity-violating one that might help understanding why weak interactions violate parity,[23] or the masslessness of a neutrino by trying to view it as a Goldstone fermion.[24] It may also be obtained by generalizing to 4 dimensions the algebra of supergauge transformations acting in the 2-dimensional string worldsheet.[25]

But what physical implications may really be extracted from the consideration of this algebraic structure? According to common knowledge, supersymmetry should relate bosons, of integer spin, with fermions, of half-integer spin, as follows:

$$\textit{bosons} \quad \overset{\text{supersymmetry}}{\longleftrightarrow} \quad \textit{fermions}. \tag{9}$$

But even this is not always valid, as there are supersymmetric theories involving only fundamental fermions, with supersymmetry transformations acting in a non-linear way.[24] Strictly speaking the algebraic structure of supersymmetry does not even require any boson at all, not to mention the superpartners that we shall introduce later. But let us leave aside such unconventional situations. Let us add, also, that supersymmetry transformations are usually expected to relate bosons and fermions with the same gauge symmetry properties.

Then, can this algebra be of any help in understanding the real world of particles and interactions? If supersymmetry is to act at the fundamental level the natural idea would be to use it to relate the known bosons and fermions in Table 1. More precisely, can one relate the bosons (gluons, $W^\pm$, Z and photon) messengers of interactions to the fermions, quarks and leptons, constituents of matter? This would lead to a sort of unification

$$\textit{Forces} \quad \overset{\text{supersymmetry?}}{\longleftrightarrow} \quad \textit{Matter}. \tag{10}$$

The idea looks attractive, even so attractive that supersymmetry is frequently presented as uniting forces with matter. This is however misleading at least at the present stage, and things do not work out that way.

Indeed the algebraic structure of supersymmetry did not seem applicable to particle physics at all, in particular as known fundamental bosons and fermions do not seem to have much in common. There are also a number of more technical reasons, dealing with: (1) the difficulties of spontaneous supersymmetry breaking, originating from the presence of the hamiltonian within the algebra; (2) the fate of the resulting Goldstone fermion, after one has succeeded in breaking supersymmetry spontaneously,[1,29–31] and as it may well continue to interact, even after getting eaten away by the spin-$\frac{3}{2}$ gravitino, according to the "equivalence theorem" of supersymmetry;[32] (3) the presence of self-conjugate Majorana fermions, unknown in Nature; (4) the requirements of baryon and lepton number conservation, which got associated with the definition of R-symmetry and the requirement of R-parity, etc.

Relating bosons and fermions, yes, but how? One has to find out which of them might be related under supersymmetry, first considering possible associations between mesons and baryons. Or, at the fundamental level, exploring as a necessary exercise tentative associations like

$$\left\{\begin{array}{ccc} \text{photon} & \overset{?}{\longleftrightarrow} & \text{neutrino} \\ W^{\pm} & \overset{?}{\longleftrightarrow} & e^{\pm} \\ \text{gluons} & \overset{?}{\longleftrightarrow} & \text{quarks} \\ & \ldots & \end{array}\right. . \tag{11}$$

But we have no chance to realize in this way systematic associations of known fundamental bosons and fermions. This is also made obvious as we know 90 fermionic field degrees of freedom (for 3 families of 15 chiral quark and lepton fields) as compared to 28 only for bosonic ones (16 + 11 + 1 including the new scalar). Furthermore these fields have different gauge and B and L quantum numbers, preventing them from being directly related.

In supersymmetry we also have to deal with the systematic appearance of self-conjugate Majorana fermions, while Nature seems to know Dirac fermions only (with a possible exception for neutrinos with Majorana mass terms). How can we obtain Dirac fermions, and attribute them conserved quantum numbers like B and L? And if we start attributing B and L also to bosons (now known as squarks and sleptons), how can we be sure that their exchanges will not spoil the B and L conservation laws, at least to a sufficiently good approximation? It is thus far from trivial to consider applying supersymmetry to the real world. But if this program can be realized and if Nature has "chosen" being supersymmetric, consequences promise being spectacular.

Addressing the difficult questions of spontaneous supersymmetry breaking, and electroweak breaking, will lead us, through the definition of a new symmetry called R symmetry with its discrete remnant known as R-parity, to the Supersymmetric Standard Model. The way to see supersymmetry now is to view it as *an extension of the standard model* that introduces a new *sparticle* for each one in the standard model,[1–4] in particular through

$$\left\{\begin{array}{ccc} \text{quarks, leptons} & \leftrightarrow & \text{spin-0 } \textit{squarks and sleptons}, \\ \text{gluons} & \leftrightarrow & \text{spin-}\tfrac{1}{2}\ \textit{gluinos}, \\ W^{\pm}, Z,\ \gamma & \leftrightarrow & \text{spin-}\tfrac{1}{2}\ \textit{charginos and neutralinos}, \end{array}\right. \tag{12}$$

with more to say about spin-0 BEH bosons, including charged and several neutral ones.

While this is now often presented as obvious, the necessity of postulating that *every known particle has its own image under supersymmetry* (SM bosons having fermionic superpartners and SM fermions bosonic ones) was long mocked as a sign of the irrelevance of supersymmetry. The introduction of a color octet of spin-$\frac{1}{2}$ Majorana fermions called *gluinos* was also, at the time, forbidden by the principle of triality.[33] This one, however, gets systematically violated within supersymmetric theories.

The necessity of *charged spin-0 BEH bosons* ($H^{\pm}$), required by the 2-doublet structure of supersymmetric theories, was also taken as an argument against supersymmetry and supersymmetric extensions of the standard model, on the grounds that even a single doublet, although possibly necessary as in the standard model, was already undesirable. These charged spin-0 bosons, which have not been discovered yet,[34,35] also appear as the spin-0 partners of the $W^{\pm}$ under *two* supersymmetry transformations, very much as the new 125 GeV boson may now also be interpreted as the spin-0 partner of the Z, up to supersymmetry-breaking effects.

3. Supersymmetry Breaking and *R* Symmetry

3.1. *Is spontaneous supersymmetry breaking possible at all?*

If bosons and fermions are directly related by supersymmetry they should have equal masses. Supersymmetry may then be only, at best, a broken symmetry. Considering terms breaking explicitly the supersymmetry (as frequently done now) would certainly make the task much easier, but may be considered only as a temporary substitute for a solution to the problem of supersymmetry breaking. If supersymmetry is to be a genuine symmetry for the theory and its equations of motion, it should be broken *spontaneously*, as for the electroweak symmetry in the standard model. This is also necessary for supersymmetry to be realized as a local fermionic gauge symmetry.[36] It must then include general relativity, leading to supergravity theories.[37,38]

To trigger a spontaneous breaking of an ordinary (global or gauge) symmetry, one simply has to arrange for the symmetric vacuum state to be unstable, e.g. by choosing a negative value for the mass^2 parameter $-\mu^2_{\rm SM}$ in the potential (1), which is easily realized.

The situation concerning supersymmetry is, however, completely different. The hamiltonian H, which governs the energy of the possible vacuum states and thus determines which one is going to be stable, may now be expressed from the squares of the four components of the supersymmetry generator, as

$$H = \frac{1}{4} \sum_{\alpha} Q_{\alpha}^2. \tag{13}$$

This implies that *a supersymmetric vacuum state* $|\,\Omega\rangle$ (verifying $Q_\alpha|\,\Omega\rangle = 0$) *must have a vanishing energy,* with $H\,|\,\Omega\rangle = 0$. On the other hand any non-supersymmetric state $|\,\Omega'\rangle$ would have, within global supersymmetry, a larger, positive, energy density, and thus would be unstable. This was originally thought to prevent any spontaneous breaking of the supersymmetry to possibly occur,[39] apparently signing the impossibility of applying supersymmetry to the real world.

3.2. *In search of a minimum of the potential breaking the supersymmetry*

In spite of this general argument however, which soon got circumvented, spontaneous supersymmetry breaking turned out to be possible, although in very specific circumstances. It is severely constrained and usually hard to obtain, at least within global supersymmetry, as any supersymmetric candidate for the vacuum state ($|\,\Omega\rangle$) is necessarily stable. Furthermore in the presence of many spin-0 fields, there are usually many opportunities for them to adjust so as to provide such a stable supersymmetric vacuum, with a vanishing value of the potential $V = 0$.

To obtain a spontaneous breaking of the global supersymmetry, one cannot just attempt to make a supersymmetric vacuum unstable. One must instead arrange for such a symmetric state to be *totally absent*, as it would otherwise be stable owing to expression (13) of the hamiltonian.

In the usual language of global supersymmetry[26,27] involving gauge superfields $V_a(x,\theta,\bar\theta)$ and (left-handed) chiral superfields $\Phi_i(x,\theta)$ with physical spin-0 and spin-$\frac{1}{2}$ components ϕ_i and $\tilde\phi_{iL}$,[40,41] the potential of scalar fields is expressed as

$$V = \frac{1}{2}\sum (D_a^2 + F_i^2 + G_i^2) = \sum_a \frac{D_a^2}{2} + \sum_i \left|\frac{\partial \mathcal{W}}{\partial \phi_i}\right|^2 . \tag{14}$$

D_a and (F_i, G_i) stand for the auxiliary components of gauge and chiral superfields. The contribution from the D terms is given by

$$V_D = \sum_a \frac{D_a^2}{2} = \frac{1}{2}\sum_a \left[\xi_a + g_a \sum_{ij} \phi_i^* \,(T_a)_{ij}\, \phi_j\right]^2 , \tag{15}$$

with the ξ_a parameters relative to abelian $U(1)$ factors in the gauge group.[29] The superpotential $\mathcal{W}(\Phi_i)$ is an analytic function of the chiral superfields.

For a supersymmetric vacuum state $|\,\Omega\rangle$ to be, not unstable but totally absent, the potential V must be *strictly positive everywhere.* One at least of these auxiliary components must then have a non-vanishing v.e.v., which is indeed the signal for a spontaneously broken supersymmetry (except for trivial situations with a free superfield). Finding a spontaneously broken supersymmetry then amounts to finding situations for which the set of equations

$$\langle D_a\rangle \;=\; \langle F_i\rangle \;=\; \langle G_i\rangle \;=0 \quad \text{must have } \textit{no solution.} \tag{16}$$

How this may be realized, as well as the definition and role of the R symmetry, leading to R-parity, to appropriately constrain the superpotential, will be further discussed in the rest of this Section. The reader mostly interested in the construction of supersymmetric extensions of the standard model (MSSM, N/nMSSM, USSM, etc.) and in the relations between massive spin-1 gauge bosons and spin-0 BEH bosons may choose to proceed directly to Sections 4 and 5.

3.3. *D and F supersymmetry breaking mechanisms, in connection with R symmetry*

To avoid having a vanishing minimum of V when all physical fields ϕ_i vanish, there are two possibilities, which may be combined:

1) The Lagrangian density may include a linear term

$$\mathcal{L}_\xi = \xi D, \tag{17}$$

associated with an abelian $U(1)$ factor in the gauge group.[1,29] This term is indeed supersymmetric, up to a derivative which does not contribute to the variation of the action integral, and gauge invariant for an abelian gauge group. It may lead to a spontaneous breaking of the supersymmetry, just by itself as in the presence of a single chiral superfield S (with a charge e such that $\xi e > 0$),[42] or by making the set of equations $\{D_a = 0\}$ without solution, as with a $SU(2) \times U(1)$ gauge group.[1] The Goldstone spinor is then a gaugino, corresponding for example to the photino in a $SU(2) \times U(1)$ theory, even if such a feature cannot persist in a physically realistic theory.[4]

One may also arrange so that the set of equations $\{D_a = 0,\ F_i = G_i = 0\}$ be without solution, as done in the presence of chiral superfields S and $\bar{S}$ with a mass term $\mu\, \bar{S}S$;[29] or with a suitable trilinear superpotential $\lambda\, H_2 H_1 S$, the electroweak gauge group being extended to an extra $U(1)$ factor, as in the USSM.[2] In all cases one has to make sure that no supersymmetric minimum of the potential exists anywhere, otherwise supersymmetry would remain (or return to) conserved.

2) The Lagrangian density may appeal to a term proportional to the auxiliary (F or G) components of a singlet chiral superfield $S(x, \theta)$. In this case the superpotential

$$\mathcal{W} = \sigma S + \cdots \tag{18}$$

includes a term linear in the singlet superfield S. One can then try to make the system of equations $\{F_i = G_i = 0\}$, i.e. $\partial \mathcal{W}/\partial \phi_i = 0$, without solution.

This first looks as an impossible task. Indeed with the superpotential

$$\mathcal{W} = \frac{\lambda_{ijk}}{3}\ \Phi_i \Phi_j \Phi_k + \frac{\mu_{ij}}{2}\ \Phi_i \Phi_j + \sigma_i\, \Phi_i, \tag{19}$$

taken for simplicity as a cubic function of the chiral superfields Φ_i for the theory to be renormalizable, the set of equations $\partial \mathcal{W}/\partial \phi_i = 0$ reads

$$\lambda_{ijk}\ \phi_j \phi_k + \mu_{ij}\ \phi_j + \sigma_i\ = 0. \tag{20}$$

With n equations for n complex field variables it is expected to have *almost always* solutions, for which supersymmetry is conserved.

Still it is possible choosing very carefully the set of interacting superfields and the superpotential $\mathcal{W}$ for spontaneous supersymmetry breaking to occur. This appeals to a new symmetry called R symmetry.[1] R transformations, further discussed in the next subsection, act on (left-handed) chiral superfields according to

$$\Phi(x,\theta) \stackrel{R}{\to} e^{i R_\Phi \alpha}\, \Phi(x, \theta\, e^{-i\alpha}). \tag{21}$$

The superpotential must transform with $R=2$ for the theory to be invariant under R. It is then said to be "R-symmetric". This symmetry can be used to select and constrain appropriately the superpotential $\mathcal{W}$ so that the set of equations (20) has no solution and the corresponding breaking of the supersymmetry is obtained *in a generic way*, not just for very specific values of the parameters.[30,31]

An interesting example is obtained with a R symmetric nMSSM-type superpotential,[1] extended to a chiral triplet T. It involves, as in the MSSM without the μ term, the two doublets H_1 and H_2 with $R=0$. They are coupled to a singlet S through a $\lambda\, H_2 H_1 S$ trilinear term as in the nMSSM, and similarly to a triplet T, both with $R=2$. The corresponding $R=2$ superpotential reads[30]

$$\mathcal{W} \;=\; \frac{1}{\sqrt{2}}\, H_2\,(g\,\tau\cdot T - g' S)\, H_1 \;+ \sigma S, \tag{22}$$

H_1 and H_2 having weak hypercharges $Y=-1$ and $+1$, respectively. This one is also, in addition, the superpotential for a $N=2$ supersymmetric gauge theory (or "hypersymmetric" theory)[43] when the $SU(2)\times U(1)$ symmetry is made again local, with the superpotential couplings fixed in terms of the electroweak gauge couplings as in (22). Its two terms may be written as proportional to

$$H_2\,\Phi H_1 \quad \text{and} \quad \text{Tr}\,\Phi, \tag{23}$$

with $\Phi = \frac{1}{2}\,(g\,\tau\cdot T - g'S)$. Or $\Phi = g\,\Lambda\cdot T - \frac{g'}{2}\,S$ where the chiral superfields T are in the adjoint representation of the gauge group, and the matrices Λ are relative to the hypermultiplet representation of the gauge group described by H_1 and H_2.

For a non-abelian $N=2$ theory the superpotential (22) reads

$$\mathcal{W} \;=\; g\sqrt{2}\; H_2\,\Lambda\cdot T\, H_1, \tag{24}$$

leading to the $N=4$ supersymmetric Yang-Mills theory when H_1 and H_2 are also taken in the adjoint representation of the gauge group.[44] The adjoint gauge superfield then interacts with 3 adjoint chiral ones, now denoted by S_1, S_2 and S_3, coupled through the trilinear superpotential

$$\mathcal{W} \;=\; g\sqrt{2}\; f_{ijk}\, S_1^i\, S_2^j\, S_3^k. \tag{25}$$

But we shall return to this later.

Let us come back to the superpotential (22) for a non-gauged $SU(2)\times U(1)$ theory involving at the moment chiral superfields only, in view of generating a spontaneous breaking of the supersymmetry through F terms.[30] The conjugates of the 4 (complex) auxiliary components of the $R=0$ superfields H_1, H_2 have $R=2$, and vanish with the 4 components of t and s (also with $R=2$). The conjugates of the 4 auxiliary components of the $R=2$ superfields T and S have $R=0$, and depend quadratically on the

4 components of h_1 and h_2. One might naively expect that they should also, "generically", be able to vanish simultaneously, so that supersymmetry would be conserved. Still this does not happen, as it would require

$$\left\{ \begin{array}{l} \dfrac{\partial W}{\partial t} = \dfrac{g}{\sqrt{2}} \, h_2 \, \tau h_1 \qquad = 0 , \\ \dfrac{\partial W}{\partial s} = -\dfrac{g'}{\sqrt{2}} \, h_2 h_1 + \sigma = 0 , \end{array} \right. \tag{26}$$

which are incompatible for $\sigma \neq 0$.

Supersymmetry is spontaneously broken, with a massless Goldstone spinor ζ_γ, taken as left-handed, carrying $R = 1$. It is described, together its $R = 2$ spin-0 partner, which remains classically massless, by a combination of the $R = 2$ chiral superfields.

This $R = 2$ chiral superfield is in this example the photon-like combination $\sin\theta \, T_3 + \cos\theta \, S$. It will be understood further by turning again the theory into a $N = 2$ supersymmetric one,[43] with a $SU(2) \times U(1)$ gauge group spontaneously broken into $U(1)_{\rm QED}$. The present Goldstone spinor ζ_γ associated with this F-breaking of the supersymmetry is described by the $R = 2$ chiral superfield $\sin\theta \, T_3 + \cos\theta \, S$. It then gets interpreted as the second photino field within $N = 2$, both photino fields λ_γ and ζ_γ, now related by a global $U(2)_R$ symmetry of $N = 2$, being the two Goldstone spinors of $N = 2$ supersymmetry.

3.4. *On the role of R symmetry to allow for supersymmetry breaking through F terms*

Without R symmetry, S^2 or S^3 terms would be allowed in the superpotential, and we would lose the benefit of having introduced a linear σS term, which may then be eliminated by a translation of S. Once σ is eliminated the potential has a vanishing minimum when all physical fields vanish, and supersymmetry is conserved. This shows *the crucial role played by R symmetry* to render possible *a generic breaking of the supersymmetry* through F terms.[30]

This mechanism leads to a classically-massless $R = 2$ spin-0 field, superpartner of the $R = 1$ Goldstone spinor (goldstino). Both are described by a $R = 2$ chiral superfield. A translation of its $R = 2$ spin-0 component, if it had to be performed, would lead to a spontaneous breaking of R symmetry (or a quasi-spontaneous breaking if R is anomalous). The imaginary part of this $R = 2$ field would then describe a massless R Goldstone boson or, for an anomalous symmetry, a classically-massless R-axion.

But is there a tighter connection between spontaneous supersymmetry breaking by this method, and the possible occurrence of spontaneous R-symmetry breaking through the v.e.v. of such a $R = 2$ scalar, superpartner of the $R = 1$ goldstino? The above example indicates that there is no need for R-symmetry to be spontaneously broken, to get spontaneous supersymmetry breaking. Conversely, for $\sigma = 0$

R symmetry may indeed be spontaneously broken, owing the $R = 2$ flat directions of the potential, with a massless R Goldstone boson, and a conserved supersymmetry. Thus spontaneous R-symmetry breaking is not a sufficient condition either, for spontaneous supersymmetry breaking to occur.

It thus appears that while spontaneous R-symmetry breaking may occur, *it is neither necessary nor sufficient* to lead to such a spontaneous breaking of the supersymmetry through F terms. What is indeed essential is the presence of R symmetry to restrict appropriately the superpotential as in (22).[30]

Furthermore, and in contrast with a current belief, spontaneous supersymmetry breaking occurs here, for $\sigma \neq 0$, in spite of having equal numbers of $R = 2$ and $R = 0$ superfields. There is thus no excess of $R = 1$ over $R = -1$ (left-handed) spinors, that would facilitate having a massless left-over $R = 1$ spinor that could become a Goldstone spinor. In fact with the same number of $R = 2$ and $R = 0$ superfields one might usually expect all spinors to acquire masses. Then there would be no candidate for a massless Goldstone spinor, and supersymmetry would have to remain conserved. Indeed the auxiliary components of the $R = 2$ superfields, which have $R = 0$, then depend on the same number of $R = 0$ physical fields, and might be expected to all vanish simultaneously.

Still this additional obstruction could be bypassed so as to render the system (26) of 4 equations for 4 variables generically without solution, and obtain spontaneous supersymmetry breaking.[30] This has been made possible, in particular, thanks to the spontaneously broken global $SU(2) \times U(1) \to U(1)$ symmetry generated by $\langle h_1 \rangle$ and $\langle h_2 \rangle$, leading to exactly massless $R = 0$ spin-0 Goldstone fields associated with $R = -1$ spinors. One of the latter balances a massless $R = 1$ spinor that is going to be the goldstino.

3.5. *Unifying D- and F-breaking mechanisms within N = 2, and going to N = 4 supersymmetry*

Beyond that, when $SU(2) \times U(1)$ is gauged again as in the nMSSM, with the gauge superfields V and V' associated to the chiral triplet and singlet T and S, the theory based on the superpotential (22) acquires an enhanced symmetry, namely extended $N = 2$ supersymmetry (or hypersymmetry), with H_1 and H_2 jointly describing a $N = 2$ hypermultiplet.[43] Of course no superpotential term proportional to S^2, S^3 as in the general NMSSM, T^2 or ST^2 may be allowed here. Such terms, which would ruin the possibility of having a $N = 2$ supersymmetry, were already excluded by means of R-symmetry, which showed the way to extended supersymmetry, and subsequently extra dimensions.

The D- and F-breaking mechanisms then become equivalent, getting unified within $N=2$ supersymmetry. Indeed the set of auxiliary components $\{-G, -F, D\}$ for a $N=2$ gauge multiplet transform as the three components of a $SU(2)_R$ isotriplet within a $U(2)_R = [SU(2) \times U(1)]_R$ global symmetry group. The ξD term for a $U(1)$ gauge superfield can then be turned into a ξF term for its associated chiral superfield through a $SU(2)_R$ transformation turning the first supersymmetry generator into the second. This $N=2$ supersymmetry breaking generates two massless Goldstone spinors, both with $R = 1$. A $SU(2) \times U(1)$, or more generally $G_{\text{non-abelian}} \times U(1)$ gauge group is then required if we intend to get a spontaneous breaking of the extended supersymmetry rather than just of the gauge symmetry.[43]

For a non-abelian $N = 2$ gauge theory the superpotential reads $\mathcal{W} = g\sqrt{2}\, H_2\, \Lambda.T\, H_1$ as in (24), with T in the adjoint representation and the Λ representing the gauge group for the hypermultiplet described by H_1 and H_2. A $N = 2$ supersymmetric theory with a massless matter hypermultiplet in the adjoint representation provides the $N = 4$ supersymmetric Yang–Mills theory.[44] The adjoint gauge superfield interacts with 3 adjoint chiral ones S_1, S_2 and S_3 coupled through the trilinear superpotential (25), $\mathcal{W} = g\sqrt{2}\, f_{ijk}\, S_1^i\, S_2^j\, S_3^k$. This provides the

$$\begin{gathered} N = 4 \textit{ supersymmetric Yang–Mills theory}, \quad \text{with} \\ \left(1 \text{ spin-1 } + 4 \text{ spin-}\frac{1}{2} + 6 \text{ spin-0}\right) \textit{ adjoint gauge fields.} \end{gathered} \tag{27}$$

The R symmetry acting chirally on the $N = 1$ supersymmetry generator (see later Eq. (35)) gets promoted from $U(1)_R$ in $N = 1$ up to $SU(2)_R$ or $U(2)_R$ in $N = 2$, and $SU(4)_R \sim O(6)_R$ in $N = 4$ supersymmetry. This corresponds to the following chain

$$\begin{aligned} R\text{-parity} \subset U(1)_R \subset\ & \underbrace{SU(2)_R \subset U(2)_R}_{N=2} \\ \subset\ & \underbrace{SU(4)_R \sim O(6)_R}_{N=4}. \end{aligned} \tag{28}$$

The spontaneous breaking of the gauge symmetry in a $N = 2$ theory will be very useful, providing larger associations between massive spin-1 gauge bosons, spin-$\frac{1}{2}$ charginos and neutralinos and spin-0 BEH bosons,[45] and leading us to a description of particle physics in a higher-dimensional space-time.[46]

3.6. *Origin of R symmetry, and of the extra $U(1)$ rotating h_1 and h_2*

R symmetry originates from an earlier Q symmetry acting within a precursor of a supersymmetric theory including two spin-0 doublets (now called h_1 and h_2), and a Dirac spin-$\frac{1}{2}$ doublet, subsequently providing the corresponding higgsinos $\tilde{h}_{1L}$ and $\tilde{h}_{2L}$.[47] The Q symmetry restricting both the possible form of the potential V and of the Yukawa couplings responsible for fermion masses was already a R-type symmetry. It acts according to

$$H_1 \overset{Q}{\rightarrow} e^{i\alpha}\, H_1(x, \theta\, e^{-i\alpha}), \quad H_2 \overset{Q}{\rightarrow} e^{i\alpha}\, H_2(x, \theta\, e^{-i\alpha}), \tag{29}$$

allowing for a $\mu\, H_2 H_1$ superpotential mass term for H_1 and H_2. It was then turned into the R symmetry familiar to us today, defined as $R = Q\, U^{-1}$ (or equivalently

$Q = RU$) and acting according to[1]

$$H_1 \overset{R}{\to} H_1(x, \theta\, e^{-i\alpha}), \quad H_2 \overset{R}{\to} H_2(x, \theta\, e^{-i\alpha}). \tag{30}$$

This R symmetry leaves h_1 and h_2 invariant so as to survive the electroweak breaking.

Here U denotes a $U(1)$ symmetry transformation commuting with supersymmetry, acting on the two electroweak doublets h_1 and h_2 according to

$$h_1 \overset{U}{\to} e^{i\alpha}\, h_1, \quad h_2 \overset{U}{\to} e^{i\alpha}\, h_2, \tag{31}$$

or in terms of superfields,

$$H_1 \overset{U}{\to} e^{i\alpha}\, H_1, \quad H_2 \overset{U}{\to} e^{i\alpha}\, H_2. \tag{32}$$

This definition was immediately extended[1] to the extra nMSSM singlet S transforming according to

$$S \overset{U}{\to} e^{-2i\alpha}\, S. \tag{33}$$

The transformation (31) was first introduced as a way to constrain the potential in a two-doublet model by *allowing for independent phase transformations of* h_1 *and* h_2, jointly with the weak hypercharge $U(1)_Y$ (h_1 and h_2 having $Y = -1$ and $+1$).[47] This does not lead to the appearance of an axion or axionlike particle as long as we are dealing with an *inert-doublet model*, keeping an unbroken symmetry combining a $U(1)$ transformation (31) with a $U(1)_Y$ transformation, under which

$$h_1 \to e^{2i\alpha}\, h_1, \quad h_2 \to h_2. \tag{34}$$

This residual $U(1)$ includes *a* Z_2 *discrete symmetry under which the inert doublet* h_1 *changes sign*, $h_1 \to -h_1$, and allows for a non-vanishing v.e.v. $\langle h_2 \rangle \neq 0$, which breaks spontaneously the electroweak symmetry. Such inert-doublet models can thus also provide, from the stability of the lightest component of h_1, a possible dark matter candidate.

In supersymmetric extensions of the standard model, however, both h_1 and h_2 must acquire non-vanishing v.e.v.'s. A classically massless particle (A) would then appear in the spectrum as a consequence of the additional $U(1)$ symmetry (31), (32), if this one remains indeed present. This particle is immediately apparent in the spectrum in the absence of the extra singlet superfield S (i.e. for $\lambda = 0$). Such a feature, considered as undesired, was *avoided from the beginning* by breaking explicitly the extra-$U(1)$ symmetry (31), (32) through the introduction of the singlet S transforming as in (33). This singlet is coupled to H_1 and H_2 by a trilinear superpotential term $\lambda\, H_2 H_1 S$, invariant under the extra-$U(1)$.

The introduction of the linear term σS in the nMSSM superpotential $\lambda\, H_2 H_1 S + \sigma S$ breaks explicitly the extra-$U(1)$ symmetry (31)–(33), providing a mass $\lambda v/\sqrt{2}$ for the would-be "axion" A.[1] Its mass vanishes with λ, the extra-$U(1)$ symmetry with its associated Goldstone (or pseudo-Goldstone) boson A getting recovered for

$\lambda = 0$. The same $U(1)$ transformation (31) acting on the two doublets h_1 and h_2 became useful later in a different context, to rotate away the CP-violating parameter θ of QCD.[48] The resulting presence of an axion A, after having escaped attention in Ref. 48, was pointed out in Refs. 49 and 50.

But no such axion has been observed yet. This may be understood if the extra-$U(1)$ symmetry is broken at a high scale through a large v.e.v. $\langle s \rangle$ for a singlet transforming non-trivially under the extra $U(1)$, as in (33). We shall return to this in Section 4.2, when dealing with the interactions of a very light neutral spin-1 gauge boson Z' (or U) as may be present in the USSM, in which the extra-$U(1)$ symmetry (32), (33) is gauged.[2] This light spin-1 boson would behave very much as the corresponding eaten-away axionlike pseudoscalar a, then mostly an electroweak singlet and interacting very weakly, thus largely "invisible".[51,52]

3.7. *Action of R symmetry*

Let us now return to R symmetry. It enlarges the initial supersymmetry algebra (8) by introducing the new symmetry generator R corresponding to an abelian group $U(1)_R$. It acts chirally on the supersymmetry generator Q according to

$$Q \overset{R}{\rightarrow} e^{-\gamma_5 \alpha}\, Q, \tag{35}$$

or equivalently $Q_L \rightarrow e^{-i\alpha}\, Q_L$, transforming gauge and (left-handed) chiral superfields according to

$$\begin{cases} V(x,\theta,\bar\theta) & \overset{R}{\rightarrow} \quad V(x,\theta\, e^{-i\alpha},\bar\theta\, e^{i\alpha}), \\ \Phi(x,\theta) & \overset{R}{\rightarrow} \quad e^{i R_\Phi \alpha}\ \Phi(x,\theta\, e^{-i\alpha}). \end{cases} \tag{36}$$

The spin-0 components $\phi = \Phi(x,0)$ of chiral superfields transform with R quantum numbers R_Φ. Their associated spin-$\frac{1}{2}$ components $\tilde\phi_L$, proportional to $[\,Q_L,\, \phi\,]$ (or equivalently to the linear term in the expansion of Φ with respect to the Grassmann coordinate θ), have $R = R_\Phi - 1$. The R-symmetry transformations (36) thus act on field components as

$$\begin{cases} V^\mu \overset{R}{\rightarrow} \quad V^\mu, & \lambda \overset{R}{\rightarrow} \quad e^{\gamma_5\alpha}\ \lambda, \\ \phi \overset{R}{\rightarrow} \quad e^{iR_\Phi\alpha}\ \phi, & \tilde\phi_L \overset{R}{\rightarrow} \ e^{i(R_\Phi-1)\alpha}\ \tilde\phi_L, \end{cases} \tag{37}$$

λ denoting the Majorana gaugino fields associated with the gauge fields V^μ. The (complex) auxiliary components $(F+iG)/\sqrt{2}$ of the chiral superfields Φ transform with $R = R_\Phi - 2$, according to

$$\frac{F+iG}{\sqrt{2}} \ \overset{R}{\rightarrow} \ e^{i(R_\Phi-2)\alpha}\ \frac{F+iG}{\sqrt{2}}\,. \tag{38}$$

The auxiliary components of $R = 2$ superfields are invariant under R. This was used in (18) and (22) to include a linear contribution σS within the superpotential $\mathcal{W}$ of a R-symmetric theory, as in the nMSSM.[1]

3.8. *Constructing* Dirac *charginos and neutralinos with a conserved R symmetry*

R symmetry (i.e. $U(1)_R$) allows in particular for R-invariant Yukawa couplings of gauginos to spin-$\frac{1}{2}$ and spin-0 fields described by chiral superfields, that may be expressed as

$$\mathcal{L}_Y = \sum_a (i)\, g_a \sqrt{2}\ \bar{\lambda}_{aR}\ \phi_i^\dagger\, (T_a)_{ij}\, \tilde{\phi}_{jL} + \text{h.c.}, \tag{39}$$

with

$$\lambda_{aR} \overset{R}{\to} e^{-i\alpha}\, \lambda_{aR},\quad \phi_i^\dagger\, (T_a)_{ij}\, \tilde{\phi}_{jL} \overset{R}{\to} e^{-i\alpha}\ \phi_i^\dagger\, (T_a)_{ij}\, \tilde{\phi}_{jL}. \tag{40}$$

The phase factor ± 1 or $\pm i$ that may appear in front of the first term in (39) is convention-dependent and may be modified by a chiral redefinition of the gaugino fields λ_a, or a relative phase redefinition of ϕ_i and $\tilde{\phi}_{iL}$.

This leads to the possibility of generating, through a spontaneous breaking of the gauge symmetry, R-invariant non-diagonal mass terms connecting gauginos with higgsinos, transforming as

$$\begin{cases} \textit{gauginos}\ \lambda & \overset{R}{\to} \quad e^{\gamma_5 \alpha}\, \lambda, \\ \textit{higgsinos}\ \psi & \overset{R}{\to} \quad e^{-\gamma_5 \alpha}\, \psi. \end{cases} \tag{41}$$

One thus gets Dirac spinors known as charginos and neutralinos — even if denominations like winos, zino, etc. could be more appropriate as we shall see. They may be expressed as[1]

$$\begin{cases} R = +1\ \ \textit{Dirac ino} & = \quad \textit{gaugino}_L + \textit{higgsino}_R\ , \\ & \qquad\quad \text{or} \\ R = -1\ \ \textit{Dirac ino} & = \quad \textit{higgsino}_L + \textit{gaugino}_R\ . \end{cases} \tag{42}$$

They have the same masses m_W, m_Z, etc. as the corresponding spin-1 gauge bosons, as long as supersymmetry is unbroken. This already hints at "gauge/BE-Higgs unification", a crucial property that may be the prime motivation for supersymmetry.[7,8]

The introduction of direct gaugino (m_1, m_2) and higgsino (μ) mass terms then modifies these R-conserving chargino and neutralino mass matrices by including supersymmetry-breaking $\Delta R = \pm 2$ contributions. The μ parameter may be considered as "supersymmetric" as a $\mu\, H_2 H_1$ mass term may be included directly in the superpotential, or regenerated from the $\lambda\, H_2 H_1 S$ coupling through the translation of the $R = 2$ spin-0 component of the singlet S, leading to

$$\mu = \lambda\, \langle s \rangle. \tag{43}$$

Still the μ term generates a supersymmetry-breaking contribution to the mass matrices when the spin-0 doublets h_1 and h_2 acquire non-vanishing v.e.v.'s, by contributing to non-vanishing v.e.v.'s for the auxiliary components of H_1 and H_2.

3.9. *From R symmetry to R parity*

R symmetry was introduced for reasons related with the triggering of the electroweak breaking induced by h_1 and h_2, which must both acquire non-vanishing v.e.v.'s. Otherwise we would stay with an unwanted massless chargino, even before thinking about introducing quarks and leptons and generating their masses. R symmetry was also introduced with the desire of defining a conserved quantum number R attributed to massless or massive Dirac spinors as in (41) and (42), with differences $\Delta R = \pm 1$ between fermions and bosons within the multiplets of supersymmetry.

This was done in a toy-model attempt at relating the photon with a "neutrino" carrying one unit of R, and the W^- with a light chargino that might have been an "electron" candidate (or even in 1976, at the time of the τ discovery, a τ candidate, with the fermionic partner of the photon as a ν_τ candidate). But the previous "neutrino", called a gaugino in modern language, must in fact be considered as a new photonic neutrino within supersymmetric extensions of the standard model.[2] It was called the photino, with, similarly, the spin-$\frac{1}{2}$ partners of the gluons called the gluinos,[32,53] so that

$$\begin{cases} \textit{photon} \quad \overset{SUSY}{\longleftrightarrow} \quad \textit{photino}, \\ \textit{gluons} \quad \overset{SUSY}{\longleftrightarrow} \quad \textit{gluinos}. \end{cases} \tag{44}$$

The parity of the new quantum number R carried by the supersymmetry generator,

$$R_p = (-1)^R, \tag{45}$$

plays an important role. It distinguishes between ordinary particles, with $R_p = +1$, and superpartners, also called *sparticles*, with $R_p = -1$, while allowing for the generation of masses for the Majorana spin-$\frac{3}{2}$ gravitino and spin-$\frac{1}{2}$ gluinos, which transform chirally under R symmetry.[32,53] Their mass terms break explicitly the continuous R symmetry, reducing it to R-parity. This one may then be identified as[2,3]

$$R_p = (-1)^R = (-1)^{2S}\,(-1)^{3B+L}. \tag{46}$$

As $R_p = (-1)^{2S}(-1)^{3(B-L)}$, its conservation follows from the conservation of $B-L$, even only modulo 2, ensuring the stability of the lightest supersymmetric particle, or LSP. This remains valid even in the presence of neutrino Majorana mass terms.

All superpartners are then expected to decay so as to ultimately provide, at the end of the decay chain, a stable LSP, usually taken to be a neutralino or a light gravitino,[32] although other possibilities may also be considered. The neutralino, in particular, turns out to be a good candidate for the non-baryonic dark matter of the Universe.

Conversely, should R-parity necessarily be conserved? A non-conservation of R-parity, as in R_p-violating theories,[54] requires B and/or L violations. It usually leads to severe difficulties with unobserved effects such as a much-too-fast proton decay mediated by squark exchanges, or too large neutrino masses, unless the corresponding products of R_p-violating couplings are taken sufficiently small. Also, if R-parity is no longer conserved, we generally lose the possibility of having a stable LSP as a candidate for the non-baryonic dark matter of the Universe.

4. N/nMSSM and MSSM Superpotentials and Potentials

4.1. *Superpotentials*

Let us precise the role of R symmetry in restricting adequately the superpotentials considered. The last component of the superpotential $\mathcal{W}$ provides a contribution to the Lagrangian density invariant under supersymmetry, up to a derivative which does not contribute to the action integral. For the theory to be invariant under R its superpotential $\mathcal{W}$ must transform according to

$$\mathcal{W}(x,\theta) \ \stackrel{R}{\rightarrow}\ e^{2\,i\,\alpha}\, \mathcal{W}(x,\theta\, e^{-\,i\,\alpha}), \tag{47}$$

so that its last component, which appears as the coefficient of the $\theta\theta$ term in its expansion and contributes to $\mathcal{L}$, be R-invariant.

A product of chiral superfields $\Pi\,\Phi_i$ transforms with $R = \sum R_{\Phi_i}$, and is allowed in the superpotential if and only if

$$\sum\ R_{\Phi_i} = 2. \tag{48}$$

The parameters λ_{ijk}, m_{ij} and σ_i in the superpotential

$$\mathcal{W} \ = \ \frac{\lambda_{ijk}}{3}\ \Phi_i\Phi_j\Phi_k + \frac{\mu_{ij}}{2}\ \Phi_i\Phi_j + \sigma_i\,\Phi_i \tag{49}$$

are required by R symmetry to vanish, unless the corresponding products of superfields verify $R_{\Phi_i} + R_{\Phi_j} + R_{\Phi_k} = 2,\ R_{\Phi_i} + R_{\Phi_j} = 2,$ or $R_{\Phi_i} = 2$.

These restrictions from R symmetry are used to select the nMSSM superpotential for the two electroweak doublets H_1 and H_2 interacting with an extra singlet S through a trilinear superpotential coupling $\lambda\, H_2H_1S$,[1]

$$\mathcal{W}_{\rm nMSSM} \ = \ S\,(\lambda\, H_2H_1 + \sigma). \tag{50}$$

The terms involving quarks and leptons will be considered later.[2] This superpotential is obtained by imposing R symmetry on the general NMSSM superpotential, also including a $\mu\, H_2H_1$ mass term as well as mass and self-interaction terms for S,

$$\mathcal{W}_{\rm NMSSM} = S\,(\lambda\, H_2H_1 + \sigma) + \mu\, H_2H_1 \ + \frac{\kappa}{3}\ S^3 + \frac{\mu_S}{2}\ S^2. \tag{51}$$

H_1 and H_2 transform as in (30) so that R symmetry can survive the electroweak breaking, extended to S according to

$$H_{1,2} \ \stackrel{R}{\rightarrow}\ H_{1,2}(x,\theta\, e^{-\,i\,\alpha}),\ \ S\ \stackrel{R}{\rightarrow}\ e^{2i\alpha}\ S(x,\theta\, e^{-\,i\,\alpha}). \tag{52}$$

Both λH_2H_1S and σS are allowed by R in the nMSSM superpotential (50). The other NMSSM terms in (51), proportional to H_2H_1, S^2 and S^3, are excluded.

Another way to restrict the general NMSSM superpotential (51) into the nMSSM one (50) is to ask for $\mathcal{W}$ to be invariant under the extra-$U(1)$ symmetry (32), (33),[1]

$$H_i \overset{U}{\to} e^{i\alpha}\, H_i, \quad S \overset{U}{\to} e^{-2i\alpha}\, S, \tag{53}$$

simply broken by the dimension-2 linear term σS, thus automatically avoiding a classically massless spin-0 "axion", before this notion was even put into light. This extra-$U(1)$ symmetry also excludes NMSSM self-interaction and mass terms proportional to S^3 and S^2 in the superpotential, as well as μH_2H_1. The latter may still be subsequently regenerated through a translation of S as in (43).

Incidentally, the μ parameter, coefficient of the μH_2H_1 superpotential mass term in the MSSM, is "supersymmetric" (in the sense that μH_2H_1 may be present in the superpotential) but comes in violation of both the R symmetry (52) and the extra-$U(1)$ symmetry (53). It may thus remain naturally small or of moderate size, as compared to very large mass scales like the grand-unification or the Planck scales.

A special version of the above general NMSSM superpotential (51) involves trilinear terms only in the superpotential,[55,56] with

$$\mathcal{W}_{\text{NMSSM}} = \lambda H_2H_1S + \frac{\kappa}{3}\, S^3, \tag{54}$$

λ and κ being dimensionless. Most of its interesting properties rely on the same trilinear λH_2H_1S coupling as in the nMSSM. In the limit $\kappa \to 0$, both the $U(1)_R$ (52) and the extra-$U(1)$ (53) would be restored. The latter being broken by $\langle h_1 \rangle$ and $\langle h_2 \rangle$ (and $\langle s \rangle$ if also present) a classically massless axionlike boson (a) would then reappear in this limit, that was precedently avoided in the nMSSM by the linear σS term (and in the above version of the NMSSM by $\frac{\kappa}{3}\, S^3$). Such a particle, which has not been observed, may also acquire a mass, possibly small, through the soft supersymmetry-breaking terms breaking explicitly the extra-$U(1)$ symmetry.

4.2. *The USSM, with a new neutral gauge boson*

Another option is to gauge the above extra-$U(1)$ symmetry (53), assuming the corresponding anomalies appropriately cancelled, usually through the introduction of extra fermion fields. These may involve, for example, mirror fermions, or exotic fermions as would be present in an $E(6)$ theory. The would-be (axionlike) Goldstone boson (a) is then "eaten away" when the additional neutral gauge boson Z' acquires a mass. This leads to the USSM, with the trilinear superpotential

$$\mathcal{W}_{\text{USSM}} = \lambda H_2H_1S, \tag{55}$$

the theory being at this stage invariant under both the R symmetry (52) and the extra-$U(1)$ symmetry (53), now promoted to a local gauge symmetry.[2]

The gauging of an additional $U(1)$, possibly appearing as a subgroup of a non-abelian grand-unification group like $E(6)$, with (anti)quark and (anti)lepton chiral superfields transforming axially according to

$$(L, Q\,;\,\bar{E}, \bar{D}, \bar{U}) \;\stackrel{U}{\to}\; e^{-\frac{i\alpha}{2}}\,(L, Q\,;\,\bar{E}, \bar{D}, \bar{U}), \tag{56}$$

requires a new spin-1 gauge boson Z'. More generally the extra-$U(1)$ symmetry generator to be gauged may involve a linear combination of the axial $U(1)$ quantum number defined from (53), (56) as

$$\begin{cases} F_A(L, Q\,;\,\bar{E}, \bar{D}, \bar{U}) \;=\; -\,\dfrac{1}{2}, \\ F_A(H_1, H_2) \;=\; 1, \quad F_A(S) \;=\; -\,2, \end{cases} \tag{57}$$

with the weak hypercharge Y and the B and L (or $B-L$) quantum numbers. A large v.e.v. for an extra singlet like s, already present in the theory and transforming as in (53), $s \to e^{-2i\alpha}\, s$, may make the new gauge boson much heavier than the $W^\pm$ and Z, giving it a large mass $\gtrsim$ TeV scale.[51] But no new heavy boson corresponding to an enlargement of the gauge group has been discovered yet.

4.3. *A new light gauge boson U? or a light pseudoscalar a?*

There is also another interesting possibility. An additional $U(1)$ factor in the gauge group, if not embedded within a grand-unification group like $O(10)$ or $E(6)$, ..., would have its own gauge coupling constant g'', next to g and g'. This one may be much smaller than g and g', in which case the mass of the new neutral gauge boson may well be small. This Z', also called a U boson, would then have, for its longitudinal polarization state, effective interactions fixed by $g'' k^\mu / m_U$. It would behave very much as the "eaten-away" Goldstone boson a, acquiring effective axionlike pseudoscalar couplings to quarks and leptons recovered from its axial couplings f_A (proportional to g''), as[52]

$$f_p \;=\; f_A\, \frac{2 m_{l,q}}{m_U}. \tag{58}$$

This is very similar to the situation for a massive but light spin-$\frac{3}{2}$ gravitino, with a very small gravitational coupling $\kappa \;=\; \sqrt{8\pi\, G_N} \;\simeq\; 4\,10^{-19}$ GeV^{-1}, and a small mass

$$m_{3/2} = \frac{\kappa d}{\sqrt{6}} \;=\; \frac{\kappa F}{\sqrt{3}}. \tag{59}$$

F, or $\sqrt{F}$, is usually referred to as the supersymmetry-breaking scale parameter. Such a light gravitino would have its $\pm 1/2$ polarization states interacting proportionally to $\kappa\, k^\mu / m_{3/2}$, or k^μ / F. It would still behave very much as the "eaten-away" spin-$\frac{1}{2}$ goldstino, according to the "equivalence theorem" of supersymmetry. The strength of its interactions then depends on the scale at which supersymmetry is spontaneously broken, getting very small if the supersymmetry-breaking scale ($\sqrt{d}$ or $\sqrt{F}$) is large enough.[32]

Let us return to a light spin-1 U boson. As it would behave very much like the corresponding equivalent Goldstone boson,[52] it would certainly be excluded if it could be produced, most notably in the radiative decays of the ψ and the Υ, much like a standard axion (A). Fortunately the singlet s already present in these theories, transforming under U as in (53) according to $s \to e^{-2i\alpha}\, s$, may acquire a large v.e.v., significantly above the weak scale. The extra-$U(1)$ symmetry is then broken "at a large scale" F_U, where the mass $m_U \propto g'' F_U$ may still be small when the extra $U(1)$ is gauged with a very small coupling. The corresponding particle (either the very light spin-1 U boson or its "equivalent" spin-0 pseudoscalar a) is then coupled effectively very weakly, proportionally to g''/m_U, or $1/F_U$.[51,52] This pseudoscalar a is mostly an electroweak singlet, largely inert.

Dealing with a spin-0 particle this also provided, as a by-product, a very early realization of the "invisible axion" mechanism that became popular later, in which the "invisible axion" is mostly an electroweak singlet.[51] Furthermore the doublet and singlet $U(1)$ quantum numbers are here appropriate to the supersymmetry framework, with an invariant $\lambda\, H_2 H_1 S$ trilinear coupling, resulting in the $U(1)$ quantum numbers $+1$ for h_1 and h_2, $-1/2$ and $+1/2$ for left-handed and right-handed quarks and leptons, and -2 for the extra singlet s as in (57).

In a similar way a light spin-1 gauge boson U, interacting very much as the eaten-away Goldstone boson a i.e. as an "invisible axion" (except for the $\gamma\gamma$ coupling of the latter), also becomes largely "invisible" if the extra-$U(1)$ symmetry is broken at a sufficiently high scale. But the hunt for such a light spin-1 U boson is another story.[57]

4.4. *N/nMSSM and MSSM potentials*

The nMSSM superpotential (50) leads to the potential[1]

$$\begin{aligned} V_{\text{nMSSM}} = & \frac{g^2+g'^2}{8}(h_1^\dagger h_1 - h_2^\dagger h_2)^2 + \frac{g^2}{2}|h_1^\dagger h_2|^2 \\ & + \frac{\xi g'}{2}\,(h_1^\dagger h_1 - h_2^\dagger h_2) + \frac{\xi^2}{2} \\ & + |\,\lambda\, h_2 h_1 + \sigma|^2 + \lambda^2\,|s|^2\,(|h_1|^2+|h_2|^2). \end{aligned} \tag{60}$$

The D-term contributions take into account an abelian $-\xi D'$ term in $\mathcal{L}$ (this sign choice, different from the usual one in (15), being made to have $\xi > 0$ for $v_2 > v_1$ i.e. $\tan\beta > 1$).

This also applies to the general NMSSM through the replacements

$$\begin{aligned} \sigma S \;\to\; f(S) \;&=\; \frac{\kappa}{3}\, S^3 + \frac{\mu_S}{2}\, S^2 + \sigma S, \quad \lambda S \;\to\; \mu + \lambda S, \\ \sigma \;\to\; \frac{df(s)}{ds} \;&=\; \kappa\, s^2 + \mu_S\, s + \sigma, \qquad\quad \lambda s \;\to\; \mu + \lambda s, \end{aligned} \tag{61}$$

in the superpotential and potential, respectively, leading to

$$\begin{aligned} V_{\rm NMSSM} = \;& \frac{g^2+g'^2}{8} (h_1^\dagger h_1 - h_2^\dagger h_2)^2 \,+\, \frac{g^2}{2}\, |h_1^\dagger h_2|^2 \\ & + \frac{\xi\, g'}{2}\, (h_1^\dagger h_1 - h_2^\dagger h_2) \,+\, \frac{\xi^2}{2} \\ & + |\, \lambda\, h_2 h_1 + \kappa\, s^2 + \mu_S\, s + \sigma|^2 + \,|\mu + \lambda s|^2\, (|h_1|^2 + |h_2|^2). \end{aligned} \tag{62}$$

The translation (43) of the singlet S restores the (N)MSSM mass term $\mu\, H_2 H_1$ from the nMSSM superpotential (50). Furthermore in the $\lambda \to 0, \sigma \to \infty$ limit, with $\lambda\sigma$ fixed, for which S decouples, we recover the MSSM potential, in the conceptually-interesting situation of a MSSM potential with dimension-2 *soft-breaking terms generated from a supersymmetric Lagrangian density.* It reads (up to a very large or infinite constant term, irrelevant at the moment)

$$\begin{aligned} V_{\rm MSSM} \;=\;& \frac{g^2+g'^2}{8}\, (h_1^\dagger h_1 - h_2^\dagger h_2)^2 \,+\, \frac{g^2}{2}\, |h_1^\dagger h_2|^2 \\ + \left(\mu^2 + \frac{\xi g'}{2}\right) h_1^\dagger h_1 \,+& \left(\mu^2 - \frac{\xi g'}{2}\right) h_2^\dagger h_2 \,+\, 2\,\lambda\sigma\, \Re\, h_2 h_1. \end{aligned} \tag{63}$$

The last term, $\propto \Re\, h_2 h_1$, forces h_1 as well as h_2 to acquire a non-vanishing v.e.v. But this does not lead to an unwanted classically-massless axion or axionlike pseudoscalar A, as this term $\propto \Re\, h_2 h_1$ breaks explicitly the extra-$U(1)$ symmetry (31), (32), $h_1 \to e^{i\alpha}\, h_1,\; h_2 \to e^{i\alpha}\, h_2$.

If the (extremely weak) interactions of the singlet S were reconsidered again, with an extremely small coupling λ, the vacuum state corresponding to (63), which then has an extremely large energy density $\simeq \sigma^2 \propto 1/\lambda^2$, would be destabilized, but still staying effectively quasi-metastable.

These expressions of the N/nMSSM and MSSM potentials illustrate how spin-0 interactions may now be viewed as *part of the electroweak gauge interactions*, with their quartic couplings fixed by

$$\frac{g^2+g'^2}{8} \quad \text{and} \quad \frac{g^2}{2}. \tag{64}$$

They lead to a spontaneous breaking of $SU(2) \times U(1)$ into $U(1)_{\rm QED}$, with non-vanishing v.e.v.'s for both h_1 and h_2,

$$\langle h_1 \rangle = \begin{pmatrix} \frac{v_1}{\sqrt{2}} \\ 0 \end{pmatrix}, \quad \langle h_2 \rangle = \begin{pmatrix} 0 \\ \frac{v_2}{\sqrt{2}} \end{pmatrix}, \tag{65}$$

where $v_1 = v\cos\beta$, $v_2 = v\sin\beta$ with $v \simeq 246$ GeV. This also leads us in the direction of gauge-Higgs unification already alluded to in (6),[7,8,42] discussed in the next Section.

The nMSSM potential (60), in particular, forces v_1 and v_2 to verify at this stage (before the introduction of extra terms breaking supersymmetry explicitly) $\sigma - \frac{1}{2}\lambda v_1 v_2 = 0$, which ensures the vanishing of the F terms in the potential. Minimizing the D terms leads (without a μ term yet) to $\langle D_Z \rangle = 0$ with $\langle D_\gamma \rangle \neq 0$ so that the photino is the Goldstone spinor, then fixing, in the absence of other soft-breaking terms,

$$m_Z^2\,(-\cos 2\beta) = \xi g'\,, \tag{66}$$

i.e. $\sqrt{\xi} \simeq m_Z/\sqrt{g'} \simeq 155$ GeV, in the large $\tan\beta$ limit.[8]

The structure of the nMSSM superpotential (50) (and resulting potential as in (60)) is useful in many circumstances, and most notably to trigger gauge symmetry breaking by rendering the gauge-symmetric vacuum state unstable. It also leads to inflationary potentials useful in the description of the very early Universe, with an initial energy density such as $\sigma^2 + \xi^2/2$, providing the necessary fuel for inflation. Additional soft-breaking terms, of dimension ≤ 3,[58] possibly induced from supergravity,[59–64] may also be added to the (N/n)MSSM potentials (60), (62) and (63).

5. Gauge/BE-Higgs Unification in the (N/n)MSSM

5.1. *An index Δ for counting massless chiral spinors*

These theories make use of gauge superfields, describing (left-handed) gaugino fields λ_L carrying $R = 1$, and $R = 2$ or 0 chiral superfields describing (left-handed) chiral spinors with $R = 1$ and -1 respectively, as seen from (37) and (42). With n_g gauge superfields and n_2 and n_0 chiral superfields with $R = 2$ or 0, we get $n_g + n_2$ and n_0 (left-handed) spinors with $R = 1$ and -1, respectively. The former are in excess by the difference

$$\Delta = \underbrace{n_g + n_2}_{R\,=\,1\ \text{spinors}} \quad - \quad \underbrace{n_0}_{R\,=\,-1\ \text{spinors}}\,. \tag{67}$$

$R = 1$ and -1 left-handed spinors may combine as in (42), in a way compatible with R symmetry, into massive Dirac spinors carrying $R = \pm 1$. For $\Delta \geq 0$, Δ chiral spinors with $R = 1$, at least, must remain massless if R symmetry is conserved, as staying unpaired with $R = -1$ counterparts.

With the nMSSM superpotential (50) $n_g = 4$ for $SU(2) \times U(1)$, $n_0 = 4$ for H_1 and H_2 and $n_2 = 1$ for S, so that $\Delta = 1$. One left-handed spinor with $R = 1$, neutral, must remain massless, which is here the photino. This one becomes the Goldstone spinor when supersymmetry is spontaneously broken. An early version of the USSM, with one additional extra-$U(1)$ gaugino, had $\Delta = 5{+}1{-}4 = 2$, leading to two massless $R=1$ spinors, with a goldstino (eaten-away by the spin-$\frac{3}{2}$ gravitino) different from the photino, superpartners ultimately decaying into gravitinos or photinos carrying away missing energy-momentum.[2,32]

5.2. *The goldstino must transform with $R=1$*

The (left-handed) massless Goldstone spinor λ_g associated with spontaneous supersymmetry breaking must have $R=1$ in a R symmetric theory, as follows from the R transformation properties (35) of the supersymmetry generator and vector-spinor current, such that

$$J^\mu_\alpha \xrightarrow{R} e^{-\gamma_5 \alpha}\, J^\mu_\alpha\,. \tag{68}$$

For spontaneously broken supersymmetry the vector-spinor current may be expressed as

$$J^\mu_\alpha = d\,\gamma^\mu \gamma_5\, \lambda_g + \cdots\,, \tag{69}$$

where $d/\sqrt{2}=F$ is the supersymmetry-breaking scale parameter which determines the gravitino mass $m_{3/2}=\kappa d/\sqrt{6}=\kappa F/\sqrt{3}$ in (59).[32] The Goldstone spinor must transform according to $\lambda_g \xrightarrow{R} e^{\gamma_5 \alpha}\, \lambda_g$, or equivalently

$$\lambda_{gL} \xrightarrow{R} e^{i\alpha}\, \lambda_{gL}, \tag{70}$$

i.e. it should transform with $R=1$.

It should be either a gaugino as in pure D-breaking,[1,29,42] or a spin-$\frac{1}{2}$ fermion field described by a $R=2$ chiral superfield as in F-breaking,[30,31] or a mixing of both.[2,29] In the nMSSM at the present stage, with an unbroken R symmetry and without any addition of soft supersymmetry-breaking terms yet, the massless goldstino field, with $R=1$, coincides with the photino field, supersymmetry remaining unbroken within neutral multiplets. This degeneracy gets broken later through terms breaking explicitly (although softly) the supersymmetry,[27,58] possibly obtained from gravity-induced supersymmetry breaking.[59–64] Still the R-symmetric nMSSM considered at the present stage is essential in the understanding of the gauge/BE-Higgs unification and of the resulting mass spectrum for the various versions of the MSSM or N/nMSSM, as we shall see.

5.3. *$U(1)_R$ symmetric nMSSM mass spectrum*

With the gauge and chiral superfields transforming under the continuous R symmetry ($U(1)_R$) according to

$$\begin{cases} V_a & \xrightarrow{R} & V_a(x,\theta e^{-i\alpha},\bar\theta e^{i\alpha}), \\ H_{1,2} & \xrightarrow{R} & H_{1,2}(x,\theta e^{-i\alpha}), \\ S & \xrightarrow{R} & e^{2i\alpha}\, S(x,\theta e^{-i\alpha}), \end{cases} \tag{71}$$

R symmetry leads to 1 chiral spinor remaining massless, with 4 massive Dirac ones. The R-symmetric and quasi-supersymmetric fermion spectrum for the nMSSM is at this stage, with $gv_1/\sqrt{2} = m_W\sqrt{2}\,\cos\beta,\ gv_2/\sqrt{2} = m_W\sqrt{2}\,\sin\beta$:[1]

$$\left\{\begin{array}{ll} 1 \text{ massless } \textit{photino:} & m = 0, \\ 2 \text{ Dirac } \textit{winos:} & m = \left\{\begin{array}{l} m_W\sqrt{2}\,\cos\beta, \\ m_W\sqrt{2}\,\sin\beta, \end{array}\right. \\ 1 \text{ Dirac } \textit{zino:} & m_Z = \sqrt{g^2+g'^2}\ v/2, \\ 1 \text{ Dirac } \textit{neutralino:} & m = \lambda v/\sqrt{2}\ , \end{array}\right. \tag{72}$$

all spinors carrying $R = \pm 1$, in agreement with their expressions from gaugino and higgsino fields as in (42). The corresponding 5×5 neutralino mass matrix expressed in a gaugino–higgsino basis will be given later in Section 6.4.

The charged and neutral spin-0 masses, obtained from the potential (60), are

$$\left\{\begin{array}{l} m_{H^\pm} = m_W, \quad m_z = m_Z, \\ m\,(\,4 \textit{ neutral spin-0 bosons}\,) = \dfrac{\lambda v}{\sqrt{2}}. \end{array}\right. \tag{73}$$

All neutral spin-0 bosons have the same mass m_Z or $\lambda v/\sqrt{2}$ as their fermionic partners in (72), thanks to the unbroken supersymmetry in the neutral sector, the photino being here the Goldstone spinor. Two of the four neutral bosons of mass $\lambda v/\sqrt{2}$ are described by the singlet superfield S, with $R = 2$. The two others are described by the $R = 0$ superfield

$$H_A = H_1^0\,\sin\beta + H_2^0\,\cos\beta. \tag{74}$$

As $\langle h_1^0\rangle = v\cos\beta/\sqrt{2},\ \langle h_2^0\rangle = v\sin\beta/\sqrt{2}$, H_A acquires the mass $\lambda v/\sqrt{2}$ from the $\lambda\, H_2H_1S$ superpotential term, by combining with the singlet S according to

$$\lambda\ H_2H_1S = -\frac{\lambda v}{\sqrt{2}}\,(H_1^0\,\sin\beta + H_2^0\,\cos\beta)\,S + \cdots\,, \tag{75}$$

in a way compatible with R symmetry (with $H_2H_1 = -\,H_1H_2 = H_2^+H_1^- - H_2^0H_1^0$).

All four scalars would return to massless for $\lambda\to 0$, for which S decouples, H_A returning to massless. Indeed in the $\lambda\to 0$ limit one recovers at the classical level a spontaneously broken extra-$U(1)$ acting as in (31) and (32), generating a classically-massless axion or axionlike particle. This one, which has here the mass $m_A = \lambda v/\sqrt{2}$, is described by the imaginary part of the spin-0 component of H_A,

$$A = \sqrt{2}\ \text{Im}\ (h_1^0\,\sin\beta + h_2^0\,\cos\beta). \tag{76}$$

5.4. *Gauge/BE-Higgs unification*

Relating gauge and BEH bosons, in spite of different electroweak properties

The superfield orthogonal to H_A in (74) is

$$H_z = -H_1^0 \cos\beta + H_2^0 \sin\beta. \tag{77}$$

The imaginary part of its spin-0 component,

$$z_g = \sqrt{2}\ \mathrm{Im}\,(-h_1^0 \cos\beta + h_2^0 \sin\beta), \tag{78}$$

orthogonal to the A field in (76), describes the would-be Goldstone boson eaten away by the Z. Indeed this Goldstone field z_g originates from the imaginary part of the SM-like combination $\varphi_{\rm sm}^{\circ} = h_1^{\circ} \cos\beta + h_2^{\circ *} \sin\beta$ responsible for the electroweak breaking, with $\langle \varphi_{\rm sm}^{\circ} \rangle = v/\sqrt{2}$.

The real part of the spin-0 component of H_z is

$$z = \sqrt{2}\ \Re\,(-h_1^0 \cos\beta + h_2^0 \sin\beta). \tag{79}$$

The signs are chosen for convenience so that $H_z \to H_2^0$ and $z \to \sqrt{2}\ \Re\ h_2^0$ in the large $\tan\beta$ limit. This field, suitably translated so that $\langle z \rangle = 0$, describes in this formalism the spin-0 boson partner of the Z within a massive gauge multiplet of supersymmetry. Its mass is obtained from the $D_Z^2/2$ contribution to the potential, as expressed in the nMSSM potential (60), with[8]

$$\begin{aligned} D_Z &= \frac{\sqrt{(g^2+g'^2)}\ v}{2}\ [\sqrt{2}\ \Re\,(-h_1^0 \cos\beta + h_2^0 \sin\beta)] + \cdots \\ &= m_Z\ z + \cdots , \end{aligned} \tag{80}$$

and

$$\frac{1}{2}\ D_Z^2 = \frac{1}{2}\ m_Z^2\ z^2 + \cdots . \tag{81}$$

One thus has

$$\boxed{\begin{gathered} m_z = m_Z \simeq 91\ \mathrm{GeV}/c^2 \\ \textit{(up to supersymmetry-breaking mass and mixing effects)}, \end{gathered}} \tag{82}$$

independently of the value of $\tan\beta$, in agreement with the unbroken supersymmetry in the neutral sector.[1]

More precisely when the BEH mechanism operates within a supersymmetric theory, it provides *massive gauge multiplets.*[42] Each of them describes a massive spin-1 gauge boson, two spin-$\frac{1}{2}$ inos constructed from gaugino and higgsino components as in (42), and a spin-0 BEH boson associated with the spontaneous breaking of the gauge symmetry. We get systematic associations between massive gauge bosons and spin-0 BEH bosons, *a quite non-trivial feature owing to their different gauge symmetry properties*, and very different couplings to quarks and leptons.[7,8,65]

We have in particular the association

$$\boxed{Z \overset{SUSY}{\longleftrightarrow} \text{2 Majorana zinos} \overset{SUSY}{\longleftrightarrow} \text{spin-0 BEH boson.}} \tag{83}$$

Independently of $\tan\beta$, and of λ in the presence of the N/nMSSM singlet S, the neutral spin-0 boson described by the z field in (79) becomes the spin-0 partner of the Z within a massive multiplet of supersymmetry. It has the same mass m_Z as long as supersymmetry is unbroken in this sector, in agreement with (6), (72), (73), (81).[1,7,8]

This also applies to the $W^\pm$, according to

$$\boxed{W^\pm \overset{SUSY}{\longleftrightarrow} \text{2 Dirac winos} \overset{SUSY}{\longleftrightarrow} \text{spin-0 boson } H^\pm.} \tag{84}$$

The $W^\pm$ is associated with two Dirac winos (usually known as charginos), obtained as in (42) with masses given in (72), and a charged spin-0 boson $H^\pm$ (or $w^\pm$), with

$$w^\pm \equiv H^\pm = \sin\beta\ h_1^\pm + \cos\beta\ h_2^\pm, \tag{85}$$

approaching $h_1^\pm$ at large $\tan\beta$. This one was originally called $w^\pm$ in Ref. 1 to emphasize its relation with the $W^\pm$, leading in a model-independent way to

$$\boxed{\begin{array}{c} m_{H^\pm} = m_{W^\pm} \simeq 80 \text{ GeV}/c^2 \\ \textit{(up to supersymmetry-breaking effects).} \end{array}} \tag{86}$$

We shall see later how these mass equalities get modified in the presence of supersymmetry-breaking effects, in the MSSM and N/nMSSM.

As seen on (83) and (84) the supersymmetry generator Q has become able to *relate bosons and fermions with different electroweak gauge symmetry properties*, a quite non-trivial feature, in contrast with the usual belief. This makes supersymmetry a very powerful symmetry, much beyond the simple replication of degrees of freedom by associating bosons and fermions with the same gauge symmetry properties.

While massive gauge bosons and spin-0 BEH bosons have different symmetry properties for the electroweak gauge group, that is spontaneously broken, they do have the same properties for the $SU(3)_{\rm QCD} \times U(1)_{\rm QED}$ symmetries, that remain unbroken.

When supersymmetry is broken the lightest neutral spin-0 boson should presumably be identified with the 125 GeV boson found at CERN[5,6] (unless a lighter one has escaped attention). This one may well correspond to the above z field (approaching $\sqrt{2}\ \Re\ h_2^0$ in the large $\tan\beta$ limit), up to a mixing angle, possibly small, induced by supersymmetry breaking.

However, the non-observation, at this stage, of a charged spin-0 BEH boson $H^\pm$ seems to indicate (unless such a boson is found, with a moderate mass) that the effects of supersymmetry breaking are more important in the $W^\pm$ than in the Z multiplet. This may be understood from the possible form of the supersymmetry-breaking terms.

5.5. *Describing spin-0 BEH bosons by massive gauge superfields*

This association between the spin-1 $W^\pm$ and Z and the spin-0 $H^\pm$ (also called $w^\pm$) and z can be made explicit in a different superfield formulation. Spin-0 BEH bosons will now be described by the spin-0 components of *massive gauge superfields*,[7,8,42] after all components of the superfields H_1^-, H_2^+ and H_z in (77), then considered as chiral Goldstone superfields, get completely gauged away through the generalized gauge choices

$$H_1^- \equiv H_2^+ \equiv 0, \quad H_z \equiv \langle H_z \rangle = -\frac{v}{\sqrt{2}} \cos 2\beta. \tag{87}$$

In this new picture these spin-0 bosons get described, in a manifestly supersymmetric formulation, by the lowest (C) spin-0 components of *massive* Z and $W^\pm$ superfields, expanded as $Z(x,\theta,\bar\theta) = C_Z + \cdots - \theta\sigma_\mu\bar\theta\ Z^\mu + \cdots$, $W^\pm(x,\theta,\bar\theta) = C_W^\pm + \cdots - \theta\sigma_\mu\bar\theta\, W^{\mu\,\pm} + \cdots$. Their spin-0 C components now describe, through *non-polynomial field transformations* linearized as $z = -m_Z C_Z + \cdots$, $w^\pm = m_W C_W^\pm + \cdots$, the same spin-0 fields z and $w^\pm$ as in the usual formalism (with signs depending on previous choices for the definitions of z and $w^\pm$).

We thus have

$$\begin{cases} Z(x,\theta,\bar\theta) = \left(\dfrac{-z}{m_Z} + \cdots\right) + \cdots - \theta\sigma_\mu\bar\theta\ Z^\mu + \cdots, \\ W^\pm(x,\theta,\bar\theta) = \left(\dfrac{w^\pm}{m_W} + \cdots\right) + \cdots - \theta\sigma_\mu\bar\theta\ W^{\mu\,\pm} + \cdots. \end{cases} \tag{88}$$

The spin-0 components of massive gauge superfields now describe spin-0 BEH bosons! Their subcanonical (χ) spin-$\frac{1}{2}$ components, instead of being gauged-away as usual, now also correspond to physical degrees of freedom describing the spin-$\frac{1}{2}$ fields usually known as higgsinos.

Supersymmetry transformations act in a linear way on the components $(C, \chi, M, N, V^\mu, \lambda, D)$ of a massive gauge superfield $V(x,\theta,\bar\theta)$, including auxiliary as well as physical components. But they act in a more complicated way when they are formulated in terms of the usual canonically-normalized spin-0 BEH and spin-$\frac{1}{2}$ higgsino fields, in particular as their expressions involve the dimensionless C components in a non-polynomial way.

6. (N/n)MSSM Mass Spectra

With gauge/BE-Higgs unification

6.1. *Spin-0 masses in the MSSM*

The non-observation, at this stage, of a charged spin-0 BEH boson seems to indicate that the effects of supersymmetry breaking should be more important in the $W^\pm$ than in the Z multiplet. This may be an effect of a significant supersymmetry-breaking term, possibly generated spontaneously from the decoupling limit of an extra singlet as indicated in (63), or from soft gravity-induced terms.

Let us define

$$\begin{cases} \varphi_{\rm sm} &= \ h_1 \cos\beta + h_2^c \sin\beta, \\ \varphi_{\rm in} &= \ h_1 \sin\beta - h_2^c \cos\beta, \end{cases} \tag{89}$$

so that $\varphi_{\rm sm}$ appears as a SM-like doublet responsible for the electroweak breaking and $\varphi_{\rm in}$ as an "inert doublet", with $\langle\varphi_{\rm sm}\rangle = v/\sqrt{2}$, $\langle\varphi_{\rm in}\rangle = 0$. Viewing for convenience β as a fixed parameter unaffected by supersymmetry-breaking terms, these terms may be viewed as providing a mass term for the "inert" doublet $\varphi_{\rm in}$, without modifying the vacuum state defined by $\langle h_1\rangle$ and $\langle h_2\rangle$.

One has, using (76) and (85),

$$\begin{aligned} |\varphi_{\rm in}|^2 &= |\, h_1 \sin\beta - \ h_2^c \cos\beta\,|^2 \\ &= |H^+|^2 + \tfrac{1}{2}\, A^2 + \tfrac{1}{2}\, |\sqrt{2}\, \Re\,(h_1^0 \sin\beta - h_2^0 \cos\beta)\,|^2. \end{aligned} \tag{90}$$

Furthermore, if these dimension-2 supersymmetry-breaking terms expressed as $m_A^2\, |\varphi_{\rm in}|^2$ were generated as in (63) from a decoupling limit of the singlet S we would have

$$m_A^2 \ = \ 2\,\mu^2. \tag{91}$$

Or in a more general way, allowing for extra soft-breaking contributions for h_1 and h_2,

$$m_A^2 \ = \ 2\,\mu^2 + \ \Delta m^2(h_1) + \Delta m^2(h_2). \tag{92}$$

The mass term for $\varphi_{\rm in}$ provides equal contributions to m_A^2 and $m_{H^\pm}^2$, and leads to a further mixing between the neutral scalars described by the real parts of φ_1^0 and φ_2^0. It provides in particular, in the large $\tan\beta$ limit for which v_1 is small, a rather large mass^2 term for h_1 contributing to $m^2(H^\pm \simeq h_1^\pm)$, m_A^2 and to a small mixing between the neutral scalars.

Specializing in the MSSM, adding the supersymmetric $(m_W^2,\ m_Z^2)$ and supersymmetry-breaking contributions to the mass^2 matrices implies immediately,

Table 2. Minimal content of the Supersymmetric Standard Model (MSSM). Neutral gauginos and higgsinos mix into a photino, two zinos and a higgsino, further mixed into four neutralinos. Ordinary particles, including additional BEH bosons have R-parity $+1$. Their superpartners have R-parity -1. The N/nMSSM includes an extra singlet with a trilinear $\lambda H_2 H_1 S$ superpotential coupling, describing a singlino and two additional neutral spin-0 bosons. The USSM also includes an extra neutral gauge boson Z' (or U) and its associated gaugino.

Spin 1	Spin 1/2	Spin 0
gluons	gluinos $\tilde{g}$	
photon	photino $\tilde{\gamma}$	
$W^\pm$	winos $\widetilde{W}^\pm_{1,2}$	$H^\pm$ } BEH bosons
Z	zinos $\widetilde{Z}_{1,2}$	h } BEH bosons
	higgsino $\widetilde{H}$	H, A } BEH bosons
	leptons l	sleptons $\tilde{l}$
	quarks q	squarks $\tilde{q}$

in this specific model,

$$m^2_{H^\pm} = m^2_W + m^2_A. \tag{93}$$

In the large $\tan\beta$ limit, $h \simeq z \simeq \sqrt{2}\,\Re\,\varphi_2^0$ and $H \simeq \sqrt{2}\,\Re\,\varphi_1^0$ have masses close to m_Z and m_A, respectively.

The mass^2 matrix for the neutral scalar fields h_1^0 and h_2^0 may be written as the sum of two supersymmetry-conserving and supersymmetry-breaking contributions. It follows from (79) involving $z = \sqrt{2}\,\Re\,(-\,h_1^0\cos\beta + h_2^0\sin\beta)$, and from the non-supersymmetric part (90) involving $\sqrt{2}\,\Re\,(h_1^0\sin\beta - h_2^0\cos\beta)$:

$$\mathcal{M}_\circ^2 = \underbrace{\begin{pmatrix} m_Z^2 & 0 \\ 0 & 0 \end{pmatrix}_{-\beta}}_{\text{SUSY-conserving}} + \underbrace{\begin{pmatrix} 0 & 0 \\ 0 & m_A^2 \end{pmatrix}_{\beta}}_{\text{SUSY-breaking}} . \tag{94}$$

The two basis denoted by $_{-\beta}$ and $_\beta$ are rotated from the (h_1^0, h_2^0) basis by angles $-\beta$ and β, and are at angle 2β. The mass of the lightest eigenstate increases from 0 to m_Z when β increases from $\pi/4$ to $\pi/2$ (or decreases from $\pi/4$ to 0), assuming $m_A > m_Z$. The second derivative of V, in the SM-like direction orthogonal to the m_A^2 eigenstate of the second matrix, at angle 2β with the direction of z, receives only a contribution $m_Z^2\cos^2 2\beta$ from the first term. This implies a mass eigenstate verifying

$$m_h \leq m_Z\,|\cos 2\beta| \quad (+\,\text{radiative corr.}). \tag{95}$$

More precisely (94) reads

$$\mathcal{M}_\circ^2 = \begin{pmatrix} c_\beta^2\, m_Z^2 + s_\beta^2\, m_A^2 & -\, s_\beta c_\beta\, (m_Z^2 + m_A^2) \\ -\, s_\beta c_\beta\, (m_Z^2 + m_A^2) & s_\beta^2\, m_Z^2 + c_\beta^2\, m_A^2 \end{pmatrix}, \tag{96}$$

and has the eigenvalues

$$m_{H,h}^2 = \frac{m_Z^2 + m_A^2}{2} \pm \sqrt{\left(\frac{m_Z^2 + m_A^2}{2}\right)^2 - m_Z^2 m_A^2 \cos^2 2\beta}. \tag{97}$$

The smallest one verifies (95), approaching m_Z in the large $\tan\beta$ limit for which the two mass eigenvalues get close to m_Z and m_A as seen from (94). The lighter scalar h becomes close to being the spin-0 partner of the Z, $h \simeq z \simeq \sqrt{2}\,\Re\, h_2^0$, with a mass close to m_Z, the heavier one $H \simeq \sqrt{2}\,\Re\, h_1^0$ having a mass close to m_A.

These formulas, leading back to $m_{H^\pm} = m_W$ and $m_H = m_Z$ when the supersymmetry-breaking parameter m_A^2 vanishes, in agreement with (82) and (86), illustrate the implications of gauge/BE-Higgs unification, even in a situation of broken supersymmetry. Large radiative corrections, involving most notably very heavy and/or strongly mixed stop quarks, are then required in the MSSM to keep a chance to get m_h sufficiently above m_Z, in view of identifying it with the 125 GeV spin-0 boson.

6.2. *Spin-0 masses in the N/nMSSM*

With heavier spin-0 bosons thanks to the extra singlet

The situation is much better in the N/nMSSM (or also in the USSM) thanks to the trilinear coupling λ in the superpotential leading to an additional quartic term in the potential $\lambda^2|h_2h_1|^2$, and to a steepest potential allowing for larger masses, already at the classical level.

Indeed starting from the R-symmetric nMSSM spectrum (73) with $m_A = \lambda v/\sqrt{2}$, $m_{H^\pm} = m_W$ and $m_h = m_Z$, independently of β,[1] the sum of supersymmetric and supersymmetry-breaking contributions leads to the mass formulas

$$\begin{cases} m_A^2 = \dfrac{\lambda^2 v^2}{2} + \delta m_A^2\ , \\ m_{H^\pm}^2 = m_W^2 + \delta m_A^2 = m_W^2 + m_A^2 - \dfrac{\lambda^2 v^2}{2}. \end{cases} \tag{98}$$

Neutral scalars are also expected to be heavier than in the MSSM. Their 2×2 mass^2 submatrix, restricted to the h_1^0, h_2^0 subspace by ignoring the singlet scalar, now reads

$$\mathcal{M}_\circ^2 = \underbrace{\begin{pmatrix} m_Z^2 & 0 \\ 0 & \dfrac{\lambda^2 v^2}{2} \end{pmatrix}_{-\beta}}_{\substack{\text{SUSY-conserving} \\ \text{(nMSSM)}}} + \underbrace{\begin{pmatrix} 0 & 0 \\ 0 & \delta m_A^2 \end{pmatrix}_{\beta}}_{\text{SUSY-breaking}} . \tag{99}$$

Its lightest mass eigenstate is immediately seen to be in the range $(m_Z,\, \lambda v/\sqrt{2})$. For

$$\lambda \geq \sqrt{\frac{g^2+g'^2}{2}} = \frac{m_Z\sqrt{2}}{v} = 2^{3/4}\, G_F^{1/2} m_Z \simeq .52 \tag{100}$$

the two mass eigenstates are already both heavier than m_Z, independently of $\tan\beta$ and before taking into account supersymmetry-breaking effects from δm_A^2. This makes it much easier to reach 125 GeV, without having to rely on large radiative corrections from very heavy stop quarks, as one must do in the MSSM.

For $\beta = \pi/4$ i.e. $v_1 = v_2$ the 2×2 matrix (99) has $\lambda^2 v^2/2$ and $m_Z^2 + \delta m_A^2$ for eigenvalues, its lightest mass eigenvalue being large if $\lambda^2 v^2/2$ and δm_A^2 are both large. We then get

$$\begin{cases} m_z^2 &= m_Z^2 + \delta m_A^2, \\ m_{H^\pm}^2 &= m_W^2 + \delta m_A^2 \end{cases} \tag{101}$$

as obtained for example with gravity-induced supersymmetry-breaking, for $\delta m_A^2 = 4\, m_{3/2}^2$.[62] This illustrates how the gauge/BE-Higgs unification may manifest on the mass spectrum. For example taking here m_z at 125 GeV would imply a $H^\pm$ around 117 GeV. In the absence of a relatively light $H^\pm$, the above relations would lead us to view, for $\tan\beta = 1$, the z as a heavier spin-0 BEH boson close in mass to the $H^\pm$, rather than the one found at 125 GeV.

More precisely the second derivative of V, in the SM-like direction orthogonal to the m_A^2 eigenstate of the second matrix, at angle 2β with the direction of z, receives a contribution only from the first matrix in (99), and is thus now equal to $m_Z^2 \cos^2 2\beta + \frac{\lambda^2 v^2}{2}\sin^2 2\beta$. This implies a neutral mass eigenstate verifying

$$m_h^2 \leq m_Z^2 \cos^2 2\beta + \frac{\lambda^2 v^2}{2}\sin^2 2\beta \quad (+\,\text{radiative corr.})\,, \tag{102}$$

This upper bound may also be obtained directly by noting that the neutral SM-like combination $\varphi_{\rm sm}^\circ = h_1^\circ \cos\beta + h_2^{\circ *}\sin\beta$ has its quartic coupling $\lambda_{\rm sm}|\varphi_{\rm sm}^\circ|^4$ in the

N/nMSSM potential (60) or (62) fixed by

$$\lambda_{\rm sm} = \frac{g^2+g'^2}{8}\,\cos^2 2\beta + \frac{\lambda^2}{4}\,\sin^2 2\beta\,. \tag{103}$$

Expanding V as a function of $\sqrt{2}\,\Re\,\varphi^\circ_{\rm sm}$ provides for this field the mass^2 parameter $2\mu^2_{\rm sm} = 2\lambda_{\rm sm}v^2 = m_Z^2\cos^2 2\beta + \frac{\lambda^2v^2}{2}\sin^2 2\beta$. Neutral spin-0 bosons cannot all be heavier, the lightest having mass $\leq \mu_{\rm sm}\sqrt{2}$ at most, leading to the mass bound (102).

Let us explicitate for completeness the mass^2 matrix (99) for the neutral scalar h_1^0, h_2^0 subspace:

$$\begin{pmatrix} c_\beta^2\, m_Z^2 + s_\beta^2\left(\dfrac{\lambda^2v^2}{2} + \delta m_A^2\right) & -\,s_\beta c_\beta\left(m_Z^2 - \dfrac{\lambda^2v^2}{2} + \delta m_A^2\right) \\ -\,s_\beta c_\beta\left(m_Z^2 - \dfrac{\lambda^2v^2}{2} + \delta m_A^2\right) & s_\beta^2\, m_Z^2 + c_\beta^2\left(\dfrac{\lambda^2v^2}{2} + \delta m_A^2\right) \end{pmatrix}, \tag{104}$$

with $m_A^2 = \frac{\lambda^2v^2}{2} + \delta m_A^2$ as in (98). m_Z^2 and $\frac{\lambda^2v^2}{2}$ correspond to the supersymmetric contributions in the R-symmetric nMSSM spectrum (73),[1] δm_A^2 being the supersymmetry-breaking contribution from the "inert doublet" $\varphi_{\rm in}$ mass term, as in (90). This 2×2 submatrix should then be included within a 3×3 matrix taking into account mixing effects with the singlet $\sqrt{2}\,\Re\, s$, involving in particular $\mu\,\lambda v_1/\sqrt{2}$ and $\mu\,\lambda v_2/\sqrt{2}$ as seen from (62).

A more instructive expression of the matrix (99), (104) is obtained in the SUSY basis $_{-\beta}$, by rotating by -2β the matrix for the supersymmetry-breaking contribution; or by writing $-\Re\varphi_{\rm in} = \Re\,(-h_1^0\sin\beta + h_2^0\cos\beta) = \sin 2\beta\;\underbrace{\Re\,(-h_1^0\cos\beta + h_2^0\sin\beta)}_{z} + \cos 2\beta\;\Re\,(h_1^0\sin\beta + h_2^0\cos\beta)$:

$$\mathcal{M}^2_\circ = \begin{pmatrix} m_Z^2 + \delta m_A^2\,\sin^2 2\beta & \delta m_A^2\,\sin 2\beta\,\cos 2\beta \\ \delta m_A^2\,\sin 2\beta\,\cos 2\beta & \dfrac{\lambda^2v^2}{2} + \delta m_A^2\,\cos^2 2\beta \end{pmatrix}_{-\beta}. \tag{105}$$

6.3. *Charginos and neutralinos mass matrices*

As understood from R symmetry and gauge/BE-Higgs unification

The higgsino fields $\tilde h_{1L}$ and $\tilde h_{2L}$, described by H_1 and H_2 with $R=0$, transform according to (37) as $\tilde h_{iL} \to e^{-i\alpha}\,\tilde h_{iL}$, $\tilde h_{2L} \to e^{-i\alpha}\,\tilde h_{2L}$. The Dirac higgsino doublet ψ constructed from $\tilde h_{1L}$ and $(\tilde h_{2L})^c$ and the Majorana gauginos λ transform chirally in opposite ways as in (41,42). Gaugino mass terms $m_{1/2}$ (denoted by m_3, m_2 and m_1 for the gluinos and $SU(2)\times U(1)$ gauginos) violate the continuous R symmetry, as for a μ term, reducing it to R-parity.[32,53]

Gaugino and higgsino fields can combine through R-invariant non-diagonal mass terms generated from the Yukawa couplings (39). The resulting charginos and neutralinos appear at this stage, with $\mu = m_i = 0$, as *Dirac particles* carrying $R = \pm 1$ as expressed in (42), leading to the $U(1)_R$ symmetric nMSSM mass spectrum (72,73).

The Dirac zino with $R = +1$ of mass m_Z, in agreement with the as-yet-unbroken supersymmetry still present in the neutral sector, is obtained by combining

$$\begin{cases} \text{the } \textit{gaugino} & \lambda_Z = \lambda_3\, c_\theta - \lambda'\, s_\theta, \\ \text{the } \textit{higgsino} & -\tilde{h}_z = \tilde{h}_1^0\, c_\beta - \tilde{h}_2^0\, s_\beta. \end{cases} \tag{106}$$

The gaugino λ_Z is directly associated with the Z, and the higgsino $\tilde{h}_z$ is described by the chiral superfield H_z in (77). The Dirac zino may be expressed as in Ref. 1 as

$$\lambda_{ZL} + (-\tilde{h}_z)_R = (\lambda_3 c_\theta - \lambda' s_\theta)_L + (\tilde{h}_1^0\, c_\beta - \tilde{h}_2^0\, s_\beta)_R, \tag{107}$$

or reexpressed in terms of the two Majorana spinors in (106). The corresponding 2×2 mass matrix in a gaugino-higgsino basis,

$$\mathcal{M}_{\text{zinos}} = \begin{pmatrix} 0 & m_Z \\ m_Z & 0 \end{pmatrix}, \tag{108}$$

may be further unpacked into a 4×4 matrix as below in (110).

Including the $\Delta R = \pm 2$ gaugino and higgsino mass terms m_i and μ breaking explicitly R symmetry, we get, with $\mu\, H_2 H_1 = \mu\, (H_2^+ H_1^- - H_2^0 H_1^0)$, the chargino and neutralino mass matrices in the MSSM,

$$\mathcal{M}_{\text{winos}} = \begin{pmatrix} m_2 & m_W \sqrt{2}\, s_\beta \\ m_W \sqrt{2}\, c_\beta & \mu \end{pmatrix}, \tag{109}$$

and

$$\mathcal{M}_{\text{inos}} = \begin{pmatrix} m_1 & 0 & -s_\theta c_\beta\, m_Z & s_\theta s_\beta\, m_Z \\ 0 & m_2 & c_\theta c_\beta\, m_Z & -c_\theta s_\beta\, m_Z \\ -s_\theta c_\beta\, m_Z & c_\theta c_\beta\, m_Z & 0 & -\mu \\ s_\theta s_\beta\, m_Z & -c_\theta s_\beta\, m_Z & -\mu & 0 \end{pmatrix}. \tag{110}$$

The part proportional to m_Z in the neutralino mass matrix is the supersymmetric contribution, in a way compatible with gauge/BE-Higgs unification, while the $\Delta R = \pm 2$ part involving m_1, m_2 and μ is the supersymmetry-breaking part. We recall that the parameter μ, although initially "supersymmetric", still leads to supersymmetry-breaking effects in the presence of v_1 and v_2.

Their eigenvalues verify relations such as

$$m^2(\textit{wino}_1) + m^2(\textit{wino}_2) = 2\, m_W^2 + \mu^2 + m_2^2, \tag{111}$$

and, similarly,

$$\sum_{1\ldots 4} m^2(\textit{neutralino}) = 2\, m_Z^2 + 2\, \mu^2 + m_1^2 + m_2^2. \tag{112}$$

Without the gaugino masses m_1, m_2, and with the dimension-2 soft-breaking mass terms for h_1 and h_2 generated spontaneously as in (63) so that $m_A^2 = 2\,\mu^2$, the average mass^2 for bosons and fermions would be the same in the multiplets considered, with

$$\left\{\begin{aligned} 3\,m_W^2 + m_{H^\pm}^2 &= 4\,m_W^2 + m_A^2 = 4\,m_W^2 + 2\mu^2 \\ &= 2\,[m^2(wino_1) + m^2(wino_2)] \\ 3\,m_Z^2 + m_h^2 + m_H^2 + m_A^2 &= 4\,m_Z^2 + 2m_A^2 = 4\,m_Z^2 + 4\mu^2 \\ &= 2\sum_{1...4} m^2(neutralino) \end{aligned}\right. \cdot \tag{113}$$

6.4. *Neutralinos in the N/nMSSM*

As understood from R symmetry and gauge/BE-Higgs unification

The N/nMSSM introduces an additional neutral singlino described by S. The chargino mass matrix (109) is simply affected by the replacement $\mu \to \mu_{\text{eff}} = \mu + \lambda\langle s\rangle$. The neutralino mass matrix (110) gets embedded into a 5×5 one. It now includes R-conserving non-diagonal mass terms corresponding to the nMSSM mass spectrum in (72), with $-\frac{\lambda v}{\sqrt{2}}\sin\beta$ and $-\frac{\lambda v}{\sqrt{2}}\cos\beta$ contributions obtained from the R-invariant $\lambda\, H_2 H_1 S$ coupling (75) mixing the doublet higgsinos $\tilde{h}_1$ and $\tilde{h}_2$ with the singlino ζ. $\tilde{h}_{1R}$, $\tilde{h}_{2R}$ and the singlino ζ_L all have $R = 1$, in agreement with the R transformation properties

$$gauginos\ \lambda \overset{R}{\to} e^{\gamma_5 \alpha}\,\lambda; \quad \tilde{h}_i \overset{R}{\to} e^{-\gamma_5\alpha}\,\tilde{h}_i \quad \zeta \overset{R}{\to} e^{\gamma_5\alpha}\,\zeta. \tag{114}$$

The R-conserving part of the mass matrix corresponds to a conserved supersymmetry. It is a rank-4 5×5 matrix obtained by unpacking the two matrices

$$\begin{pmatrix} 0 & m_Z \\ m_Z & 0 \end{pmatrix} \quad \text{and} \quad \begin{pmatrix} 0 & -\dfrac{\lambda v}{\sqrt{2}} \\ -\dfrac{\lambda v}{\sqrt{2}} & 0 \end{pmatrix}. \tag{115}$$

This provides as in (72) 5 neutralinos organized as a Dirac zino of mass m_Z, a Dirac neutralino mixing the singlino with the left-over higgsino, carrying $R = \pm 1$, and a massless chiral photino with $R = 1$. The massive spinors involve, as in (107)[1]

$$\left\{\begin{array}{lcl} \lambda_Z = \lambda_3 c_\theta - \lambda' s_\theta & \text{combined with} & -\tilde{h}_z = \tilde{h}_1^0\, c_\beta - \tilde{h}_2^0\, s_\beta, \\ \tilde{s} & \text{combined with} & -(\tilde{h}_1^0\, s_\beta + \tilde{h}_2^0\, c_\beta). \end{array}\right. \tag{116}$$

The neutralino mass spectrum (72) for the $U(1)_R$-invariant nMSSM[1] is reexpressed into the 5×5 neutralino mass matrix

$$\begin{pmatrix} 0 & 0 & -s_\theta c_\beta m_Z & s_\theta s_\beta m_Z & 0 \\ 0 & 0 & c_\theta c_\beta m_Z & -c_\theta s_\beta m_Z & 0 \\ -s_\theta c_\beta m_Z & c_\theta c_\beta m_Z & 0 & 0 & -\frac{\lambda v}{\sqrt{2}} s_\beta \\ s_\theta s_\beta m_Z & -c_\theta s_\beta m_Z & 0 & 0 & -\frac{\lambda v}{\sqrt{2}} c_\beta \\ 0 & 0 & -\frac{\lambda v}{\sqrt{2}} s_\beta & -\frac{\lambda v}{\sqrt{2}} c_\beta & 0 \end{pmatrix}. \tag{117}$$

One completes this R-symmetric 5×5 mass matrix by the $\Delta R = \pm 2$ contributions, re-introducing the μ and μ_S doublet and singlet mass parameters previously discarded from the nMSSM superpotential[1] to get R symmetry. If a translation on S has to be performed, μ and μ_S get modified into $\mu_{\text{eff}} = \mu + \lambda \langle s \rangle$ and $\mu_{S\,\text{eff}} = \mu_S + 2\kappa \langle s \rangle$. The gaugino mass parameters m_1 and m_2 may be generated by radiative corrections, or from gravity-induced supersymmetry breaking as e.g. in Ref. 62.

The resulting 5×5 neutralino mass matrix reads

$$\begin{pmatrix} m_1 & 0 & -s_\theta c_\beta m_Z & s_\theta s_\beta m_Z & 0 \\ 0 & m_2 & c_\theta c_\beta m_Z & -c_\theta s_\beta m_Z & 0 \\ -s_\theta c_\beta m_Z & c_\theta c_\beta m_Z & 0 & -\mu_{\text{eff}} & -\frac{\lambda v}{\sqrt{2}} s_\beta \\ s_\theta s_\beta m_Z & -c_\theta s_\beta m_Z & -\mu_{\text{eff}} & 0 & -\frac{\lambda v}{\sqrt{2}} c_\beta \\ 0 & 0 & -\frac{\lambda v}{\sqrt{2}} s_\beta & -\frac{\lambda v}{\sqrt{2}} c_\beta & \mu_{S\,\text{eff}} \end{pmatrix}. \tag{118}$$

It involves 12 non-diagonal terms $\propto m_Z$ and $\lambda v/\sqrt{2}$, originating from the supersymmetric and $U(1)_R$-invariant nMSSM mass spectrum (72), (117), next to the 5 additional $\Delta R = \pm 2$ gaugino and doublet + singlet higgsino mass terms. The chargino and neutralino masses now verify

$$m^2(wino_1) + m^2(wino_2) = 2\, m_W^2 + \mu_{\text{eff}}^2 + m_2^2, \tag{119}$$

and

$$\sum_{1\ldots 5} m^2(neutralino) = 2\left(m_Z^2 + \frac{\lambda^2 v^2}{2} + \mu_{\text{eff}}^2\right) + m_1^2 + m_2^2 + \mu_{S\,\text{eff}}^2, \tag{120}$$

reducing to (111, 112) for $\lambda = 0$.

6.5. *Squarks, sleptons and supersymmetry breaking*

Left-handed quark and lepton fields are described by the left-handed chiral quark and lepton doublet superfields, Q and L. Right-handed ones, viewed as the conjugates of antiquark and antilepton fields, are described by the left-handed singlet superfields $\bar{U}$, $\bar{D}$ and $\bar{E}$. H_1 and H_2 generate charged-lepton and down-quark masses, and up-quark masses, from the trilinear superpotential couplings $\mathcal{W}_{lq}$ of lepton and quark superfields to H_1 and H_2,[2]

$$\mathcal{W}_{lq} = \lambda_e \, H_1 \, \bar{E} L + \lambda_d \, H_1 \, \bar{D} Q - \lambda_u \, H_2 \, \bar{U} Q. \tag{121}$$

H_1 and H_2 are separately responsible, from $\langle h_1^0 \rangle = v_1/\sqrt{2}$, $\langle h_2^0 \rangle = v_2/\sqrt{2}$, for charged-lepton and down-quark masses, and up-quark masses, respectively, with

$$m_e = \frac{\lambda_e v_1}{\sqrt{2}}, \quad m_d = \frac{\lambda_d v_1}{\sqrt{2}}, \quad m_u = \frac{\lambda_u v_2}{\sqrt{2}}. \tag{122}$$

This tends to favor a smaller v_1 as compared to v_2, i.e. a large $\tan\beta = v_2/v_1$, in view of the large mass of the t quark as compared to the b.

The superpotential interactions resulting from (121) are invariant under the continuous $U(1)_R$ symmetry (36), (37), with $R = +1$ for left-handed (anti)quark and (anti)lepton superfields, so that leptons and quarks carry $R = 0$ and sleptons ans squarks $+1$ (for $\tilde{l}_L, \tilde{q}_L$) or -1 (for $\tilde{l}_R, \tilde{q}_R$). Gauge and chiral superfields transform under R according to (71) so that, altogether

$$\begin{cases} V_a & \stackrel{R}{\to} & V_a(x, \theta e^{-i\alpha}, \bar{\theta} e^{i\alpha}), \\ (L, Q; \bar{E}, \bar{D}, \bar{U}) & \stackrel{R}{\to} & e^{i\alpha} \, (L, Q; \bar{E}, \bar{D}, \bar{U})(x, \theta e^{-i\alpha}), \\ H_{1,2} & \stackrel{R}{\to} & H_{1,2}(x, \theta e^{-i\alpha}), \\ S & \stackrel{R}{\to} & e^{2i\alpha} \; S(x, \theta e^{-i\alpha}), \end{cases} \tag{123}$$

the superpotential $\mathcal{W}$, including $\mathcal{W}_{lq}$, having $R = 2$.

A mixing between left-handed doublet and right-handed singlet squarks, e.g. $\tilde{t}_L$ and $\tilde{t}_R$, with $R = +1$ and -1 respectively, can be generated by the $\Delta R = \pm 2$ terms in the Lagrangian density. It leads for the $\tilde{t}$ squarks to a non-diagonal term $A_t - \mu m_t \cot\beta$, combining the contributions from the dimension-3 soft-breaking terms involving h_2 with those from F terms.

One essential question is the way by which squarks and sleptons can acquire very large masses. A spontaneous breaking of the global supersymmetry generates a massless chiral Goldstone spinor, carrying $R = 1$ as seen in (70). It may describe the photino, or more generally a neutral gaugino possibly involving an extra-$U(1)$ gaugino, combined with a neutral spin-$\frac{1}{2}$ field ζ described by a $R = 2$ chiral superfield.[2]

An extension of the gauge group to include an extra $U(1)$ factor is necessary if one intends to generate large masses for all squarks and sleptons, at the tree approximation, in a globally supersymmetric theory. Indeed with $SU(3) \times SU(2) \times U(1)$ or $SU(5)$ as the gauge group the squarks of the first generation would verify, at the classical level,[4]

$$\sum_{1,2} m^2(\tilde{u}_i) + m^2(\tilde{d}_i) \;=\; 2\,(m_u^2 + m_d^2). \tag{124}$$

One at least should then have a very small or even negative mass^2, leading to a charge and color-breaking vacuum. This leads either to consider an extension of the gauge group to include an extra-$U(1)$ factor, or to generate large masses from radiative corrections as in the "gauge-mediated" supersymmetry-breaking (GMSB) models, or to go to local supersymmetry. Or, finally, to move to extra dimensions to generate the required breaking of the supersymmetry using *discrete boundary conditions involving R-parity*, as discussed in Section 9.

Making all squarks and sleptons heavy was first done through the v.e.v. of the D component from an extra $U(1)$ with non-vanishing axial couplings to quarks and leptons, as in the USSM briefly introduced in Section 4.2. This may be done by taking advantage of the U transformations (32), (33), (53), (56), under which

$$\begin{cases} (L,Q;\bar{E},\bar{D},\bar{U}) & \stackrel{U}{\rightarrow} \quad e^{-\frac{i\alpha}{2}}\,(L,Q;\bar{E},\bar{D},\bar{U}), \\ H_{1,2} & \stackrel{U}{\rightarrow} \quad e^{i\alpha}\,H_{1,2}, \\ S & \stackrel{U}{\rightarrow} \quad e^{-2i\alpha}\,S. \end{cases} \tag{125}$$

This is in particular a symmetry of the trilinear superpotential $\mathcal{W}_{lq}$ responsible for quarks and lepton masses.[2] It acts axially on quarks and leptons, the corresponding (axionlike) Goldstone boson (a) getting eliminated as the extra neutral boson Z' (also called U) acquires a mass. This still leaves us with the question of how to generate a mass for gluinos, possibly from radiative corrections,[53] or from supergravity as anticipated in Ref. 32, since supergravity requires abandoning the continuous R symmetry in favor of the discrete R-parity.

Supersymmetry breaking is now usually realized by generating soft supersymmetry-breaking terms[27,58] from radiative corrections, or supergravity. In the first case, gauge-mediated models are generally characterized by the possibility of a light or very light gravitino LSP, behaving very much like a goldstino.[32]

When the local supersymmetry is spontaneously broken, it generates a massive spin-$\frac{3}{2}$ Majorana gravitino. Its mass term $m_{3/2}$ breaks explicitly the continuous R symmetry ($U(1)_R$), reducing it to R-parity. This reallows, in the supergravity framework, direct gaugino mass parameters, whose mass scale may be naturally fixed from the gravitino mass $m_{3/2}$. Soft supersymmetry-breaking terms may then be generated from supergravity,[59] leading to gravity-induced supersymmetry-breaking models,[60–64] where the gravitino is generally taken to be heavy.

In all these cases (MSSM, N/nMSSM, USSM, ...) supersymmetry-breaking and R-symmetry breaking contributions are added to the Lagrangian density.

This includes the reintroduction of the "supersymmetric" parameter μ (with $\Delta R = \pm 2$ and also breaking the extra-$U(1)$ symmetry), possibly regenerated from a translation of S, and the inclusion of the $\Delta R = \pm 2$ gaugino mass parameters, and of other terms of dimension ≤ 3 breaking supersymmetry explicitly but softly.

This would be a natural place to stop this presentation of the Supersymmetric Standard Model, hoping for supersymmetric particles to be discovered soon at LHC.

Still there may be more, and this is likely not to be the end of the story. More symmetries may be jointly operating to provide a better understanding of the electroweak and grand-unification breakings, opening the way to new compact dimensions of space-time, next to the quantum anticommuting dimensions of supersymmetry.

7. Beyond the N/nMSSM

7.1. *Towards a $N = 2$ supersymmetric spectrum*

Let us return to the R-symmetric superpotential (50),[1] $\mathcal{W}_{\text{nMSSM}} = S\,(\lambda\, H_2 H_1 + \sigma)$ and resulting spectrum (72), (73), (115), (117). For λ equal to the limiting value (100),

$$\lambda = \sqrt{\frac{g^2 + g'^2}{2}} \simeq .52 \tag{126}$$

(up to a possible convention-dependent sign for a real coupling), so that

$$m_Z = \frac{\sqrt{g^2 + g'^2}\; v}{2} = \frac{\lambda v}{\sqrt{2}}, \tag{127}$$

the theory has in its Z sector, with the superpotential

$$\mathcal{W}_\circ = -\sqrt{\frac{g^2 + g'^2}{2}}\; H_2^0 H_1^0\, S + \sigma S, \tag{128}$$

an unbroken $N = 2$ supersymmetry, independently of the value of the mixing angle β.[43,66] Its effects may be observed in the two 2×2 neutralino mass matrices (115), which participate equally in the 5×5 neutralino mass matrix (117). The Z gets associated with 2 Dirac zinos (or 4 Majorana ones) and 5 neutral spin-0 bosons within a massive gauge multiplet of $N = 2$ supersymmetry, according to

$$\boxed{\begin{array}{c} nMSSM \ \text{ with } \ \lambda = \sqrt{\dfrac{g^2 + g'^2}{2}} \quad \Rightarrow \\ Z \ \overset{SUSY}{\Longleftrightarrow} \ 4 \text{ Majorana zinos } \overset{SUSY}{\Longleftrightarrow} \ 5 \text{ spin-0 bosons.} \end{array}} \tag{129}$$

This is valid as long as supersymmetry remains unbroken in this sector. This massive $U(1)_R$ symmetric nMSSM spectrum even presents, for the $W^\pm$ and Z multiplets, an effective $N = 2$ supersymmetry, with in this case the electric charge acting as a central charge,[44] this $N = 2$ supersymmetry being broken in the $W^\pm$ multiplet for $\tan\beta \neq 1$.

The $N = 2$ supersymmetry may be extended to gluons and gluinos, if the latter are turned into Dirac particles, and accompanied by a complex octet of spin-0 "sgluons".[53] If we want to pursue in this direction of $N = 2$ supersymmetry we must introduce, next to the singlet S, adjoint $SU(2)$ and $SU(3)$ chiral superfields T and O with trilinear super-Yukawa couplings fixed by the gauge couplings,

$$\lambda_i = g_i\,\sqrt{2}, \quad \lambda' = \frac{g'}{\sqrt{2}}. \tag{130}$$

The electroweak couplings of H_1 and H_2 to the singlet S and triplet T are given by the superpotential (22) already encountered for the F-breaking of the supersymmetry,[30,43]

$$\mathcal{W} = \frac{1}{\sqrt{2}}\, H_2\,(g\,\tau.T - g'S)\,H_1 + \sigma S. \tag{131}$$

This $N = 2$ superpotential includes in particular, precisely, the nMSSM-type one

$$\mathcal{W}_\circ = -\sqrt{\frac{g^2+g'^2}{2}}\; H_2^0 H_1^0\, S_Z + \sigma S \tag{132}$$

for the chiral superfield $S_Z = \cos\theta\, T_3 - \sin\theta\, S$ associated with the Z, as written earlier in (128). This provides an $N = 2$ interpretation for the mass degeneracy occurring at the classical level, for unbroken supersymmetry, between the 5 neutral spin-0 bosons of the nMSSM, all of mass m_Z for $\lambda = \sqrt{(g^2+g'^2)/2}$.[66]

One must also consider 4 doublet chiral superfields (or $SU(5)$ quintuplets) instead of 2, by introducing H'_1 and H'_2 next to H_1 and H_2 to provide the required degrees of freedom for constructing 4 Dirac winos, so that

$$W^\pm \overset{SUSY}{\Longleftrightarrow} \text{4 Dirac winos} \overset{SUSY}{\Longleftrightarrow} \text{5 charged spin-0 bosons}. \tag{133}$$

The $W^\pm$ and Z are then associated with 5 charged and 5 neutral spin-0 bosons, all of masses m_W and m_Z as long as the $N = 2$ supersymmetry remains unbroken. This ultimately provides the spectrum for the gauge-and-BE-Higgs sector of $N = 2$ supersymmetric grand-unified theories,[45] which may then be formulated in a 5 or 6-dimensional space-time.[46]

7.2. *Radiative gluino masses from messenger quarks*

Gluinos being Majorana particles transforming chirally as in (37), a continuous R symmetry would forbid gluino masses, except if gluinos are turned into Dirac particles, as would be the case, precisely, within a $N = 2$ theory as we discussed.

But let us return to $N = 1$ for a moment. Gluinos are massless at lowest order, within global supersymmetry. Still if one abandons the continuous R symmetry one may consider generating *radiatively* gluino masses from their couplings to a new set of massive *messenger quarks* described by the chiral superfields $\mathcal{Q}$ and $\bar{\mathcal{Q}}$, *vectorially coupled* to standard model particles and sensitive to the source of supersymmetry-breaking, for messenger squarks and quarks to have different masses.[53] One also needs to introduce a source of breaking for the continuous R-symmetry, otherwise gluinos would still remain massless.

This requires some care especially if we intend to appeal to the F-mechanism for breaking spontaneously supersymmetry,[30,31] as the presence of an R symmetry is needed for a generic breaking of the supersymmetry as discussed in Section 3.4. At the same time however, R symmetry must be broken to get gluino masses, then leading to a massless *R-Goldstone-boson* or light *R-axion.* This may lead to prefer generating the spontaneous breaking of the supersymmetry in the messenger sector through the extra-$U(1)$ gauge interactions of the messenger (s)quarks with a non-vanishing $\langle D \rangle$, as done in Ref. 53.

Let us consider a second octet of paragluinos ζ_a, described by a chiral octet superfield O with $R = 0$. One may then generate radiatively a gluino mass in a R-symmetric way, with the two Majorana octets transforming in opposite ways according to

$$\begin{cases} \textit{gluinos}\ \lambda_a & \overset{R}{\to} \quad e^{\gamma_5 \alpha}\ \lambda_a, \\ \textit{paragluinos}\ \zeta_a & \overset{R}{\to} \quad e^{-\gamma_5 \alpha}\ \zeta_a. \end{cases} \tag{134}$$

These two octets could originate, together with their associated spin-0 gluons, now often called *sgluons*, from an underlying $N = 2$ supersymmetry.[43,44] They are described by a chiral octet superfield O with $R = 0$ (rather than 2, so as to lead to (134)), coupled to *massive messenger quark superfields* $\mathcal{Q}$ and $\bar{\mathcal{Q}}$ with $R = 1$, themselves vectorially coupled to standard model particles.

These messenger quark superfields interact with the octet O through the $R = 2$ superpotential[53]

$$\mathcal{W}_{\text{mess.}} = m_{\mathcal{Q}}\, \bar{\mathcal{Q}}\mathcal{Q} + \lambda_O\, \bar{\mathcal{Q}}\, O \mathcal{Q}, \tag{135}$$

color indices being omitted for simplicity. This generates a R-conserving Dirac mass term (m_D) for the Dirac gluino octet with $R = 1$,

$$\textit{Dirac gluinos} \ = \lambda_{aL} + \zeta_{aR}, \tag{136}$$

through one-loop diagrams involving the massive messenger quarks and squarks, sensitive to the source of spontaneous supersymmetry breaking, e.g. through an extra $U(1)$.

We still have to pay attention to the two additional real octets of spin-0 gluons described by O (sgluons), as one tends to acquire a negative mass^2 from quantum corrections. This instability may be avoided by introducing some amount of

R-symmetry breaking, to locally stabilize the vacuum through a superpotential Majorana mass term $\frac{1}{2}\,\mu_O\, O^2$ for the second gluino octet ζ. The resulting explicit breaking of R symmetry may be, again, a reason to prefer breaking spontaneously supersymmetry in the messenger sector through D terms, rather than through the F-breaking mechanism making use of R symmetry.

This leads for gluinos to both Dirac and Majorana mass terms, with a *see-saw* type 2×2 mass matrix[53]

$$\mathcal{M}_{\text{gluinos}} = \begin{pmatrix} 0 & m_D \\ m_D & \mu_O \end{pmatrix}, \qquad (137)$$

a mechanism introduced for gluinos even before it started getting widely considered for neutrinos. To be general a direct Majorana mass term m_3 for ordinary gluinos could still be added, although the purpose of this study was to discuss how an effective m_3 could be generated radiatively from the above see-saw type mass matrix.

This still leaves us, however, with a vacuum state that is locally-stable but only *metastable*, with a lower energy vacuum state for which color would be spontaneously broken.[53] Fortunately this metastable vacuum is in practice effectively stable. The possible interest of such metastable vacuum states, that had escaped attention at the time, was brought back to consideration more recently.[67] To generate radiatively in this way a significant mass for the gluinos, which must now be $\sim$ TeV scale at least as the result of LHC experiments,[68,69] it is necessary to consider quite high values for the messenger quark masses.

One may also imagine gauging the R symmetry, eliminating the corresponding Goldstone boson if R is spontaneously broken. This would lead to *a new force acting only on supersymmetric particles* (still to be discovered), and therefore, presumably, on dark matter. This new force may even be long-ranged if R symmetry stayed unbroken, otherwise it would have a finite range $\hbar/mc$ where m is the mass of the corresponding gauge boson associated with R symmetry.

8. $N = 2$ Supersymmetric Grand-Unified Theories

8.1. *Moving to higher dimensions*

We saw that a $N = 2$ supersymmetric theory with a massless matter hypermultiplet in the adjoint representation provides the $N = 4$ supersymmetric Yang–Mills theory, the adjoint gauge superfield interacting with 3 adjoint chiral ones coupled through the trilinear superpotential (25) $\mathcal{W} = g\,\sqrt{2}\, f_{ijk}\, S_1^i\, S_2^j\, S_3^k$, describing 1 spin-1 + 4 spin-$\frac{1}{2}$ + 6 spin-0 adjoint fields, with a $SU(4)_R \sim O(6)$ group acting on the 4 supersymmetry generators.[43,44] These theories, also obtained from the low-energy region of the dual spinor model,[70] or from the dimensional reduction of a supersymmetric Yang–Mills theory in 10 dimensions,[71] are remarkably elegant but very constrained, and more difficult to apply to fundamental particles and interactions. Still the extra dimensions of space-time may well be at the origin of the breaking of both the supersymmetry and the grand-unification symmetry, as we shall see.

Starting again from $N = 1$ supersymmetry in 4 dimensions, with gaugino and higgsino mass terms related by $m_1 = m_2 = m_3 = -\mu$, possibly also equal to the gravitino mass $m_{3/2}$, and assuming $\tan\beta = 1$ for simplicity, we get, directly or from (109), (110), remarkable mass relations like[72]

$$\left\{ \begin{array}{rcl} m^2(\text{winos}) &=& m_W^2 + m_{3/2}^2, \\ m^2(\text{zinos}) &=& m_Z^2 + m_{3/2}^2, \\ m(\text{photino}) = m(\text{gluinos}) &=& m_{3/2}, \end{array} \right. \tag{138}$$

up to radiative corrections.

These formulas, obtained in 4 dimensions, already point to a higher-dimensional origin of the $m_{3/2}^2$ contributions to the 4d mass^2, with supersymmetric particles carrying momenta $\pm\, m_{3/2}$ along an extra compact dimension. This leads us to consider again theories with a $N = 2$ extended supersymmetry applied to electroweak and strong interactions, or with a grand-unification symmetry like $SU(5)$, $O(10)$ or $E(6)$, before moving to higher dimensions.[45,65]

8.2. *Are the $N = 2$ "central charges" really central?*

Within $N = 2$ supersymmetry the particles get again organized within massless or massive multiplets, the $W^\pm$ and Z, $X^{\pm 4/3}$ and $Y^{\pm 1/3}$ gauge bosons belonging to different kinds of massive gauge multiplets.

To start with, $N = 2$ theories involve a new sort of massive multiplet known as an "hypermultiplet", describing massive spin-$\frac{1}{2}$ and spin-0 particles, with the two spin-0 fields transforming as the two components of a $SU(2)_R$ isodoublet while the Dirac spinor is an isosinglet.[43] This massive multiplet with maximum spin-$\frac{1}{2}$ looks at first intriguing as in principle it should not exist, not being a representation of the $N = 2$ supersymmetry algebra $\{\, Q^i, \bar{Q}^j \,\} = -\,2\,\not{P}\,\delta^{\,ij}$.

The theory being nevertheless invariant under the two supersymmetry generators Q_1 and Q_2, and thus under $\{Q^1, \bar{Q}^2\}$, the above $N = 2$ algebra must be modified, to allow for additional bosonic symmetry generators within the expression of $\{Q^1, \bar{Q}^2\}$. How it gets modified is quite interesting, especially in view of the spontaneous breakings of the grand-unification symmetry, in a way allowing for the electroweak breaking to occur.

We are led to consider, in a way compatible with Lorentz symmetry (and the symmetry property of the anticommutators) the extended algebra

$$\{\, Q^i, \bar{Q}^j \,\} \;=\; -\,2\,\not{P}\;\delta^{ij} \;+\; 2\,\epsilon^{ij}\,(Z + \gamma_5 Z'), \tag{139}$$

where one should still identify correctly the two symmetry generators Z and Z'. They are often referred to as "central charges", meaning that they ought to commute with all other symmetry generators of the theory.

Let us consider, however, a R-symmetric theory under which Q^1 and Q^2 transform chirally according to (35),

$$Q^i \overset{R}{\to} e^{-\gamma_5\alpha}\, Q^i, \tag{140}$$

or equivalently $Q^i_R \to e^{i\alpha}\, Q^i_R$. The operators Z and Z' appearing in (139) get rotated according to

$$Z - iZ' \overset{R}{\to} e^{2i\alpha}\,(Z - iZ'), \tag{141}$$

so that[45]

$$[\,R,\, Z\,] = -\,2i\,Z', \quad [\,R,\, Z'\,] = 2i\,Z. \tag{142}$$

These operators Z and Z', although commonly referred to as "central charges", *do not belong to the center of the symmetry algebra!*

This may seem surprising in view of the study of all possible supersymmetries of the S matrix, according to which Z and Z' must commute with all symmetry generators.[73] This analysis, however, disregards massless particles and symmetry breaking. It is thus not directly applicable here, both massless particles and symmetry-breaking effects playing an essential role. The Z and Z' generated from the anticommutation relations (139) do not necessarily commute with all symmetry generators, as seen in (141), (142) for R symmetry.[45] Z and Z' do not in general commute between themselves nor with gauge symmetry generators, in the non-abelian case, and as such do not qualify as "central charges".

Indeed in a gauge theory the anticommutators of the two supersymmetry generators may be expressed as in (139) but only *up to* (non-abelian or abelian) *field-dependent gauge transformations* (and modulo field equations of motion). For a $N = 2$ supersymmetric Yang–Mills theory equation (139) reads[44]

$$\{\,Q^1_R,\, \overline{Q^2_L}\,\} \;=\; 2\,\frac{1+i\gamma_5}{2}\,\underbrace{g\,T_i\,(a_i - ib_i)}_{Z\,-\,iZ'}, \tag{143}$$

a_i and b_i being the spin-0 partners of the adjoint or singlet gauge fields V_i^μ, so that

$$Z = g\,T_i\,a_i, \quad Z' = g\,T_i\,b_i. \tag{144}$$

The T_i denote the generators of the gauge symmetry group. a_i and b_i, described by adjoint or singlet chiral superfields with $R = 2$ (as for the nMSSM singlet S in (52)), transform under R according to

$$a_i - ib_i \overset{R}{\to} e^{2i\alpha}\,(a_i - ib_i), \tag{145}$$

so that Z and Z' do actually transform according to (141).

The spin-0 adjoints a_i and b_i will soon be interpreted as originating from the 5th and 6th components $V^5 = a$ and $V^6 = b$ of the 6d gauge fields $V^{\hat\mu}$. The R transformation (145) gets then associated with a 6d rotation $\mathcal{R}_{56}$ in compact space, under which

$$V_i^5 - iV_i^6 \overset{\mathcal{R}_{56}}{\to} e^{2i\alpha}\,(V_i^5 - iV_i^6), \tag{146}$$

$$P^5 - iP^6 \overset{\mathcal{R}_{56}}{\to} e^{2i\alpha}\,(P^5 - iP^6), \tag{147}$$

and similarly for the extra components of the covariant translation operator $\mathcal{P}^{\hat\mu}$ in 6 dimensions.

The translation of these adjoint gauge scalars (constrained to $f_{ijk}\langle a_j\rangle\langle b_k\rangle = 0$ for the potential to be minimum) leads to a spontaneous breaking of the non-abelian gauge symmetry. It generates in the anticommutation relations (139) finite field-independent parts

$$\langle Z\rangle = g\,T_i\langle a_i\rangle \quad \text{and} \quad \langle Z'\rangle = g\,T_i\langle b_i\rangle. \tag{148}$$

These now truly deserve the name of central charges, commuting between themselves and with all unbroken symmetry generators[44,45]

$$\langle Z\rangle \text{ and } \langle Z'\rangle \textit{ are the central charges.} \tag{149}$$

This leads to the *spontaneous generation of central charges* in the anticommutation relations of the $N = 2$ supersymmetry algebra, i.e. to *a spontaneous modification of the graded symmetry algebra.* Note that a central charge $Z_\circ$ may already be present before spin-0 fields get translated. These symmetry operators act as abelian, commuting in particular with all gauge symmetry generators surviving the spontaneous breaking. In practice we shall often drop the symbols $\langle\ \rangle$ and simply refer for convenience to Z and Z', instead of $\langle Z\rangle$ and $\langle Z'\rangle$, as being the central charges.

8.3. *Solving the "doublet-triplet splitting problem"*

Central charges can thus be "spontaneously generated" in a $N = 2$ supersymmetry algebra, one of them at least being closely connected with the spontaneous breaking of the grand-unification symmetry.[44,45] It could be the weak-hypercharge operator Y, spontaneously generated in the algebra through the symmetry breaking $SU(5) \to SU(3) \times SU(2) \times U(1)$, in a way which preserves the rank of the gauge group.

But Y is not conserved by the electroweak breaking. And, conversely, this breaking cannot occur in a $N = 2$ theory with a central charge proportional to Y. This is easily seen as initially massless spin-0 BEH doublets with $Y = \pm 1$ would acquire large masses $\frac{3}{5}\,m_X$, getting unable to trigger the electroweak breaking.

But we can start with initially massive quintuplets of mass m before the grand-unification breaking, and take advantage of the *flat directions for the adjoint or singlet gauge scalars* a_i *and* b_i *in a* $N = 2$ *theory.* Indeed the adjoint mass parameter vanishes in the superpotential, already as a consequence of R symmetry, as for the singlet S in the nMSSM. The magnitude of the adjoint v.e.v., denoted by V, can then freely adjust in the weak-hypercharge direction preserving $SU(3)\times SU(2)\times U(1)$, so that the resulting doublet mass parameter $m_D = m-\frac{3}{2}\,gV$ vanishes. This vanishing allows for the electroweak breaking, with the triplet mass parameter $m+\,gV$ getting identical to $m_X = \frac{5}{2}\,gV$:[45]

$$\boxed{m_{\text{quint.}} \quad \overset{\textit{GUT breaking}}{\longrightarrow} \quad \begin{cases} m_D \;=\; m - \dfrac{3}{2}\,gV \;\equiv\; 0, \\[2ex] m_T \;=\; m \,+\, gV \;=\; \frac{5}{2}\,gV \equiv m_X. \end{cases}} \tag{150}$$

This mechanism provides an automatic and natural solution to the "doublet-triplet splitting problem". This one is usually considered as a serious difficulty for the electroweak breaking in a $N=1$ supersymmetric grand-unified theory (and even more in a non-supersymmetric one), requiring a very large and unnatural adjustment for parameters of the order of the grand-unification scale. This severe fine-tuning problem is solved by moving to $N=2$, and from there to higher-dimensional theories. The vanishing of the doublet mass parameter m_D allows for their translation generating the electroweak breaking. Even better, this translation requires, conversely, that the doublet mass parameter m_D vanishes exactly *by locking it to 0, for the energy to be minimum.*

8.4. *The massive* $X^{\pm 4/3}$, $Y^{\pm 1/3}$, $W^{\pm}$ *and* Z *multiplets, within* $N=2$ *supersymmetry*

All spin-1 gauge bosons must then belong to massless or massive multiplets of $N=2$ supersymmetry. But there are *three different types of massive gauge multiplets*, in contrast with $N=1$. Type I multiplets are appropriate to describe the $W^{\pm}$ and Z as in (129), (133). They involve no central charge, and may be complex or real. Type II and type III multiplets, on the other hand, have a non-vanishing value of the central charge $\langle Z\rangle$ (from now on simply denoted by Z) and are necessarily complex. They differ by their field content, and are appropriate to the description of grand-unification gauge bosons such as $X^{\pm 4/3}$ and $Y^{\pm 1/3}$, in a $SU(5)$ theory.

These grand-unification bosons, which have the same weak hypercharge $Y=\pm\,5/3$, belong to two multiplets with the same value of the central charge

$$Z(X^{\pm 4/3}) \;=\; Z(Y^{\pm 1/3}) \;=\; \pm\; m_X \tag{151}$$

(up to a possible convention-dependent sign). This one includes a contribution $\pm\frac{3}{5}\;m_X Y$, spontaneously generated[44] in the $N\;=\;2$ algebra when the grand-unification symmetry is spontaneously broken to $SU(3)\times SU(2)\times U(1)$. This central charge reads

$$Z = Z_\circ + \frac{3}{5}\, m_X\, Y, \tag{152}$$

and is such that

$$Z(\textit{spin-0 doublets}) \equiv 0, \tag{153}$$

as in (150), naturally allowing for the doublet translations responsible for the electroweak breaking.

The $X^{\pm 4/3}$ belongs to a smaller multiplet, of type II, including a single spin-0 boson $x^{\pm 4/3}$ and two Dirac xino (anti)triplets, with $m_X = |Z| > 0$. The $Y^{\pm 1/3}$ belongs to a larger multiplet, of type III, with 5 spin-0 bosons and 4 Dirac yino (anti)triplets, verifying altogether[74]

$$\begin{cases} \text{type II}: & m_X = |Z| > 0, \\ \text{type III}: & m_Y > |Z| = m_X > 0. \end{cases} \tag{154}$$

The smaller character of the $X^{\pm 4/3}$ multiplet as compared to the $Y^{\pm 1/3}$ one is associated with the mass equality $m_X = |Z|$, in contrast with $m_Y > |Z|$. This may be easily understood when moving to 6 dimensions, where the $X^{\pm 4/3}$ is massless in relation with an unbroken $SU(4)$ *electrostrong symmetry* in 6 dimensions while the $Y^{\pm 1/3}$ is already massive (with mass m_W) in 6 dimensions.

The $Y^{\pm 1/3}$ multiplet, larger than the $X^{\pm 4/3}$ one, also accommodates the 4 triplet components from the 4 quintuplets. All of them have the same mass^2 m_Y^2, including a m_X^2 contribution in agreement with (150). These 4 quintuplets describe in particular the 4 spin-0 doublets responsible for the electroweak breaking in a $N = 2$ theory, also providing spin-0 partners for the $W^\pm$ and Z bosons.

The $Y^{\pm 1/3}$ (and associated partner) mass^2 originates from the two contributions generated by the adjoint and doublet v.e.v.'s V and v, respectively, so that

$$m_Y^2 = \underbrace{m_X^2}_{Z^2} + m_W^2. \tag{155}$$

The $W^\pm$ and Z, on the other hand, carry no central charge Z and belong to massive gauge multiplets of type I, describing 4 inos and 5 spin-0 bosons for every spin-1 particle, as in (129), (133).

When supersymmetry is broken mass relations similar to (138), like

$$\begin{cases} m^2(\text{xinos}) = m_X^2 + m_{3/2}^2, \\ m^2(\text{yinos}) = m_Y^2 + m_{3/2}^2 = m_X^2 + m_W^2 + m_{3/2}^2, \end{cases} \tag{156}$$

are obtained for xinos and yinos, and interpreted in terms of momenta $\pm m_{3/2}$ carried along an extra compact dimension.[46]

9. Supersymmetry and Grand-Unification in Extra Dimensions

9.1. *From $N = 2$ supersymmetry to 6 dimensions*

These $N = 2$ theories may then be formulated in a 5 or 6 dimensional space-time,[46,65] with the "central charges" Z and Z' getting turned into the 5th and 6th components of the (covariant) momentum along the compact dimensions.[44] The two spin-0 photons and spin-0 "sgluon" octets present in $N = 2$ supersymmetry get described by the fifth and sixth components of the photon and gluon fields $V_i^{\hat{\mu}}$ in 6 dimensions. *The $W^\pm$ and Z masses are already present in 6 dimensions*, where the photon and gluons are coupled with the same strength, in a grand-unified theory.

Viewing $Q\sqrt{3/8}$ as one of the $SU(4)$ *electrostrong symmetry* generators, suitably normalized in the same way as for the $SU(3)$ generators, provides in 6d the $SU(4)$ relation between the electromagnetic and strong couplings,

$$\begin{gathered} \underline{\textit{electrostrong symmetry}} \implies \\ e_{6d} = \sqrt{\frac{3}{8}}\; g_{3\,6d}, \quad \text{i.e.} \quad \alpha_{6d} = \frac{3}{8}\; \alpha_{3\,6d}. \end{gathered} \tag{157}$$

This relation is exact in 6d as long as we do not introduce the grand-unification breaking through antiperiodic boundary conditions for GUT-odd particles, discussed in the next section. We also have, by returning to $SU(5)$ to include weak in addition to electrostrong interactions, $\sin^2\theta = e^2/g^2 = 3/8$ for the electroweak angle in 6d, at the classical level.

The central charge Z of the $N = 2$ algebra in 4 dimensions, essential to the discussion of the grand-unification breaking, originates from the fifth component of the (covariant) momentum along a compact dimension, according to

$$\mathcal{P}^5 = -\left(Z_\circ + \frac{3}{5}\, m_X\, Y\right) \tag{158}$$

(up to a possible convention-dependent sign). Once we are in 5 or 6 dimensions, we only have to refer to the extra components of the covariant momenta, $\mathcal{P}^5$ and $\mathcal{P}^6$, rather than to the corresponding central charges Z and Z' of the 4d $N = 2$ theory.

The particle content of the $N = 2$ multiplets are given in Refs. 45 and 46. The massive gauge multiplet (129) describing the Z in 4 dimensions originates from the massive Z gauge multiplet in 6 dimensions, such that

$$\boxed{Z \;\overset{\substack{6d \\ SUSY}}{\Longleftrightarrow}\; \text{8-comp. Dirac zino} \;\overset{\substack{6d \\ SUSY}}{\Longleftrightarrow}\; \text{3 spin-0 bosons.}} \tag{159}$$

This multiplet reduces to (129) in 4 dimensions, in which the Z is associated with 4 Majorana zinos and 5 spin-0 BEH bosons. Similar expressions hold for the $W^\mp$ and $Y^{\pm 1/3}$ multiplets. This set of 5 neutral spin-0 BEH bosons associated with the Z in 4 dimensions, before the breaking of the $N = 2$ supersymmetry, is similar to the nMSSM one in (129) for $\lambda = \sqrt{(g^2 + g'^2)/2} \simeq .52$, and presumably includes the 125 GeV boson found at CERN.

$N = 2$ theories in 4 dimensions, being vectorlike, also include mirror partners for quarks and leptons, to which they are coupled through the exchanges of spin-0 gluons and photons, in particular. But no such particles have been observed yet. Their presence at low energy may be avoided by considering a mirror-parity operator M_p under which mirror fields, as well as spin-0 gluon and photon fields, ..., are M_p-odd. The Z multiplet then gets further reduced, to include just a single spin-0 boson z associated with the Z.

Indeed among the 4 spin-0 doublets (originating from 4 quintuplets) the usual ones h_1 and h_2 are taken as M_p-even so as to survive in the low-energy theory. Their $N = 2$ partners h'_1 and h'_2, being M_p-odd, disappear from the low-energy theory. The definition of this M_p operator involves in particular, as seen from (145-147), a rotation $\mathcal{R}_{56}(\pi)$, equivalent to a reflection symmetry in compact space, $x^5 \to -x^5$, $x^6 \to -x^6$, under which

$$V_i^\mu \to V_i^\mu ; \quad V_i^5 \to -V_i^5, \; V_i^6 \to -V_i^6 . \tag{160}$$

Anticipating on the supersymmetry breaking discussed in the next section, the R-odd zinos are present only at the compactification scale associated with the sixth dimension, starting with two Dirac zinos (combining a Dirac gaugino with a Dirac higgsino),[46] at mass^2

$$m^2(zinos) = m_Z^2 + \pi^2/L_6^2 . \tag{161}$$

We then simply remain, in the low-energy 4d theory below the compactification scales, with the field-content of the standard model but for the presence of the two spin-0 doublets h_1 and h_2, with quartic doublet couplings fixed by $(g^2 + g'^2)/8$ and $g^2/2$ as in (5,6,63). This is crucial for the gauge-BEH unification according to which[1,7,8]

$$\boxed{\textit{spin-1 } Z \quad \overset{SUSY}{\longleftrightarrow} \; \overset{SUSY}{\longleftrightarrow} \quad \textit{spin-0 BEH boson}\, z,} \tag{162}$$

with the spin-0 z having the same mass as the Z before supersymmetry breaking effects get taken into account. We also expect, in the same way, the following association

$$\textit{spin-1 } W^\pm \quad \overset{SUSY}{\longleftrightarrow} \; \overset{SUSY}{\longleftrightarrow} \quad \textit{spin-0 BEH boson } w^\pm , \tag{163}$$

for the charged spin-0 boson in (85),

$$w^\pm \equiv H^\pm = \sin\beta \; h_1^\pm + \cos\beta \; h_2^\pm . \tag{164}$$

In a grand-unified theory, with for example $SU(5)$ as the gauge group, this one gets spontaneously broken down to an $SU(4)$ *electrostrong symmetry* group. The

$X^{\pm 4/3}$ (anti)triplet remains massless in 6d, where the $Y^{\pm 1/3}$ (anti)triplet has the same mass m_W as the $W^{\mp}$, with which they form a

$$SU(4) \textit{ electrostrong antiquartet } \begin{pmatrix} Y^{+1/3} \\ W^- \end{pmatrix}. \tag{165}$$

In this higher-dimensional space-time, the $SU(5)$ symmetry is broken through the BEH-quintuplet v.e.v.'s, providing in 6 dimensions equal masses to the $Y^{\pm 1/3}$ and $W^{\mp}$ gauge fields, according to

$$\boxed{SU(5) \overset{\substack{EW\ breaking \\ in\ 6d}}{\longrightarrow} SU(4) \textit{ electrostrong gauge group,}} \tag{166}$$

leading to the $SU(4)$ relation $\alpha_{6d} = (3/8)\ \alpha_{s\,6d}$ in 6 dimensions, and to $\sin^2\theta = 3/8$, at the classical level.

This electrostrong-weak breaking in 6d *separating weak from electrostrong interactions* leads in 4d to the mass relation (155)

$$m_Y^2 = m_X^2 + m_W^2 \tag{167}$$

found previously, with $m_X^2 = P_5^2 + P_6^2$, for the X and Y gauge bosons and their susy partners. These relations are valid for each excitation level of the extra compact dimensions, for the $X^{\pm 4/3}$ and $Y^{\pm 1/3}$ gauge fields in 6 dimensions.

9.2. *Grand-unification and supersymmetry breaking from* discrete *boundary conditions in compact space*

The extra compact space dimensions may play an essential role in the breaking of the supersymmetry and grand-unification symmetries, together with the mirror-parity operator M_p allowing to avoid mirror quarks and leptons, spin-0 gluons and photons, and additional spin-0 BEH bosons, in the low-energy spectrum. This may be done through boundary conditions involving, in an interesting way, *discrete* rather than continuous symmetries. They include R-parity, a GUT-parity G_p and mirror parity M_p acting as translation and reflection symmetries in the compact space, thanks to its topological properties.[46,65]

These three discrete symmetries naturally allow for the presence at low energy of the two spin-0 doublets h_1 and h_2, even under R_p, G_p and M_p. They can thus generate the same spontaneous electroweak breaking in 4 dimensions as already resulting in (159), (166) from the grand-unification breaking into the $SU(4)_{es}$ electrostrong symmetry subgroup in 6 dimensions, with the $W^{\pm}$ and Z masses in 4 dimensions directly originating from the 6d theory.

The breaking of the supersymmetry may be obtained by identifying the action of traveling along a complete loop $\mathcal{L}_6$ in compact space (i.e. a translation $x^6 \to x^6 + L_6$, in the simplest example of a flat torus) with a discrete R-parity transformation, $R_p = (-1)^{3(B-L)}\,(-1)^{2S} = \pm 1$:

$$\boxed{\begin{array}{c}\textit{traveling along a complete loop } \mathcal{L}_6 \textit{ in compact space}\\ \equiv \quad R\textit{-parity transformation.}\end{array}} \tag{168}$$

This makes all superpartners naturally "very heavy", i.e. at the compactification scale:

$$m(R\textit{-odd superpartners}) \;\approx\; \textit{compactification scale.} \tag{169}$$

This compactification scale is unknown but has to be $\gtrsim$ TeV scale, at least. We may thus be lucky enough to see superpartners, together with the opening of extra space dimensions, in a not-too-distant future. But we may also have to face the eventuality that superpartner masses be considerably larger than the presently accessible $\approx$ TeV scale, especially if the compactification of extra dimensions also sets the scale for the grand-unification breaking. The latter scale, however, may be substantially reduced as compared to usual expectations, as we shall see.

In a similar way, the breaking of the $SU(4)$ *electrostrong symmetry* group in 6d may be obtained by identifying traveling along a complete loop $\mathcal{L}_5$ (e.g. a translation $x^5 \to x^5 + L_5$ on a flat torus) with a discrete Z_2 GUT-parity transformation G_p,

$$\boxed{\begin{array}{c}\textit{traveling along another loop } \mathcal{L}_5 \textit{ in compact space}\\ \equiv \quad GUT\textit{-parity transformation.}\end{array}} \tag{170}$$

This one is defined from expression (152) of the central charge Z, as $e^{i\pi Z/m_X}$, or more precisely as

$$GUT\textit{-parity } G_p \;=\; G' \,\times\, e^{\,i\pi\frac{3}{5}Y} \;=\; (-1)^{Z/m_X} \;=\; \pm 1. \tag{171}$$

Here $G' = e^{i\pi Z_\circ/m_X}$ is a global symmetry operator commuting with both $SU(5)$ and supersymmetry, acting in particular on quark and lepton grand-unification multiplets, and spin-0 BEH multiplets. G_p may be expressed in terms of the central charge Z present in the $N=2$ supersymmetry algebra in 4 dimensions, as in (139), (151)–(153), part of which, proportional to the weak-hypercharge Y, is generated spontaneously during the breaking of the grand-unification symmetry.[44,45] Taking the fifth dimension as cyclic, of size L_5, with periodic and antiperiodic boundary conditions for GUT-even and GUT-odd fields, we can identify the action of a GUT-parity transformation with the one of a translation of L_5 on the torus, so that

$$\begin{array}{l}\textit{action of } GUT\textit{-parity}\\ \quad\equiv \textit{action of } e^{\,i\,P_5 L_5} \;=\; (-1)^{P_5/\frac{\pi}{L_5}}.\end{array} \tag{172}$$

The $X^{\pm 4/3}$ and $Y^{\pm 1/3}$ gauge bosons, with $Y = \pm 5/3$, are odd under GUT-parity, and carry momenta $(2n_5 + 1)\ \pi/L_5$ along the fifth dimension. This is also the case for the spin-0 triplet partners of the electroweak doublets within $SU(5)$ quintuplets. These belong to the same massive gauge multiplet of type III as the $Y^{-1/3}$, in agreement with (151), (171).[45,46,74]

This *triplet-doublet splitting mechanism,* already operating within $N = 2$ supersymmetric GUTs as in (150),[45] is of the same nature as the one splitting the $X^{\pm 4/3}$ and $Y^{\pm 1/3}$ masses away from the gluon, photon, $W^{\mp}$ and Z masses. It provides for their physical spin-0 triplet and doublet components the same masses as for the $Y^{\pm 1/3}$, $W^{\mp}$ and Z, before the supersymmetry breaking. The massless components associated with the would-be Goldstone bosons are eliminated when the $Y^{\pm 1/3}$, $W^{\mp}$ and Z acquire their 6d masses m_W, m_W and m_Z.

The $X^{\pm 4/3}$, $Y^{\pm 1/3}$ and color-(anti)triplet spin-0 bosons, being G_p-odd, have no direct couplings between two ordinary (anti)quarks or (anti)leptons, even under GUT-parity. This is in contrast with ordinary GUTs, and implies that

$$\textit{the proton is expected to be stable,} \tag{173}$$

at least in the simplest situations.[65,75] The corresponding compactification scale associated with the grand-unification breaking might then be lower, and possibly significantly lower, than the $\approx 10^{16}$ GeV usually considered.

Altogether the spontaneous breaking of the supersymmetry and grand-unification symmetries may both be induced through the compactification of the extra dimensions. This leads to the possibility of fixing the scales associated with these breakings in terms of the compactification scales for the extra dimensions.[46] In the simplest case of two flat extra dimensions and for the lowest-lying excited states, we would get relations like

$$\left\{ \begin{array}{l} m_{3/2} = \dfrac{\pi}{L_6} = \dfrac{1}{2R_6} \\ \qquad \text{(from } R\textit{-parity} \equiv \text{ translation of } L_6 \text{)}, \\ m_X = \dfrac{\pi}{L_5} = \dfrac{1}{2R_5} \\ \qquad \text{(from } GUT\textit{-parity} \equiv \text{translation of } L_5 \text{)}, \end{array} \right. \tag{174}$$

up to radiative corrections. The lowest-lying superpartners, or grand-unification particles, are expected to be present at these mass scales determined by $m_{3/2}$ and m_X, respectively.

This use of *discrete* boundary conditions associated with a non-trivial topology, involving for supersymmetry R-parity rather than a continuous symmetry, allows to link rigidly these fundamental supersymmetry and grand-unification breaking parameters $m_{3/2}$ and m_X to the compactification scales. This approach contrasts with the initial one in Ref. 76 disregarding fields corresponding to excited states that become infinitely massive when the size of the compact space is shrunk to zero (in particular states of masses proportional to π/L, essential here). We obtain instead *quantized mass parameters* fixed in terms of the compactification scales, with the *geometry* now determining the masses of the new particles in which we are interested.

9.3. *Implications for the compactification scales*

The resulting 4d theory has, in its simplest version, *the same content as the standard model* at low-energy *but for the two spin-0 doublets h_1 and h_2*, while still allowing for the gauge/BE-Higgs unification that is one of the most interesting features of supersymmetric theories. The new (sixth) dimension opens up at the compactification scale $m_{3/2}$, i.e. π/L_6 in the simplest case. Superpartners, mirror particles, spin-0 gluons etc., as for a $N = 2$ theory in 4 dimensions,[45] would then appear at or above this threshold. They now originate from a $N = 1$ theory in 5 dimensions, its mass spectrum resulting from the discrete boundary conditions involving the R_p and M_p symmetries.

Let us assume $m_{3/2}$ smaller than m_X. Below $m_{3/2}$ the theory has in its simplest version the same field content as the standard model, but for the second spin-0 doublet. The evolution of the gauge couplings (or simply of the differences $g_i^{-2} - g_j^{-2}$) in the 4d theory between m_W and $m_{3/2}$ is only slightly modified as compared to the standard model. It does not lead a grand-unification of these three couplings below the compactification scale.

Above $m_{3/2}$ the theory gets 5-dimensional, non-renormalizable, with gauge couplings having the dimension of mass$^{-1/2}$. We can no longer discuss as usual the running of the gauge couplings. One may still feel tempted to continue evaluating an evolution of effective couplings with energy, taking into account only a finite number of states up to a cut-off mass Λ, but one should be cautious before drawing conclusions. In addition the asymptotic freedom of QCD is expected to be ruined owing to the extra mirror families of quarks and leptons, unless one considers that quarks, leptons and mirror partners do not have excited states for the compact dimensions.

We recall that the dimension (x^6) responsible for the evolution of (effective) gauge couplings between $m_{3/2}$ and m_X is distinct from the one (x^5) responsible for the breaking of the GUT symmetry at the higher grand-unification scale. The latter, $m_X = \pi/L_5$, however, may be only slightly larger than $m_{3/2} = \pi/L_6$.

$\sin^2\theta$, evaluated in a 4d theory, is in particular sensitive to the number of spin-0 doublets and associated higgsinos, usually h_1, h_2 and $\tilde{h}_1, \tilde{h}_2$ in a $N = 1$ theory (counting very much as for 6 spin-0 doublets) with the field content of the (N/n)MSSM. Only the 2 doublets h_1 and h_2, without their accompanying higgsinos, are here present in the 4d theory below $m_{3/2}$. The evolution of the three gauge couplings, if extrapolated up to a unified value within the 4d theory, would necessitate a too small $\sin^2\theta$, much as for the standard model.

This may be more than compensated, however, by the 4 spin-0 doublets h_1, h_2, h'_1, h'_2 + associated higgsinos (twice as much as in the MSSM) present in the 5d theory between $m_{3/2}$ and m_X. These extra doublet degrees of freedom (taking also into account the extra adjoint gaugino and spin-0 fields) tend to lead to a too large value of $\sin^2\theta$. This indicates that the correct value of $\sin^2\theta$ may be obtained from a balance between these two effects, presumably with $m_{3/2}$ not far below m_X.

If the proton is indeed stable m_X may be much lower than the usual $\approx 10^{16}$ GeV scale, especially with a faster evolution of effective gauge couplings in the 5d theory between $m_{3/2}$ and m_X. Their unification may then occur for a rather low value of the grand-unification scale m_X. One may even imagine that these unification and compactification scales

$$m_{3/2} = \frac{\pi}{L_6} < m_X = \frac{\pi}{L_5} \tag{175}$$

may be not so far above the $\approx$ few TeV scale soon accessible at LHC. This would provide new perspectives for a possible discovery of superpartners, new space dimensions and maybe grand-unification particles, in a not-too-distant future.

This set of jointly-operating mechanisms, based on supersymmetry, extra dimensions and discrete symmetries, allows for the electroweak breaking to already occur in 6d dimensions, where it leaves unbroken an electrostrong symmetry group. It provides in 4d the electroweak breaking induced by h_1 and h_2 at low energies, even in the presence of significantly larger scales associated with grand-unification and possibly (in a more remarkable way) supersymmetry breaking. In particular

no fine-tuning between GUT-scale parameters is required,

(176)

and the electroweak breaking in the low-energy theory appears largely insensitive to the behavior of the higher-dimensional theory. In addition, stable particles are associated with the excitation of the compact fifth and sixth dimensions, and are possible candidates for the non-baryonic dark matter of the Universe.

10. Conclusion

In addition to superpartners, supersymmetric theories lead to an extended set of spin-0 bosons $H^{\pm}, H, h, A, \dots$. Some of them appear as extra states for massive spin-1 gauge bosons, providing a relation between spin-1 mediators of gauge interactions and spin-0 particles associated with symmetry breaking and mass generation.

Searches for supersymmetric particles started in the late seventies, first looking for light gluinos and associated R-hadrons, light charged sleptons, etc., often relying on the missing-energy momentum carried away by unobserved neutralino or gravitino LSP's, at the modest energies accessible at the time.[3,77] Considerable work has been done since throughout the world, most notably at PETRA (DESY) and PEP (SLAC), LEP (CERN) and at the Tevatron (Fermilab). These searches are now at the forefront of particle physics with the restart of LHC experiments at CERN.

All this could not be discussed here, nor the status of the lightest supersymmetric particle, presumably a neutralino, as a possible dark matter candidate in a R-parity conserving theory. We know now that strongly-interacting squarks and gluinos should be heavier than about 1 TeV at least. We refer the reader to the original results from the ATLAS and CMS collaborations at LHC,[34,35,68,69] and to the other articles in this book to complete this theoretical description with the presentation of experimental results and constraints on supersymmetric particles and additional spin-0 BEH bosons.

The next run of LHC experiments, with an energy increased from 8 to 13 TeV, may well allow for the direct production of supersymmetric particles, and of an extended system of spin-0 bosons including a charged $H^{\pm}$. Will this energy be sufficient, and at which energy scale should the new superpartners be found ? Is it indeed not too far from the electroweak scale, and accessible at LHC ? Or still significantly larger, as it could happen for superpartner masses determined by the very small size of an extra dimension ($L \lesssim 10^{-17}$ cm corresponding to $\pi\hbar/Lc \gtrsim 6$ TeV/c^2)?

In any case the 125 GeV boson observed at CERN may well be interpreted, up to a mixing angle induced by supersymmetry breaking, as the spin-0 partner of the Z under two supersymmetry transformations,

$$\textit{spin-1 } Z \quad \overset{SUSY}{\longleftrightarrow} \; \overset{SUSY}{\longleftrightarrow} \quad \textit{spin-0 BEH boson}, \tag{177}$$

i.e. as a Z that would be deprived of its spin. This provides within a theory of electroweak and strong interactions the first example of two known fundamental particles of different spins that may be related by supersymmetry.

Even if R-odd superpartners were still to remain out of reach for some time, possibly due to large momenta along very small space dimensions, supersymmetry could still be tested in the gauge-and-BEH sector at present and future colliders, in particular through the properties of the new spin-0 boson and the search for additional ones.

References

1. P. Fayet, *Nucl. Phys. B* **90**, 104 (1975).
2. P. Fayet, *Phys. Lett. B* **64**, 159 (1976); *ibid.* **69**, 489 (1977).
3. G. Farrar and P. Fayet, *Phys. Lett. B* **76**, 575 (1978).
4. P. Fayet, *Phys. Lett. B* **84**, 416 (1979).
5. ATLAS Coll., *Phys. Lett. B* **716**, 1 (2012).
6. CMS Coll., *Phys. Lett. B* **716**, 30 (2012).

7. P. Fayet, *Nucl. Phys. B* **237**, 367 (1984).
8. P. Fayet, *Phys. Rev. D* **90**, 015033 (2014).
9. S. Glashow, *Nucl. Phys.* **22**, 579 (1961).
10. A. Salam and J. Ward, *Phys. Lett.* **13**, 168 (1964).
11. S. Weinberg, *Phys. Rev. Lett.* **19**, 1264 (1967).
12. A. Salam, in *Proc. Nobel Symp. on Elementary Particle Theory*, ed. Svartholm, p. 367 (1968).
13. J. Goldstone, *Nuovo Cim.* **19**, 154 (1961).
14. F. Englert and R. Brout, *Phys. Rev. Lett.* **13**, 321 (1964).
15. P. Higgs, *Phys. Rev. Lett.* **13**, 508 (1964).
16. G. Guralnik, C. Hagen and T. Kibble, *Phys. Rev. Lett.* **13**, 585 (1964).
17. H. Georgi and S. Glashow, *Phys. Rev. Lett.* **32**, 438 (1974).
18. H. Georgi, H. Quinn and S. Weinberg, *Phys. Rev. Lett.* **33**, 451 (1974).
19. S. Weinberg, *Phys. Rev. D* **13**, 974 (1976); *ibid.* **19**, 1277 (1979).
20. L. Susskind, *Phys. Rev. D* **20**, 2619 (1979).
21. S. Dimopoulos and L. Susskind, *Nucl. Phys. B* **155**, 237 (1979).
22. E. Eichten and K. Lane, *Phys. Lett. B* **90**, 125 (1980).
23. Yu. Gol'fand and E. Likhtman, *ZhETF Pis. Red.* **13**, 452 (1971) [*JETP Lett.* **13**, 323 (1971)].
24. D. Volkov and V. Akulov, *Phys. Lett. B* **46**, 109 (1973).
25. J. Wess and B. Zumino, *Nucl. Phys. B* **70**, 39 (1974); *Phys. Lett. B* **49**, 52 (1974).
26. P. Fayet and S. Ferrara, *Phys. Rep.* **32**, 249 (1977).
27. S. Martin, A Supersymmetry Primer, http://arxiv.org/abs/hep-ph/9709356 (2016).
28. P. Ramond, *Eur. Phys. J. C* **74**, 2698 (2014).
29. P. Fayet and J. Iliopoulos, *Phys. Lett. B* **51**, 461 (1974).
30. P. Fayet, *Phys. Lett. B* **58**, 67 (1975).
31. L. O'Raifeartaigh, *Nucl. Phys. B* **96**, 331 (1975).
32. P. Fayet, *Phys. Lett. B* **70**, 461 (1977); *ibid.* **86**, 272 (1979).
33. M. Gell-Mann, P. Ramond and R. Slansky, *Rev. Mod. Phys.* **50**, 721 (1978).
34. https://twiki.cern.ch/twiki/bin/view/AtlasPublic/HiggsPublicResults
35. https://twiki.cern.ch/twiki/bin/view/CMSPublic/PhysicsResultsHIG
36. D. Volkov and V. Soroka, *Pisma Zh. Eksp. Teor. Fiz.* **18**, 529 (1973) [*JETP Lett.* **18**, 312 (1973)].
37. S. Ferrara, D. Freedman and P. van Nieuwenhuizen, *Phys. Rev. D* **13**, 3214 (1976).
38. S. Deser and B. Zumino, *Phys. Lett. B* **62**, 335 (1976).
39. J. Iliopoulos and B. Zumino, *Nucl. Phys. B* **76**, 310 (1974).
40. A. Salam and J. Strathdee, *Nucl. Phys. B* **76**, 477 (1974).
41. S. Ferrara, J., Wess and B. Zumino, *Phys. Lett. B* **51**, 239 (1974).
42. P. Fayet, *Nuovo Cim. A* **31**, 626 (1976).
43. P. Fayet, *Nucl. Phys. B* **113**, 135 (1976).
44. P. Fayet, unpublished (1976); *Nucl. Phys. B* **149**, 137 (1979).
45. P. Fayet, *Phys. Lett. B* **142**, 263 (1984); *Nucl. Phys. B* **246**, 89 (1984).
46. P. Fayet, *Phys. Lett. B* **159**, 121 (1985); *Nucl. Phys. B* **263**, 649 (1986).
47. P. Fayet, *Nucl. Phys. B* **78**, 14 (1974).
48. R. Peccei and H. Quinn, *Phys. Rev. D* **16**, 1791 (1977).
49. F. Wilczek, *Phys. Rev. Lett.* **40**, 279 (1978).
50. S. Weinberg, *Phys. Rev. Lett.* **40**, 223 (1978).
51. P. Fayet, *Phys. Lett. B* **95**, 285 (1980).
52. P. Fayet, *Nucl. Phys. B* **187**, 184 (1981).
53. P. Fayet, *Phys. Lett. B* **78**, 417 (1978).

54. R. Barbier *et al.*, *Phys. Rep.* **420**, 1 (2005).
55. U. Ellwanger, C. Hugonie and A. Teixeira, *Phys. Rep.* **496**, 1 (2010).
56. M. Maniatis, *Int. J. Mod. Phys. A* **25**, 3505 (2010).
57. P. Fayet, *Phys. Rev. D* **75**, 115017 (2007).
58. L. Girardello and M. Grisaru, *Nucl. Phys. B* **194**, 65 (1982).
59. E. Cremmer *et al.*, *Nucl. Phys. B* **147**, 105 (1979).
60. A. Chamseddine, R. Arnowitt and P. Nath, *Phys. Rev. Lett.* **49**, 970 (1982).
61. R. Barbieri, S. Ferrara and C. Savoy, *Phys. Lett. B* **119**, 343 (1982).
62. E. Cremmer, P. Fayet and L. Girardello, *Phys. Lett. B* **122**, 41 (1983).
63. H. Nilles, M. Srednicki and D. Wyler, *Phys. Lett. B* **120**, 346 (1983).
64. L. Hall, J. Lykken and S. Weinberg, *Phys. Rev. D* **27**, 2359 (1983).
65. P. Fayet, *Eur. Phys. J. C* **74**, 2837 (2014).
66. P. Fayet, *AIP Conf. Proc.* **1241**, 78 (2010).
67. K. Intriligator, N. Seiberg and D. Shih, *JHEP* **04**, 021 (2006).
68. https://twiki.cern.ch/twiki/bin/view/AtlasPublic/SupersymmetryPublicResults
69. https://twiki.cern.ch/twiki/bin/view/CMSPublic/PhysicsResultsSUS
70. F. Gliozzi, J. Scherk and D. Olive, *Nucl. Phys. B* **122**, 253 (1977).
71. L. Brink, J. Scherk and J. Schwarz, *Nucl. Phys. B* **121**, 77 (1977).
72. P. Fayet, *Phys. Lett. B* **125**, 178 (1983).
73. R. Haag, J. Lopuszanski and M. Sohnius, *Nucl. Phys. B* **88**, 257 (1975).
74. P. Fayet, *Phys. Lett. B* **146**, 41 (1984).
75. P. Fayet, *Phys. Lett. B* **153**, 397 (1985).
76. J. Scherk and J. Schwarz, *Phys. Lett. B* **82**, 60 (1979); *Nucl. Phys. B* **153**, 61 (1979).
77. G. Farrar and P. Fayet, *Phys. Lett. B* **79**, 442 (1978); *ibid.* **89**, 191 (1980).

Chapter 21

Future Direction Beyond the Standard Theory

Alex Pomarol

Dept. de Física, Universitat Autònoma de Barcelona,
08193 Bellaterra, Barcelona, Spain
alex.pomarol@uab.cat

With the Higgs discovery, the full Standard Model (SM) has been experimentally established, and most of its sectors accurately tested. Nevertheless, the SM deeply relies in the presence of the Higgs, a spin-zero field, whose mass term is not expected, on theoretical grounds, to be much smaller than the Planck scale. This problem of naturalness demands a modification of the SM around the weak scale, making the exploration of the TeV-energy regime a top experimental priority. In this chapter we briefly review the most well-motivated scenarios beyond the SM that can accommodate a light Higgs, mainly centering in two ideas: Supersymmetry and compositeness.

1. Motivations to go Beyond

How wonderful that we have met with a paradox;
now we have some hope of making progress.
Niels Bohr

The most important discoveries in physics have come out in physical regimes in which the existent theory had failed to give sensitive predictions, leading, in most of the cases, to a change of paradigm. Sometimes it was foreseen which new theory had to replace the old paradigm, and dedicated experimental searches were put forward. This has been the case for particle physics in recent times, as for example, in the search for the top-quark and Higgs. Nevertheless, sometimes we do not even realize that our current theory has a flaw, not making sense in certain energy regimes, as it happened with classical physics in the subatomic domain. In these circumstances experiments have led the new-physics searches, with the theory, in this case quantum mechanics, coming behind.

The Higgs discovery was led by a "no-lose theorem" for new physics:[1] Theories of massive vectors, as the W and Z bosons, did not make sense if no extra degrees of

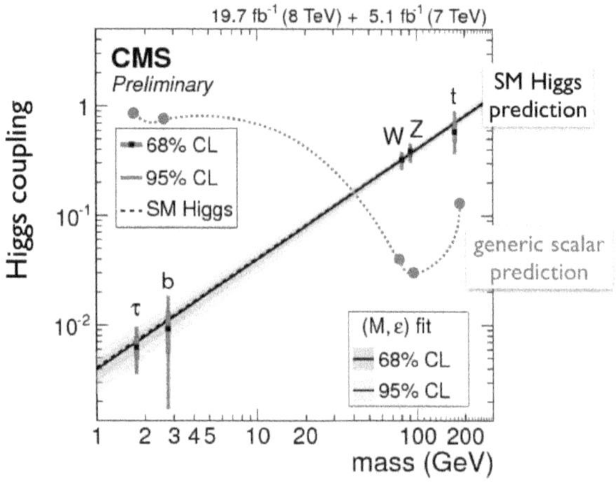

Fig. 1. Fit of the couplings of the newly discovered state at the LHC to the SM particles as a function of their masses.[3] The predictions from the SM Higgs are given by a straight black line. A generic scalar would have couplings to the SM particles laying in any point of this plane, as in the example shown with a dashed line. The experimental data clearly favors a SM Higgs.

freedom were added, since they were becoming strongly-coupled at energies above $\approx 4\pi m_W/g$. Therefore either new particles or new strong-dynamics were guaranteed to be discovered at the LHC below or around TeV energies. One possibility to make theories of massive vector bosons consistent was the Brout–Englert–Higgs (BEH) mechanism,[2] that predicted a new particle below the TeV, the Higgs, whose couplings were fully determined. At the LHC Run 1, a new state with these properties was indeed observed,[3,4] as shown in Fig. 1, giving the first experimental confirmation of the BEH-mechanism and a full validation of the Standard Model (SM) of elementary particles.[5] We can be proud of this outstanding achievement. For the first time we are confident to have a theory for the fundamental interactions of the universe that allows to make sensible predictions for physics all the way to very small scales. Indeed, being the Higgs mass $m_h \simeq 125$ GeV, the SM is a consistent effective field theory valid up to energies of order the Planck-mass $M_P = \sqrt{1/G_N} \sim 10^{19}$ GeV.[6] At these energies, of course, quantum gravitational effects are expected to be large, entering a new unpredictable strong-coupling regime, that asks again for a new paradigm. Either string theory or something else awaits there to be discovered.

The Planck-mass scale is however too large to be fully explored with present facilities. We can only hope to detect "echoes" arising from physics at those energies. For example, neutrino Majorana masses can be expected to arise from Planckian physics, incarnated in dimension-five operators suppressed by the heavy scale.[7] This can lead to a rate of neutrinoless double-beta decays observable within near future experiments.[8] Also processes mediated by Planckian states could lead to proton decay rates relatively close to present sensitivities. Dark Matter could also be a remnant of this very high-energy new physics, for example, as a very weakly interacting particle, such as the axion.[9] If this is the case, we will face a big challenge to detect DM beyond its gravitational presence already observed. Though very limited, it is clear that these types of searches must have a high priority in any experimental physics agenda.

With this in mind, TeV colliders seem hopeless to make any important discovery. The SM is a perfectly consistent theory at the TeV, giving us precise predictions. Nevertheless, we encounter, for the first time, a different motivation to expect new physics to show up at the TeV regime. We have reasons to believe that the BEH-mechanism cannot be the full story. It is true that a virtue of the BEH-mechanism for electroweak symmetry breaking (EWSB) is its simplicity. But, as it is well-known from daily life, simplicity is in conflict with stability.[a] The fact that the Higgs is a scalar, a spin-zero state, makes it difficult to keep it light ($m_h \ll M_P$), as Kenneth Wilson denounced long time ago:[10] "*Scalar particles are the only kind of free particles whose mass term does not break either an internal or a gauge symmetry.*"

The seed of the problem can be easily understood just by looking at the propagating degrees of freedom (DOF) of a massless and massive state of spin zero, as compared with those of a state of spin equal to 1/2, 1, or higher. A massless vector boson, like the photon, has two polarizations (2 DOF), while a massive vector has 3 polarizations. The simple formula $2 \neq 3$ guarantees that a massless vector can never get a mass by continuous variations of parameters (or quantum fluctuations); only a discrete change in the theory, increasing its DOF, can make vector massive. Similarly for fermions, we have that a charged massless fermion has 2 DOF, while a massive one has the double (the left- and right-handed chiral states), and therefore, for the same reason, massless fermions are safe from getting masses under small fluctuations.[b] Now, massless scalars have the same DOF as massive scalars: 1 DOF for neutral ones. Even if we start with a massless scalar at tree-level, it is not guaranteed that quantum corrections will not give it a mass. Natural expectations for this mass is then just dictated by dimensional analysis and approximate symmetries of the theory. In the SM, for example, in the presence of a large threshold scale M_P in which gravitational interactions are large and new physics must be present, the Higgs mass is expected to be

$$m_h^2 = aM_P^2, \tag{1}$$

where a is a number close to one, as no accidental symmetries in the SM dictate additional suppressions (at most a loop suppression, $\sim g^2/16\pi^2$, due to its unavoidable quantum nature). Therefore, the empirical evidence $m_h \ll M_P$ appears to be, for quantum field theorists, very unnatural.

---

[a]We learn this at the very early stages of our lives when at the kindergarten we become knowledgable with *The Three Little Pigs* fable.

[b]If a fermion has no charge, it can get a Majorana-type mass without increasing the DOF. For this reason, to keep naturally massless fermions, we must take the extra assumption that the fermion has some type of charge.

2. New Paradigms Awaiting at the TeV

Idealized models have a useful role to play,
as ways to clarify your thinking.
Paul Krugman

Theorists contemplate mainly two possible explanations for the above described Higgs-mass oddity. One is to consider that the SM is upgraded at TeV energies to incorporate an extra symmetry which could relate the Higgs, a scalar, to a fermion, whose mass can be protected. This is the case of supersymmetry. An alternative option is to assume that the Higgs is not really an elementary particle but a composite state made of elementary fermions, as pions in QCD. In this case, we must postulate a new strong-sector at the TeV from whose dynamics must emerge a Higgs-like state.

Recently, a third possibility has started to be seriously considered in the physics community. This relies in the quite controversial possibility that our universe is only one among a vast number of other universes in which the laws of physics can be different. If so, we can *naturally* expect that in most of the universes the Higgs mass is close to the Planck scale, with only few in which $m_h \ll M_P$. Nevertheless, it is in these few universes where the laws of physics can lead to observers.[11] The fact that we live in a special universe would be similar to the fact that we live in a special planet, the Earth, different from most of the planets. Where else could we live? This idea goes by the name of the "Multiverse", and has recently found a solid theoretical framework: Eternal inflation and the huge landscape of string theory vacua. The Multiverse approach is often criticized of being non-scientific, as if it could be used to explain anything. Nevertheless, this is far from being true. Scientifically speaking, this approach is in very similar footing as Darwin's Theory of Evolution: Though it cannot predict what species can arise on Earth, it affords a mechanism to explain their varieties and sophistications. And more importantly, it can be experimentally dismissed! Finding human's fossils in the Cambrian epoch would be enough to throw Darwin's theory away. We could also dismiss the Multiverse solution to the hierarchy problem if another scalar is discovered at the LHC. It can be argued that $m_h \ll M_P$ is crucial to have the chemistry of our universe,[12] but why on earth another unnaturally light scalar would be needed for? This would definitely point towards an alternative solution to the hierarchy problem. A special motivation for the Multiverse paradigm is that it is the only one that properly addresses the smallness of the cosmological constant. As Weinberg pointed out long ago,[13] a cosmological constant close to its present value could be anthropically selected in the landscape. We will not further discuss the Multiverse idea. It is clear that this will receive an important boost if no new physics is discovered at the LHC.

It is interesting to point out that all scenarios for new-physics discussed above predicted a Higgs with a mass around the observed one. For instance, in a minimal supersymmetric version of the SM, what is called the MSSM, the lightest-Higgs mass was predicted to be in the range[14]

$$m_h \lesssim 135 \text{ GeV}, \tag{2}$$

while in minimal versions of composite Higgs (MCHM) the predictions[15] were

$$115 \text{ GeV} \lesssim m_h \lesssim 185 \text{ GeV}. \tag{3}$$

Finally, to have the SM valid all the way up to Planck energies,[6] as could be natural in a Multiverse, the Higgs mass had to lie within

$$110 \text{ GeV} \lesssim m_h \lesssim 170 \text{ GeV}. \tag{4}$$

Finding the Higgs at around 125 GeV did not discriminate among these three possibilities. In the MSSM, it can be accommodated close to the upper limit, having the important implication though that the supersymmetry-breaking scale must be beyond the TeV.[16] On the contrary, in the MCHM a 125 GeV Higgs can be accommodated close to the lower limit, implying that color fermionic resonances must be below the TeV,[15,17] at the reach then of the LHC.

Supersymmetry

Supersymmetry allows to relate the properties of scalars with those of fermions such that the stability of the fermion masses can guarantee the stability of the scalar masses. The most economical realization would be to impose supersymmetry to relate the Higgs to any of the SM left-handed lepton, either the tau, muon or electron. Interestingly, this is doable since all these fields have the same quantum numbers under the SM. This possibility was proposed long ago,[18] but only recently realistic models have been explored.[19] The major difficulty comes however from the up-quark masses that can only be generated if supersymmetry is broken, requiring then extra dynamics at the TeV.

On the other hand, if we demand that all fermion masses must arise from supersymmetry-preserving terms, the minimal supersymmetric version of the SM is the MSSM. This requires to double the full spectrum of the SM, a new fermion for each SM boson and vice versa, implying a new layer of particles, the superpartners, to show up at TeV energies. We should not dismay with this doubling of the spectrum. We already came across before when Dirac predicted an anti-particle for each existing particle in the venue of relativistic quantum theories, a story with a successful ending. One of the most interesting prediction of the MSSM is the physical Higgs mass, that can be fully calculated as a function of the mass spectrum of the theory. For this reason, learning at the LHC the Higgs-mass value was the most relevant piece of information for the MSSM. In particular, a 125 GeV Higgs leads to the requirement that the breaking of supersymmetry in the stops, the partners of the top in the MSSM, must be large, above the TeV scale. This obviates many direct searches for superpartners at the LHC! Furthermore, this implies, based on naturalness, that the MSSM starts to be disfavored by the experimental data, and one must look for non-minimal versions such as the NMSSM in which an extra singlet field is added. This is one of the main lessons from the LHC Run 1. We stop here as further discussions on supersymmetry can be found in Chapter 20 of this book.

Compositeness

Probably the easiest solution to the Higgs-mass problem is to renounce of elementary scalars. This was one of the main motivation for Technicolor models[20] where a mechanism for EWSB similar to QCD was postulated at the TeV. In QCD the breaking of the electroweak symmetry comes from the condensate of quarks due to the strong-interacting gluons at GeV energies. This phenomena is close related to superconductivity in which the Cooper pair plays the role of the quark condensate. Following the same idea, Technicolor models consist of gauge theories strongly coupled at the TeV in which a condensate of Techni-fermions leads to EWSB. Nevertheless, Technicolor models do not predict a light Higgs-like state. In fact, in Technicolor models we expect many heavy resonances with different spins, but all with masses around the TeV and none of them with couplings to SM states necessarily proportional to their mass, as LHC data suggests (see Fig. 1). For this reason these type of models are at present dismissed.

A different, but close related possibility, is to look for theories at the TeV whose strong dynamics does not break the electroweak symmetry but instead deliver a composite BEH-mechanism.[21] At first glance, this might seem a tough demand. But this is in fact not the case, if we wisely make use of the so-called Goldstone theorem, a theorem inspired by the work of Nambu,[22] conjectured by Goldstone,[23] and proven by Goldstone, Salam and Weinberg.[24] This theorem states that whenever a spontaneous breaking of global symmetry occurs, massless bosons must appear. These are called Goldstone bosons. For example, if the global-symmetry breaking pattern of a quantum field theory is $SU(3) \to SU(2)_L \times U(1)_Y$, the Goldstone theorem tells us that a massless scalar field, transforming as a $SU(2)$-doublet with $Y = 1$, must be contained in the theory. This is a beautiful result! If we postulate to have at the TeV a new strong-sector with this breaking pattern, we are guaranteed to have a composite scalar with the right quantum numbers to be identified with the Higgs. Nevertheless, this simple incarnation does not fully work by many reasons, as it predicts, for instance, large deviations from the relation $m_W^2 \simeq m_Z^2 \cos^2\theta_W$ due to the absence of a custodial symmetry in the TeV strong-sector. The minimal realistic version of a composite Higgs model was given in Ref. 25 based on a TeV strong-sector with the global-symmetry breaking pattern

$$SO(5) \to SO(4) \simeq SU(2)_L \times SU(2)_R, \tag{5}$$

with the SM $U(1)_Y$ embedded in $SU(2)_R$. The Higgs appears as a Goldstone boson and the "custodial" $SO(4)$ symmetry preserves the relation $m_W^2 \simeq m_Z^2 \cos^2\theta_W$. There is an additional requirement to make the model realistic. SM fermions and gauge bosons must couple to the Higgs to get masses, that implies that they must have direct couplings to the TeV strong-sector. These couplings however break explicitly the global $SO(5)$ symmetry, making the Higgs a "Pseudo" Goldstone boson (PGB), as Weinberg pointed out long ago.[26] What this means is that the Higgs is not massless anymore, as a Higgs potential is generated by one-loop quantum corrections involving SM particles. The main contribution comes from the top-quark loop due to its large Yukawa coupling, which forces the Higgs to get a vacuum expectation value and trigger EWSB. This one-loop contribution can also allow to naturally accommodate a Higgs mass of 125 GeV.[15,17,c]

It is therefore clear that the top-quark is one of the main players in composite Higgs models. Had the top-quark been lighter, the SM gauge-boson one-loop contributions would have dominated the Higgs potential, and no EWSB would have occurred. Since the top-quark must have sizable linear couplings to the TeV strong sector, in order to get its large mass, the top can be used as a portal to this sector. Measuring then the properties of the top-quark can be as important as those of the Higgs.

The above are the generic features of the composite PGB Higgs idea. Nevertheless, one could wonder about which concrete underlying theory at the TeV could implement these properties, i.e., which are the UV degrees of freedom of the new TeV sector, such as quarks and gluons are to QCD hadrons and mesons. This quite ambitious question is however very difficult to address, mainly due to our limitation to deal with strong dynamics. We must recall that it took us many years, and plenty of experimental data, to discover that the underlying theory behind protons, neutrons and pions was QCD. One can find a theoretical handle in the work of Seiberg,[28] that conjectured some dualities between strongly-coupled gauge theories and weakly-coupled ones that are much easier to treat. Using these dualities, it was found in[29] different UV completed models of composite Higgs. It is also recently receiving some interest UV completed composite-Higgs models with enough (though not all) ingredients to be explored in the lattice.[30]

Alternatively, one can use the AdS/CFT correspondence[31] as a playground for these ideas. Composite Higgs models can be easily realized as weakly-coupled five-dimensional (5D) models in Anti-de Sitter (AdS),[32] in which the Higgs corresponds to the fifth-component of the 5D gauge bosons.[25] The Higgs mass is protected by 5D gauge invariance and can only get a nonzero value from non-local one-loop effects.[33] The AdS/CFT correspondence allows to built composite Higgs models where the mass spectrum of resonances, corresponding to the Kaluza–Klein modes, can be determined.[25]

---

[c] Variations on the composite PGB Higgs idea have also been put forward under the name of Little Higgs models.[27] In these models however the SM gauge and fermion sector is extended in order to guarantee that Higgs-mass corrections involving the new strong-sector arise at the two-loop level instead of one-loop, allowing for a better insensitivity of the electroweak scale to the new strong dynamics.

On the other hand, interestingly, many predictions of composite Higgs models do not require at all the full knowledge of the strong TeV theory, but only the symmetry breaking pattern. For instance, many Higgs properties can be model-independently derived in an equivalent way as pions in QCD can be very well described at low-energies by the Chiral Lagrangian. Following this approach, it was shown in Ref. 34 which Higgs couplings are expected to deviate from the SM predictions if the Higgs is composite. We will come back on this issue later on.

Up to now we have been independently discussing the two main ideas beyond the SM, supersymmetry and composite Higgs. Nevertheless nature could well be using, in a non-trivial way, a blend of these two ideas to deliver naturally a light Higgs. We must be aware that the territory of supersymmetric and strongly-coupled theories is still uncharted, so nature could surprise us with some unexpected new-physics. For example, if the strong-sector at the TeV is also supersymmetric, a light Higgs of 125 GeV could emerge due to supersymmetry, instead of its Goldstone nature.[35] The main crucial difference here with respect to the MSSM is that, beyond the Higgs, the rest of the SM would not need to be supersymmetric, implying that only the Higgs would have a supersymmetric partner, the Higgsino. An alternative option is to consider models in which the EWSB is triggered by a Technicolor sector coupled to the MSSM. These models can accommodate a 125 GeV Higgs without the need of very heavy stops, apart from also solving other difficulties of the MSSM such as the μ-problem.[36] More drastically, following the suggestion of Ref. 37, it could also be that string theory comes in at the TeV, as can occur if we allow gravity to propagate in large extra dimensions (LED). The lightness of the Higgs is not a big problem anymore, as quantum gravity does not appear in this case at the huge scale M_P but at energies around the corner, $\sim$ TeV. This is a dream scenario for experimental physics, as we could in the next decades address most of the fundamental questions of particle physics. The LED scenarios however still lack for an explanation of why $m_h \ll$ TeV; of course, a much milder requirement, but important to understand which new physics we could discover first (new string excitations, black holes, gravitons, ...).

3. Looking for Experimental Evidences of TeV New Physics

The great tragedy of science,
the slaying of a beautiful hypothesis by an ugly fact.
Thomas Huxley

If indeed new physics is lurking around TeV energies, this could show up either indirectly, by slightly modifying the SM predictions for physical processes, or in a direct way, as new states at colliders. In the first case, we must advocate for searches in the intensity frontier, while for the second it is clear that it is more convenient to explore the energy frontier. Though both approaches can provide relevant information on physics beyond the SM (BSM), it is clear that the search for indirect deviations is more limited as their interpretations are often unclear (for example, the anomalous rate of precession of the perihelion Mercury's orbit was first recognized in 1859, but it was not till 1915 with Einstein's General Relativity that its origin was determined). Therefore to fully determine which new theory will replace the SM, the exploration at the energy frontier will be really essential.

Let us start by understanding which indirect searches are the most interesting ones. For this purpose it is convenient to make use of Effective Quantum Field Theories (EQFT), as these give us a model-independent parametrization of all the indirect BSM effects. Assuming that the characteristic scale of new physics, Λ, is heavier than the electroweak scale, the SM EQFT is obtained as an expansion in SM fields and derivatives over Λ:

$$\mathcal{L}_{\rm EQFT} = \mathcal{L}_4 + \mathcal{L}_6 + \cdots , \tag{6}$$

where $\mathcal{L}_4$ is made of dimension-four operators and defines what we call the SM Lagrangian,[5] while $\mathcal{L}_6$, that contains dimension-six operators,[38] gives the leading indirect BSM effects.[d] In principle, one can expect that flavor and CP-violating processes are the most sensitive to $\mathcal{L}_6$, as they are suppressed in the SM by the smallness of the fermion masses of the first families.[39] — see also Chapter 17 in this book. Nevertheless, a similar suppression as in the SM is expected in certain BSM. For example, in supersymmetric theories where supersymmetry-breaking is mediated from a "hidden sector" to the SM by gauge interactions, the so-called gauge mediated supersymmetry breaking models (GMSB),[40] all squark masses are family-universal up to small corrections involving the SM fermion masses. Hence their effects to flavor observables are as suppressed as in the SM. Similar scenarios can be found in composite Higgs models. It is therefore not fully guaranteed that flavor and CP-violating effects are the most relevant ones.

Assuming family-universality, a full classification of the physical effects arising from $\mathcal{L}_6$ was provided in Ref. 41. These can be encoded in 59 *primary couplings* between SM fields (for one family) that can be chosen to be:

- 11 Higgs couplings.
- 7 Z-couplings to fermions.
- 1 W-coupling to right-handed fermions.
- 7 triple-gauge couplings.
- 8 fermion dipole-moments.
- 25 four-fermion interactions.

A first complete global fit to all these BSM effects was given in Ref. 42 (except for the four-fermion interactions), showing which sectors of the SM were very well-tested and which require more experimental examination.

---

[d] We are assuming lepton and baryon number conservation.

Among the 59 primaries, the most interesting ones are the Higgs couplings, as the Higgs is the most sensitive SM particle to BSM corrections. For this reason, as we will show below, today determination of the Higgs couplings, even if not very accurate, can place important bounds on new physics. From the EQFT analysis, one can deduce[43] that among the Higgs couplings, the most relevant ones are the primary couplings,[41] that for the case of CP-conservation correspond to

$$\begin{aligned}\mathcal{L}_h^{\text{primary}} &= g^h_{VV}\, h\left[W^{+\,\mu}W^-_\mu + \frac{1}{2\cos^2\theta_W}Z^\mu Z_\mu\right]\\ &\quad+\frac{1}{6}\,g_{3h}\,h^3 + g^h_{ff}(h\bar{f}_L f_R + \text{h.c.})\\ &\quad+\kappa_{GG}\,\frac{h}{2v}G^{A\,\mu\nu}G^A_{\mu\nu} + \kappa_{\gamma\gamma}\,\frac{h}{2v}A^{\mu\nu}A_{\mu\nu} + \kappa_{Z\gamma}\frac{h}{v}A^{\mu\nu}Z_{\mu\nu},\end{aligned} \tag{7}$$

where $G^{A\,\mu\nu}$, $A_{\mu\nu}$, $Z_{\mu\nu}$ are the field-strengths of the gluon, photon and Z. Equation (7) gives the set of SM Higgs couplings whose effects from $\mathcal{L}_6$ are independent from effects to other SM observables.[43] On the contrary, the rest of the Higgs couplings can be always written as a function of other SM couplings, as explicitly shown in Ref. 44. At the LHC one can combine the different Higgs-production mechanisms and branching ratios to determine six of the primary Higgs couplings: g^h_{ff} ($f = t, b, \tau$), g^h_{VV}, κ_{GG} and $\kappa_{\gamma\gamma}$.[e] The CMS and ATLAS fit of these six primary Higgs couplings can be found in Refs. 3 and 4, and for a combination of the ATLAS and CMS data see, for example, Ref. 45. Though the primary coupling $\kappa_{Z\gamma}$ has not yet been determined, one can use the experimental bound[46] $BR(h \to Z\gamma)/BR(h \to Z\gamma)_{\text{SM}} \lesssim 10$ to derive the constraint[42] $-0.01 \lesssim \kappa_{Z\gamma} \lesssim 0.02$. The fact that in the SM $h \to Z\gamma$ arises at the one-loop level, and therefore has a small branching fraction, $BR(h \to Z\gamma) \sim 0.15\%$, makes this decay channel very sensitive to new physics; it probably provides the last chance to find large BSM effects in SM Higgs couplings. Among the remaining primary Higgs couplings to be measured we also have g_{3h}. Its determination however will be very difficult, since it requires to search for double-Higgs production $pp \to hh$ that has small rates.[47] Also Higgs couplings to light fermions g^h_{ff} (beyond the 3rd family) are going to be difficult to measure, since we expect these couplings to be proportional to m_f/m_W, giving then very small branching ratios. For example, for the case of the muon, that is probably the most accessible, we have in the SM $BR(h \to \mu\mu) \sim 0.02\%$. Therefore high luminosities will be needed to measure these Higgs couplings at the LHC Run 2.

The experimental fit of the Higgs primary couplings shows a good agreement with the SM predictions,[3,4] with no signal of new physics. This leads to important implications for BSM. In the MSSM, for example, the Higgs couplings to

[e] We note that g^h_{tt} and g^h_{VV} also affect $BR(h \to \gamma\gamma/Z\gamma)$ and $\sigma(GG \to h)$ at the one-loop level.[34]

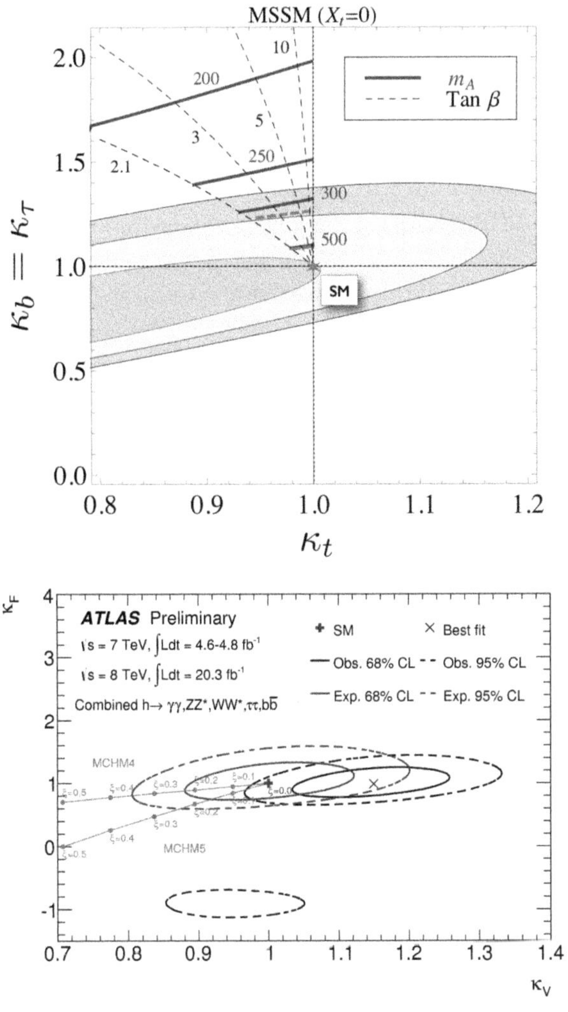

Fig. 2. Two-dimensional fit of Higgs couplings and predictions from the MSSM (top plot),[49] and composite Higgs models MCHM4 and MCHM5 (bottom plot).[50] Other Higgs couplings are put to their SM value. We follow the notation $\kappa_V \equiv g^h_{VV}/g^{h\,\text{SM}}_{VV}$ and $\kappa_F \equiv g^h_{ff}/g^{h\,\text{SM}}_{ff}$.

fermions receive sizable tree-level corrections due to the extra heavy Higgs.[f] The main effects[48,49] are then expected to be for g^h_{tt} and $g^h_{bb,\tau\tau}$, with a pattern of deviations with respect to the SM shown in Fig. 2 (top plot). The absence of deviations leads then to a lower-bound on the heavy Higgs mass of approximately $m_A \gtrsim 400$ GeV.[50] Similarly, in strongly-interacting BSM in which the Higgs is composite,

[f]Also g^h_{VV} and g_{3h} are modified at tree-level by the heavy Higgs exchange. Nevertheless, the corrections to g^h_{VV} are suppressed by extra powers of the heavy Higgs mass, while g_{3h} is, as we mentioned above, very difficult to be measured in the near future.

effects on the Higgs coupling to fermions and $V = W/Z$ can be enhanced by a strong-coupling factor g_ρ^2/g^2,[34] that can be as large as $g_\rho^2/g^2 \sim 16\pi^2$ with respect to effects in other SM sectors. The pattern of deviations is shown in Fig. 2 (bottom plot) for several MCHM and as a function of $\xi = (v/f)^2$ where f is the Higgs decay-constant, related to the composite scale by $\Lambda = g_\rho f$. Bounds on the scale of compositeness Λ coming from LHC Higgs physics are starting to to be as competitive as those from LEP, even that we have produced much less Higgs at the LHC than Z at LEP.[50]

The most compelling way to discover new physics is, without doubt, by direct detection of new particles. Both, supersymmetry and composite Higgs models, predict a bunch of new states lurking around the TeV. Specialized searches are on the way by LHC experimentalists and a large number of different analysis have been already pursued. If we had to prioritize few of them, we would select the hunting for color particles, specially those dedicated searches for the partners of the top, either in supersymmetric or composite Higgs models. The reason for this priority is the following. If TeV new-physics must explain the lightness of the Higgs versus M_P, the loop corrections to the Higgs mass must be "tamed" by the new particles. The most relevant one is the top-quark loop, since it has the largest coupling to the Higgs. This loop contribution is controlled by the top-quark partners, that generate a Higgs mass of order the electroweak scale if these new states are around 500 GeV. For larger masses, the parameters of the model must be tuned to keep the Higgs light. Therefore, based on naturalness arguments, the top partners are expected to be below the TeV. Furthermore, being color particles, we are guaranteed to have sizable production cross-section at the LHC to be easily discovered. In the case of supersymmetric models, the top partners correspond to scalars with the quantum numbers of the top and bottom, the so-called stops and sbottoms. They are supposed to mostly decay into tops and bottoms, and into the lightest supersymmetric particle that in most natural scenarios is the Higgsino, or the Gravitino for the case of GMSB models. For composite Higgs models, the top partners are color fermionic resonances with electric charges $Q = 5/3, 2/3, -1/3$,[15] and a phenomenology described in detail in Ref. 51. This is depicted in Fig. 3 where it is shown the mass spectrum of a natural supersymmetric and composite Higgs model. Present limits on top partners from the LHC Run 1 are around 500–800 GeV,[52] scratching at present the most natural region of the parameter space of the MSSM and MCHM. Nevertheless, it will not be until the LHC Run 2 where the naturalness of these BSM will be really at stake.

Clues for cosmological conundrums

Could TeV physics be behind other fundamental questions in particle physics and cosmology, such as the origin of Dark Matter (DM), the abundance of matter over anti-matter in our universe (Baryogenesis), the origin of inflation or neutrino masses? Though not necessary the case, as the mandatory new-physics at the Planck

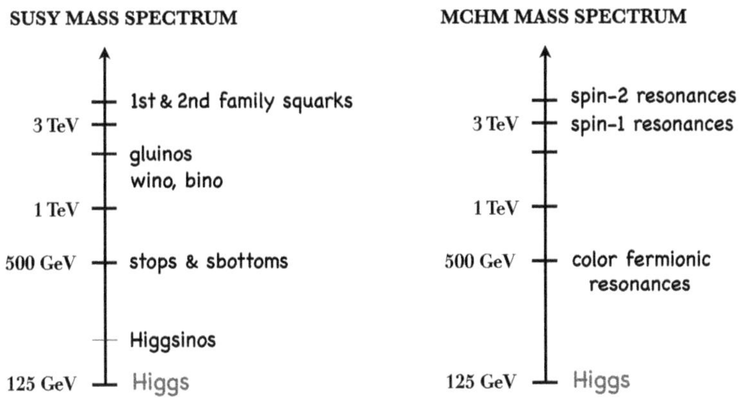

Fig. 3. Natural expectations for the mass spectrum in supersymmetric models (left) and composite Higgs models (right).

scale could be the true responsible for these phenomena, it is well possible that some of these questions are addressed by TeV physics, opening an exciting possibility of resolving these mysteries in well controlled experiments, such as TeV colliders. The most likely of the above important questions to be addressed by TeV new-physics is the DM origin. This hope arises from the so-called "WIMP miracle": A stable particle with mass of order the electroweak scale and $O(1)$ renormalizable-interactions is in the ballpark of the needed relic abundance for a DM candidate. In the MSSM, as well as in the MCHM, we find many DM candidates.[53] For instance, the lightest superpartner, if neutral, as the neutralinos (superpartners of the Z, photon or Higgs), can be a good candidate for DM in certain "well-tempered" region of the parameter space.[54] Similarly, DM can arise in composite Higgs models as an extra Goldstone boson,[55] or as the "baryons" of the TeV strong-sector which can be stable, as ordinary baryons in QCD, by an accidental symmetry.[56] Detecting these types of DM candidates is possible, but not guaranteed, as they could be too heavy, around few TeV, to have impact in present detectors. "Blind" searches at the LHC, i.e., model-independently looking for missing energy (from the undetected DM) plus a jet/vector-boson, can also be performed to scan over a large variety of models. A lot of effort has been put behind these searches that, with a little bit of luck, could give us significant rewards.

Baryogenesis is another interesting phenomena that could have its origin at the TeV. The universe, as it cools down, undergoes a phase transition from a symmetric vacuum to an EWSB one, at a critical temperature of $T_c \sim m_h$. If this transition is strongly first-order and bubbles of EWSB phase are produced as the universe gets cooler, there is the possibility to create the needed baryons that populate our universe. Of course, we still need that new physics at the TeV afford baryon number violating processes, apart from CP violation.[57] Unfortunately, if the electroweak phase transition is driven only by the SM Higgs, the transition is close to second-order and baryons cannot be produced. Different possibilities could change this behavior. If extra scalars, such as an extra singlet,[58] coexist with the SM Higgs, the electroweak phase transition could be of first-order. Therefore detecting this singlet at the LHC, though not an easy task,[59] can be of fundamental importance. Another possibility is to have Baryogengesis, not at the electroweak phase transition, but at a $T_c \sim$ TeV phase-transition arising from a new strong-sector at the TeV. As commented, this strong-sector must be there if the Higgs is composite, or even in the MSSM if supersymmetry is broken at low-energies. This option is quite interesting, but again will be difficult to be fully explored at the LHC, probably needing a more energetic collider.

4. Epilogue

Doubt is not a pleasant condition,
but certainty is absurd.
Voltaire

We are facing a new era in particle physics in which discoveries at the energy frontier are not anymore fully guaranteed. The times of "no-lose theorems" for discovery are gone for experimental physics at TeV energies and we have the risk of not finding any new physics at the LHC. In fact, the most radical change of the present paradigm, the Multiverse idea, gives the dramatic possibility to find nothing new at TeV colliders. We encompassed this situation before: For example, the Michelson–Morley experiment gave an unexpected null result. But in spite of the frustration of knowing that experimentally we could not learn anything about the properties of the medium in which electromagnetic waves were propagating, we were able to contemplate the birth of a new paradigm, Einstein's theory of relativity. We therefore should not fear as we always learn things from well-motivated experiments.

On the other hand, a different type of motivation is coming forth in the search for new physics at TeV energies: The unnaturalness of the SM. Understanding the smallness of the electroweak scale versus the Planck scale can require new physics to be present at the TeV. This gives us confidence to believe that the LHC has high chances to discover new physics. We have well-defined proposals and a loaded agenda for LHC searches. And, of course, we must be opened to whatever surprises nature can bring us beyond our expectations. As Thomas Henry Huxley once advised us, we must "*be prepared to give up every preconceived notion, follow humbly wherever and to whatever abysses nature leads, or you shall learn nothing.*"

Acknowledgments

We acknowledge support from the Catalan ICREA Academia Program, and also the grants FPA2011-25948, 2014SGR1450.

References

1. M. S. Chanowitz, in *Proceedings of the 23rd International Conference on High Energy Physics*, Berkeley, California, ed. by S. Loken (World Scientific, Singapore, 1987).
2. F. Englert and R. Brout, *Phys. Rev. Lett.* **13**, 321 (1964); P. W. Higgs, *Phys. Rev. Lett.* **13**, 508 (1964).
3. The CMS Collaboration, CMS-PAS-HIG-14-009.
4. The ATLAS Collaboration, ATLAS-CONF-2015-007, ATLAS-COM-CONF-2015-011.
5. S. L. Glashow, *Nucl. Phys.* **22**, 579 (1961); S. Weinberg, *Phys. Rev. Lett.* **19**, 1264 (1967); A. Salam, *Conf. Proc. C* **680519**, 367 (1968).
6. See for example, J. Ellis, J. R. Espinosa, G. F. Giudice, A. Hoecker and A. Riotto, *Phys. Lett. B* **679**, 369 (2009).
7. S. Weinberg, *Phys. Rev. D* **22**, 1694 (1980).
8. See for example, T. Johnson's talk at 50th Rencontres de Moriond EW 2015.
9. J. Preskill, M. B. Wise and F. Wilczek, *Phys. Lett. B* **120**, 127 (1983); L. F. Abbott and P. Sikivie, *Phys. Lett. B* **120**, 133 (1983); M. Dine and W. Fischler, *Phys. Lett. B* **120**, 137 (1983).
10. K. G. Wilson, *Phys. Rev. D* **3**, 1818 (1971).
11. See for example, S. Weinberg, *Universe or Multiverse?*, ed. B. Carr (Cambridge University Press, 2009), pp. 29–42 [hep-th/0511037]; J. Garriga, D. Schwartz-Perlov, A. Vilenkin and S. Winitzki, *JCAP* **0601**, 017 (2006).
12. V. Agrawal, S. M. Barr, J. F. Donoghue and D. Seckel, *Phys. Rev. D* **57**, 5480 (1998).
13. S. Weinberg, *Phys. Rev. Lett.* **59**, 2607 (1987).
14. See for example, A. Djouadi, *Phys. Rept.* **459**, 1 (2008).
15. R. Contino, L. Da Rold and A. Pomarol, *Phys. Rev. D* **75**, 055014 (2007).
16. See for example, L. J. Hall, D. Pinner and J. T. Ruderman, *JHEP* **1204**, 131 (2012).
17. O. Matsedonskyi, G. Panico and A. Wulzer, *JHEP* **1301**, 164 (2013); M. Redi and A. Tesi, *JHEP* **1210**, 166 (2012); D. Marzocca, M. Serone and J. Shu, *JHEP* **1208**, 013 (2012); A. Pomarol and F. Riva, *JHEP* **1208**, 135 (2012).
18. P. Fayet, *Phys. Lett. B* **64**, 159 (1976).
19. F. Riva, C. Biggio and A. Pomarol, *JHEP* **1302**, 081 (2013).
20. S. Weinberg, *Phys. Rev. D* **13**, 974 (1976); *Phys. Rev. D* **19** (1979) 1277; L. Susskind, *Phys. Rev. D* **20** (1979) 2619.
21. D. B. Kaplan and H. Georgi, *Phys. Lett. B* **136**, 183 (1984); B **136**, 187 (1984); H. Georgi, D. B. Kaplan and P. Galison, *Phys. Lett. B* **143**, 152 (1984); H. Georgi and D. B. Kaplan, *Phys. Lett. B* **145**, 216 (1984); M. J. Dugan, H. Georgi and D. B. Kaplan, *Nucl. Phys. B* **254**, 299 (1985).
22. Y. Nambu, *Phys. Rev.* **117**, 648 (1960).
23. J. Goldstone, *Nuovo Cim.* **19**, 154 (1961).
24. J. Goldstone, A. Salam and S. Weinberg, *Phys. Rev.* **127**, 965 (1962).
25. K. Agashe, R. Contino and A. Pomarol, *Nucl. Phys. B* **719**, 165 (2005).
26. S. Weinberg, *Phys. Rev. Lett.* **29**, 1698 (1972).
27. N. Arkani-Hamed, A. G. Cohen, H. Georgi, *Phys. Lett. B* **513** 232–240 (2001); N. Arkani-Hamed, A. G. Cohen, T. Gregoire *et al.*, *JHEP* **0208**, 020 (2002).
28. N. Seiberg, *Nucl. Phys. B* **435**, 129 (1995).
29. F. Caracciolo, A. Parolini and M. Serone, *JHEP* **1302**, 066 (2013).
30. See for example, M. Golterman and Y. Shamir, arXiv:1502.00390 [hep-ph].
31. J. M. Maldacena, *Adv. Theor. Math. Phys.* **2**, (1998) 231; S. S. Gubser, I. R. Klebanov and A. M. Polyakov, *Phys. Lett. B* **428**, 105 (1998); E. Witten, *Adv. Theor. Math. Phys.* **2**, 253 (1998).

32. L. Randall and R. Sundrum, *Phys. Rev. Lett.* **83**, 3370 (1999).
33. Y. Hosotani, *Phys. Lett. B* **126**, 309 (1983).
34. G. F. Giudice, C. Grojean, A. Pomarol and R. Rattazzi, *JHEP* **0706**, 045 (2007).
35. T. Gherghetta and A. Pomarol, *Phys. Rev. D* **67**, 085018 (2003); R. Sundrum, *JHEP* **1101**, 062 (2011).
36. For models with this property, see for example, M. Dine, A. Kagan and S. Samuel, *Phys. Lett. B* **243**, 250 (1990), and more recently, A. Azatov, J. Galloway and M. A. Luty, *Phys. Rev. Lett.* **108**, 041802 (2012); *Phys. Rev. D* **85**, 015018 (2012); T. Gherghetta and A. Pomarol, *JHEP* **1112**, 069 (2011).
37. N. Arkani-Hamed, S. Dimopoulos and G. Dvali, *Phys. Lett. B* **429** 263 (1998); *Phys. Rev. D* **59**, 086004 (1999).
38. B. Grzadkowski, M. Iskrzynski, M. Misiak and J. Rosiek, *JHEP* **1010**, 085 (2010) [arXiv:1008.4884 [hep-ph]].
39. See for example, Y. Nir, *Phys. Scripta T* **158**, 014005 (2013).
40. For a review see, G. F. Giudice and R. Rattazzi, *Phys. Rept.* **322**, 419 (1999).
41. R. S. Gupta, A. Pomarol and F. Riva, *Phys. Rev. D* **91**, 3 (2015), 035001; E. Masso, *JHEP* **1410**, 128 (2014).
42. A. Pomarol and F. Riva, *JHEP* **1401**, 151 (2014); A. Falkowski and F. Riva, *JHEP* **1502**, 039 (2015).
43. J. Elias-Miro, J. R. Espinosa, E. Masso and A. Pomarol, *JHEP* **1311**, 066 (2013).
44. A. Pomarol, in *Proceedings of the 2014 European School of High-Energy Physics (ESHEP 2014)*, arXiv:1412.4410 [hep-ph]; M. Gonzalez-Alonso, A. Greljo, G. Isidori and D. Marzocca, *Eur. Phys. J. C* **75**, 3, 128 (2015); A. Falkowski, arXiv:1505.00046 [hep-ph].
45. A. Falkowski, F. Riva and A. Urbano, *JHEP* **1311**, 111 (2013).
46. CMS Collaboration, arXiv:1307.5515 [hep-ex]; ATLAS Collaboration, ATLAS-CONF-2013-009.
47. See for example, A. Azatov, R. Contino, G. Panico and M. Son, arXiv:1502.00539 [hep-ph].
48. L. Maiani, A. D. Polosa and V. Riquer, *Phys. Lett. B* **718**, 465 (2012).
49. R. S. Gupta, M. Montull and F. Riva, *JHEP* **1304**, 132 (2013).
50. ATLAS Collaboration, ATLAS-CONF-2014-010.
51. A. De Simone, O. Matsedonskyi, R. Rattazzi and A. Wulzer, *JHEP* **1304**, 004 (2013).
52. Particle Data Group Collaboration, *Chin. Phys. C* **38**, 090001 (2014).
53. G. Jungman, M. Kamionkowski and K. Griest, *Phys. Rept.* **267**, 195 (1996).
54. See for example, N. Arkani-Hamed, A. Delgado and G. F. Giudice, *Nucl. Phys. B* **741**, 108 (2006).
55. See for example, M. Frigerio, A. Pomarol, F. Riva and A. Urbano, *JHEP* **1207**, 015 (2012); D. Marzocca and A. Urbano, *JHEP* **1407**, 107 (2014).
56. S. Nussinov, *Phys. Lett. B* **165**, 55 (1985); S. M. Barr, R. S. Chivukula and E. Farhi, *Phys. Lett. B* **241**, 387 (1990).
57. A. G. Cohen, D. B. Kaplan and A. E. Nelson, *Ann. Rev. Nucl. Part. Sci.* **43**, 27 (1993); M. Quiros, hep-ph/9901312.
58. See for example, J. R. Espinosa, T. Konstandin and F. Riva, *Nucl. Phys. B* **854**, 592 (2012).
59. D. Curtin, P. Meade and C. T. Yu, *JHEP* **1411**, 127 (2014).

www.ingramcontent.com/pod-product-compliance
Ingram Content Group UK Ltd.
Pitfield, Milton Keynes, MK11 3LW, UK
UKHW052114170926
8624IPUK00026B/377

9 789814 733502